# Lecture Notes in Compu  )8

Commenced Publication in 1973
Founding and Former Series Editors:
Gerhard Goos, Juris Hartmanis, and Jan van Leeuwen

Frank Dehne   Alejandro López-Ortiz
Jörg-Rüdiger Sack (Eds.)

# Algorithms and Data Structures

9th International Workshop, WADS 2005
Waterloo, Canada, August 15-17, 2005
Proceedings

 Springer

Volume Editors

Frank Dehne
Jörg-Rüdiger Sack
Carleton University
School of Computer Science
Ottawa, Ontario, K1S 5B6, Canada
E-mail: frank@dehne.net, sack@scs.carleton.ca

Alejandro López-Ortiz
University of Waterloo
School of Computer Science
Waterloo, Ontario, N2L 3G1, Canada
E-mail: alopez-o@uwaterloo.ca

Library of Congress Control Number: 2005929608

CR Subject Classification (1998): F.2, E.1, G.2, I.3.5, G.1

ISSN        0302-9743
ISBN-10     3-540-28101-0 Springer Berlin Heidelberg New York
ISBN-13     978-3-540-28101-6 Springer Berlin Heidelberg New York

Springer is a part of Springer Science+Business Media

springeronline.com

© Springer-Verlag Berlin Heidelberg 2005
Printed in Germany

Typesetting: Camera-ready by author, data conversion by Scientific Publishing Services, Chennai, India
Printed on acid-free paper        SPIN: 11534273        06/3142        5 4 3 2 1 0

# Preface

The papers in this volume were presented at the 9th Workshop on Algorithms and Data Structures (WADS 2005). The workshop took place during August 15 – 17, 2005, at the University of Waterloo, Waterloo, Canada. The workshop alternates with the Scandinavian Workshop on Algorithm Theory (SWAT), continuing the tradition of SWAT and WADS starting with SWAT 1988 and WADS 1989. From 90 submissions, the Program Committee selected 37 papers for presentation at the workshop. In addition, invited lectures were given by the following distinguished researchers: Allan Borodin and Max J. Egenhofer.

On behalf of the Program Committee, we would like to express our sincere appreciation to the many persons whose effort contributed to making WADS 2005 a success. These include the invited speakers, members of the steering and Program Committees, the authors who submitted papers, and the many referees who assisted the Program Committee. We are indebted to Robert Kane for installing and modifying the submission software, maintaining the submission server and interacting with authors as well as for helping with the preparation of the program.

August 2005        Frank Dehne, Alejandro López-Ortiz, and Jörg-Rüdiger Sack

# WADS Organization

## Organizing Institutions

## Steering Committee

Frank Dehne     Carleton University, Canada
Ian Munro     University of Waterloo, Canada
Jörg-Rüdiger Sack     Carleton University, Canada
Roberto Tamassia     Brown University, Canada

## Program Co-chairs

Frank Dehne     Carleton University, Canada
Alejandro López-Ortiz     University of Waterloo, Canada
Jörg-Rüdiger Sack     Carleton University, Canada

## Conference Chair

Alejandro López-Ortiz     University of Waterloo, Canada

## Program Committee

Pankaj Agarwal     Duke University, USA
Michael Atkinson     University of Otago, New Zealand
Gill Barequet     Technion, Israel
Mark de Berg     Tech. Universiteit Eindhoven, The Netherlands
Gilles Brassard     University of Montreal, Canada
Leah Epstein     Tel Aviv University, Israel
Rolf Fagerberg     University of Southern Denmark, Denmark
Sandor Fekete     Tech. Universität Braunschweig, Germany
Faith Fich     University of Toronto, Canada

| | |
|---|---|
| Komei Fukuda | ETH Zürich, Switzerland |
| Anupam Gupta | Carnegie Mellon University, USA |
| Susanne Hambrusch | Purdue University, USA |
| Valerie King | University of Victoria, Canada |
| Rolf Klein | Universität Bonn, Germany |
| Peter Bro Miltersen | University of Aarhus, Denmark |
| Alistair Moffat | University of Melbourne, Australia |
| Ian Munro | University of Waterloo, Canada |
| Venkatesh Raman | The Institute of Mathematical Sciences, India |
| Rajeev Raman | University of Leicester, UK |
| Andrew Rau-Chaplin | Dalhousie University, Canada |
| Michiel Smid | Carleton University, Canada |
| Srinivasa Rao | University of Waterloo, Canada |
| Stefan Szeider | Durham University, UK |
| Peter Widmayer | ETH Zürich, Switzerland |
| Norbert Zeh | Dalhousie University, Canada |

## Referees

| | |
|---|---|
| Eyal Ackerman | Shlomo Hoory |
| Spyros Angelopoulos | Riko Jacob |
| Sunil Arya | Rohit Khandekar |
| Franz Aurenhammer | Christina Knauer |
| Mihai Badoiu | Stephen Kobourov |
| Amotz Bar-Noy | Peter Kornerup |
| Hajo Broersma | Sven O. Krumke |
| Hervé Brönnimann | Elmar Langetepe |
| Paz Carmi | Asaf Levin |
| Luc Devroye | Florent Madelaine |
| David Eppstein | Avner Magen |
| Esther (Eti) Ezra | Anil Maheshwari |
| Rudolf Fleischer | Daniel Marx |
| Greg Frederickson | Catherine McCartin |
| Tom Friedetzky | Ulrich Meyer |
| Stefan Funke | David Mount |
| Joachim Giesen | Kamesh Munagala |
| Ansgar Grüne | Jonathan Naor |
| Joachim Gudmundsson | Daniel Paulusma |
| Sariel Har-Peled | Leon Peeter |
| Refael Hassin | David Peleg |
| Mathias Hauptmann | Seth Pettie |
| Herman Haverkort | Jeff Phillips |
| Klaus Hinrichs | Helmut Prodinger |
| Winfried Hochstättler | Prabhakar Ragde |

Dror Rawitz
Vera Rosta
Nathan Rountree
Gerth S. Brodal
Peter Sanders
Saket Saurabh
Yuval Scharf
Avishay Sidlesky
Somnath Sikdar

Sagi Snir
Jan Vahrenhold
Yusu Wang
Birgitta Weber
Tony Wirth
Alik Zamansky
Uri Zwick
Rob van Stee

# Table of Contents

## Session 4A

## Session 4B

## Session 5

## Session 6A

## Session 6B

## Session 7A

## Session 7B

## Session 8A

# Towards a Theory of Algorithms

Allan Borodin

Department of Computer Science,
University of Toronto

**Abstract.** Undergraduate (and graduate) algorithms courses and texts often organize the material in terms of "algorithmic paradigms", such as greedy algorithms, backtracking, dynamic programming, divide and conquer, local search, primal-dual, etc. (but not etc. etc.). We seem to be able to intuitively describe what we have in mind when we discuss these classes of algorithms but rarely (if ever) do we attempt to define precisely what we mean by such terms as greedy, dynamic programming, etc. Clearly, a precise definition is required if we want to defend a statement such as "This is a difficult optimization problem.... in particular, there is no greedy algorithm that provides a good approximation for this problem".

In the context of combinatorial search and optimization problems, I will present precise models for some common basic paradigms. In a sense what we would really like to have are algorithmic models (e.g. for a greedy optimization algorithm) that are as widely accepted as the Church-Turing definition for "computable function". While this goal is probably much too ambitious, we would at least like to have models that capture most of the algorithms that fit within these common paradigms.

This talk is based on results from a number of papers. In particular, I will present precise definitions for greedy and greedy-like algorithms [2], simple dynamic programming and backtracking [3], and basic primal-dual/local ratio algorithms [1].

## References

1. A. Borodin, D. Cashman, and A. Magen. How well can primal dual and local ratio algorithims perform? In *Proceedings of the 32nd International Colloquium on Automata, Languages and Programming*, 2005.
2. A. Borodin, M. N. Nielsen, and Rackoff. (Incremental) priority algorithms. In *Proceedings of the 13th Annual ACM-SIAM Symposium on Discrete Algorithms*, 2002.
3. M. Alekhnovich, A. Borodin, J. Buresh-Oppenheim, R. Impagliazzo, A. Magen, and T. Pitassi. Toward a model for backtracking and dynamic programming. *Unpublished manuscript*, 2004.

F. Dehne, A. López-Ortiz, and J.-R. Sack (Eds.): WADS 2005, LNCS 3608, p. 1, 2005.
© Springer-Verlag Berlin Heidelberg 2005

# k-Restricted Rotation with an Application to Search Tree Rebalancing

Alejandro Almeida Ruiz[1], Fabrizio Luccio[2], Antonio Mesa Enriquez[3], and Linda Pagli[4]

[1] Universidad de Matanzas, Cuba
alejandro.almeida@umcc.cu
[2] Università di Pisa, Italy
luccio@di.unipi.it
[3] Universidad de la Habana, Cuba
tonymesa@matcom.uh.cu
[4] Università di Pisa, Italy
pagli@di.unipi.it

**Abstract.** The *restricted rotation distance* $d_R(S,T)$ between two binary trees $S$, $T$ of $n$ vertices is the minimum number of rotations by which $S$ can be transformed into $T$, where rotations can only take place at the root of the tree, or at the right child of the root. A sharp upper bound $d_R(S,T) \leq 4n - 8$ is known, based on the word metric of Thompson's group. We refine this bound to a sharp $d_R(S,T) \leq 4n-8-\rho_S-\rho_T$, where $\rho_S$ and $\rho_T$ are the numbers of vertices in the rightmost vertex chains of the two trees, by means of a very simple transformation algorithm based on elementary properties of trees. We then generalize the concept of restricted rotation to *k-restricted* rotation, by allowing rotations to take place at all the vertices of the highest $k$ levels of the tree. For $k = 2$ we show that not much is gained in the worst case, although the classical problem of rebalancing an AVL tree can be solved efficiently, in particular rebalancing after vertex deletion requires $O(\log n)$ rotations as in the standard algorithm. Finding significant bounds and applications for $k \geq 3$ is open.

**Keywords:** Rotation, Rotation distance, Binary tree, Search tree, AVL tree, Rebalancing, Data structures, Design of algorithms.

## 1 Tree Transformation via Rotations

We consider rooted binary trees of $n$ vertices, simply called *trees* in the following. These trees are relevant in different fields of computing, in particular they are used as *search trees* storing $n$ keys at the vertices in *infix order*, i.e. the keys in the left (respectively, right) subtree of each vertex $v$ precede (respectively, follow) the key in $v$ according to a given ordering [3]. Although completely general the following considerations are carried out having search trees in mind.

Rotations are well known local changes in the tree structure, preserving key ordering in search trees. A *right rotation* at a vertex $y$ raises its left child $x$ to

F. Dehne, A. López-Ortiz, and J.-R. Sack (Eds.): WADS 2005, LNCS 3608, pp. 2–13, 2005.

the place of $y$ while $y$ becomes the right child of $x$, and the original right child of $x$ becomes the left child of $y$ (the rest of the tree remains unchanged). A *left rotation* is symmetrical. If applied to vertex $x$, after the right rotation, it reconstructs the original tree. Rotations maintain the infix ordering of the keys, and are used to keep search trees *balanced*. That is, the height of a tree, hence the search time for a key, remains of $O(\log n)$ after insertions or deletions of vertices [3].

The *rotation distance* $d(S,T)$ between two trees $S$, $T$ of $n$ vertices is the minimum number of rotations by which $S$ can be transformed into $T$, where the rotations can take place at any vertex. As right and left rotation are one the inverse of the other, we have $d(S,T) = d(T,S)$. In a scholarly paper [9], Sleator, Tarjan, and Thurston proved that $d(S,T) \leq 2n - 6$ for any pair $S$, $T$ using hyperbolic geometry, and that this bound is sharp for large values of $n$. Mäkinen [6] showed that slightly weaker results can be obtained by simple arguments, and Luccio and Pagli [4] then gave an elementary constructive proof for the upper bound of [9]. No polynomial-time algorithm is known to compute $d(S,T)$, but estimates of this value were given by Pallo [7] and Rogers [8].

The new *restricted rotation distance* $d_R(S,T)$ was introduced by Cleary [1], where rotations can take place only at the root of the tree, or at the right child of the root. Cleary and Taback [2] then strongly improved the results of [1], proving a sharp upper bound $d_R(S,T) \leq 4n - 8$ for any pair $S$, $T$, and a stricter lower bound if $S$ and $T$ satisfy a special condition. Both these works are based on the word metric of Thompson's group.

The works of Cleary and Taback are remarkable for several reasons. First, the upper bound for $d_R(S,T)$ is only about twice the one for $d(S,T)$ "despite of the fact that it may take many rotations to move a vertex to one of the places where rotations are allowed" [1]. Second, the properties of an abstract algebric structure are transferred into the theory of data structures with brillant results. Finally, and important from an information processing point of view, allowing rotations only at selected vertices is relevant for example in handling distributed files where only a fraction of the vertices reside in a location (e.g., a server) where data can be updated.

Given an arbitrary tree $T$, let the *left arm* (respectively, *right arm*) of a vertex $v$ be the longest sequence of vertices reached from $v$ when descending $T$ via left (respectively, right) edges only. The arms of the root will be simply called the left and right arm of $T$, and the number of vertices in such arms will be denoted by $\lambda_T$ and $\rho_T$, respectively (see figure 1). The union of the left arm, the right arm, and the root, is called the *border* of $T$. In this work we do the following:

1. We refine the upper bound $d_R(S,T) \leq 4n - 8$ of [2] to $d_R(S,T) \leq 4n - 8 - \rho_S - \rho_T$, and prove that the new bound is sharp.[1] The bound of [2] can then be seen as a special case of the new one. The two bounds coincide for $\rho_S = \rho_T = 0$, but we shall see in the next section that the probability of this case to occur goes to zero for increasing $n$, and is strictly equal to zero for balanced search trees.

---

[1] This bound was anticipated in [5] without a matching lower bound.

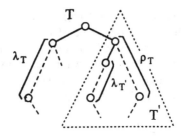

**Fig. 1.** The left and right arms of a tree $T$, of lengths $\lambda_T$ and $\rho_T$. $T'$ is a subtree of $T$, with left arm of length $\lambda_{T'}$

We underline that the new upper bound is the result of a very simple algorithm to transform $S$ into $T$, based on elementary properties of trees.

2. The concept of restricted rotation is generalized by performing rotations at all the vertices in the highest $k$ levels of the tree, as motivated by searching into large files. We study the new distance for $k = 2$, that is the rotations are allowed only at the root and at both its children, showing that not much is necessarily gained over the previous case. The case $k \geq 3$ will be left substantially open.

3. We finally show how restricted rotations are sufficient for rebalancing AVL trees efficiently, giving an insertion and a deletion algorithms that use rotations at the highest $k = 2$ levels only.

For brevity the proofs not crucial for following the development of our arguments are omitted.

## 2   The Basic Algorithm

To present our rotation algorithm we pose:

**Definition 1.** *A tree $T$ is an* (i,j)-*chain, with $i, j \geq 0$ and $i + j + 1 = n$, if all the vertices of $T$ lie on the border, with $\lambda_T = i$ and $\rho_T = j$.*

First we give a basic algorithm to transform any given tree into a fixed reference tree $R$ consisting of an (n-3,2)-chain. As the transformation process can be reverted by exchanging right and left rotations, transforming $S$ into $T$ will amount to transforming both $S$ and $T$ into $R$, then reverting the second transformation to get $S \to R \to T$.

Let *rot1-right, rot1-left* respectively denote the right and left rotation at the root, and let *rot2-right, rot2-left* denote the same rotations at the right child of the root. We have:

**Algorithm 1.** *Transforming a tree $T$ with $n \geq 3$ into an (n-3,2)-chain. $T'$ is the subtree rooted at the right child of the root of $T$ (see figure 1). $\overline{T}$ and $\overline{T}'$ denote the trees generated at each step of the transformation, starting with $\overline{T} = T$ and $\overline{T}' = T'$.*

**while** $\lambda_{\overline{T}} > 0$ **do** rot1-right;

**while** $\overline{T}' \neq (0,1)$-chain **do**

    **if** $\lambda_{\overline{T}'} > 0$ **then** rot2-right **else** rot1-left.

**Theorem 1.** *Algorithm 1 transforms any tree $T$ with $n \geq 3$ into an $(n\text{-}3,2)$-chain performing $2n - 4 - \rho_T$ rotations. (Proof omitted.)*

A tree $S$ can be transformed into another tree $T$ by applying algorithm 1 to transform $S$ into an $(n\text{-}3,2)$-chain $R$, then transforming $R$ into $T$ by reverting the rotations of algorithm 1 that would transform $T$ into $R$. We have:

**Corollary 1.** *For two arbitrary trees $S$, $T$ with $n \geq 3$ we have: $d_R(S,T) \leq 4n - 8 - \rho_S - \rho_T$.*

The upper bound of corollary 1 coincides with the one of [2] for $\rho_S = \rho_T = 0$, that is the roots of the two trees have no right descendants. In fact, this is the only condition under which the bound of [2] is sharp, and is very unlikely or even impossible to occur. For trees built at random, the probability $P_0$ of having $\rho_S = \rho_T = 0$ decreases exponentially with $n$. For search trees the major distinction is between trees built by random key insertions from the root, and balanced trees. In the first case we have $P_0 = 1/n^2$, as this corresponds to having the maximum key being inserted as the first item in both trees. For *AVL* balanced trees [3] it can be easily shown that both $\rho_S$ and $\rho_T$ must fall into the positive interval $[h/2, h-1]$, where $h$ is the index of the $h$-th Fibonacci number $F_h$ such that $F_h \leq n+1 < F_{h+2}$. Then we have the stricter upper bound: $d_R(S,T) \leq 4n - 8 - \Omega(\lg n)$.

Due to the result of [2], the bound of Corollary 1 is sharp because, for any $n$, there are trees with $\rho_S = \rho_T = 0$. However we shall proof that the bound is also sharp for arbitrary values of $\rho_S$ and $\rho_T$. To this end we start with a basic result. Identify the vertices of a tree $S$ with the integers 1 to $n$ in infix order, and let $\Pi^S(v)$ be the sequence of vertices in the path from the root to a vertex $v$. We have:

**Lemma 1.** *Let $S, T$ be two trees to be transformed into one another. And let $v$, $w = v + 1$ be two vertices such that $v \in \Pi^S(w)$ and $w \in \Pi^T(v)$. The transformation process must pass through an intermediate tree whose border contains $v$ and $w$. (Proof omitted.)*

It is interesting to note that the only way of inverting the relative ordering of $v$ and $w$ in a path $\Pi$, as required in the transformation of $S$ into $T$, is passing through an intermediate tree $Q$ with $v$ and $w$ on the border, where either $v$ is the root, $w$ is the right child of $v$, and rot1-left is applied to $Q$; or $v$ is the right child of the root, $w$ is the right child of $v$, and rot2-left is applied to $Q$.

We now prove that the bound of Corollary 1 is sharp for arbitrary values of $\rho_S$ and $\rho_T$. We have:

**Theorem 2.** *For any $n \geq 3$ there are trees $S$, $T$ with arbitrary values of $\rho_S$, $\rho_T$ such that $d_R(S,T) = 4n - 8 - \rho_S - \rho_T$.*

**Proof.** For $n = 3$ the proof trivially goes by exhaustion. For $n \geq 4$, consider the two trees $S$, $T$ of figure 2, with arbitrary values $\rho_S = s$ and $\rho_T = t$. The two vertices $n - 2$, $n - 1$ are in the same condition as $v$, $w$ of Lemma 1 because $(n-2) \in \Pi^S(n-1)$ and $(n-1) \in \Pi^T(n-2)$. Therefore $S$ must be transformed into an intermediate tree $Q$ with both $n - 2$ and $n - 1$ on the border. Since the subsequent rotations for transforming $Q$ into $T$ can be inverted, $d_R(S,T)$ can be computed as the minimum number of rotations for transforming both $S$ and $T$ into the same $Q$.

For bringing vertex $n - 1$ to the border of $S$, this vertex must be lifted at least up to level 3. This requires an initial sequence of at least $\rho_S - 1$ rot1-left rotations to lift the chain $s + 1, s + 2, \ldots, n - 1$, bringing $s + 1$ at level 3. The vertices of the chain can be moved from it, one after the other, only if the chain itself resides in the right subtree of the root, and a rot2-right rotation is applied to each of its vertices. For each vertex $x$ in $s+1, s+2, \ldots, n-2$ such a rotation must be followed by at least one rot1-left, because rot2-right does not lift the vertices following $x$ in the chain which are attached to the new right child of $x$. As there are $n - \rho_S - 2$ vertices in $s + 1, s + 2, \ldots, n - 2$, at least $2(n - \rho_S - 2)$ rotations are needed to bring vertex $n - 1$ at level 3, and a final rot2-right is needed to bring $n-1$ into the right arm, thus obtaining a tree with $n-2$ and $n-1$ on the border. The total number of rotations mentioned above is $2n - \rho_S - 4$. If only these rotations are performed, an (n-3,2)-chain $Q_1$ is obtained from $S$, that is the tree at minimum distance from $S$ with the vertices $n - 2$ and $n - 1$ on the border.

A similar reasoning holds for bringing vertex $n-2$ to the border of $T$. Only one rot2-right is now required for vertex $n-1$ as this lifts the chain $t+1, t+2, \ldots, n-2$ one level up. After $2n - \rho_T - 3$ rotations $T$ is then transformed into an (n-4,3)-chain $Q_2$, that is the tree at minimum distance from $T$ with vertices $n - 2$ and $n - 1$ on the border. Another rotation is now required to transform $Q_1$ into $Q_2$, for a total number of $4n - 8 - \rho_S - \rho_T$ rotations.                     □

## 3   Rotating at Different Tree Levels

Allowing rotations only at selected vertices may be relevant for handling very large files that mainly reside in secondary storage, while the root, and the vertices in its neighborhood, are brought into main memory where the rotations are performed. We then introduce the *k-restricted rotation distance* $d_{k-R}(S,T)$ with $k \geq 2$, where rotations (now called "k-restricted") can take place at the highest $k$ levels of the tree.

For $k = 2$, that is the rotations are allowed only at the root and at both its children. In addition to the notation *rot1-right, rot1-left, rot2-right, rot2-left* introduced before, we shall denote by *rot3-right, rot3-left* the right and left rotation at the left child of the root. The transformation process can now be

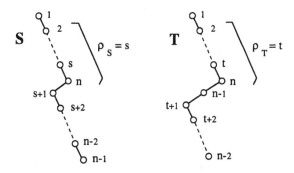

**Fig. 2.** Two trees for which the bound of Theorem 2 is met

refined by first transferring all the vertices to the right subtree of the root as in Algorithm 1, then inserting all the vertices into the border, with the leaves not already in the border brought into the left arm by only one rot3-left rotation. The algorithm is as follows:

**Algorithm 2.** *Transforming a tree $T$ with $n \geq 3$ into an (n-3,2)-chain, using 2-restricted rotations. $T'$ is the right subtree of the root of $T$. $T''$ is the left subtree of the root of $T'$. Start with $\overline{T} = T$, $\overline{T}' = T'$, $\overline{T}'' = T''$.*

    **while** $\lambda_{\overline{T}} > 0$ **do** rot1-right;

    **while** $(\overline{T}' \neq (0,1)\text{-chain}$ **and** $\overline{T}' \neq (1,0)\text{-chain}$ **and** $\overline{T}' \neq (1,1)\text{-chain})$ **do**

        **if** $\overline{T}'' = $ empty **then** rot1-left **else**

            **if** $\overline{T}'' = $ leaf **then** (rot1-left, rot3-left) **else** rot2-right;

    **if** $(\overline{T}' = (1,0)\text{-chain}$ **then** rot2-right;

    **if** $(\overline{T}' = (1,1)\text{-chain}$ **then** (rot2-right, rot1-left).

For a given tree $T$, let $F_T$ be the set of leaves not in the border of $T$, and let $\phi_T = |F_T|$. We have:

**Theorem 3.** *Algorithm 2 transforms any tree $T$ with $n \geq 3$ into an (n-3,2)-chain performing at most $2n - \rho_T - \phi_T - 3$ 2-restricted rotations. (Proof omitted.)*

Algorithm 2 admits a symmetric version requiring $2n - \lambda_T - \phi_T - 3$ 2-restricted rotations for transforming $T$ into a (2,n-3)-chain. To transform a tree $S$ into $T$, one or the other intermediate chain may be chosen whichever requires less rotations. We then have:

**Corollary 2.** *For two arbitrary trees $S$, $T$ with $n \geq 3$, we have: $d_{2-R}(S,T) \leq 4n - \mu - \phi_S - \phi_T - 6$, where $\mu = max(\lambda_S + \lambda_T, \rho_S + \rho_T)$.*

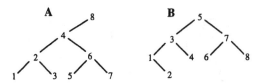

**Fig. 3.** Two opposite trees $A$, $B$ for the set $\{1, 2, ..., 8\}$

Corollary 2 gives an upper bound of $d_{2-R}(S,T) \leq 4n - \phi_S - \phi_T - 7$ since $\mu \geq 1$. Not much is gained with the introduction of the new rotations if $F_S$ and $F_T$ are small, in fact we now prove a lower bound for 2-restricted rotation, showing that only marginal room for improvement remains over the result of Corollary 2 in the worst case. First note that Lemma 1 holds unchanged since its proof can be immediately extended to the new allowed rotations. We also have:

**Lemma 2.** *Let a tree $T$ contain a subtree $C$ whose root is not in the border of $T$. Let $C$ contain $m$ vertices, of which $c \geq 0$ have two children. Then at least $m + c$ 2-restricted rotations are needed to move all the vertices of $C$ into the border of $T$. (Proof omitted.)*

To apply Lemma 1 and 2 to large portions of the trees we pose:

**Definition 2.** *Two opposite trees $A$, $B$ are built as search trees on a set of consecutive integers $\{h, h + 1, ..., h + 2^r - 1\}$, with $r \geq 1$, such that $A$ has vertex $h + 2^r - 1$ at the root, and the other integers of the set follow, arranged in a complete binary tree; and $B$ has vertex $h + 2^{r-1}$ at the root of a complete binary tree containing all the integers of the set, except for $h + 1$ that is displaced in the $(r + 1)$-th level.*

Figure 3 shows two opposite trees for the set $\{1, 2, ..., 8\}$, that is for $h = 1$ and $r = 3$. Due to Lemma 1, if two opposite trees $A$, $B$ are respectively part of two trees to be transformed into one another, all the vertices of $A$ and $B$ must be brought into the border of the tree under transformation. This consideration allows to define two trees $S$, $T$ for which a significant lower bound on $d_{2-R}(S,T)$ can be found. We have:

**Theorem 4.** *For any $n \geq 7$ there are two trees $S$, $T$ with non-empty arms such that $d_{2-R}(S,T) \geq 4n - \lambda_S - \lambda_T - \rho_S - \rho_T - \phi_S - \phi_T - 12$.*
**Proof.** Let $A_1$, $B_1$ and $A_2$, $B_2$ be two pairs of opposite trees containing $m_1$ and $m_2$ vertices, respectively. Note that $m_1/2 - 1$ and $m_2/2 - 1$ of these vertices have two children, and $m_1/2$ and $m_2/2$ are leaves. Let $S$, $T$ be two trees of $n \geq m_1 + m_2 + 3$ vertices (i.e., $n \geq 7$), composed of a border with non-empty arms, with $A_1$, $B_1$ attached to the right of the extremes of the left arms of $S$ and $T$, and $A_2$, $B_2$ attached to the left of the extremes of the right arms of $S$

and $T$. Note that vertices 1 and $n$ are respectively the extremes of the left and the right arms, and all the vertices of $A_1$, $B_1$, $A_2$, $B_2$ are not in the borders of $S$, $T$ and must be brought into such borders by Lemma 1. By Lemma 2 this requires a total of at least $2(m_1 + m_1/2 - 1) + 2(m_2 + m_2/2 - 1) = 3m_1 + 3m_2 - 4$ 2-restricted rotations. To start such rotations, however, the roots of $A_1$, $B_1$, $A_2$, $B_2$ must be brought to level 3, requiring at least $\lambda_S - 1 + \lambda_T - 1 + \rho_S - 1 + \rho_T - 1$ rotations. Since $n = m_1 + m_2 + \lambda_S + \rho_S + 1 = m_1 + m_2 + \lambda_T + \rho_T + 1$, and $\phi_S = \phi_t = (m_1 + m_2)/2$, the result follows with straightforward calculations.  □

The lower bound of Theorem 4 is not sharp, however, the difference with the upper bound of Corollary 2 depends only on the values of the arm lengths, besides a constant, and is independent of $n$. In the limit case $\lambda_S = \rho_S = \lambda_T = \rho_T = 1$ we have: $4n - \phi_S - \phi_t - 16 \leq d_{2-R}(S, T) \leq 4n - \phi_S - \phi_t - 8$, i.e. the difference between the bounds is a constant 8.

For $k \geq 3$ the situation is much more complex. For example, for $k = 3$ the left and the right subtrees of the root can be transfored independently, making use of the 2-restricted rotations in their upper two levels where the case $k = 2$ applies (note, however, that $S$ and $T$ must have the same root). Such a divide and conquer approach can then be extended to greater values of $k$. The study of ($k \geq 3$)-restricted rotation is fully open, some ideas can be found in [5].

## 4   Rebalancing AVL Trees

A natural question arises under the restricted rotations model, namely, can a balanced search tree, unbalanced after insertion or deletion of a vertex, be rebalanced efficiently? The answer is postive at least for AVL trees, in fact we will show how rebalancing one such a tree after vertex insertion or deletion with $O(\log n)$ 2-restricted rotations.[2]

Consider rebalancing after insertion. While the standard algorithm performs a unique (simple or double) rotation around the critical vertex, the key idea in the restricted rotation model is lifting the critical vertex up to the root with a sequence of rotations in the two highest levels of the tree, then performing the required AVL rotation at the root to rebalance the subtree locally unbalanced, and then moving this subtree down to its original position. In fact, when the critical vertex is at the root, simple and double AVL rotations can be expressed in terms of restricted rotations. In particular a simple right rotation and a simple left rotation correspond to rot1-right and rot1-left, respectively; a double right rotation corresponds to rot3-left followed by rot1-right; and a double left rotation corresponds to rot2-right followed by rot1-left.

In an AVL tree $T$, let $\Pi = a_0, a_1, ..., a_{k+1}$ be a subsequence of vertices encountered in the search path for insertion of a vertex. $\Pi$ starts with the root $a_0$ of $T$, reaches the critical vertex $a_{k-1}$ and its child $a_k$, and ends with the child

---

[2] The same result can be obtained with rotations at the root and at one of its children only, using a slightly more complicated procedure.

$a_{k+1}$ of $a_k$. Note that $a_{k-1}$ and $a_k$ enter in a simple AVL rotation, while $a_{k-1}$, $a_k$, and $a_{k+1}$ enter in a double AVL rotation. For performing at the root the required AVL rotation, all the vertices in $\Pi$, starting from $a_1$ and up to $a_k$ (for simple rotation) or to $a_{k+1}$ (for double rotation) are lifted to the root one after the other, then the AVL rotation is done, and then the vertices are moved down to their original posititions. These up and down movements are better expressed with an *ad hoc* notation.

For a vertex $x$ residing in level 2 or 3 of the tree we define $UP(x)$ as the operation of lifting $x$ up to the root, and $UP^{-1}(x)$ as the one of moving $x$ from the root down to its original position. $UP(x)$ must be performed before $UP^{-1}(x)$, and both operations depend on the position of $x$ in the tree when $UP(x)$ is performed. However, since the two operations may be separated by several time steps during which the tree may change, the actions taken by $UP(x)$ must be recorded for later execution of $UP^{-1}(x)$. The two operations are defined as follows:

if $x$ is the left child of the root **then** $UP(x)$=rot1-right, $UP^{-1}(x) =$ rot1-lef;
if $x$ is the right child of the root **then** $UP(x)$=rot1-left, $UP^{-1}(x)$=rot1-right;
if $x$ is the left child of the right child of the root **then**
  $UP(x)=$ rot2-right, rot1-left, $UP^{-1}(x)$=rot1-right, rot2-left;
if $x$ is the right child of the left child of the root **then**
  $UP(x)=$ rot3-left, rot1-right, $UP^{-1}(x)$=rot1-left, rot3-right.

Note that $UPs$ are equivalent to the AVL rotations executed in the upper two levels of the tree, and $UP^{-1}s$ are their inverses.

We now define the rebalancing algorithm, whose functioning can be followed on the example of figure 4 (integer keys not in the range $1:n$). Recall that $\Pi = a_0, a_1, ..., a_{k+1}$ is the search sequence. Note that the vertices in $\Pi$ get 'marked'.

**Algorithm 3.** *Rebalancing an AVL tree after insertion, using 2-restricted rotations. In the search sequence, $A = a_{k-1}$ is the critical vertex, $B = a_k$ is the child of $A$, and $C = a_{k+1}$ is the child of $B$.*

**if** an AVL simple rotation on $A$ is required **then**

1. mark $a_0$;
   **for** $i = 1$ **to** $k$ **do** mark $a_i$, $UP(a_i)$;
2. **if** $A$ is the left-child of $B$ **and** the left-child of $A$ is marked **then** rot3-right;
   **if** $A$ is the right-child of $B$ **and** the right-child of $A$ is marked **then** rot2-left;
   $a_{k-1} \leftarrow B$;
3. **for** $i = k - 1$ **downto** 1 **do** $UP^{-1}(a_i)$;

**if** an AVL double rotation on $A$ is required **then**

1. mark $a_0$;
   **for** $i = 1$ **to** $k + 1$ **do** mark $a_i$, $UP(a_i)$;

2. **if** the left-child of the left-child of $C$ is marked **then** rot3-right;
   **if** the right-child of the right-child of $C$ is marked **then** rot2-left;
   $a_{k-1} \leftarrow C$;
3. **for** $i = k - 1$ **downto** $1$ **do** $UP^{-1}(a_i)$.

We have:

**Theorem 5.** *Algorithm 3 correctly rebalances an AVL tree, unbalanced after vertex insertion, with $O(\log n)$ 2-restricted rotations. (Proof omitted.)*

In AVL trees rebalancing after deletion is more complicated than for insertion, because the height of the subtree rooted at the critical vertex always decreases by one after rebalancing at that vertex, thus possibly unbalancing the parent of the critical vertex. The unbalancing can then propagate up to the root so that $O(\log n)$ rotations are required in the worst case. However we can detect in advance the vertices on the search path requiring a rotation, called *candidates*, as the ones whose balance factor is different from zero and is of opposite sign with respect to the search path [3].

The stategy for rebalancing a tree after deletion, using a restricted set of rotations, is similar to that of insertion. As before all the vertices of the critical path are lifted one by one to the root. In addition, when each such a vertex is lifted, it is also rotated if satisfies the candidate condition. A rebalancing after deletion algorithm can then be derived with minor modifications from Algorithm 3, still requiring $O(\log n)$ rotations. This result seems to be interesting because the same number of rotations, in order of magnitude, is required by the standard AVL deletion algorithm where rotations are performed at all levels.

As before $a_0, \ldots, a_{k-1}$ is the search subsequence from the root to the critical vertex. Unlike for insertion, $a_k$ is now the child of $a_{k-1}$ *not* in the search path, and $a_{k+1}$ is the left child (respectively, right child) of $a_k$ if $a_k$ is the right child (respectively, left child) of $a_{k-1}$. The rebalancing algorithm is as follows. As usual we assume that the candidate vertices have already been detected during the search for the node to be deleted.

**Algorithm 4.** *Rebalancing an AVL tree after deletion, using 2-restricted rotations. In the search sequence, $A = a_{k-1}$ is the critical vertex, $B = a_k$ is the child of $A$, and $C = a_{k+1}$ is the child of $B$.*

**if** an AVL simple rotation on $A$ is required **then**
    1. **for** $i = 0$ **to** $k$ **do**
        **if** $a_i$ is a candidate **then**
            **if** $a_{i+1}$ is the left-child of $a_i$ **then** **let** $b_i$ = right-child of $a_i$;
            **if** $a_{i+1}$ is the right-child of $a_i$ **then** **let** $b_i$ = left-child of $a_i$;
            mark $b_i$, $UP(b_i)$;
        mark $a_i$, **if** $i \neq 0$ **then** $UP(a_i)$;

    **2. if**  $A$  is the left-child of  $B$  **and**  the left-child of  $A$  is marked  **then** rot3-right;

        **if**  $A$  is the right-child of  $B$  **and**  the right-child of  $A$  is marked  **then** rot2-left;

         $a_{k-1} \leftarrow B$ ;

    **3. for**  $i = k - 1$  **downto**  0 **do**

        **if**  $i \neq 0$  **then**  $UP^{-1}(a_i)$ ;

        **if**  $a_i$  is a candidate **then**  $UP^{-1}(b_i)$ ;

**if**  an AVL double rotation on  $A$  is required  **then**

    **1. for**  $i = 0$  **to**  $k + 1$  **do**

        **if**  $a_i$  is a candidate **then**

            **if**  $a_{i+1}$  is the left-child of  $a_i$  **then**  **let**  $b_i =$  right-child of  $a_i$ ;

            **if**  $a_{i+1}$  is the right-child of  $a_i$  **then**  **let**  $b_i =$  left-child of  $a_i$ ;

            mark  $b_i$ ,  $UP(b_i)$ ;

        mark  $a_i$ , **if**  $i \neq 0$  **then**  $UP(a_i)$ ;

    **2. if**  the left-child of the left-child of  $C$  is marked  **then**  rot3-right;

        **if**  the right-child of the right-child of  $C$  is marked  **then**  rot2-left;

         $a_{k-1} \leftarrow C$ ;

    **3. for**  $i = k - 1$  **downto**  0 **do**

        **if**  $i \neq 0$  **then**  $UP^{-1}(a_i)$ ;

        **if**  $a_i$  is a candidate **then**  $UP^{-1}(b_i)$ .

We then have the following corollary as a simple extension of Theorem 5:

**Corollary 3.** *Algorithm 4 rebalances an AVL tree after vertex deletion with*  $O(\log n)$  *2-restricted rotations.*

## 5  Concluding Remarks

Much work has to be done to complete this work. First, a sharp lower bound and/or a better algorithm for 2-restricted rotation have yet to be found. Moreover the present study is directed to worst case performance, while an efficient transformation algorithm for the average case is still unknown. ($k \geq 3$)-restricted rotations open a promising new line of investigation. More importantly, other applications of restricted rotations should be studied besides the AVL rebalancing problem.

    Another interesting approach for further studies could be investigating the relationship between group theory and our new results, along the line opened by Cleary and Taback. Not only this could lead to improve the bounds, but new significant interpretations of tree distance could possibly arise in group theory.

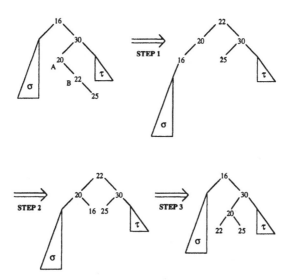

**Fig. 4.** Rebalancing an AVL tree after the insertion of the unbalancing vertex 25 (simple rotation), using 2-restricted rotations,. We have: $\Pi = a_0, a_1, a_2, a_3 = 16, 30, 20, 22$ and $k = 3$, $a_{k-1} = A = 20$ (critical vertex), $B = 22$. Step 1 of Algorithm 3 makes: $UP(a_1 = 30) = $ rot1-left; $UP(a_2 = 20) = $ rot3-left, rot1-right; $UP(a_3 = 22) = $ rot2-right, rot1-left. Step 2 makes: rot3-right, $a_2 = 22$. Step 3 makes: $UP^{-1}(a_2 = 22) = $ rot1-left, rot3-right; $UP^{-1}(a_1 = 30) = $ rot1-right

# References

1. S. Cleary, Restricted rotation distance between binary trees, *Information Processing Letters* 84 (2002) 333-338.
2. S. Cleary and J. Taback, Bounding restricted rotation distance, *Information Processing Letters* 88 (5) (2003) 251-256.
3. D.E. Knuth, *The Art of Computer Algorithms: Sorting and Searching*, Vol. 3, Addison-Wesley, Reading, MA, 1973.
4. F. Luccio and L. Pagli, On the upper bound on the rotation distance of binary trees, *Information Processing Letters* 31 (2) (1989) 57-60.
5. F. Luccio and L. Pagli, General restricted rotation distance, *Università di Pisa, Dipartimento di Informatica*, Tech. Rep. TR-04-22 (2004).
6. E. Mäkinen, On the rotation distance of binary trees, *Information Processing Letters* 26 (5) (1988) 271-272.
7. J. Pallo, An efficient upper bound of the rotation distance of binary trees, *Information Processing Letters* 73 (3-4) (2000) 87-92.
8. R. Rogers, On finding shortest paths in the rotation graph of binary trees, in: *Proc. Southeastern Internat. Conf. on Combinatorics, Graph Theory, and Computing*, Vol. 137 (1999) 77-95.
9. D.D. Sleator, R.E. Tarjan, W.R. Thurston, Rotation distance, triangulations, and hyperbolic geometry, *J. Amer. Math. Soc.* 1 (3) (1988) 647-681.

# Heap Building Bounds

Zhentao Li[1] and Bruce A. Reed[2]

[1] School of Computer Science, McGill University
zhentao.li@mail.mcgill.ca
[2] School of Computer Science, McGill University
breed@cs.mcgill.ca

**Abstract.** We consider the lower bound for building a heap in the worst case and the upper bound in the average case. We will prove that the supposedly fastest algorithm in the average case[2] does not attain its claimed bound and indeed is slower than that in [6]. We will then prove that the adversarial argument for the claimed best lower bound in the worst case[1] is also incorrect and the adversarial argument used yields a bound which is worse than that in [5] given by an information theory argument. Finally, we have proven a lower bound of $1.37n + o(n)$ for building a heap in the worst case.

## 1 Introduction

Heaps are a classical and commonly used implementation of priority queues. They are so fundamental that computer science students typically learn about them in their first year of university study. In this paper, we discuss bounds on building heaps.

We will prove that the supposedly fastest algorithm in the average case[2] does not attain its claimed bound and indeed is slower than that in [6]. We will then prove that the adversarial argument for the claimed best lower bound in the worst case[1] is also incorrect and the adversarial argument used yields a bound which is worse than that in [5] given by an information theory argument. Finally, we have proven a lower bound of $1.37n + o(n)$ for building a heap in the worst case. Forthwith the details.

A heap [7, 4, 3] is a binary tree in each node of which we have stored a key. The tree has a special shape. All of its levels are full except the last one. The nodes on the last level are all as much to the left of the tree as possible. A min-heap has the property that every node has value less than or equal to its children. All heaps in this paper are min-heaps. A perfect heap is a heap whose last level is full.

The height of a tree is defined as the number of arcs of the longest path from the root to a leaf. Therefore, a perfect heap of height $k$ has $2^{k+1} - 1$ nodes.

One of the attractive features of heaps is that they can be implemented using an array where the children of a node at A[i] are located at A[2i] and A[2i+1].

We will only consider building heaps in the comparisons model. That is, at each step, an algorithm chooses two keys and compares them to determine

F. Dehne, A. López-Ortiz, and J.-R. Sack (Eds.): WADS 2005, LNCS 3608, pp. 14–23, 2005.

which is bigger. Since we are dealing with min-heaps, we will call the winner of a comparison the key that is smaller and the loser of a comparison the key that is bigger.

The first heap-building algorithm due to Williams [7] runs in $O(n \log(n))$ by inserting the keys one by one. More precisely, it's worst case running time is $\sum_{i=1}^{n-1} \lfloor \log(i+1) \rfloor$. The key is added in a new leaf node so that the tree remains heap-shaped and "bubbled up" the tree until the heap order is restored.

A classical algorithm of Floyd [4] for building heaps from the bottom up yields an upper bound of $2n$ comparisons. This algorithm builds a heap on $n = 2^k - 1$ nodes by first recursively builds 2 heaps of size $2^{k-1} - 1$. It then "trickle down" another node to merge these two heaps. An information theory lower bound of $1.364n$ comparisons to build a heap on $n$ keys is shown in [5]. An algorithm which uses $1.521n$ comparisons on average is developed in [6] by combining ideas from Floyd's and Williams algorithm. Faster algorithms for building heaps in the worst case were developed with the aid of binomial trees.

The binomial tree of size 1 (height 0) is a single node. A binomial tree of height $k$ is defined as follows: It has a root that has $k$ children. The subtrees rooted at the children of the root are binomial trees of each of heights 1 to $k-1$. As in the min-heap, every node has a key whose value less or equal to that of its children. Clearly a binomial tree of height $k$ has $2^k$ nodes.

A binomial tree on $2^k$ nodes can be built recursively using $2^k - 1$ comparisons by first building two binomial trees of $2^{k-1}$ nodes and then comparing the keys at their roots. This is clearly best possible since we know that the root contains the min, any key that has not lost at least once could still be the min and each comparison can only make one key, that has not yet lost, lose.

Faster algorithms for building heaps on $2^k$ elements first build a binomial tree and then recursively convert this into a heap. As discussed in [5], this approach can be used to build a heap on $n$ nodes in $1.625n + o(n)$ comparisons in the worst case.

The contributions of this paper are threefold:

1. The algorithm shown in [2] claims to have an average case running time of $1.5n + o(n)$ or faster. We will show that the analysis gives a lower bound of $\frac{43}{28}n + o(n)$ which is slower than the algorithm shown in [6].
2. The authors of [1] claim that their adversary yields a lower bound of $1.5n + o(n)$ comparisons in the worst case. We will show that this adversary yields a lower bound which is at best $\frac{5}{4}n + o(n)$ comparisons. This is worse than that of the information theory lower bound of $1.364n$ comparisons [5].
3. We have proven a new lower bound of $1.3701 \ldots n + o(n)$ for building heaps.

In what follows, we consider only heaps of size $2^k$ and $2^k - 1$. This is not really a restriction. For example, to build a heap with 23 elements, we can first build the 15 element heap rooted at the left child of the root, then build the 7 elements heap rooted at the right child of the root and then "trickle down" the remaining element from the root using $2 \log(n)$ comparisons (see [3]) to construct our heap. In the same vein, if we can construct heaps of size $n = 2^k - 1$ in

$an + o(n)$ comparisons for all $k$, then we can build heaps of any size in $an + o(n)$ comparisons.

A pseudo-binomial tree is a binomial tree with one leaf missing somewhere in the tree.

## 2   Average Case Algorithm

The algorithm described by Carlsson and Chen [2] is as follows:

1. To build a perfect heap of size $n = 2^k - 1$, first build a binomial trees of height $i$ for $i = 1, 2, 4, \ldots, 2^{k-1}$.
2. Repeatedly compare the keys of the roots of two smallest trees until a pseudo-binomial tree of size $2^k - 1$ is created. Note that these first two steps take $n - 1$ comparisons in total.
3. Let $\bar{T}(2^k - 1)$ denote the number of comparisons required on average by this algorithm to transform a pseudo-binomial tree of size $2^k - 1$ into a heap. Note that $\bar{T}(1) = \bar{T}(3) = 0$ since these pseudo-binomial trees are heaps.

    Note that the subtree rooted at the children of the root are all binomial trees except for one which is a pseudo-binomial tree. If $k < 2$ then we have constructed the desired heap. otherwise we proceed depending on where the missing leaf is:

    **Case 1.** If the largest subtree of the root is the pseudo binomial tree, recurse on it. Then compare the keys of the roots of the other subtrees of the root to create a pseudo-binomial tree (i.e.:same as step 2) and recurse on it.

    This takes $\bar{T}(2^{k-1} - 1) + k - 2 + \bar{T}(2^{k-1} - 1)$ comparisons. It happens only if the min was in the largest binomial tree (before step 2) and this occurs $\frac{n+1}{2n}$ of the time as discussed in [2].

    **Case 2.** Otherwise, transform the largest subtree, $T$, of the root $R$, which is now a binomial tree, into a heap plus an extra element $x$. We do this as follows: The root $r$ of $T$ will be the root of the heap. The subheap rooted at the right child of $r$ will be formed from the union of the elements of the subtrees rooted at the children of $r$ except for the largest. The largest subtree of $T$ rooted at the left child of $r$ will be used to form the subheap rooted at the left child of $r$ (and will yield an extra element). To form the right subheap, we build a pseudo-binomial tree from the union of the trees under consideration and we recursively apply this algorithm starting at step 2. To form the subheap rooted at the left child of $r$, we recursively apply the procedure described in this paragraph.

    At this point, we have built a heap on $T - x$. We now need to build a heap on the elements not in $T - x + r$. To do so, we consider $x$ as a child of $R$. Recurse on the children of $R$, excluding $T$ but including $x$, starting from step 2.

    According to Carlsson and Chen, this takes $\sum_{i=2}^{k-2} \left( (i-1) + \bar{T}(2^i - 1) \right) + k - 1 + \bar{T}(2^{k-1} - 1)$ comparisons on average. Since $\bar{T}(3) = 0$ this is just

$2 + \sum_{i=3}^{k-1} \left( (i-1) + \bar{T}(2^i - 1) \right)$ comparisons. This happens $\frac{n-1}{2n}$ of the time.

4. Stop the recursion at the heaps of size 7 and build them in $\frac{9}{7}$ comparisons as discussed in [2].

It seems to us that Carlsson and Chen's analysis is faulty as they ignore important conditioning on $x$. We show now that their analysis is faulty even assuming their conditioning assumptions are correct. Accepting their hypothesis, we have:

**Theorem 1.** *For the algorithm in [2],* $\bar{T}(2^k - 1) \geq \frac{15}{28}2^k - k \quad \forall k \geq 3$

*Proof.* First note that $\bar{T}(7) = \frac{9}{7} \geq \frac{15}{28}2^k - k$ for this algorithm.
If $\bar{T}(2^i - 1) \geq \frac{15}{28}2^i - i$ for $i = 3, \ldots, k-1$ and $k \geq 4$ then

$$
\begin{aligned}
\bar{T}(2^k - 1) &= \frac{n-1}{2n}\left( 2 + \sum_{i=3}^{k-1} \left( (i-1) + \bar{T}(2^i - 1) \right) \right) \\
&\quad + \frac{n+1}{2n}\left( \bar{T}(2^{k-1} - 1) + k - 2 + \bar{T}(2^{k-1} - 1) \right) \\
&= \frac{n-1}{2n}\left( 2 + \sum_{i=3}^{k-1} \left( (i-1) + \bar{T}(2^i - 1) \right) \right) \\
&\quad + \frac{n+1}{2n}\left( k - 2 + 2\bar{T}(2^{k-1} - 1) \right) \\
&\geq \frac{n-1}{2n}\left( 2 + \sum_{i=3}^{k-1} \left( (i-1) + \frac{15}{28}2^i - i \right) \right) \\
&\quad + \frac{n+1}{2n}\left( k - 2 + 2\left( \frac{15}{28}2^{k-1} - k + 1 \right) \right) \\
&= \frac{n-1}{2n}\left( 2 + \frac{15}{28}\left( 2^k - 8 \right) - (k-3) \right) \\
&\quad + \frac{n+1}{2n}\left( \frac{15}{28}2^k - k \right) \\
&= \frac{n-1}{2n}\left( \frac{15}{28}2^k - k + \frac{5}{7} \right) + \frac{n+1}{2n}\left( \frac{15}{28}2^k - k \right) \\
&\geq \frac{n-1}{2n}\left( \frac{15}{28}2^k - k \right) + \frac{n+1}{2n}\left( \frac{15}{28}2^k - k \right) \\
&= \frac{15}{28}2^k - k
\end{aligned}
$$

$\therefore \bar{T}(2^k - 1) \geq \frac{15}{28}2^k - k$
   By induction, $\bar{T}(2^k - 1) \geq \frac{15}{28}2^k - k \quad \forall k \geq 3$

Note that this implies that the algorithm in [2] is worse than the algorithm in [6].

## 3   Worst Case Adversary

The adversary described by Carlsson and Chen [1] does the following: For all keys $x$, define $Up(x) = \{y|y < x\}$ and $Down(x) = \{y|y \geq x\}$. When comparing two keys $x$ and $y$, the adversary will answer as follows:

If $x \in Up(y)$, we must answer $x < y$. If $y \in Up(x)$, we must answer $y < x$.

If $x \notin Up(y)$ and $y \notin Up(x)$ then answer $x < y$ according to the first rule that can apply:

**Rule 1.** If $\|Down(x)\| > \|Down(y)\|$ then $x$ is the winner, otherwise
**Rule 2.** if $\|Up(x)\| < \|Up(y)\|$ then $x$ is the winner.
**Rule 3.** For all other cases, answer $x < y$.

We will now show a counter-example for which the adversarial argument given in [1] fails to attain the claimed bound of $1.5n$.

**Theorem 2.** *Given a complete heap $H$ of height $k \geq 2$, in which the key at its leaves have never won, a key Loser which has never won, and a set $S$ of $2^{k+1}+2^{k+2}$ keys which have not yet been compared, we can build in $\frac{5}{4}(2^{k+1}+2^{k+2})$ comparisons, against this adversary, a heap $H'$ of height $k+2$ containing $S$ and all the nodes of $H$ such that no leaf of $H'$ contains a key which has won a comparison.*

We proceed in the following way: We consider a node P with both children being leaves. We call the key at the left child L and the key at the right child R.

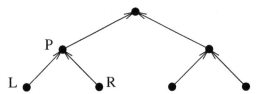

It is enough to prove the theorem for heaps of size 7 as we can treat the $2^{k-2}$ heaps of size 7 at the bottom of $H$ separately. In order to add 12 nodes to this heap, we will do the following:

**Step 1.**   – *Compare Loser to P. Loser will lose since $\|Down(Loser)\|=1$ (and $\|Down(Loser)\|$ remains 1 after this comparison).*

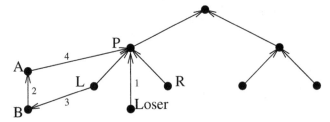

*A number $n$ on an edge is the $n$th comparison that we are making.*

**Step 2.**  – *Compare two keys in S and call the winner A and the loser B.*
- *Compare B to L. B will win since $\|Down(L)\| = \|Down(B)\| = 1$ and $\|Up(L)\| \geq 2$ while $\|Up(B)\| = 1$.*
- *Compare P to A. P will win since $\|Down(P)\| \geq 4$ and $\|Down(A)\| = 3$. We can redraw the tree to record the current information:*

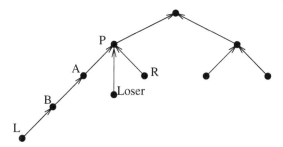

- *Compare two more keys in S and call the winner C and the loser D.*
- *Compare C to A. C will lose since $\|Down(C)\|=2$ and $\|Down(A)\|=3$.*
- *Compare a key $N_1$ in S to B and a key $N_2$ in S to C. The new keys will lose since $\|Down(B)\|=2$, $\|Down(C)\|=2$ and $\|Down(N_1)\|=\|Down(N_2)\|=1$.*

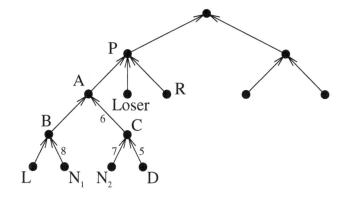

**Step 3.** *Do step 2 on R and P instead of L and P.*

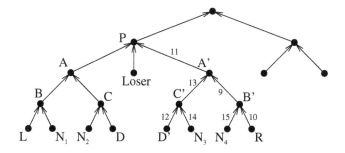

We have taken 15 comparisons to add 12 keys (Loser doesn't count as an added key). We can repeat this process using the same Loser key.

Note that we did not use the fact that the adversary chooses arbitrarily if both $\|Up\|$ and $\|Down\|$ are equal.

Also note that the only property that we used was that $\|Down(L)\| = 1$ (and $\|Down(R)\| = 1$), $\|Up(L)\| > 1$ (and $\|Up(R)\| > 1$) and $\|Down(P)\| \geq 2$). These properties are kept for the keys on the last two levels after we have inserted the new keys.

Here is a possible way of building the initial 7 nodes heap:

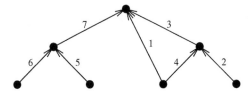

Note that we put the winner of the 3rd comparison at the root. This allows us to build perfect heaps of odd height. To build perfect heaps of even height, we can just start with a heap of 15 nodes instead.

Here is a possible way of building the 15 nodes heap:

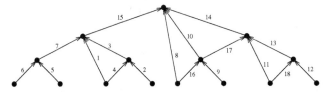

Now we can use Gonnet and Munro's algorithm [5] to build heaps of any height from perfect heaps.

Therefore, the lower bound that the adversary provides is at most $\frac{5}{4}n + O(\log^2 n)$ which is worse than the information theory lower bound of $1.362n$.

## 4    A Simple Adversary

We now describe an adversary which yields a lower bound of $1.3701\ldots n + o(n)$ comparisons for building a heap $H$ on $2^k - 1$ elements.

Since we are dealing with min-heaps, we will call the winner of a comparison the key that is smaller and the loser of a comparison the key that is bigger.

The adversary decides how to answer comparisons by looking at the first loss graph. This is a directed acyclic graph which contains, for every node $x$ which has lost, an edge from $x$ to the first node to which it lost. These are the only edges of the graph. Note that each component of this graph is a tree all of whose edges are directed towards some root. Note further that this graph changes as the algorithm progresses. Initially, it is empty and when the algorithm terminates, it has $n - 1$ edges.

There are $n - 1$ comparisons which are first losses and hence correspond to edges of the final first loss graph. We bound the number of comparisons used in building the heap which are not part of the final first loss graph. If in the final first loss graph, everybody but the minimum lost to the minimum, then since all but the top three nodes of the heap lose to somebody who is not the minimum, there must be at least $n - 3$ such extra comparisons in total.

More generally, our approach is to try to ensure that there are many vertices of large indegree high up (i.e. close to the root) in the final first loss graph.

Ideally, we would like the indegree in the first loss tree of a node to be less than the corresponding value for its parent. This is difficult to ensure as the indegree of $x$ may increase after its first loss. So instead, we colour an edge $xy$ of the first loss graph red if $x$ lost to $y$ before $y$ had lost and blue otherwise. We let $a(x)$ be the red indegree of $x$. We note that $a(x)$ can only increase during the heap building process and that after $x$ loses, $a(x)$ does not change. So if we insist that:

1. When comparing two nodes, the node with the higher $a$ value wins
2. If the two nodes have equal $a$ value then the node which has not yet lost wins

then the $a(x)$ value of any node is indeed strictly less than that of its parent in the final first loss tree.

We use $b(x)$ to denote $a(y)$ where $y$ is the parent of $x$ in the first loss graph. If $x$ has not yet lost, $b(x)$ is undefined.

We analyze this adversary using a LP. We define some variables such as $p_0 = \|\{x | a(x) = 0\}\|/n$ and $p_{(0,i)} = \|\{x | a(x) = 0, b(x) = i\}\|/n$ by looking at the first loss graph. We also define variables depending on the shape of the final heap that is built. They includes $q_{(0,1)} = \|\{x | a(x) = 0, b(x) = 1$ and $x$ is a leaf $\}\|/n$ as well as $q_{(0,1,4,5,2,3)} = \|\{x | a(x) = 4, b(x) = 5, x$ is not a leaf, $x$ has children $c_1, c_2$ and $a(c_1) = 0, b(c_1) = 1, a(c_2) = 2, b(c_2) = 3\}\|/n$. Our LP has a total of 4099 variables.

Recall that we consider only heaps on $n = 2^k - 1$ nodes. Thus every internal vertex has two children.

With these variables, we define some constraints by simply counting the nodes. An example of such constraint is $\sum p_i = 1$. We also have constraints due to the structure of a heap such as $\sum q_{(i,j)} = 0.5 + \frac{1}{2n}$ since there are $2^{k-1}$ leaves in a heap on $2^k - 1$ elements. We note that instead of an equation with RHS $= 0.5 + \frac{1}{2n}$, we use two inequalities which this implies. One with RHS $\leq \frac{1}{2}$ and the other with RHS $\geq \frac{1}{2} + \epsilon$ for a small but fixed $\epsilon$. Our final analysis uses 209 constraints.

To give a flavour of the LP, we close this section by proving here a lower bound of $(1 + \frac{1}{13})n - 1$ on the number of comparisons needed to build a heap. We actually consider the number of comparisons not in the first loss graph. We denote this number by $Extra$. We also use $p_1^*$ to denote $\|\{x | a(x) = 1, x$ has a unique child in the first loss graph$\}\|/n$

Now, consider a node $y$ with $a(y) \geq 1$. When $a(y)$ increases from 0 to 1, $y$ won a comparison against a node $x$. Since $y$ won, $a(x)$ had value 0. Since $a(y)$ increased, $x$ had not yet lost and $xy$ is a red edge of the first loss graph. It follows that for all $i$, $p_{(0,i)} \geq p_i$. Thus, $p_0 \geq \frac{1}{2}$ and more strongly

$$p_0 \geq \frac{1}{2} + \frac{1}{2}(p_{(0,i)} - p_i) \tag{1}$$

If $y$ has $a(y) \geq 2$ then for $a(y)$ to become 2, it either has to have at least two children in the first loss graph with $a$ value 0, or one child with $a$ value 0 and another with $a$ value 1. It follows easily that:

$$p_0 + p_1^* \geq \frac{3}{5} \qquad (2)$$

$$p_0 + p_1^* \geq \frac{2}{3} - \frac{1}{3}(p_{(0,1)} - p_1) \qquad (3)$$

Now, if a vertex $x$ with $a(x) = 0$ is a non-leaf of the heap, then $x$ must win a comparison and this is not a comparison of the first loss graph. If a vertex $x$ with $a(x) = 1$ which has a unique child $z$ in the first loss graph, is a non-leaf of the heap, then one of $z$ or $x$ must win a comparison which is not a comparison of the first loss graph. Since there are at most $\frac{n}{2} + \frac{1}{2}$ leaves, amortizing over $x$ and $z$ in the second case, we have:

$$Extra \geq \frac{1}{2}(p_0 n + p_1^* n - (\frac{n}{2} + \frac{1}{2})) \qquad (4)$$

Combining this with (2) gives

$$Extra \geq \frac{n}{20} - \frac{1}{4} \qquad (5)$$

The point of our LP is that we can combine one such argument with others. For example, if we consider only $x$ with $a(x) = 0$, for each such $x$ which is not a leaf in the heap, each child $y$ of $x$ in the heap has $a(x) = 0$, and must have lost twice so we have:

$$Extra \geq 2(p_0 n - \frac{n}{2} - \frac{1}{2}) \qquad (6)$$

Combining (3) with (4) gives:

$$Extra \geq \frac{n}{12} - \frac{1}{6}(p_{(0,1)} - p_1)n - \frac{1}{4} \qquad (7)$$

Combining (6) with (1) gives:

$$Extra \geq (p_{(0,1)} - p_1)n - 1 \qquad (8)$$

Finally, combining (7) and (8) which arise via two similar but different argument give

$$Extra \geq \frac{n}{13} - 1 \qquad (9)$$

Our LP of course involves much more sophisticated arguments and many more than two of them. Details can be found at

www.cs.mcgill.ca/~zli47/heaps.html

# 5   Conclusions

We have shown that the analysis of the average case running time of algorithm
[2] is incorrect. We have also shown that the lower bound on heap building from
[1] is incorrect. The full proof of our adversary which yields a lower bound of
$1.37n + o(n)$ is available at `www.cs.mcgill.ca/~zli47/heaps.html` .

# References

1. Svante Carlsson and Jingsen Chen. The complexity of heaps. In *Proceedings of the third annual ACM-SIAM symposium on Discrete algorithms*, pages 393–402. SIAM, 1992.
2. Svante Carlsson and Jingsen Chen. Heap construction: Optimal in both worst and average cases? In *Algorithms and Computation*, pages 254–263. Springer, 1995.
3. Thomas H. Cormen, Charles E. Leiserson, Ronald L. Rivest, and Clifford Sten. *Introduction to Algorithms*. The MIT Press, McGraw-Hill Book Company, 2 edition, 2001.
4. Robert W. Floyd. Algorithm 245: Treesort. *Commun. ACM*, 7(12):701, 1964.
5. Gaston H Gonnet and J Ian Munro. Heaps on heaps. *SIAM Journal of Computing*, 15(4):964–971, 1986.
6. C. J. McDiarmid and B. A. Reed. Building heaps fast. *J. Algorithms*, 10(3):352–365, 1989.
7. J. W. J. Williams. Algorithm 232: Heapsort. *Commun. of the ACM*, 7(6):347–348, 1964.

# The Multi-radius Cover Problem

Refael Hassin and Danny Segev

School of Mathematical Sciences,
Tel-Aviv University, Tel-Aviv 69978, Israel
{hassin, segevd}@post.tau.ac.il

**Abstract.** Let $G = (V, E)$ be a graph with a non-negative edge length $l_{u,v}$ for every $(u, v) \in E$. The vertices of $G$ represent locations at which transmission stations are positioned, and each edge of $G$ represents a continuum of demand points to which we should transmit. A station located at $v$ is associated with a set $R_v$ of allowed transmission radii, where the cost of transmitting to radius $r \in R_v$ is given by $c_v(r)$. The *multi-radius cover* problem asks to determine for each station a transmission radius, such that for each edge $(u, v) \in E$ the sum of the radii in $u$ and $v$ is at least $l_{u,v}$, and such that the total cost is minimized.

In this paper we present LP-rounding and primal-dual approximation algorithms for discrete and continuous variants of multi-radius cover. Our algorithms cope with the special structure of the problems we consider by utilizing greedy rounding techniques and a novel method for constructing primal and dual solutions.

## 1 Introduction

### 1.1 Problem Definition

In this paper we consider the following generalization of the vertex cover problem. Let $G = (V, E)$ be a graph with a non-negative edge length $l_{u,v}$ for every $(u, v) \in E$. The vertices of $G$ represent locations at which transmission stations are positioned, and each edge of $G$ represents a continuum of demand points to which we should transmit. A station located at $v$ is associated with a set of allowed transmission radii $R_v = \{r_0^v, \ldots, r_{k_v}^v\}$, $0 = r_0^v < \cdots < r_{k_v}^v$, where the cost of transmitting to radius $r_i^v$ is $c_{v,i}$. Without loss of generality, $0 = c_{v,0} < \cdots < c_{v,k_v}$ for every $v \in V$. The *multi-radius cover* problem (MRC) asks to determine for each station a transmission radius, such that for each edge $(u, v) \in E$ the sum of the radii in $u$ and $v$ is at least $l_{u,v}$, and such that the total cost is minimized.

When all edges are of unit length and the set of transmission radii from each station is $\{0, 1\}$, MRC reduces to the vertex cover problem. Therefore, hardness results regarding vertex cover extend to MRC. In particular, it is NP-hard to approximate MRC to within any factor smaller than 1.3606 [4]. We note that the currently best approximation guarantee for vertex cover in general graphs is $2 - o(1)$ [2, 6].

The MRC problem introduces two new difficulties when compared to the vertex cover problem. Consider an edge $(u, v) \in E$. The first difficulty in MRC is the dependence between the transmission radii we choose for $u$ and $v$, as their sum must be at

F. Dehne, A. López-Ortiz, and J.-R. Sack (Eds.): WADS 2005, LNCS 3608, pp. 24–35, 2005.

least $l_{u,v}$. In vertex cover, the edge $(u,v)$ is covered if we pick one of its endpoints, regardless of whether or not we pick the other. The second difficulty in MRC is that unlike vertex cover, in which we either pick a vertex or not, there are several "degrees" to which we can pick each vertex, corresponding to the set of allowed transmission radii.

We also study the *expensive-cheap vertex cover* problem (EC-VC). Given a graph $G = (V, E)$, at each vertex $v$ we can locate an expensive facility with cost $\beta_v$, or a cheap facility with cost $\alpha_v$. An edge $(u, v)$ is said to be covered if an expensive facility is located at $u$ or at $v$, or if cheap facilities are located at both $u$ and $v$. The objective is to locate expensive and cheap facilities with minimum total cost, such that all edges of $G$ are covered.

EC-VC is a special case of MRC that is obtained when all edges are of unit length, and for each $v \in V$ we have $R_v = \{0, \frac{1}{2}, 1\}$ with costs $\alpha_v$ and $\beta_v$ for transmitting to radii $\frac{1}{2}$ and 1, respectively. Even this restricted problem is a generalization of vertex cover, since when $\alpha_v = \beta_v$ for every $v \in V$ there is no motivation to locate a cheap facility at any vertex.

We also discuss the *piecewise linear multi-radius cover* problem (PL-MRC), a continuous variant of MRC in which the set of allowed transmission radii is $\mathbb{R}_+$, and we are given a transmission cost function $c_v : \mathbb{R}_+ \to \mathbb{R}_+$ for each station $v$. Each $c_v$ is a non-decreasing piecewise linear function, and we assume that it is given by specifying the pairs $\langle r, c_v(r) \rangle$ for every $r \in BP_v$, where $BP_v$ is the set of breakpoints of $c_v$.

## 1.2   Our Results

In Section 2 we present an LP-rounding algorithm for EC-VC. It uses the optimal fractional solution to identify a special set of vertices, at which facilities are located. It then determines the type of facility to locate at each special vertex based on a greedy rule: Always locate a cheap facility, unless it would lead to an infeasible solution. We prove that by applying this simple rule, the algorithm constructs a feasible solution whose cost is at most twice the optimum.

We then develop in Section 3 a linear time primal-dual algorithm with a similar approximation guarantee. This algorithm simultaneously constructs an integral primal solution and a feasible dual solution, so that their values are provably close to each other. We prove that the total cost of the solution we obtain is at most twice the optimum, by showing how to separately charge the cost of locating cheap and expensive facilities to the dual variables. A crucial property of the algorithm, that guarantees a linear run time, is that in each step we locate an expensive facility or a pair of cheap facilities that cover at least one edge.

We proceed in Section 4 to provide an LP-relaxation of the MRC problem, in which the covering constraint is specified in terms of vertices and transmission radii. Our formulation hides the actual lengths of edges and radii using a special indexing scheme. We present a primal-dual algorithm that is guaranteed to obtain a feasible integral solution whose cost is at most twice that of the optimal fractional solution. Our algorithm initially transmits to radius $r_0^v$ from each station $v$, and in each step it slightly increases the transmission radius of at least one station. This process eventually results in a feasible solution, and is guided by a novel method for constructing the primal and dual solutions. This method enables us to easily analyze the algorithm, although it appears to be relatively challenging to ensure that feasibility of the dual solution is maintained in each step.

Finally, in Section 5 we discuss the hardness of approximating PL-MRC, and in particular prove that this problem is as hard to approximate as vertex cover. We then present a polynomial time $(2 + \epsilon)$-approximation algorithm for every fixed $\epsilon > 0$. This algorithm is based on a polynomial time reduction to MRC by constructing a discrete set of transmission radii for each station, that depends on the cost function, $\epsilon$, and additional parameters. This immediately gives the stated result, by using the primal-dual algorithm in Section 4 and guessing the remaining parameters.

## 2 An LP-Rounding Algorithm for EC-VC

In this section we first describe a natural LP-relaxation of the EC-VC problem. We then present an algorithm that rounds the optimal fractional solution to a feasible integral solution, while increasing the cost by a factor of at most 2.

### 2.1 A Linear Program

We suggest the following LP-relaxation of EC-VC[1]:

$$\text{minimize} \quad \sum_{v \in V} \alpha_v z_v + \sum_{v \in V} \beta_v y_v \qquad \text{(LP)}$$

$$\text{subject to} \quad z_v \geq x_e \qquad \forall e \in E, v \in e \qquad (2.1)$$

$$y_u + y_v + x_e \geq 1 \qquad \forall e = (u, v) \in E \qquad (2.2)$$

$$x_e, y_v, z_v \geq 0 \qquad \forall e \in E, v \in V \qquad (2.3)$$

In an integral solution, the variable $z_v$ indicates whether a cheap facility is located at $v$, and the variable $y_v$ indicates whether an expensive facility is located at $v$. In addition, the variable $x_e$ indicates whether the edge $e$ is covered by cheap facilities located at its endpoints. Constraint (2.1) ensures that no edge is covered by cheap facilities unless we indeed locate them. Constraint (2.2) ensures that each edge $(u, v)$ is covered by locating an expensive facility at $u$ or at $v$, or by locating cheap facilities at both $u$ and $v$.

### 2.2 The Algorithm

Let $(x^*, y^*, z^*)$ be an optimal fractional solution to (LP). We can assume without loss of generality that for each $(u, v) \in E$ we have $x^*_{u,v} = \min\{z^*_u, z^*_v\}$. We identify a special set of vertices $S = \{v \in V : y^*_v + z^*_v \geq \frac{1}{2}\}$, and locate facilities as follows. For each $v \in S$, we locate a cheap facility at $v$ if and only if $N(v) \subseteq S$, and otherwise we locate an expensive facility there. We do not locate a facility at any $v \notin S$.

In Lemma 1 and Theorem 2 we show that the solution constructed by the algorithm is feasible, and prove that its cost is at most twice the optimum of (LP), which is a lower bound on the cost of any feasible solution to the EC-VC problem.

---

[1] Although EC-VC can be formulated as an integer covering LP, we prefer to use the current formulation to simplify the presentation.

**Lemma 1.** *The facilities located by the LP-rounding algorithm cover all edges.*

*Proof.* We first prove that $S$ is a vertex cover in $G$. Suppose that there is an edge $(u, v) \in E$ such that $u, v \notin S$. By definition of $S$, $y_u^* + z_u^* < \frac{1}{2}$ and $y_v^* + z_v^* < \frac{1}{2}$. It follows that

$$y_u^* + y_v^* + x_{u,v}^* = y_u^* + y_v^* + \min\{z_u^*, z_v^*\} \leq (y_u^* + z_u^*) + (y_v^* + z_v^*) < 1 \;,$$

and $(x^*, y^*, z^*)$ is not a feasible solution to (LP), a contradiction.

Now consider an edge $(u, v) \in E$. Since we locate a facility at each vertex in $S$, this edge is clearly covered if $u, v \in S$. Otherwise, since $S$ is a vertex cover, we can assume without loss of generality that $u \in S$. As $v \notin S$, we locate an expensive facility at $u$, and $(u, v)$ is covered.    □

**Theorem 2.** *The cost of the solution constructed by the LP-rounding algorithm is at most* $2 \cdot \mathrm{OPT(LP)}$.

*Proof.* Let $v \in S$. We claim that the cost of the facility we locate at $v$ is at most twice the cost of the optimal fractional facilities located at $v$. There are two cases:

1. $N(v) \subseteq S$. In this case the cost of the cheap facility we locate at $v$ is

$$\alpha_v \leq 2\alpha_v(z_v^* + y_v^*) \leq 2(\alpha_v z_v^* + \beta_v y_v^*) \;.$$

2. $N(v) \not\subseteq S$, that is, there is a vertex $u \in N(v)$ such that $u \notin S$. Since $(x^*, y^*, z^*)$ is feasible for (LP),

$$1 \leq y_u^* + y_v^* + x_{u,v}^* = y_u^* + y_v^* + \min\{z_u^*, z_v^*\} \leq y_u^* + y_v^* + z_u^* \;,$$

and $y_v^* \geq 1 - (y_u^* + z_u^*) > \frac{1}{2}$. Therefore, the cost of the expensive facility we locate at $v$ is

$$\beta_v \leq 2\beta_v y_v^* \leq 2(\alpha_v z_v^* + \beta_v y_v^*) \;.$$

This implies that the combined cost of the facilities we locate is

$$\sum_{\substack{v \in S: \\ N(v) \subseteq S}} \alpha_v + \sum_{\substack{v \in S: \\ N(v) \not\subseteq S}} \beta_v \leq 2 \sum_{v \in S} (\alpha_v z_v^* + \beta_v y_v^*) \leq 2 \cdot \mathrm{OPT(LP)} \;. \qquad \square$$

## 3    A Primal-Dual Algorithm for EC-VC

In this section we present a linear time primal-dual algorithm with an approximation ratio of 2. Our algorithm is based on the vertex cover primal-dual algorithm due to Bar-Yehuda and Even [1]. However, it departs from their approach, as we use the linear program (LP) and its dual, that do not form a covering-packing pair.

## 3.1   The Dual Program

The dual of the linear program (LP) is:

$$\text{maximize} \quad \sum_{e \in E} t_e \qquad \qquad \text{(DLP)}$$

$$\text{subject to} \quad \sum_{e \in \delta(v)} s_{e,v} \le \alpha_v \qquad \forall v \in V \qquad (3.1)$$

$$\sum_{e \in \delta(v)} t_e \le \beta_v \qquad \forall v \in V \qquad (3.2)$$

$$t_e - s_{e,u} - s_{e,v} \le 0 \qquad \forall e = (u,v) \in E \qquad (3.3)$$

$$s_{e,v}, t_e \ge 0 \qquad \forall e \in E, \, v \in e \qquad (3.4)$$

By weak duality, the cost of any feasible solution to (DLP) provides a lower bound on the cost of the optimal solution to (LP), which is a lower bound on the cost of the EC-VC problem.

## 3.2   The Algorithm

Given a feasible dual solution $(s, t)$, a vertex $v$ is called $\alpha$-*tight* if $\sum_{e \in \delta(v)} s_{e,v} = \alpha_v$, and $\beta$-*tight* if $\sum_{e \in \delta(v)} t_e = \beta_v$. We define an associated primal solution by locating expensive facilities at $\beta$-tight vertices and cheap facilities at $\alpha$-tight remaining vertices. Therefore, the primal solution covers an edge $(u, v)$ if at least one of $u$ and $v$ is $\beta$-tight, or if both $u$ and $v$ are $\alpha$-tight.

We begin with the trivial dual solution $s = t = 0$. We then consider the edges of $G$ in an arbitrary order, and for each edge $e = (u, v)$ we perform a *maximal increment step*. The goal of this step is to increase the value of $t_e$, which is currently 0, as much as possible. If we increase $t_e$ by $\epsilon > 0$, to maintain feasibility we also have to increase $s_{e,u} + s_{e,v}$ by at least $\epsilon$. Therefore, there are two limiting bounds:

1. In addition to (3.3), the variable $t_e$ appears twice in (3.2): $\sum_{e' \in \delta(u)} t_{e'} \le \beta_u$ and $\sum_{e' \in \delta(v)} t_{e'} \le \beta_v$, since $e \in \delta(u) \cap \delta(v)$. This implies that we cannot increase $t_e$ by more than $\epsilon_1 = \min\{\beta_u - \sum_{e' \in \delta(u)} t_{e'}, \beta_v - \sum_{e' \in \delta(v)} t_{e'}\}$.

2. In addition to (3.3), the variable $s_{e,u}$ appears once in (3.1), $\sum_{e' \in \delta(u)} s_{e',u} \le \alpha_u$, and we cannot increase $s_{e,u}$ by more than $\epsilon_2 = \alpha_u - \sum_{e' \in \delta(u)} s_{e',u}$. Similarly, for $s_{e,v}$ we have $\sum_{e' \in \delta(v)} s_{e',v} \le \alpha_v$, and we cannot increase $s_{e,v}$ by more than $\epsilon_3 = \alpha_v - \sum_{e' \in \delta(v)} s_{e',v}$.

Let $\epsilon = \min\{\epsilon_1, \epsilon_2 + \epsilon_3\}$. We increase $t_e$ by $\epsilon$, and balance this by increasing $s_{e,u}$ by $\epsilon'_2$ and $s_{e,v}$ by $\epsilon'_3$, for some $\epsilon'_2, \epsilon'_3 \ge 0$ that satisfy $\epsilon'_2 \le \epsilon_2$, $\epsilon'_3 \le \epsilon_3$ and $\epsilon'_2 + \epsilon'_3 = \epsilon$.

Let $(s', t')$ be the final dual solution, and let $(x', y', z')$ be the indicator vector of the associated primal solution: $y'_v = 1$ and $z'_v = 0$ for every $\beta$-tight vertex; $y'_v = 0$ and $z'_v = 1$ for every $\alpha$-tight remaining vertex; $y'_v = z'_v = 0$ for all other vertices; $x'_e = \min\{z'_u, z'_v\}$ for every $e = (u, v) \in E$.

### 3.3    Analysis

In the following we first prove that the integral primal solution we obtained is indeed feasible for (LP). We then suggest a charging scheme that specifies how the cost of the facilities we locate is paid for using the dual variables. Specifically, we show that the cost of these facilities can be paid for, such that no edge $e \in E$ is charged more than twice the value of its dual variable $t'_e$. It follows that the approximation ratio of our algorithm is 2.

**Lemma 3.** $(x', y', z')$ *is a feasible solution to* (LP).

*Proof.* By definition of $(x', y', z')$, the constraints (2.1) and (2.3) are clearly satisfied. Suppose that the constraint (2.2) is not satisfied for some edge $e = (u, v)$, that is, $y'_u + y'_v + x'_e < 1$. Since the solution is integral, $y'_u = y'_v = x'_e = 0$. By definition of $y'$, we have $\sum_{e' \in \delta(u)} t'_{e'} < \beta_u$ and $\sum_{e' \in \delta(v)} t'_{e'} < \beta_v$. Let $\epsilon'_1 = \min\{\beta_u - \sum_{e' \in \delta(u)} t'_{e'}, \beta_v - \sum_{e' \in \delta(v)} t'_{e'}\} > 0$. By definition of $x'$, $x'_e = \min\{z'_u, z'_v\} = 0$, and we assume without loss of generality that $z'_u = 0$. Finally, by definition of $z'$ and since $y'_u = 0$, we have $\sum_{e' \in \delta(u)} s'_{e', u} < \alpha_u$. Let $\epsilon'_2 = \alpha_u - \sum_{e' \in \delta(u)} s'_{e', u} > 0$ and $\epsilon'_3 = \alpha_v - \sum_{e' \in \delta(v)} s'_{e', v} \geq 0$.

Since the values of the dual variables never decrease during the construction of the dual solution, it follows that $t_e$ could have been further increased in the maximal increment step of $e$ by at least $\min\{\epsilon'_1, \epsilon'_2 + \epsilon'_3\} > 0$. This contradicts the fact that $t_e$ was maximally increased.                                                    □

In Theorem 4 we separately charge the cost of locating cheap and expensive facilities to the dual variables. We remark that the dual solution $(s', t')$ satisfies constraint (3.3) with equality, for every $e \in E$. This follows from observing that in each maximal increment step we increase $t_e$ and $s_{e,v} + s_{e,u}$ by exactly the same value, and these variables are not changed later.

**Theorem 4.** *The primal-dual algorithm constructs in linear time a solution whose cost is at most* $2 \cdot \mathrm{OPT(LP)}$.

*Proof.* Let $A = \{v \in V : z'_v = 1\}$ and $B = \{v \in V : y'_v = 1\}$ be the sets of vertices at which we locate cheap and expensive facilities, respectively. Clearly, $A$ and $B$ are disjoint. The combined cost of the cheap and expensive facilities we locate is

$$
\begin{aligned}
\sum_{v \in V} \alpha_v z'_v + \sum_{v \in V} \beta_v y'_v &= \sum_{v \in A} \alpha_v + \sum_{v \in B} \beta_v \\
&= \sum_{v \in A} \sum_{e \in \delta(v)} s'_{e,v} + \sum_{v \in B} \sum_{e \in \delta(v)} t'_e \\
&\leq \sum_{v \in A} \sum_{e = (u,v) \in \delta(v)} (s'_{e,v} + s'_{e,u}) + \sum_{v \in B} \sum_{e \in \delta(v)} t'_e \\
&= \sum_{v \in A} \sum_{e \in \delta(v)} t'_e + \sum_{v \in B} \sum_{e \in \delta(v)} t'_e \ ,
\end{aligned}
$$

where the second equality holds since each $v \in A$ is $\alpha$-tight and each $v \in B$ is $\beta$-tight, and the last equality holds since the dual solution $(s', t')$ satisfies constraint (3.3) with equality. Since each dual variable $t_e$ appears at most twice in the right-hand-side of this inequality,

$$\sum_{v \in V} \alpha_v z'_v + \sum_{v \in V} \beta_v y'_v \leq 2 \sum_{e \in E} t'_e \leq 2 \cdot \text{OPT(LP)} \ ,$$

where the last inequality holds since $(s', t')$ is a feasible dual solution, and its cost is a lower bound on OPT(LP).

Our algorithm runs in time $O(|V| + |E|)$, since it consists of exactly $|E|$ maximal increment steps. By keeping track of the sums in each constraint of (DLP), we can compute the $\epsilon$'s and update the variables in a single step by performing a constant number of operations.                                                                      □

## 4     A Primal-Dual Algorithm for MRC

In this section we suggest an asymmetric LP-relaxation of the MRC problem. In contrast with the relaxation of EC-VC, in which the covering constraint is symmetric with respect to the endpoints of each edge, here we make use of a relaxation in which the covering constraint is specified in terms of vertices and transmission radii. We then present a primal-dual algorithm that is guaranteed to obtain a feasible integral solution whose cost is at most twice that of the optimal fractional solution. Our algorithm is based on a novel method for constructing the primal and dual solutions. This method shifts the effort in the analysis from proving the approximation guarantee of the algorithm to ensuring that feasibility of the dual solution is maintained in each step.

### 4.1     A Linear Program and Its Dual

Our formulation is based on the following observation. Suppose we transmit to radius $r_i^v$ from the station located at $v$. This covers all demand points on the edges adjacent to $v$ that are within distance of at most $r_i^v$ from $v$. Therefore, from each $u \in N(v)$ we must transmit to radius at least $l_{u,v} - r_i^v$. This enables us to hide the actual lengths of edges and radii in the linear program by using a special indexing scheme.

For every $v \in V, i = 0, \ldots, k_v$ and $u \in N(v)$, if we transmit from $v$ to radius $r_i^v$, we must transmit from $u$ to radius $r_j^u$, $j \geq \mathbb{I}_u^v(i)$, where $\mathbb{I}_u^v(i) = \min\{j : r_j^u \geq l_{u,v} - r_i^v\}$ if there exists some $0 \leq j \leq k_u$ such that $r_j^u \geq l_{u,v} - r_i^v$, and $\mathbb{I}_u^v(i) = \infty$ otherwise. Note that if $\mathbb{I}_u^v(i) = \infty$, we cannot transmit from $v$ to radius $r_i^v$. In addition, we assume that $r_{k_u}^u + r_{k_v}^v \geq l_{u,v}$ for every $(u, v) \in E$, or otherwise there is no feasible solution.

Using this notation, we suggest the following LP-relaxation of MRC:

$$\text{minimize} \quad \sum_{v \in V} \sum_{i=0}^{k_v} c_{v,i} x_{v,i} \tag{LP'}$$

$$\text{subject to} \quad \sum_{i=0}^{k_v} x_{v,i} \geq 1 \qquad \forall v \in V \tag{4.1}$$

$$\sum_{j=\mathbb{I}_u^v(i)}^{k_u} x_{u,j} \geq \sum_{j=0}^{i} x_{v,j} \qquad \forall v \in V, \, i = 0, \ldots, k_v, \, u \in N(v) \qquad (4.2)$$

$$x_{v,i} \geq 0 \qquad \forall v \in V, \, i = 0, \ldots, k_v \qquad (4.3)$$

In an integral solution, the variable $x_{v,i}$ indicates whether we transmit to radius $r_i^v$ from $v$. Constraint (4.1) ensures that we choose at least one radius for each vertex. Clearly, there is no motivation to choose multiple radii. Constraint (4.2) ensures that if we transmit from $v$ to radius at most $r_i^v$, then we transmit from $u \in N(v)$ to radius at least $r_{\mathbb{I}_u^v(i)}^u$. The dual of the linear program (LP′) is:

maximize $\quad \displaystyle\sum_{v \in V} y_v \qquad\qquad\qquad\qquad\qquad\qquad\qquad$ (DLP′)

subject to $\quad y_v + \displaystyle\sum_{\substack{u \in N(v) \\ i:t \geq \mathbb{I}_v^u(i)}} z_{u,v,i} - \sum_{\substack{u \in N(v) \\ i:i \geq t}} z_{v,u,i} \leq c_{v,t} \qquad \begin{array}{l} \forall v \in V, \\ t = 0, \ldots, k_v, \end{array} \qquad (4.4)$

$\qquad\qquad y_v, \, z_{v,u,t} \geq 0 \qquad\qquad\qquad\qquad\quad \begin{array}{l} \forall v \in V, \\ t = 0, \ldots, k_v, \\ u \in N(v) \end{array} \qquad (4.5)$

## 4.2   The Algorithm

**The High-Level Idea.** Suppose $(y, z)$ is a feasible dual solution. For every $v \in V$, let $T(v)$ be the set of indices $t$ for which the constraint (4.4) is tight:

$$T(v) = \left\{ t : y_v + \sum_{\substack{u \in N(v) \\ i:t \geq \mathbb{I}_v^u(i)}} z_{u,v,i} - \sum_{\substack{u \in N(v) \\ i:i \geq t}} z_{v,u,i} = c_{v,t} \right\}.$$

Furthermore, suppose that $T(v) \neq \emptyset$ for every $v \in V$, and let $t_v = \max T(v)$. We define an associated primal solution by: From each vertex $v$ transmit to radius $r_{t_v}^v$.

The initial dual solution is $y = z = 0$. Note that $0 \in T(v)$ at the beginning of the algorithm, since we have $c_{v,0} = 0$ for every $v \in V$. In each step we convert the current solution $(y, z)$ to a new feasible solution $(y', z')$, such that

1. For every $v \in V$, $T'(v) \neq \emptyset$ and $t'_v \geq t_v$.
2. There exists $v \in V$ for which $t'_v > t_v$.

It follows that our algorithm obtains a feasible primal solution within $\sum_{v \in V} k_v$ steps, since if this number of steps was already performed, we have $t_v = k_v$ for every $v \in V$, and all edges are covered by the associated primal solution.

**Implementing a Single Step.** We assume that when a step begins we are given a feasible dual solution $(y, z)$ such that $T(v) \neq \emptyset$ for every $v \in V$, and an edge $(p, q)$ that is not covered by the associated primal solution, that is, $r_{t_p}^p + r_{t_q}^q < l_{p,q}$. Clearly, $t_p < k_p$

or $t_q < k_q$, since $r^p_{k_p} + r^q_{k_q} \geq l_{p,q}$. Without loss of generality, $t_p \leq k_p$ and $t_q < k_q$. Let

$$
\epsilon_p = \min_{r=t_p+1,\ldots,k_p} \left\{ c_{p,r} - \left( y_p + \sum_{\substack{u \in N(p) \\ i:r \geq \mathbb{I}^u_p(i)}} z_{u,p,i} - \sum_{\substack{u \in N(p) \\ i:i \geq r}} z_{p,u,i} \right) \right\},
$$

$$
\epsilon_q = \min_{r=t_q+1,\ldots,k_q} \left\{ c_{q,r} - \left( y_q + \sum_{\substack{u \in N(q) \\ i:r \geq \mathbb{I}^u_q(i)}} z_{u,q,i} - \sum_{\substack{u \in N(q) \\ i:i \geq r}} z_{q,u,i} \right) \right\}.
$$

If $t_p = k_p$, we simply define $\epsilon_p = \infty$. The maximality of $t_p$ and $t_q$ implies $\epsilon_p > 0$ and $\epsilon_q > 0$, and we assume without loss of generality that $\epsilon_q \leq \epsilon_p$. We also assume that $\epsilon_q$ is attained at the constraint $r$ of $q$.

We first increase $y_q$ by $\epsilon_q$, and the constraint $r$ of $q$ now becomes tight. However, it is possible that due to this increase in $y_q$ some of the constraints of $q$ are violated. By definition of $\epsilon_q$, it follows that the constraints $t_q + 1, \ldots, k_q$ of $q$ are still satisfied, since the margin in these constraints was at least $\epsilon_q$. We now increase $z_{q,p,t_q}$ by $\epsilon_q$. This variable does not appear in the constraints $t_q + 1, \ldots, k_q$ of $q$, but appears in the constraints $1, \ldots, t_q$ (with negative sign), and therefore this increase balances the increase in $y_q$, and these constraints are also satisfied.

The variable $z_{q,p,t_q}$ also appears in some of the constraints of $p$ (with positive sign). Specifically, it appears in all the constraints $j$ of $p$ for $j \geq \mathbb{I}^q_p(t_q)$. We claim that $\mathbb{I}^q_p(t_q) > t_p$, since when the step began we had the radius $r^q_{t_q}$ in $q$ and $r^p_{t_p}$ in $p$, but these radii did not cover the edge $(p,q)$. Since the margin in the constraints $t_p + 1, \ldots, k_p$ of $p$ was at least $\epsilon_p \geq \epsilon_q$, these constraints are still satisfied after this increase in $z_{q,p,t_q}$.

Note that $T(v)$ does not change in this step for $v \neq p, q$. In addition, the constraint $t_p$ of $p$ is still tight and therefore $t'_p \geq t_p$. Finally, since the constraint $r$ of $q$ is now tight and $r > t_q$, we have $t'_q > t_q$.

## 4.3    Analysis

Let $(y', z')$ be the dual solution constructed by the primal-dual algorithm. By construction, its associated primal solution is feasible. In Theorem 5 we prove that the cost of the primal solution is at most $2 \sum_{v \in V} y'_v$. Since the cost of the dual solution is a lower bound on the cost of any feasible primal solution, it follows that the approximation ratio of our algorithm is 2.

**Theorem 5.** *The primal-dual algorithm constructs a solution whose cost is at most* $2 \cdot \mathrm{OPT}(\mathrm{LP}')$ *in time* $O(|E|K + |V|K^2)$, *where* $K = \max_{v \in V} k_v$.

*Proof.* To prove the claim, it is sufficient to show that the cost of the primal solution is at most $2 \sum_{v \in V} y'_v$. We first observe that since the initial dual solution is $y = z = 0$, and in each step we increase a single $y$ and a single $z$ by the same value,

$$
\sum_{v \in V} y'_v = \sum_{v \in V} \sum_{u \in N(v)} \sum_{i=0}^{k_v} z'_{v,u,i} .
$$

Therefore, the cost of the primal solution is

$$\sum_{v \in V} c_{v,t_v} = \sum_{v \in V} \left( y'_v + \sum_{\substack{u \in N(v) \\ i:t_v \geq I^u_v(i)}} z'_{u,v,i} - \sum_{\substack{u \in N(v) \\ i:i \geq t_v}} z'_{v,u,i} \right)$$

$$\leq \sum_{v \in V} y'_v + \sum_{v \in V} \sum_{u \in N(v)} \sum_{i=0}^{k_v} z'_{v,u,i}$$

$$= 2 \sum_{v \in V} y'_v \ ,$$

where the first equality follows from the observation that since $t_v \in T(v)$, the dual constraint $t_v$ of $v$ is tight.

By keeping track of the sums in each constraint of (DLP$'$), the representation of this linear program can be constructed in time $O(|E|K)$. In addition, our algorithm performs at most $\sum_{v \in V} k_v \leq |V|K$ steps, where the number of operations in each step is $O(K)$. □

## 5    The PL-MRC Problem

In this section we first prove that PL-MRC is as hard to approximate as vertex cover. We then present a polynomial time $(2 + \epsilon)$-approximation algorithm for any fixed $\epsilon > 0$.

### 5.1    Hardness Results

**Lemma 6.** *A polynomial time approximation algorithm for the PL-MRC problem with factor $\alpha$ would imply a polynomial time approximation algorithm for the vertex cover problem with the same factor.*

*Proof.* We show how to reduce vertex cover to PL-MRC by an approximation preserving reduction. Given a vertex cover instance $G = (V, E)$, we construct an instance of PL-MRC on $G$. The length of each edge is 1 and the cost function is

$$c(r) = \begin{cases} 2r, & \text{if } r \in [0, 1/2) \\ 1, & \text{if } r \in [1/2, \infty) \end{cases} \ .$$

Let $S^* \subseteq V$ be a minimum cardinality vertex cover in $G$. Suppose we can approximate PL-MRC to within factor $\alpha$ of optimum. We show how to find a vertex cover with cardinality at most $\alpha |S^*|$.

Since $S^*$ is a vertex cover, when we transmit from each $v \in S^*$ to radius 1 all demand points are covered, and the cost of this transmission is $|S^*|$. It follows that the cost of an optimal transmission is at most $|S^*|$, and we can find in polynomial time a feasible transmission with cost at most $\alpha |S^*|$. Let $S$ be the set of vertices from which we transmit to radius at least $\frac{1}{2}$. $S$ is clearly a vertex cover, or otherwise there is an edge $(u, v) \in E$ for which the sum of transmission radii from $u$ and $v$ is less than 1. In addition, $|S| \leq \alpha |S^*|$, since the cost of transmitting from $S$ is $|S|$ and the total cost is at most $\alpha |S^*|$. □

## 5.2    The Algorithm

Given a PL-MRC instance $I$ and a fixed parameter $\epsilon > 0$, we show how to construct in polynomial time an instance $\rho(I)$ of MRC such that any feasible MRC solution for $\rho(I)$ is also a feasible PL-MRC solution for $I$ with identical cost, and such that $\mathrm{OPT}_{MRC}(\rho(I)) \leq (1+\epsilon)^2 \mathrm{OPT}_{PL}(I)$. We immediately get a $(2+\epsilon)$-approximation, by using the primal-dual algorithm in Section 4 and choosing an appropriate $\epsilon$.

Without loss of generality, we assume that $\mathrm{OPT}_{PL}(I) > 0$, since we can easily verify if there is a feasible solution with cost 0. We also assume that we are given an additional parameter $\Delta > 0$ that satisfies $\mathrm{OPT}_{PL}(I) \leq \Delta \leq (1+\epsilon)\mathrm{OPT}_{PL}(I)$, since we can try every $\Delta = (1+\epsilon)^k LB \leq UB$, where

$$LB = \max_{(u,v)\in E} \min_{r\in[0,l_{u,v}]} \left( c_u(r) + c_v(l_{u,v} - r) \right) \ ,$$

$$UB = \sum_{(u,v)\in E} \min_{r\in[0,l_{u,v}]} \left( c_u(r) + c_v(l_{u,v} - r) \right) \ .$$

We remark that

1. $0 < LB \leq \mathrm{OPT}_{PL}(I) \leq UB$.
2. Since $\frac{UB}{LB} \leq |E|$, there are $O(\log_{1+\epsilon}|E|)$ values of $\Delta$ to be tested.
3. $\min_{r\in[0,l_{u,v}]} \left( c_u(r) + c_v(l_{u,v} - r) \right)$ is attained at a breakpoint of $u$ or $v$.

Let $r_{\max}^v = \max\{l_{u,v} : u \in N(v)\}$. Since the cost function $c_v$ is non-decreasing and the radius $r_{\max}^v$ is sufficient to cover all edges adjacent to $v$, there is no motivation to transmit from $v$ to a greater radius. In addition, let

$$r_{\min}^v = \begin{cases} \max\{r \in [0, r_{\max}^v] : c_v(r) \leq \frac{\epsilon\Delta}{n}\}, & \text{if } c_v(r_{\max}^v) \geq \frac{\epsilon\Delta}{n} \\ r_{\max}^v, & \text{otherwise} \end{cases} \ .$$

We assume that $\{0, r_{\min}^v, r_{\max}^v\} \subseteq BP_v$, where $BP_v$ is the set of breakpoints of $c_v$.

To construct an instance of MRC, we define for each $v \in V$ a discrete set $R_v$ of transmission radii. If $r_{\min}^v = r_{\max}^v$, we simply define $R_v = \{0, r_{\max}^v\}$. Otherwise, let $r_0, \ldots, r_k$ be the breakpoints in $BP_v$ such that $r_{\min}^v = r_0 < \cdots < r_k = r_{\max}^v$. For each interval $[r_i, r_{i+1}]$, $0 \leq i \leq k - 1$, we define a set of points $P_v[i, i+1] = \{r_i^1, \ldots, r_i^t\}$ as follows. The $j$th point $r_i^j$ satisfies $c_v(r_i^j) = (1+\epsilon)^j c_v(r_i)$, and the index $t$ is the maximal integer for which $c_v(r_i^t) = (1+\epsilon)^t c_v(r_i) < c_v(r_{i+1})$. Note that $t$ is polynomial in the input length, since $t = O(\log_{1+\epsilon}(\frac{c_v(r_{i+1})}{c_v(r_i)}))$. We define

$$R_v = \left( \bigcup_{i=0}^{k-1} P_v[i, i+1] \right) \bigcup BP_v \ .$$

**Lemma 7.** $\mathrm{OPT}_{MRC}(\rho(I)) \leq (1+\epsilon)^2 \mathrm{OPT}_{PL}(I)$.

*Proof.* Consider the optimal solution for the instance $I$ of PL-MRC, and for each $v \in V$ let $r_v^* \in [0, r_{\max}^v]$ be the radius to which we transmit from $v$. Let $A = \{v \in V : r_v^* \leq r_{\min}^v\}$ and $B = V \setminus A$. Using $r^*$ we define the following solution for the instance $\rho(I)$

of MRC: Transmit from each $v \in V$ to radius $\hat{r}_v = \min\{r \in R_v : r \geq r_v^*\}$. Clearly, this solution is feasible.

Let $\text{cost}(A)$ and $\text{cost}(B)$ be the new costs of transmitting from $A$ and $B$, respectively. Since for each vertex $v \in A$ we have $\hat{r}_v \leq r_{\min}^v$ and $c_v(r_{\min}^v) \leq \frac{\epsilon \Delta}{n}$,

$$\text{cost}(A) = \sum_{v \in A} c_v(\hat{r}_v) \leq \sum_{v \in A} c_v(r_{\min}^v) \leq \epsilon \Delta \frac{|A|}{n} \leq \epsilon \Delta \leq \epsilon(1+\epsilon)\text{OPT}_{PL}(I) .$$

In addition, by construction of $R_v$ for $v$ such that $r_{\min}^v < r_{\max}^v$, it follows that $c_v(\hat{r}_v) \leq (1+\epsilon)c_v(r_v^*)$. Therefore,

$$\text{cost}(B) = \sum_{v \in B} c_v(\hat{r}_v) \leq (1+\epsilon) \sum_{v \in B} c_v(r_v^*) \leq (1+\epsilon)\text{OPT}_{PL}(I) .$$

These inequalities show that the total cost of the new transmission is

$$\text{cost}(A) + \text{cost}(B) \leq \epsilon(1+\epsilon)\text{OPT}_{PL}(I) + (1+\epsilon)\text{OPT}_{PL}(I) = (1+\epsilon)^2\text{OPT}_{PL}(I) .$$

$\square$

# References

[1] R. Bar-Yehuda and S. Even. A linear-time approximation algorithm for the weighted vertex cover problem. *Journal of Algorithms*, 2:198–203, 1981.

[2] R. Bar-Yehuda and S. Even. A local-ratio theorem for approximating the weighted vertex cover problem. *Annals of Discrete Mathematics*, 25:27–46, 1985.

[3] J. Chuzhoy and J. Naor. Covering problems with hard capacities. In *Proceedings of the 43rd Annual Symposium on Foundations of Computer Science*, pages 481–489, 2002.

[4] I. Dinur and S. Safra. The importance of being biased. In *Proceedings of the 34th Annual ACM Symposium on Theory of Computing*, pages 33–42, 2002.

[5] S. Guha, R. Hassin, S. Khuller, and E. Or. Capacitated vertex covering. *Journal of Algorithms*, 48:257–270, 2003.

[6] E. Halperin. Improved approximation algorithms for the vertex cover problem in graphs and hypergraphs. *SIAM Journal on Computing*, 31:1608–1623, 2002.

[7] R. Hassin and A. Levin. The minimum generalized vertex cover problem. In *Proceedings of the 11th Annual European Symposium on Algorithms*, pages 289–300, 2003.

[8] D. S. Hochbaum. Approximation algorithms for the set covering and vertex cover problems. *SIAM Journal on Computing*, 11:555–556, 1982.

[9] D. P. Williamson. The primal-dual method for approximation algorithms. *Mathematical Programming, Series B*, 91:447–478, 2002.

# Parameterized Complexity
# of Generalized Vertex Cover Problems

Jiong Guo[*], Rolf Niedermeier[*], and Sebastian Wernicke[**]

Institut für Informatik, Friedrich-Schiller-Universität Jena,
Ernst-Abbe-Platz 2, D-07743 Jena, Fed. Rep. of Germany
{guo, niedermr, wernicke}@minet.uni-jena.de

**Abstract.** Important generalizations of the VERTEX COVER problem
(CONNECTED VERTEX COVER, CAPACITATED VERTEX COVER, and MAX-
IMUM PARTIAL VERTEX COVER) have been intensively studied in terms
of approximability. However, their parameterized complexity has so far
been completely open. We close this gap here by showing that, with
the size of the desired vertex cover as parameter, CONNECTED VER-
TEX COVER and CAPACITATED VERTEX COVER are both fixed-parameter
tractable while MAXIMUM PARTIAL VERTEX COVER is W[1]-hard. This
answers two open questions from the literature. The results extend to
several closely related problems. Interestingly, although the considered
generalized VERTEX COVER problems behave very similar in terms of
constant-factor approximability, they display a wide range of different
characteristics when investigating their parameterized complexities.

## 1   Introduction

Given an undirected graph $G = (V, E)$, the NP-complete VERTEX COVER prob-
lem is to find a set $C \subseteq V$ with $|C| \leq k$ such that each edge in $E$ has at least one
endpoint in $C$. In a sense, VERTEX COVER could be considered the *Drosophila*
of fixed-parameter algorithmics [17, 25]:

1. There is a long list of continuous improvements on the combinatorial explosion with respect to the parameter $k$ when solving the problem exactly. The
   currently best exponential bound is below $1.28^k$ [8, 26, 14, 28, 12].
2. VERTEX COVER has been a benchmark for developing sophisticated data
   reduction and problem kernelization techniques [1, 19].
3. It was the first parameterized problem where the usefulness of interleaving
   depth-bounded search trees and problem kernelization was proven [27].
4. Restricted to planar graphs, it was—besides DOMINATING SET—one of the
   central problems for the development of "subexponential" fixed-parameter
   algorithms and the corresponding theory of relative lower bounds [2, 4, 11].

---

[*] Supported by the Deutsche Forschungsgemeinschaft, Emmy Noether research group
PIAF (fixed-parameter algorithms), NI 369/4.
[**] Supported by Deutsche Telekom Stiftung and Studienstiftung des deutschen Volkes.

F. Dehne, A. López-Ortiz, and J.-R. Sack (Eds.): WADS 2005, LNCS 3608, pp. 36–48, 2005.
© Springer-Verlag Berlin Heidelberg 2005

5. VERTEX COVER served as a testbed for algorithm engineering in the realm of fixed-parameter algorithms [1, 3, 13].
6. Studies of VERTEX COVER led to new research directions within parameterized complexity such as counting [7], parallel processing [13], or using "vertex cover structure" as a general strategy to solve parameterized problems [30].

This probably incomplete list gives an impression of how important VERTEX COVER was and continues to be for the whole field of parameterized complexity. However, research in this field to date appears to have neglected a closer investigation of recent significant generalizations and variants of VERTEX COVER. These appear in various application scenarios such as drug design [22] and have so far only been studied in the context of their polynomial-time approximability. We close this gap here by providing several first-time parameterized complexity results, which also answers two open questions from the literature.

We are only aware of two papers that perform somewhat related research. First, Nishimura, Ragde, and Thilikos [29] also study generalizations of VERTEX COVER. However, they follow a completely different route: Whereas we study concrete problems such as CAPACITATED VERTEX COVER or MAXIMUM PARTIAL VERTEX COVER on general graphs, their interest lies in recognizing general classes of graphs with a very special case of interest being the class of graphs with bounded vertex cover (refer to [29] for details). Second, Bläser [9] shows that some partial covering problems are fixed-parameter tractable when the parameter is the number of objects covered instead of the size of the covering set. (In this paper, as well as in the abovementioned studies, the parameter is always the size of the covering set.)

We deal with a whole list of vertex covering problems, all of them possessing constant-factor (mostly 2) polynomial-time approximation algorithms. Deferring their formal definitions to the next section, we now informally describe the studied problems and the known and new results. In the presentation of our results, $n$ denotes the number of vertices and $m$ denotes the number of edges of the input graph. The parameter $k$ always denotes the size of the vertex cover.

1. For CONNECTED VERTEX COVER one demands that the vertex cover set is connected. This problem is known to have a factor-2 approximation [6]. We show that it can be solved in $O(6^k n + 4^k n^2 + 2^k n^2 \log n + 2^k nm)$ time. In addition, we derive results for the closely related variants TREE COVER and TOUR COVER.
2. For CAPACITATED VERTEX COVER, the "covering capacity" of each graph vertex is limited in that it may not cover all of its incident edges. This problem has a factor-2 approximation [22]. Addressing an open problem from [22], we show that CAPACITATED VERTEX COVER can be solved in $O(1.2^{k^2} + n^2)$ time using an enumerative approach. We also provide a problem kernelization. Altogether, we thus show that CAPACITATED VERTEX COVER—including two variants with "hard" and "soft" capacities—is fixed-parameter tractable.
3. For MAXIMUM PARTIAL VERTEX COVER, one only wants to cover a specified number of edges (that is, not necessarily all) by at most $k$ vertices. This

**Table 1.** New parameterized complexity results for some NP-complete generalizations of VERTEX COVER shown in this work. The parameter $k$ is the size of the desired vertex cover, $m$ denotes the number of edges, and $n$ denotes the number of vertices

| Problem | Result | |
|---|---|---|
| CONNECTED VERTEX COVER | $6^k n + 4^k n^2 + 2^k n^2 \log n + 2^k nm$ | Thm. 2 |
| TREE COVER | $(2k)^k \cdot km$ | Cor. 3 |
| TOUR COVER | $(4k)^k \cdot km$ | Cor. 3 |
| CAPACITATED VERTEX COVER | $1.2^{k^2} + n^2$ | Thm. 5 |
| SOFT CAPACITATED VERTEX COVER | $1.2^{k^2} + n^2$ | Thm. 10 |
| HARD CAPACITATED VERTEX COVER | $1.2^{k^2} + n^2$ | Thm. 10 |
| MAXIMUM PARTIAL VERTEX COVER | W[1]-hard | Thm. 11 |
| MINIMUM PARTIAL VERTEX COVER | W[1]-hard | Cor. 12 |

problem is known to have a factor-2 approximation [10]. Answering an open question from [5], we show that this problem appears to be fixed-parameter intractable—it is W[1]-hard. The same is proven for its minimization version.

Summarizing, we emphasize that our main focus is on deciding between fixed-parameter tractability and W[1]-hardness for all of the considered problems. Interestingly, although all considered problems behave in more or less the same way from the viewpoint of polynomial-time approximability—all have factor-2 approximations—the picture becomes completely different from a parameterized complexity point of view: MAXIMUM PARTIAL VERTEX COVER appears to be intractable and CAPACITATED VERTEX COVER appears to be significantly harder than CONNECTED VERTEX COVER. Table 1 surveys all of our results.

## 2    Preliminaries and Previous Work

Parameterized complexity is a two-dimensional framework for studying the computational complexity of problems.[1] One dimension is the input size $n$ (as in classical complexity theory) and the other one the *parameter* $k$ (usually a positive integer). A problem is called *fixed-parameter tractable* (fpt) if it can be solved in $f(k) \cdot n^{O(1)}$ time, where $f$ is a computable function only depending on $k$. A core tool in the development of fixed-parameter algorithms is polynomial-time preprocessing by *data reduction rules*, often yielding a *reduction to a problem kernel*. Here the goal is, given any problem instance $x$ with parameter $k$, to transform it into a new instance $x'$ with parameter $k'$ such that the size of $x'$ is bounded by some function only depending on $k$, $(x, k)$ has a solution iff $(x', k')$ has a solution, and $k' \leq k$. A formal framework to show *fixed-parameter intractability* was developed by Downey and Fellows [17] who introduced the concept of *parameterized reductions*. A parameterized reduction from a parameterized lan-

---

[1] For a more detailed introduction see, e.g., [17, 19, 24].

guage $L$ to another parameterized language $L'$ is a function that, given an instance $(x, k)$, computes in time $f(k) \cdot n^{O(1)}$ an instance $(x', k')$ (with $k'$ only depending on $k$) such that $(x, k) \in L \Leftrightarrow (x', k') \in L'$. The basic complexity class for fixed-parameter intractability is W[1] as there is good reason to believe that W[1]-hard problems are not fixed-parameter tractable [17].

In this work, we consider three directions of generalizing VERTEX COVER (VC), namely demanding that the vertices of the cover must be *connected* (Section 3), introducing *covering capacities* for the vertices (Section 4), and relaxing the condition that *all* edges in the graph must be covered (Section 5). Our corresponding parameterized complexity results are summarized in Table 1, the formal definitions of the problems follow.

CONNECTED VERTEX COVER: Given a graph $G = (V, E)$ and an integer $k \geq 0$, determine whether there exists a vertex cover $C$ for $G$ containing at most $k$ vertices such that the subgraph of $G$ induced by $C$ is connected.

This problem is NP-complete and approximable within 2 [6]. Two variants are derived by introducing a weight function $w : E \to \mathbb{R}^+$ on the edges and requiring that the cover must induce a subgraph with a certain structure and minimum weight.

TREE COVER: Given a graph $G = (V, E)$ with edges weighted with positive real numbers, an integer $k \geq 0$, and a real number $W > 0$, determine whether there exists a subgraph $G' = (V', E')$ of $G$ with $|V'| \leq k$ and $\sum_{e \in E'} w(e) \leq W$ such that $V'$ is a vertex cover for $G$ and $G'$ is a tree.[2]

The closely related problem TOUR COVER differs from TREE COVER only in that the edges in $G'$ should form a closed walk instead of a tree. Note that a closed walk can contain repeated vertices and edges. Both TREE COVER and TOUR COVER were introduced in [6] where it is shown that TREE COVER is approximable within 3.55 and TOUR COVER within 5.5. Könemann et al. [23] improved both approximation factors to 3.

Section 4 considers the CAPACITATED VERTEX COVER (CVC) problem and related variants. Here, each vertex $v \in V$ is assigned a *capacity* $c(v) \in \mathbb{N}^+$ that limits the number of edges it can cover when being part of the vertex cover.

**Definition 1.** *Given a capacitated graph $G = (V, E)$ and a vertex cover $C$ for $G$. We call $C$ capacitated vertex cover if there exists a mapping $f : E \to C$ which maps each edge in $E$ to one of its two endpoints such that the total number of edges mapped by $f$ to any vertex $v \in C$ does not exceed $c(v)$.*

CAPACITATED VERTEX COVER: Given a vertex-weighted (with positive real numbers) and capacitated graph $G$, an integer $k \geq 0$, and a real number $W \geq 0$, determine whether there exists a capacitated vertex cover $C$ for $G$ containing at most $k$ vertices such that $\sum_{v \in C} w(v) \leq W$.

---

[2] TREE COVER is equivalent to CONNECTED VERTEX COVER for unweighted graphs.

The CVC problem was introduced by Guha et al. [22] who also give a factor-2 approximation algorithm. Two special flavors of CVC exist in the literature that arise by allowing "copies" of a vertex to be in the capacitated vertex cover [22, 15, 20]. In that context, taking a vertex $l$ times into the capacitated vertex cover causes the vertex to have $l$ times its original capacity. The number of such copies is unlimited in the SOFT CAPACITATED VERTEX COVER (SOFT CVC) problem while it may be restricted for each vertex individually in the HARD CAPACITATED VERTEX COVER (HARD CVC) problem. For unweighted HARD CVC, the best known approximation algorithm achieves a factor of 2 [20]. The weighted version HARD CVC is at least as hard to approximate as Set Cover [15].

Section 5 considers a third direction of VC generalizations besides connectedness and capacitation. In the MAXIMUM PARTIAL VERTEX COVER problem, we relax the condition that *all* edges must be covered.

> MAXIMUM PARTIAL VERTEX COVER: Given a graph $G = (V, E)$ and two integers $k \geq 0$ and $t \geq 0$, determine whether there exists a vertex subset $V' \subseteq V$ of size at most $k$ such that $V'$ covers at least $t$ edges.

This problem was introduced by Bshouty and Burroughs [10] who showed it to be approximable within 2. Further improvements can be found in [21]. Note that MAXIMUM PARTIAL VERTEX COVER is fixed-parameter tractable with respect to the parameter $t$ [9]. In case of MINIMUM PARTIAL VERTEX COVER we are asked for a vertex subset with *at least* $k$ vertices covering *at most* $t$ edges.

# 3      Connected Vertex Cover and Variants

In this section we show that CONNECTED VERTEX COVER is fixed-parameter tractable with respect to the size of the connected vertex cover. More precisely, it can be solved by an algorithm running in $O(6^k n + 4^k n^2 + 2^k n^2 \log n + 2^k nm)$ time where $n$ and $m$ denote the number of vertices and edges in the input graph and $k$ denotes the size of the connected vertex cover. We modify this algorithm to also show the fixed-parameter tractability for two variants of CONNECTED VERTEX COVER, namely TREE COVER and TOUR COVER.

We solve CONNECTED VERTEX COVER by using the Dreyfus-Wagner algorithm as a subprocedure for computing a Steiner minimum tree in a graph [18]. For an undirected graph $G = (V, E)$, a subgraph $T$ of $G$ is called a *Steiner tree* for a subset $K$ of $V$ if $T$ is a tree containing all vertices in $K$ such that all leaves of $T$ are elements of $K$. The vertices of $K$ are called the *terminals* of $T$. A *Steiner minimum tree* for $K$ in $G$ is a Steiner tree $T$ such that the number of edges contained in $T$ is minimum. Finding a Steiner minimum tree leads to an NP-complete problem. The Dreyfus-Wagner algorithm computes a Steiner minimum tree for a set of at most $l$ terminals in $O(3^l n + 2^l n^2 + n^2 \log n + nm)$ time [18].

Our algorithm for CONNECTED VERTEX COVER consists of two steps:

1. Enumerate all minimal vertex covers with at most $k$ vertices. If one of the enumerated minimal vertex covers is connected, then output it and terminate.

2. Otherwise, for each of the enumerated minimal vertex covers $C$, use the Dreyfus-Wagner algorithm to compute a Steiner minimum tree with $C$ as the set of terminals. If one minimal vertex cover has a Steiner minimum tree $T$ with at most $k - 1$ edges, then return the vertex set of $T$ as output; otherwise, there is no connected vertex cover with at most $k$ vertices.

**Theorem 2.** CONNECTED VERTEX COVER *can be solved in* $O(6^k n + 4^k n^2 + 2^k n^2 \log n + 2^k nm)$ *time.*

*Proof.* The first step of the algorithm is correct since each connected vertex cover (covc) contains at least one minimal vertex cover. For a given graph, there are at most $2^k$ minimal vertex covers with at most $k$ vertices. We can enumerate all such minimal vertex covers in $O(2^k \cdot m)$ time. Then, the running time of the first step is $O(2^k \cdot m)$.

The correctness of the second step follows directly from the following easy to prove observation: For a set of vertices $C$, there exists a connected subgraph of $G$ with at most $k$ vertices which contains all vertices in $C$ iff there exists a Steiner tree in $G$ with $C$ as the terminal set and at most $k - 1$ edges. By applying the Dreyfus-Wagner algorithm on $G$ with $C$ as the terminal set, we can easily find out whether there are $k - |C|$ vertices from $V \setminus C$ connecting $C$ and, hence, whether there is a covc with at most $k$ vertices and containing $C$. Since $|C| < k$, the second step can be done in $O(2^k \cdot (3^k n + 2^k n^2 + n^2 \log n + nm)) = O(6^k n + 4^k n^2 + 2^k n^2 \log n + 2^k nm)$ time.    □

The algorithm for CONNECTED VERTEX COVER can be modified to solve TREE COVER and TOUR COVER. The proof is omitted.

**Corollary 3.** TREE COVER *and* TOUR COVER *can be solved in* $O((2k)^k \cdot km)$ *and* $O((4k)^k \cdot km)$ *time, respectively.*

# 4    Capacitated Vertex Cover and Variants

In this section we present fixed-parameter algorithms for the CVC problem and its variants HARD CVC and SOFT CVC. In the case of CVC, the easiest way to show its fixed-parameter tractability is to give a reduction to a problem kernel. This is what we begin with here, afterwards complementing it with an enumerative approach for further improving the overall time complexity.

**Proposition 4.** *Given an $n$-vertex graph $G = (V, E)$ and an integer $k \geq 0$ as part of an input instance for CVC, then it is possible to construct an $O(4^k \cdot k^2)$-vertex graph $\tilde{G}$ such that $G$ has a size-$k$ solution for CVC iff $\tilde{G}$ has a size-$k$ solution for CVC. In the special case of uniform vertex weights, $\tilde{G}$ has only $O(4^k \cdot k)$ vertices. The construction of $\tilde{G}$ can be performed in $O(n^2)$ time.*

*Proof.* We first assume uniform vertex weights, generalizing the approach to weighted graphs at the end of the proof.

Let $u, v \in V$, $u \neq v$, and $\{u, v\} \notin E$. The simple observation that lies at the heart of the data reduction rule needed for the kernelization is that if the open neighborhoods of $u$ and $v$ coincide (i.e., $N(u) = N(v)$) and $c(u) < c(v)$, then $u$ is part of a minimum capacitated vertex cover only if $v$ is as well. We can generalize this finding to a data reduction rule: Let $\{v_1, v_2, \ldots, v_{k+1}\} \subseteq V$ with the induced subgraph $G[\{v_1, v_2, \ldots, v_{k+1}\}]$ being edgeless, and $N(v_1) = N(v_2) = \cdots = N(v_{k+1})$. Call this the *neighbor set*. Then delete from $G$ a vertex $v_i \in \{v_1, v_2, \ldots, v_{k+1}\}$ which has minimum capacity. This rule is correct because any size-$k$ capacitated vertex cover $C$ containing $v_i$ can be modified by replacing $v_i$ with a vertex from $\{v_1, v_2, \ldots, v_{k+1}\}$ which is not in $C$.

Based on this data reduction rule, $\tilde{G}$ can be computed from $G$ as claimed by the following two steps:

1. Use the straightforward linear-time factor-2 approximation algorithm to find a vertex cover $S$ for $G$ of size at most $2k'$ (where $k'$ is the size of a minimum vertex cover for $G$ and hence $k' \leq k$). If $|S| > 2k$, then we can stop because then no size-$k$ (capacitated) vertex cover can be found. Note that $V \setminus S$ induces an edgeless subgraph of $G$.
2. Examining $V \setminus S$, check whether there is a subset of $k+1$ vertices that fulfill the premises of the above rule. Repeatedly apply the data reduction rule until it is no longer applicable. Note that this process continuously shrinks $V \setminus S$.

The above computation is clearly correct. The number of all possible neighbor sets can be at most $2^{2k}$ (the number of different subsets of $S$). For each neighbor set, there can be at most $k$ neighboring vertices in $V \setminus S$; otherwise, the reduction rule would apply. Hence, in the worst case we can have at most $2^{2k} \cdot k$ vertices in the remaining graph $\tilde{G}$. The generalization to non-uniform vertex weights works as follows: We have $|S| \leq 2k$. Hence, the vertices in $V \setminus S$ may have maximum vertex degree $2k$ and the capacity of a vertex in $V \setminus S$ greater than $2k$ without any harm can be replaced by capacity $2k$. Therefore, without loss of generality, one may assume that the maximum capacity of vertices in $V \setminus S$ is $2k$. We then have to modify the reduction rule as follows. If there are vertices $v_1, v_2, \ldots, v_{2k^2+1} \in V$ with $N(v_1) = N(v_2) = \ldots = N(v_{2k^2+1})$, partition them into subsets of vertices with equal capacity. There are at most $2k$ of these sets. If such a set contains more than $k$ vertices, delete the vertex with maximum weight. Altogether, we thus end up with a problem kernel of $2^{2k} \cdot 2k^2 = O(4^k \cdot k^2)$ vertices.

It remains to justify the polynomial running time. First, note that the trivial factor-2 approximation algorithm runs in time $O(|E|) = O(n^2)$. Second, examining the common neighborhoods can be done in $O(n^2)$ time by successively partitioning the vertices in $V \setminus S$ according to their neighborhoods. □

Clearly, a simple brute-force search within the reduced instance (with a size of only $O(4^k \cdot k^2)$ vertices) already yields the fixed-parameter tractability of CVC, albeit in time proportional to $\binom{4^k \cdot k^2}{k}$. As the next theorem shows, we can do much better concerning the running time.

**Theorem 5.** *The* CVC *problem can be solved in* $O(1.2^{k^2} + n^2)$ *time.*

The theorem is proved by first giving an algorithm to solve CAPACITATED VER-
TEX COVER and then proving its running time. The basic idea behind the algorithm is as follows: We start with a minimal vertex cover $C = \{c_1, \ldots, c_i\} \subseteq V$
for the input graph $G = (V, E)$. Due to lack of capacities, $C$ is not necessarily
a *capacitated* vertex cover for $G$. Hence, if $C$ is not a capacitated vertex cover,
we need to add some additional vertices from $V \setminus C$ to $C$ in order to provide
additional capacities. More precisely, since for each vertex $v \in (V \setminus C)$ all of
its neighbors are in $C$, adding $v$ can be seen as "freeing" exactly one unit of
capacity for as many as $c(v)$ neighbors of $v$. The algorithm uses an exhaustive
search approach based on this observation by enumerating all possible patterns
of capacity-freeing and for each pattern computing the cheapest set of vertices
from $V \setminus C$ (if one exists) that matches it.

**Definition 6.** *Given a graph* $G = (V, E)$ *and a vertex cover* $C = \{c_1, \ldots, c_i\} \subseteq$
$V$ *for* $G$. *A capacity profile of length* $i$ *is a binary string* $s = s[1] \ldots s[i] \in \{0, 1\}^i$.
*A vertex* $w \in V \setminus C$ *is said to* match *a capacity profile* $s$ *if it is incident to each
vertex* $c_j \in C$ *with* $s[j] = 1$ *and its capacity is at least the number of ones in* $s$.

Using Definition 6, the following pseudocode gives an algorithm for CVC.

**Algorithm:** CAPACITATED VERTEX COVER
**Input:** A capacitated and vertex-weighted graph $G = (V, E)$, $k \in \mathbb{N}^+$, $W \in \mathbb{R}^+$
**Output:** "YES" if $G$ has a capacitated vertex cover of size at most $k$
          with weight $\leq W$; "NO" otherwise

*01*   Perform the kernelization from Proposition 4 on $G$
*02*   **for** every minimal vertex cover $C$ of $G$ with size $i \leq k$ **do**
*03*       **if** $C$ is a cap. vertex cover with weight $\leq W$ **then return** "YES"
*04*       **for** every multiset $M$ of $(k - i)$ capacity profiles of length $i$ **do**
*05*           remove the all-zero profiles from $M$
*06*           find the cheapest set $\hat{C} \subseteq (V \setminus C)$ so that there exists a
               bijective mapping $f : \hat{C} \to M$ where each $\hat{c} \in \hat{C}$ matches
               the capacity profile of $f(\hat{c})$. Set $\hat{C} \leftarrow \emptyset$ if no such set exists
*07*           **if** $\hat{C} \neq \emptyset$, the weight of $\hat{C}$ is $\leq W$, and $C \cup \hat{C}$ is a
               capacitated vertex cover for $G$ **then return** "YES"
*08*   **return** "NO"

**Lemma 7.** *The given algorithm for* CAPACITATED VERTEX COVER *is correct.*

*Proof.* Preprocessing the graph in line *01* is correct according to Proposition 4.
Since a capacitated vertex cover for a graph $G = (V, E)$ is also a vertex cover,
its vertices can be partitioned into two sets $C$ and $\hat{C}$ such that $C$ is a minimal
vertex cover for $G$. Each vertex in $\hat{C}$ gives additional capacity to a subset of the
vertices in $C$, i.e., for every $\hat{c} \in \hat{C}$, we can construct a capacity profile $s_{\hat{c}}$ where
$s_{\hat{c}}[j] = 1$ if and only if $\hat{c}$ uses its capacity to cover the edge to the $j$-th vertex
in $C$. The correctness of the algorithm follows from its exhaustive nature: It tries

all minimal vertex covers, all possible combinations of capacity profiles and for each combination determines the cheapest possible set $\hat{C}$ such that $C \cup \hat{C}$ is a capacitated vertex cover for $G$.                                                                    □

**Lemma 8.** *The given algorithm for* CVC *runs in* $O(1.2^{k^2} + n^2)$ *time.*

*Proof.* The preprocessing in line *01* can be carried out in $O(n^2)$ time according to Proposition 4. This leads to a new graph containing at most $\tilde{n} = O(4^k \cdot k^2)$ vertices. Line *02* of the algorithm can be executed in $O(2^k) \cdot \tilde{n}^{O(1)}$ time and causes the subsequent lines *03–07* to be called at most $2^k$ times. Due to [15–Lemma 1], we can decide in $\tilde{n}^{O(1)}$ time whether a given vertex cover is also a capacitated vertex cover (lines *03* and *07*). For line *04*, note that for a given $0 \le i \le k$ there exist $2^i$ different capacity profiles of that length. Furthermore, it is well-known that given a set $A$ where $|A| = a$, there exist exactly $\binom{a+b-1}{b}$ $b$-element multisets with elements drawn from $A$. Hence, line *04* causes lines *05–07* to be executed $\binom{2^i+(k-i)-1}{k-i}$ times. The delay between enumeration of two multisets can be kept constant. As it will be shown in Lemma 9, line *06* takes $\tilde{n}^{O(1)}$ time. Overall, the running time of the algorithm is bounded from above by

$$O(n^2) + 2^k \cdot \max_{1 \le i \le k} \binom{2^i + (k-i) - 1}{k-i} \cdot \tilde{n}^{O(1)}.$$

With some effort, we can bound this number by $O(n^2 + 1.2^{k^2})$.                    □

It remains to show the running time for line *06* of the algorithm.

**Lemma 9.** *Given a weighted, capacitated graph* $G = (V, E)$, *a vertex cover* $C$ *of* $G$ *of size* $i \le k$ *for* $G$, *and a multiset* $M$ *of* $k - i$ *capacity profiles of length* $i$. *Then, it takes* $n^{O(1)}$ *time to find the cheapest set* $\hat{C} \subseteq (V \setminus C)$ *so that there exists a bijective mapping* $f : \hat{C} \to M$ *where each* $\hat{c} \in \hat{C}$ *matches the capacity profile of* $f(\hat{c})$ *or determine that no such set* $\hat{C}$ *exists.*

*Proof.* Finding $\hat{C}$ is equivalent to finding a minimum weight maximum bipartite matching on the bipartite graph $G' = (V_1', V_2', E')$ where each vertex in $V_1'$ represents a capacity profile from $M$, $V_2' = V \setminus C$, and two vertices $v \in V_1', u \in V_2'$ are connected by an edge in $E'$ if and only if the vertex represented by $u$ matches the profile represented by $v$ (the weight of the edge is $w(u)$). Finding such a matching is well-known to be solvable in polynomial time [16].                    □

It is possible to solve SOFT CVC and HARD CVC by adapting the above algorithm for CVC: Observe that if we choose multiple copies of a vertex into the cover, each of these copies will have its own individual capacity profile. Thus, only line *06* of the CVC algorithm has to be adapted to solve SOFT CVC and HARD CVC. The proof is omitted.

**Corollary 10.** SOFT CVC *and* HARD CVC *are solvable in* $O(1.2^{k^2} + n^2)$ *time.*

# 5    Maximum and Minimum Partial Vertex Cover

All the VERTEX COVER variants we studied in the previous sections are known to have a polynomial-time constant-factor approximation (mostly factor 2). All of them were shown to be fixed-parameter tractable. By way of contrast, we now present a result where a variant that has a polynomial-time factor-2 approximation is shown to be fixed parameter intractable. More precisely, we show that MAXIMUM PARTIAL VERTEX COVER (MAxPVC) is W[1]-hard with respect to the size $k$ of the partial vertex cover by giving a parameterized reduction from the W[1]-complete INDEPENDENT SET problem [17] to MAxPVC. With the solution size as parameter, we also show the W[1]-hardness of its minimization version MINPVC by a reduction from CLIQUE.

> INDEPENDENT SET: Given a graph $G = (V, E)$ and an integer $k \geq 0$, determine whether there is a vertex subset $I \subseteq V$ with at least $k$ vertices such that the subgraph of $G$ induced by $I$ contains no edge.

An *independent set* in a graph is a set of pairwise nonadjacent vertices.

**Theorem 11.** MAXIMUM PARTIAL VERTEX COVER *is W[1]-hard with respect to the size of the cover.*

*Proof.* We give a parameterized reduction from INDEPENDENT SET to MAx-PVC. Given an input instance $(G = (V, E), k)$ of INDEPENDENT SET. For every vertex $v \in V$, let $\deg(v)$ denote the degree of $v$ in $G$. We construct a new graph $G' = (V', E')$ in the following way: For each vertex $v \in V$ we insert $|V| - \deg(v)$ new vertices into $G$ and connect each of these new vertices with $v$. In the following, we show that a size-$k$ independent set in $G$ one-to-one corresponds to a size-$k$ partial vertex cover in $G'$ which covers $t := k \cdot |V|$ edges.

Firstly, a size-$k$ independent set in $G$ also forms a size-$k$ independent set in $G'$. Moreover, each of these $k$ vertices has exactly $|V|$ incident edges. Then, these $k$ vertices form a partial vertex cover covering $k \cdot |V|$ edges. Secondly, if we have a size-$k$ partial vertex cover in $G'$ which covers $k \cdot |V|$ edges, then we know that none of the newly inserted vertices in $G'$ can be in this cover. Hence, this cover contains $k$ vertices from $V$. Moreover, a vertex in $G'$ can cover at most $|V|$ edges and two adjacent vertices can cover only $2|V| - 1$ edges. Therefore, no two vertices in this partial vertex cover can be adjacent, which implies that this partial cover forms a size-$k$ independent set in $G$.                                          □

In MINIMUM PARTIAL VERTEX COVER (MINPVC), we wish to choose *at least* $k$ vertices such that *at most* $t$ edges are covered. Through a parameterized reduction from the W[1]-complete CLIQUE problem [17], it is possible to show analogously to MAxPVC that MINPVC is also W[1]-hard. This reduction works in a similar way as the reduction in the proof above. The proof is omitted.

**Corollary 12.** MINIMUM PARTIAL VERTEX COVER *is W[1]-hard with respect to the size of the cover.*

## 6 Conclusion

We extended and completed the parameterized complexity picture for natural variants and generalizations of VERTEX COVER. Notably, whereas the fixed-parameter tractability of VERTEX COVER immediately follows from a simple search tree strategy, this appears not to be the case for all of the problems studied here. Table 1 in Section 2 summarizes our results, all of which, to the best of our knowledge, are new in the sense that no parameterized complexity results have been known before for these problems. Our fixed-parameter tractability results clearly generalize to cases where the vertices have real weights $\geq c$ for some given constant $c > 0$ and the parameter becomes the weight of the desired vertex cover (see [28] for corresponding studies for VERTEX COVER). Our work also complements the numerous approximability results for these problems. It is a task for future research to significantly improve on the presented worst-case running times (exponential factors in parameter $k$). In particular, it would be interesting to learn more about the amenability of the considered problems to problem kernelization by (more) efficient data reduction techniques.

Besides the significant interest (with numerous applications behind) in the studied problems on their own, we want to mention one more feature of our work that lies a little aside. Adding our results to the already known large arsenal of facts about VERTEX COVER, this problem can be even better used and understood as a seed problem for parameterized complexity as a whole: New aspects now related to vertex covering by means of our results are issues such as enumerative techniques or parameterized reduction. This might be of particular use when learning or teaching parameterized complexity through basically *one* natural and easy to grasp problem—VERTEX COVER—and its "straightforward" generalizations.

*Acknowledgment.* We are grateful to an anonymous referee of WADS 2005 for spotting a flaw in a previous proof of the fixed-parameter tractability of CONNECTED VERTEX COVER.

## References

1. F. N. Abu-Khzam, R. L. Collins, M. R. Fellows, M. A. Langston, W. H. Suters, and C. T. Symons. Kernelization algorithms for the Vertex Cover problem: theory and experiments. In *Proc. ALENEX04*, pages 62–69. ACM/SIAM, 2004.
2. J. Alber, H. L. Bodlaender, H. Fernau, T. Kloks, and R. Niedermeier. Fixed parameter algorithms for Dominating Set and related problems on planar graphs. *Algorithmica*, 33(4):461–493, 2002.
3. J. Alber, F. Dorn, and R. Niedermeier. Experimental evaluation of a tree decomposition based algorithm for Vertex Cover on planar graphs. *Discrete Applied Mathematics*, 145(2):219–231, 2005.
4. J. Alber, H. Fernau, and R. Niedermeier. Parameterized complexity: exponential speed-up for planar graph problems. *Journal of Algorithms*, 52:26–56, 2004.

5.  J. Alber, J. Gramm, and R. Niedermeier. Faster exact algorithms for hard problems: a parameterized point of view. *Discrete Mathematics*, 229:3–27, 2001.
6.  E. M. Arkin, M. M. Halld´orsson, and R. Hassin. Approximating the tree and tour covers of a graph. *Information Processing Letters*, 47(6):275–282, 1993.
7.  V. Arvind and V. Raman. Approximation algorithms for some parameterized counting problems. In *Proc. 13th ISAAC*, volume 2518 of *LNCS*, pages 453–464. Springer, 2002.
8.  R. Balasubramanian, M. R. Fellows, and V. Raman. An improved fixed-parameter algorithm for Vertex Cover. *Information Processing Letters*, 65(3):163–168, 1998.
9.  M. Bläser. Computing small partial coverings. *Information Processing Letters*, 85(6):327–331, 2003.
10. N. H. Bshouty and L. Burroughs. Massaging a linear programming solution to give a 2-approximation for a generalization of the Vertex Cover problem. In *Proc. 15th STACS*, volume 1373 of *LNCS*, pages 298–308. Springer, 1998.
11. L. Cai and D. Juedes. On the existence of subexponential parameterized algorithms. *Journal of Computer and System Sciences*, 67(4):789–807, 2003.
12. L. S. Chandran and F. Grandoni. Refined memorisation for Vertex Cover. *Information Processing Letters*, 93(3):125–131, 2005.
13. J. Cheetham, F. Dehne, A. Rau-Chaplin, U. Stege, and P. J. Taillon. Solving large FPT problems on coarse-grained parallel machines. *Journal of Computer and System Sciences*, 67(4):691–706, 2003.
14. J. Chen, I. A. Kanj, and W. Jia. Vertex Cover: further observations and further improvements. *Journal of Algorithms*, 41:280–301, 2001.
15. J. Chuzhoy and J. S. Naor. Covering problems with hard capacities. In *Proc. 43rd FOCS*, pages 481–489. IEEE Computer Society Press, 2002.
16. T. H. Cormen, C. E. Leiserson, R. L. Rivest, and C. Stein. *Introduction to Algorithms, 2nd Edition*. MIT Press, 2001.
17. R. G. Downey and M. R. Fellows. *Parameterized Complexity*. Springer, 1999.
18. S. E. Dreyfus and R. A. Wagner. The Steiner problem in graphs. *Networks*, 1:195–207, 1972.
19. M. R. Fellows. New directions and new challenges in algorithm design and complexity, parameterized. In *Proc. 8th WADS*, volume 2748 of *LNCS*, pages 505–520. Springer, 2003.
20. R. Gandhi, E. Halperin, S. Khuller, G. Kortsarz, and A. Srinivasan. An improved approximation algorithm for Vertex Cover with hard capacities. In *Proc. 30th ICALP*, volume 2719 of *LNCS*, pages 164–175. Springer, 2003.
21. R. Gandhi, S. Khuller, and A. Srinivasan. Approximation algorithms for partial covering problems. In *Proc. 28th ICALP*, volume 2076 of *LNCS*, pages 225–236. Springer, 2001.
22. S. Guha, R. Hassin, S. Khuller, and E. Or. Capacitated vertex covering. *Journal of Algorithms*, 48(1):257–270, 2003.
23. J. Könemann, G. Konjevod, O. Parekh, and A. Sinha. Improved approximations for tour and tree covers. *Algorithmica*, 38(3):441–449, 2004.
24. R. Niedermeier. Ubiquitous parameterization-invitation to fixed-parameter algorithms. In *Proc. MFCS*, volume 3153 of *LNCS*, pages 84–103. Springer, 2004.
25. R. Niedermeier. *Invitation to Fixed-Parameter Algorithms*. Oxford University Press, forthcoming, 2005.
26. R. Niedermeier and P. Rossmanith. Upper bounds for Vertex Cover further improved. In *Proc. 16th STACS*, volume 1563 of *LNCS*, pages 561–570. Springer, 1999.

27. R. Niedermeier and P. Rossmanith. A general method to speed up fixed-parametertractable algorithms. *Information Processing Letters*, 73:125–129, 2000.

28. R. Niedermeier and P. Rossmanith. On efficient fixed-parameter algorithms for Weighted Vertex Cover. *Journal of Algorithms*, 47(2):63–77, 2003.

29. N. Nishimura, P. Ragde, and D. M. Thilikos. Fast fixed-parameter tractable algorithms for nontrivial generalizations of Vertex Cover. In *Proc. 7th WADS*, volume 2125 of *LNCS*, pages 75–86. Springer, 2001.

30. E. Prieto and C. Sloper. Either/or: using vertex cover structure in designing FPTalgorithms—the case of $k$-Internal Spanning Tree. In *Proc. 8th WADS*, volume 2748 of *LNCS*, pages 474–483. Springer, 2003.

# The Complexity of Implicit and
# Space Efficient Priority Queues

Christian W. Mortensen* and Seth Pettie**

[1] IT University of Copenhagen
cworm@itu.dk
[2] Max Planck Institut für Informatik
pettie@mpi-sb.mpg.de

**Abstract.** In this paper we study the time-space complexity of implicit priority queues supporting the *decreasekey* operation. Our first result is that by using *one* extra word of storage it is possible to match the performance of Fibonacci heaps: constant amortized time for insert and decreasekey and logarithmic time for deletemin. Our second result is a lower bound showing that that one extra word really is necessary. We reduce the decreasekey operation to a cell-probe type game called the *Usher's Problem*, where one must maintain a simple data structure without the aid of any auxiliary storage.

## 1 Introduction

An *implicit* data structure on $N$ elements is one whose representation consists simply of an array $A[0..N-1]$, with one element stored in each array location. The most well known implicit structure is certainly Williams's binary heap [26], which supports the priority queue operations *insert* and *delete-min* in logarithmic time. Although the elements of Williams's heap are conceptually arranged in a fragment of the infinite binary tree, the tree edges are not explicitly represented. It is understood that the element at $A[i]$ is the parent of $A[2i+1]$ and $A[2i+2]$. The practical significance of implicit data structures is that they are, in certain circumstances, maximally space efficient. If the elements can be considered *atomic* then there is no better representation than a packed array.

A natural suspicion is that by insisting on an implicit representation one may be sacrificing asymptotic time optimality. After 40 years of sporadic research on implicit structures [26, 17, 21, 27, 20, 5, 8, 7, 28, 9, 10, 15] we can say that this suspicion is almost completely misguided. In various dictionary & priority queue problems, for instance, there are either optimal implicit structures or ones that can be made optimal with *a couple extra words* of storage.

---

* This work was performed while visiting the Max-Planck-Institut für Informatik as a Marie Curie Doctoral Fellow.
** Supported by an Alexander von Humboldt Postdoctoral Fellowship.

F. Dehne, A. López-Ortiz, and J.-R. Sack (Eds.): WADS 2005, LNCS 3608, pp. 49–60, 2005.

In this paper we study the complexity of implicit priority queues that support the decrease-key operation. Our main positive result is that given *one extra word of storage* there is an implicit priority queue matching the performance of Fibonacci heaps. It supports delete-min in logarithmic time and insert and decrease-key in constant time, all amortized. That one extra word obviously has no practical consequences but it is a thorn in our side. We propose a variation on our data structure that uses no additional storage and supports decrease-key in $O(\log^* n)$ time, without affecting the other operations. This $O(\log^* n)$ bound is a theoretical burden. Is it natural? We prove that it is, in the following sense: if any implicit priority queue uses zero extra space and supports decrease-key in $o(\log^* n)$ amortized time then the amortized cost of insert/delete-min jumps dramatically, from logarithmic to $\Omega(n^{1/\log^{(k)} n})$, for any $k$.

We reduce the decrease-key operation to the *Absent Minded Usher's Problem*, a game played in a simplified cell-probe model. Imagine an usher seating indistinguishable theater patrons one-by-one in a large theater. The usher is equipped with two operations: he can *probe* a seat to see if it is occupied or not and he can *move* a given patron to a given unoccupied seat. (Moving patrons after seating them is perfectly acceptable.) The catch is this: before seating each new patron we wipe clean the usher's memory. That is, he must proceed without any knowledge of which seats are occupied or the number of patrons already seated. We prove that any deterministic ushering algorithm must seat $m$ patrons with $\Omega(m \log^* m)$ probes and moves, and that this bound is asymptotically tight.

Our lower bound proof attacks a subtle but fundamental difficulty in implicit data structuring, namely, orchestrating the movement of elements within the array, given little auxiliary storage. In its present form the ushering problem is limited to proving small time-space tradeoffs. However it is likely that generalizations of the ushering method could yield more impressive lower bounds.

**Organization.** In the remainder of this section we define what an implicit priority queue is, survey previous work and discuss our contributions. In Section 2 we present our new data structure. Section 3 is devoted to the Usher's Problem and its relationship with implicit priority queues.

*Implicit Priority Queues.* We first give a specification for an abstract implicit priority queue which is suitable for theoretical analysis but impractical. We then propose a particularly space efficient method for implementing such a data structure.

An implicit priority queue of size $n$ consists of an array $A[0..n-1]$ (plus, possibly, a little extra storage) where $A[0], \ldots, A[n-1]$ contain distinct elements (or *keys*) from a total order. We also use $A$ to denote the set of elements in the priority queue. The following operations are supported.

insert$(\kappa)$         : $A := A \cup \{\kappa\}$
deletemin$()$       : Return $\min A$ and set $A := A \backslash \{\min A\}$
decreasekey$(i, \kappa)$ : Set $A[i] := \min\{A[i], \kappa\}$

An operation decides what to do based on the auxiliary information and any comparisons it makes between elements. Before returning it is free to alter the auxiliary information and permute the contents of $A$, so long as its $n$ elements lie in $A[0, \ldots, n-1]$. We assume that "$n$" is known to all operations.

The definition above sweeps under the rug a few issues that are crucial to an efficient and useful implementation. First, applications of priority queues store not only elements from a total order but *objects* associated with those elements. For example, Dijkstra's shortest path algorithm repeatedly asks for the *vertex* with minimum tentative distance; the tentative distance alone is useless. Any practical definition of a priority queue must take the application's objects into account. The second issue relates to the peculiar arguments to decreasekey. The application must tell decreasekey the *index* in $A$ of the element to be decreased, which is necessarily a moving target since the data structure can permute the elements of $A$ at will.

We propose a priority queue interface below that addresses these and other issues. Let us first sketch the normal (real world) interaction between application and priority queue. To insert the object $v$ the application passes the priority queue an identifier $id(v)$ of its choosing, together with $key(v)$, drawn from some total order. In return the priority queue gives the application a $pq\_id(v)$, also of its choosing, which is used to identify $v$ in later calls to decreasekey. When $v$ is removed from the queue, due to a deletemin, the application receives both $id(v)$ and $key(v)$.

In our interface we give the data structure an extra degree of freedom, without placing any unreasonable demands on the governing application. Whereas the standard interface forces the data structure to assign $pq\_ids$ once and for all, we let the data structure update $pq\_ids$ as necessary. We also let the application maintain control of the keys. This is for two reasons, both concerning space. First, the application may not want to explicitly represent keys at all if they can be deduced in constant time. Second, the application can now use the same key in multiple data structures without producing a copy for each one. Below $\mathcal{Q}$ represents the contents of the data structure, which is initially empty. (Observe that this interface is modular. Neither the application nor the data structure needs to know any details about the other.)

*The priority queue implements:*

**insert**$(id(v))$      : Sets $\mathcal{Q} := \mathcal{Q} \cup \{id(v)\}$ and returns a $pq\_id(v)$
**deletemin**()      : Return, and remove from $\mathcal{Q}$, the $id(v)$ minimizing
                       $key(v)$
**decreasekey**$(pq\_id(v))$ : A notification that $key(v)$ has been reduced
*The application implements:*
**update**$(id(v), x)$    : Set $pq\_id(v) := x$
**compare**$(id(v), id(w))$ : True iff $key(v) < key(w)$

Using this interface it is simple to implement an abstract implicit priority queue. The data structure would consist of an array of *ids* and we maintain, with appropriate update operations, that $pq\_id(v)$ indexes the position of $id(v)$ in $A$. For example, if we implemented a $d$-ary heap [17] with this interface every priority queue operation would end with at most $\log_d n$ calls to update, which is the maximum number of elements that need to be permuted in $A$.

Our interface should be contrasted with a more general solution formalized by Hagerup and Raman [14], in which $pq\_ids$ would be fixed once and for all.

In their schema the application communicates with the data structure through an intermediary *quasidictionary*, which maps each *pq_id* (issued by the quasidictionary) to an identifier that can be updated by the data structure. Since saving space is one of the main motivations for studying implicit data structures we prefer our interface to one based on a quasidictionary.

*Defining "Extra Storage."* All the priority queues cited in this paper store $n$ *ids* and $n$ *keys*, and if decreasekey is supported, $n$ *pq_ids* as well. We consider any further storage *extra*. For the sake of simplicity we ignore temporary space used by the individual operations and any overhead involved in memory allocation. Here "memory allocation" means simulating the array $A$, whose length varies as elements are inserted and deleted. Brodnik et al. [3] proved that the standard solution—array doubling/halving—can be improved so that only $\Theta(\sqrt{n})$ extra words of space are used, where $n$ is the current size of the array. In pointer-based structures (like Fibonacci heaps) the cost of memory allocation is more severe. There is a measurable overhead for each allocated block of memory.

*Previous work.* Much of the work on implicit data structures has focussed on the dictionary problem, in all its variations. The study of dynamic implicit dictionaries in the comparison model was initiated by Munro & Suwanda [21]. They gave a specific partial order (à la Williams's binary heap) that allowed inserts, deletes, and searches in $O(\sqrt{n})$ time, and showed, moreover, that with any partial order $\Omega(\sqrt{n})$ time is necessary for some operation. Munro [20] introduced a novel pointer encoding technique and showed that all dictionary operations could be performed in $O(\log^2 n)$ time, a bound that stood until just a few years ago. After a series of results Franceschini & Grossi [9] recently proved that all dictionary operations could be performed in worst-case logarithmic time with no extra storage. Franceschini & Grossi [10] also considered the static dictionary problem, where the keys consist of a vector of $k$ characters. Their implicit representation allows for searches in optimal $O(k + \log n)$ time, which is somewhat surprising because Andersson et al. [1] already proved that optimal search is impossible if the keys are arranged in *sorted order*.

Yao [27] considered a static version of the dictionary problem where the keys are integers in the range $\{1, \ldots, m\}$. He proved that if $m$ is sufficiently large relative to $n$ then no implicit representation can support $o(\log n)$ time queries. In the same paper Yao proved that *one-probe* queries are possible if $m \leq 2n-2$. Fiat et al. [8,7] gave an implicit structure supporting constant time queries for any universe size, provided only $O(\log n + \log \log m)$ bits of extra storage. Zuckerman [28], improving a result of [8,7], showed that $O(1)$ time queries are possible with zero extra storage, for $m$ slightly superpolynomial in $n$. In the integer-key model we need to qualify the term "extra storage." An implicit representation occupies $n \lceil \log m \rceil$ bits, which is roughly $n \log n$ more than the information bound of $I = \lceil \log \binom{m}{n} \rceil$ bits. See [4, 22, 23, 24] for $I + o(I)$ space dictionaries.

*Implicit Priority Queues.* Williams's binary heap [26] uses zero extra storage and supports inserts and deletemins in worst-case logarithmic time. It was generalized by Johnson [17] to a $d$-ary heap, which, for any fixed $d$, supports inserts and decreasekeys in $O(\log_d n)$ time and deletemins in $O(d \log_d n)$ time. Carlsson et al.'s implicit binomial heap [5] supports inserts in constant time and deletemins in logarithmic time, both worst case. Their data structure uses $O(1)$ extra words of storage. Harvey and Zatloukal's postorder heap [15] also supports insert in $O(1)$ time; the time bound is amortized but they use zero extra storage. Their data structure can be generalized to a postorder $d$-ary heap.

*General Priority Queues.* The $d$-ary implicit heap has an undesirable tradeoff between decreasekeys and deletemins. In many applications of priority queues the overall performance depends crucially on a fast decreasekey operation. Fredman and Tarjan's Fibonacci heap [13] supports all operations in optimal amortized time: $O(\log n)$ for deletemin and $O(1)$ for the rest. Aside from the space taken for *keys*, *ids*, and *pq_ids*, Fibonacci heaps require $4n$ pointers and $n(\log \log n + 2)$ bits. Each node keeps a $(\log \log n + 1)$-bit rank, a mark bit, and references its parent, child, left sibling, and right sibling. Kaplan and Tarjan [18] shaved the space requirements of Fibonacci heaps by $n$ pointers and $n$ bits. The Pairing heap [12] can be represented with $2n$ pointers though it does not handle decreasekeys in constant time [11].

*Our Contributions.* In this paper we show that it is possible to match the performance of Fibonacci heaps using *one* extra word of storage, and that no deterministic implicit priority queue can achieve the same bounds without one extra word; see Figure 1. Our data structure may be of separate interest because it uses a completely new method for supporting decreasekeys in constant amortized time. Whereas Fibonacci-type heaps [13, 6, 18, 25] organize the elements in heap-ordered trees and link trees based on their *ranks*, our priority queue is conceptually composed of a set of unordered lists, whose elements adhere to a particular partial order. We do not tag elements with ranks. The primitives of our data structure are simply the concatenation of lists and the division of lists, using any linear-time selection algorithm [2].

| | Decreasekey | Deletemin | Extra Storage | Ref |
|---|---|---|---|---|
| Fibonacci | $O(1)$ | $O(\log n)$ | $4n$ ptrs, $n(\log \log n + 2)$ bits | [13] |
| Thin | $O(1)$ | $O(\log n)$ | $3n$ ptrs, $n(\log \log n + 1)$ bits | [18] |
| Pairing | $\Omega(\log \log n)$ $O(\log n)$ | $O(\log n)$ | $2n$ ptrs, $n$ bits | [11] [16] |
| Post. $d$-ary | $O(\log_d n)$ | $O(d \log_d n)$ | zero | [15] |
| **New** | $O(1)$ | $O(\log n)$ | 1 ptr | |
| **New** | $O(\log^* n)$ | $O(\log n)$ | zero | |
| **New l.b.** | if $o(\log^* n)$ | $\Omega(n^{1/\log^{(k)} n})$ | zero | |

**Fig. 1.** All priority queues support inserts in amortized constant time. The Fibonacci & Thin Heaps support amortized constant time melds. See [19] for results on worst-case bounds

Our lower bounds reinforce a theme in implicit data structures [27, 5, 8, 15], that it takes only a couple extra words of storage to alter the complexity of a problem. Although our results depend on small amounts of extra memory, the *ushering* technique is new and abstract enough to be applied elsewhere.

## 2   A New Priority Queue

In this section we design an *abstract* implicit priority queue. We cover the high-level invariants of the structure, some very low-level encoding issues, then sketch the operational "flow" of the priority queue. Many details are omitted; see [19].

*Encoding Bits.* We encode bits using the standard technique: for two given indices $i, j$ we equate $A[i] < A[j]$ with zero and $A[j] < A[i]$ with one. In this way we can encode small integers and pointers with $O(\log n)$ elements.

*Junk Elements.* All elements are tagged either *normal* or *junk*, where the tags are represented implicitly. We divide the entire array $A$ into consecutive *triplets*. The first two elements of any triplet are junk, and their relative order encodes whether the third element is junk or normal. A junk (normal) triplet is one whose third element is junk (normal). We maintain the invariant that the minimum element in the queue is normal.

*L-lists and I-lists.* At a high level the data structure consists of a sequence of $O(\log n)$ $L$-lists and a set of $O(\log^2 n)$ $I$-lists, each of which is associated with a distinct interval of $L$-lists. For any list $T$ we let $T^\star$ denote the set of normal elements in $T$ and $|T|$ its length in triplets, including junk triplets. The relation $S < T$ holds when $\max S^\star < \min T^\star$, or if either $S^\star$ or $T^\star$ is empty.
    The list $L_{ij}$ belongs in slot $j$ of zone $i$, where $i \in [0, \log_4 n]$, $j \in [0, 6]$, and the length of $L_{ij}$ is roughly exponential in $i$. The $L$-lists are internally unordered but, as a whole, in sorted order. That is, if $ij < kl$ (lexicographically) then $L_{ij} < L_{kl}$. Loosely speaking, the list $I_{ij,kl}$ contains elements that, *were* they to be included in $L$-lists, could be assigned to some list in the interval $L_{ij}, \ldots, L_{kl}$. We let $L_{s(ij)}$ and $L_{p(ij)}$ be the non-empty successor and predecessor of list $L_{ij}$, respectively. The $L$- and $I$-lists obey the following *order* and *size* invariants. Some invariants refer to parameters $N \geq n$, $\omega = \lceil \log N \log \log N \rceil$ and $\gamma = \log^4 N$. In addition to $L$- and $I$-lists there is a *buffer* $B$ which is discussed later.

**Inv. O1** If $ij < kl$ then $L_{ij} < L_{kl}$.
**Inv. O2** If $ij \leq kl$ then $L_{p(ij)} < I_{ij,kl} < L_{s(kl)}$.
**Inv. S1** $|L_{ij}| \in [\gamma 4^i, 2\gamma 4^i]$ and $|L_{ij}|$ is a multiple of $\omega$. $|L_{ij}|$ is non-empty only if $L_{i0}, \ldots, L_{i(j-1)}$ are also non-empty, i.e., $L$-lists are packed in each zone.
**Inv. S2** For $ij > 00$, $|L_{ij}^\star| \geq \frac{1}{2}|L_{ij}|$.
**Inv. S3** For any $ij \leq kl$, $|I_{ij,kl}|$ is a multiple of $\omega$ and $I_{ij,kl}$ is non-empty only if $L_{ij}$ and $L_{kl}$ are also non empty. $I_{00,kl}$ is empty for all $kl$.
**Inv. S4** $|B| = |L_{00}| = 2\gamma$, and $|L_{00}^\star| \geq 1$.

Assuming that the minimum element is normal, it follows from Invariants O1, O2, S3, and S4 that the minimum always lies in $L_{00}$ or $B$. In our data structure

$L_{00}$ and $B$ are treated as fixed size mini priority queues that support findmin in constant time, deletemin in $O(\log n)$ time, and decreasekey in either $O(1)$ or $O(\log^* n)$ time, depending on whether we are allowed to store one extra pointer.

*Periodic Rebuilding.* We use a few mechanisms to guarantee that min $A$ is normal. First, any element that is inserted or decreasekey'd is made normal. Every insertion takes two otherwise unused junk elements to form a normal triplet and every decreasekey can require up to three unused junk elements. Our data structure rebuilds itself whenever there may be a shortage of unused junk elements. First it finds the $n/6$ smallest elements, designates them normal, and assigns them to triplets. The remaining $n/2$ junk elements constitute the *junk reservoir*. We maintain an implicit operation counter that is initially set to $n/6$. Every operation decrements the counter and when it reaches zero (i.e, when the junk reservoir may be empty) the data structure is rebuilt in linear time. It follows that the minimum element is always normal. We charge the cost of rebuilding the structure to the $n/6$ preceding operations. Some parameters are w.r.t. an $N \geq n$. Upon rebuilding we let $N = 2^{\lceil \log(7n/6) \rceil}$. Since $n$ (the size of the queue) does not uniquely determine $N$ we dedicate two junk triplets to indicate the correct value.

*Block Structure.* The entire array $A$ is divided into a sequence of *blocks*, each containing $\omega = \lceil \log N \log \log N \rceil$ triplets. The primary purpose of blocks is to allow a dynamic implicit representation of $L$- and $I$-lists, which are circular and doubly-linked. We dedicate $O(\log N)$ triplets of each block to represent successor and predecessor pointers. Thus, given two blocks in different $L$- or $I$-lists it is possible to splice them together in $O(\log N)$ time by altering 4 pointers. Blocks also contain other counters and pointers; see [19].

*The Structure of $B$ and $L_{00}$.* Recall that both $B$ and $L_{00}$ have fixed size and are located in $A$ at fixed locations depending on $N$. We keep the normal triplets of $L_{00}$ packed together and arranged in a $(\log N)$-ary implicit heap. Thus $L_{00}$ supports deletemins in $O(\log N \log_{\log N} \omega) = O(\log N)$ time and decreasekeys on elements in $L_{00}^\star$ in $O(\log_{\log N} \omega) = O(1)$ time. It allows bulk insertion of $k$ elements in $O(k + \log \gamma)$ time, where the $\log \gamma = O(\log \log N)$ term is the time to determine $|L_{00}^\star|$. When dealing with $B$ and $L_{00}$ a deletemin accepts a junk element to plug up the hole and an insert returns a displaced junk element. The buffer $B$ consists of, in this order, a triplet containing min $B^\star$ followed by fixed size mini-buffers $B^{\log^* N}, \ldots, B^1$, where $|B^1| = \Theta(\gamma) = \Theta(\log^4 N)$ and $|B^\ell| = \Theta(\log^{(\ell)} N)$. The normal triplets of each mini-buffer are packed together, and furthermore, the normal triplets of $B^1$ are arranged in a $(\log N)$-ary implicit heap. To insert the element $e \notin B$ into $B$ we tag it normal, if not already. We identify the first junk triplet of $B^{\log^* N}$ and swap $e$ with the third element of this triplet. If $e = \min B$ we swap it with the old minimum. At this point $B^{\log^* N}$ may be full, that is, it contains only normal triplets. In general, whenever $B^{\ell+1}$ is full we perform a binary search to determine the first junk triplet in $B^\ell$, say $B^\ell[j]$. We then swap the whole mini-buffer $B^{\ell+1}$ with the junk triplets in $B^\ell[j..j+|B^{\ell+1}|-1]$, which may cause $B^\ell$ to be full. If $\ell = 1$ an artificial decreasekey operation is performed on each of these elements in order to restore the heap order of $B^1$. The cost of relieving an overflow of $B^{\ell+1}$ is $O(|B^{\ell+1}| + \log |B^\ell|) = O(|B^{\ell+1}|)$, i.e., $O(1)$ per element. Since an element appears in each buffer at most once (per insertion) the amortized cost of an insert is $O(\log^* N)$. See [19] for $B$'s deletemin and decreasekey routines.

If $B$ can store an extra pointer (outside of the array $A$) then it can support inserts and decreasekeys in $O(1)$ time. We get rid of $B^{\log^* N}, \ldots, B^2$ and let the pointer index the first junk triplet in $B^1$.

*Memory Layout.* The array $A$ is partitioned into five parts. The first part consists entirely of junk and the first three occupy fixed locations depending on $N$.

1. Preamble. Contains a flag indicating $N$, the operation counter, and the free block pointer, which points to the location of item (5), below.
2. The Buffer $B$ and the list $L_{00}$.
3. Representative blocks: one block from each $L$- and $I$-list is kept in a fixed location. If the list is empty an all-junk block is kept in its spot. The representative blocks contain additional statistics about their respective lists.
4. An array of blocks in use, by either $L$- or $I$-lists.
5. An array of unused blocks, a.k.a. the junk reservoir.

*High-Level Operations.* The priority queue operations only deal with $L_{00}$ and $B$. However, such an operation can induce a number of low level operations if one of the invariants is violated, for instance, if $|L_{00}^\star|$ reaches zero or $|B| > 2\gamma$. The asterixes below mark places where a sequence of low level operations may be necessary. Rebuilding the structure is considered a low level operation.

**Insert($\kappa$):** Insert* $\kappa$ into $B$. Put the displaced junk element at the end of the junk reservoir. Decrement* the global operation counter.

**Decreasekey($i, \kappa$):** If $A[i]$ lies in $B$ or $L_{00}$ then perform the decreasekey there. Otherwise insert* the element $A[i]$ into $B$ with the new key $\kappa$, changing its status to normal if it was junk. Put the displaced junk element at $A[i]$. Decrement* the global operation counter.

**Deletemin():** Return $\min(L_{00}^\star \cup \min B^\star)$ using the deletemin operation* provided either by $B$ or $L_{00}$. Use the last junk element in the junk reservoir to plug up the hole. Decrement* the global operation counter.

*A Sketch of the Rest.* Ignoring other parts of the data structure, the behavior of our $L$-lists is straightforward. Whenever $L_{00}^\star$ becomes empty we find its successor $L' = L_{s(00)}$. If $L'$ is in zone 0 we simply rename it $L_{00}$ and if $L'$ is in zone $i > 0$ we *divide* it, with a linear-time selection algorithm, into $O(i)$ shorter lists, which are distributed over zones 0 through $i - 1$. Similarly, when $L_{00}^\star$ becomes full we divide it into smaller lists, which are inserted into zone 0. In general, whenever a zone $i$ contains more than 7 lists (a violation) we concatenate some of the lists to form one whose size is suitable to be inserted into zone $i + 1$. Our main difficulty is filing newly inserted/decreasekey'd elements into the correct $L$-list. Any direct method, like performing a binary search over the $L$-lists, is doomed to take $\Omega(\log \log n)$ time. The purpose of the $I$-lists is to direct unfiled elements to their correct $L$-lists, at constant amortized time per element.

Inserts and decreasekeys are directed toward the buffer $B$. When $B$ is full we divide it into geometrically increasing sets $J_0 < J_1 < J_2 \ldots$, with $|J_{\ell+1}| = 2|J_\ell|$.

Each set $J_\ell$ is concatenated with the narrowest possible $I$-list that satisfies Invariant O2. That is, we pick $O(\log n)$ normal representatives $e_{ij} \in L^\star_{ij}$ and concatenate $J_\ell$ with $I_{ij,kl}$ such that $e_{ij} < J_\ell < e_{kl}$, where $ij$ is maximal and $kl$ minimal. The procedure for dividing $I$-lists is identical to dividing the buffer. (We basically consider $B$ to be a special $I$-list.) In particular, whenever $I_{ij,kl}$ *might* contain the minimum element—if $L_{ij}$ is the first non-empty $L$-list—we divide it into $J$-sets of geometrically increasing size and concatenate them with the proper $I$-lists. It is not obvious why this method should work. Every time an $I$-list is processed its elements have a good chance of ending up in another $I$-list, which presumably covers a narrower interval of $L$-lists. That is, it looks like we're just implementing binary search in slow motion. The full analysis of our data structure relies on a complicated potential function; see [19] for the details.

# 3     The Absent Minded Usher's Problem

In [19] we show that in any implicit priority queue the decreasekey operation must be prepared to solve a version of the Absent Minded Usher's Problem, which is the focus of this section. Refer to [19] for the complete lower bound proof.

Let $A$ be an array of infinite length, where each location of $A$ can contain a *patron* (indistinguishable from other patrons) or be empty. An *usher* is a deterministic program to insert a new patron into $A$ without any knowledge of its contents. That is, the usher does not know how many patrons it has already inserted. The usher can probe a location of $A$ to see if it is occupied and move any patron to an empty location of $A$. We are interested in the complexity of the best ushering program, that is, the number of probes and moves needed to seat $N$ patrons in an initially empty array. There exists a simple $O(N \log^* N)$ time usher; it is based on the same *cascading buffers* technique used in $B$ from the previous section. We prove that any usher requires $\Omega(N \log^* N)$ time.

We imagine an infinite graph whose vertices are layed out in a grid. The $x$-axis corresponds to time (number of insertions) and the $y$-axis corresponds to the array $A$. An usher is partially modeled as a set of $x$-monotone paths through the grid, with each path representing where a particular patron was at each moment in time. We assign each edge a cost, which represents in an amortized sense the time taken to bring that patron to its current position. By reasoning about the usher's decision tree we are able to derive a recurrence relation describing the costs of edges. The solution to this recurrence is then used to lower bound the complexity of the ushering problem.

We put the patron to be inserted in the artificial position $A[-1]$. The usher's algorithm is modeled as a binary decision tree. At each internal node is an array position to be probed and at each leaf is a list of pairs of the form $(j_1, j_2)$, indicating that the patron at $A[j_1]$ should be moved to $A[j_2]$. Each leaf is called an *operation* and the cost of executing an operation $o$ is its depth in the decision tree, $d(o)$, plus the number of patron moves, $m(o)$.

Consider an infinite graph with vertex set $\{A_i[j] : i \geq 0, j \geq -1\}$, where $A_i[j]$ represents $A[j]$ *after* $i$ insertions. There exists an edge $(A_{i-1}[j_1], A_i[j_2])$ exactly

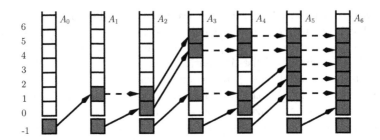

**Fig. 2.** This (infinite) two dimensional grid depicts the flow of patrons over time for a particular ushering algorithm. Probes made by the algorithm are not represented. Solid edges are fresh, dashed ones leftover

when $A_{i-1}[j_1]$ and $A_i[j_2]$ contain the same patron. Note that the graph is composed exclusively of paths. An edge is *leftover* if it is of the form $(A_{i-1}[j], A_i[j])$ and *fresh* otherwise, i.e., fresh edges correspond to patron movements and leftover edges correspond to unmoved patrons; see Figure 2. Let $\mathrm{pred}(A_{i-1}[j_1], A_i[j_2])$ denote the edge $(A_{i-2}[j_3], A_{i-1}[j_1])$ if such an edge exists.

We now define a cost function $c$. If the edge $(u, v)$ does not exist then $c(u, v) = 0$. Any edge of the form $(A_{i-1}[j], A_i[j])$ has cost $c(\mathrm{pred}(A_{i-1}[j], A_i[j]))$: leftover edges inherit the cost of their predecessor. Let $C_i = \sum_{j_1, j_2} c(A_{i-1}[j_1], A_i[j_2])$ be the total cost of edges into $A_i$, and let $P_i = \sum_{j_1 \neq j_2} c(\mathrm{pred}(A_{i-1}[j_1], A_i[j_2]))$ be the cost of the *predecessors* of the fresh edges into $A_i$. Let $o_i$ be the operation performed at the $i$th insertion. (For example, in the ushering algorithm partially depicted in Figure 2, $o_2 = o_4 = o_6$ and $o_1$, $o_3$, and $o_5$ are distinct.) Each fresh edge $e$ between $A_{i-1}$ and $A_i$ is assigned the same cost:

$$c(e) \stackrel{\mathrm{def}}{=} \frac{P_i + d(o_i) + m(o_i)}{m(o_i)}$$

That is, the total cost assigned to these $m(o_i)$ fresh edges is their inherited cost, $P_i$, plus the actual time of the operation $o_i$: $d(o_i) + m(o_i)$. It follows from the definitions that $C_i$ is exactly the time to insert $i$ patrons. Let $T(m)$ be the *minimum* possible cost of a fresh edge associated with an operation performing $m$ movements. We will show $T(m) = \Omega(\log^* m)$ and that this implies the amortized cost of $N$ insertions is $\Omega(N \log^* N)$.

For the remainder of this section we consider the $i$th insertion. Suppose that the operation $o_i$ moves patrons *from* locations $A_{i-1}[j_1, j_2, \ldots, j_{m(o_i)}]$. The patrons in these locations were placed there at various times in the past. Define $i_q < i$ as the index of the insertion that *last* moved a patron to $A[j_q]$.

**Lemma 1.** *Let $p, q$ be indices between 1 and $m(o_i)$. If $i_p \neq i_q$ then $o_{i_p} \neq o_{i_q}$.*

We categorize all operations in the patron's decision tree w.r.t. $m = m(o_i)$— recall that $i$ is fixed in this section. An operation $o$ is *shallow* if $d(o) < \lfloor \log m \rfloor /2$ and *deep* otherwise. It is *thin* if $m(o) < \lfloor \sqrt{\log m} \rfloor$ and *fat* otherwise.

**Lemma 2.** $\left| \{q : o_{i_q} \text{ is shallow and thin}\} \right| < \frac{1}{2} \sqrt{m \log m}$

**Lemma 3.** *If $(u, v)$ is a fresh edge, where $v \in A_k$ and $o_k$ is deep and thin, then $c(u, v) \geq \frac{1}{2}\sqrt{\log m}$.*

In summary, Lemma 2 implies that at least $m - \sqrt{m \log m}/2$ of the fresh edges between $A_{i-1}$ and $A_i$ can be traced back to earlier edges that are either in deep and thin operations or fat operations. The cost of edges in deep and thin operations is bounded by Lemma 3 and the cost of edges in fat operations is bounded inductively. Recall that $T(m)$ is the minimum cost of a fresh edge associated with an operation performing $m$ moves.

**Lemma 4.** $T(m) \geq (\log^* m)/4$

*Proof.* Let $e$ be a fresh edge into $A_i$ with $m = m(o_i)$ and let $\beta = \min\{\frac{1}{2}\sqrt{\log m}, T(\sqrt{\log m}), T(\sqrt{\log m} + 1), \ldots\}$. Then:

$$c(e) = \frac{m(o_i) + P_i + d(o_i)}{m(o_i)} \geq \frac{m + (m - \frac{1}{2}\sqrt{m \log m}) \cdot \beta}{m}$$

Since the only property of $e$ that we used in the above inequalities is $m(o_i) = m$, any lower bound on $c(e)$ implies the same lower bound on $T(m)$. We assume inductively that $T(r) \geq \frac{1}{4}\log^* r$, which holds for $r \leq 2^{16}$ since $T(r) \geq 1$ and $\log^* 2^{16} = 4$. For $m > 2^{16}$ we have:

$$T(m) \geq 1 + \left(1 - \sqrt{\log m/4m}\right) \cdot \beta \geq 1 + \left(1 - \sqrt{\log m/4m}\right) \cdot \frac{\log^*(\sqrt{\log m})}{4}$$

$$> \left(1 - \sqrt{\log m/4m}\right) \left(\frac{\log^* m - 2}{4} + 1\right) > \frac{\log^* m}{4}$$

**Theorem 1.** *Any usher seating $N$ patrons must perform $\Omega(N \log^* N)$ operations. For some patron it must perform $\Omega(\log N)$ operations.*

See [19] for the proof of Theorem 1. Our lower bound on implicit priority queues shows that the decreasekey operation can be forced to behave like an usher, seating $m$ patrons for *some* $m$ of its choosing. If the data structure has no extra storage and if $m > \log^{(k)} n$ (for some fixed $k$), then by Theorem 1 the amortized cost per decreasekey is $\Omega(\log^*(\log^{(k)} n)) = \Omega(\log^* n - k)$. If $m$ is smaller, i.e., decreasekeys were performed quickley, then we show that a further sequence of $O(m \log^* n)$ decreasekeys, inserts, and deletemins must take $\Omega(n^{1/\log^{(k)} n})$ time.

**Acknowledgment.** We thank Rasmus Pagh for many helpful comments.

# References

1. A. Andersson, T. Hagerup, J. Hastad, and O. Petersson. Tight bounds for searching a sorted array of strings. *SIAM J. Comput.*, 30(5):1552–1578, 2000.
2. M. Blum, R. W. Floyd, V. Pratt, R. L. Rivest, and R. E. Tarjan. Time bounds for selection. *J. Comput. Syst. Sci.*, 7(4):448–461, 1973.

3. A. Brodnik, S. Carlsson, E. Demaine, J. I. Munro, and R. Sedgewick. Resizable arrays in optimal time and space. In *WADS*, pages 37–??, 1999.
4. A. Brodnik and J. I. Munro. Membership in constant time and almost-minimum space. *SIAM J. Comput.*, 28(5):1627–1640, 1999.
5. S. Carlsson, J. I. Munro, and P. V. Poblete. An implicit binomial queue with constant insertion time. In *SWAT*, pages 1–13, 1988.
6. J. R. Driscoll, H. N. Gabow, R. Shrairman, and R. E. Tarjan. Relaxed heaps: an alternative to Fibonacci heaps with applications to parallel computation. *Comm. ACM*, 31(11):1343–1354, 1988.
7. A. Fiat and M. Naor. Implicit $O(1)$ probe search. *SIAM J. Comput.*, 22(1):1–10, 1993.
8. A. Fiat, M. Naor, J. P. Schmidt, and A. Siegel. Nonoblivious hashing. *J. ACM*, 39(4):764–782, 1992.
9. G. Franceschini and R. Grossi. Optimal worst-case operations for implicit cache-oblivious search trees. In *Proc. 8th WADS*, 2003.
10. G. Franceschini and R. Grossi. No sorting? better searching! In *Proc. 45th IEEE Symp. on Foundations of Computer Science (FOCS)*, pages 491–498, 2004.
11. M. L. Fredman. On the efficiency of pairing heaps and related data structures. *J. ACM*, 46(4):473–501, 1999.
12. M. L. Fredman, R. Sedgewick, D. D. Sleator, and R. E. Tarjan. The pairing heap: a new form of self-adjusting heap. *Algorithmica*, 1(1):111–129, 1986.
13. M. L. Fredman and R. E. Tarjan. Fibonacci heaps and their uses in improved network optimization algorithms. *J. ACM*, 34(3):596–615, 1987.
14. T. Hagerup and R. Raman. An efficient quasidictionary. In *SWAT*, 2002.
15. N. J. A. Harvey and K. Zatloukal. The post-order heap. In *Proc. FUN*, 2004.
16. J. Iacono. Improved upper bounds for pairing heaps. In *SWAT*, pages 32–43, 2000.
17. D. B. Johnson. Priority queues with update and finding minimum spanning trees. *Info. Proc. Lett.*, 4(3):53–57, 1975.
18. H. Kaplan and R. E. Tarjan. New heap data structures. Technical Report TR-597-99, Computer Science Dept., Princeton University, March 1999.
19. C. W. Mortensen and S. Pettie. The complexity of implicit and space-efficient priority queues. Manuscript, http://www.mpi-sb.mpg.de/ pettie/, 2005.
20. J. I. Munro. An implicit data structure supporting insertion, deletion, and search in $O(\log^2 n)$ time. *J. Comput. Syst. Sci.*, 33(1):66–74, 1986.
21. J. I. Munro and H. Suwanda. Implicit data structures for fast search and update. *J. Comput. Syst. Sci.*, 21(2):236–250, 1980.
22. Rasmus Pagh. Low redundancy in static dictionaries with constant query time. *SIAM J. Comput.*, 31(2):353–363, 2001.
23. R. Raman, V. Raman, and S. S. Rao. Succinct indexable dictionaries with applications to encoding $k$-ary trees and multisets. In *SODA*, pages 233–242, 2002.
24. R. Raman and S. S. Rao. Succinct dynamic dictionaries and trees. In *Proc. 30th Int'l Colloq. on Automata, Languages and Programming*, pages ??–??, 2003.
25. T. Takaoka. Theory of 2–3 heaps. *Discrete Appl. Math.*, 126(1):115–128, 2003.
26. J. W. J. Williams. Algorithm 232 (heapsort). *Comm. ACM*, 7:347–348, 1964.
27. A. C. C. Yao. Should tables be sorted? *J. ACM*, 28(3):615–628, 1981.
28. D. Zuckerman. *Computing Efficiently Using General Weak Random Sources*. Ph.D. Thesis, The University of California at Berkeley, August 1991.

# Analysis of a Class of Tries with Adaptive Multi-digit Branching

Yuriy A. Reznik

RealNetworks, Inc.,
2601 Elliott Avenue, Seattle, WA 98121
yreznik@acm.org

**Abstract.** We study a class of adaptive multi-digit tries, in which the numbers of digits $r_n$ processed by nodes with $n$ incoming strings are such that, in memoryless model (with $n \to \infty$):

$$r_n \to \frac{\log n}{\eta} \quad (pr.)$$

where $\eta$ is an algorithm-specific constant. Examples of known data structures from this class include LC-tries (Andersson and Nilsson, 1993), "relaxed" LC-tries (Nilsson and Tikkanen, 1998), tries with logarithmic selection of degrees of nodes, etc. We show, that the average depth $D_n$ of such tries in asymmetric memoryless model has the following asymptotic behavior (with $n \to \infty$):

$$D_n = \frac{\log \log n}{- \log (1 - h/\eta)} \left(1 + o\left(1\right)\right)$$

where $n$ is the number of strings inserted in the trie, and $h$ is the entropy of the source. We use this formula to compare performance of known adaptive trie structures, and to predict properties of other possible implementations of tries in this class.

## 1 Introduction

Radix search trees or *tries*, introduced in late 1950's by R. E. de la Briandeis [3] and E. Fredkin [14] have long become one of the most basic and much appreciated concepts in computer science. It is well known, that they enable access to $n$ variable-length strings in $O(\log n)$ time (on average), while using only $O(n)$ of space [15]. These fundamental properties remain in force for a large class of statistical models [8, 5, 28], and are inherited by most of their modifications, such as *digital search trees* [6, 13], *Patricia tries* [16], *bucket tries* [26, 15, 12], and others [4]. Most of the original research on these structures has been done in 1960s–70s.

Nevertheless, in recent years, tries have seen a resurgence of interest in connection with several new modifications, allowing search operations to be executed much faster, typically in $O(\log \log n)$ time on average [1, 2, 18, 20, 10, 22, 25, 11].

F. Dehne, A. López-Ortiz, and J.-R. Sack (Eds.): WADS 2005, LNCS 3608, pp. 61–72, 2005.

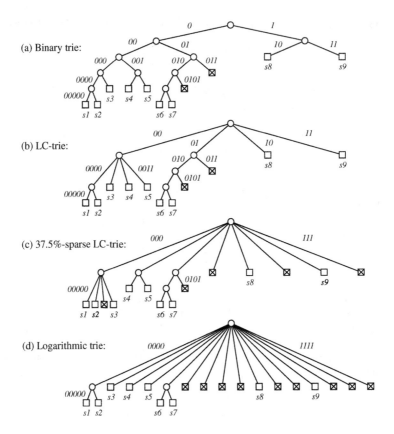

(a) Binary trie:

(b) LC-trie:

(c) 37.5%-sparse LC-trie:

(d) Logarithmic trie:

**Fig. 1.** Examples of tries built from 9 binary strings: $s1 = 00000\ldots$, $s2 = 00001\ldots$, $s3 = 00011\ldots$, $s4 = 0010\ldots$, $s5 = 0011\ldots$, $s6 = 01000\ldots$, $s7 = 01001\ldots$, $s8 = 1010\ldots$, $s9 = 1100\ldots$

The basic technique that enables such a speed up is called *adaptive multi-digit branching* [1, 22].

The best known example of such a structure is a *level compressed trie* (or *LC-trie*) of Andersson and Nilsson [1], which simply replaces all complete subtrees in a trie with larger multi-digit nodes[1] (see Fig. 1.a,b). Other known implementations include "sparse" LC-tries of Nilsson and Tikkanen [19, 20], which allow the resulting multi-digit nodes to have a certain percentage of empty positions (see Fig. 1.c), tries with logarithmic branching [22, 25], etc.

To the best of our knowledge, only LC-tries have been thoroughly analyzed in the past. First results, suggesting that the average depth $D_n^{LC}$ of LC-tries in asymmetric memoryless model is $O(\log \log n)$ and only $\Theta(\log^* n)$ if model is sym-

---

[1] It is assumed that multi-digit nodes can be represented by lookup tables, so the time required to parse such a structure is still proportional to the number of nodes in each particular path.

metric, were reported by Andersson and Nilsson [1, 2]. A more refined estimate in the symmetric case: $D_n^{LC} \sim \log^* n$ has been obtained by Devroye [10]. In the asymmetric case, Devroye and Szpankowski [11], and Reznik [24] have independently arrived at the following expression: $D_n^{LC} = \frac{\log \log n}{-\log(1-h/h_{-\infty})}(1+o(1))$ where $h$ and $h_{-\infty}$ are Shannon and Rényi entropies correspondingly [28]. Analysis of the expected height (longest path length) of LC-tries in memoryless model can be found in [10, 11].

At the same time, not much is known about other implementations of adaptive tries. Most relevant results in this direction include analysis of basic statistics of multi-digit nodes [24, 25], and experimental data collected for some specific implementations [19, 20]. Why useful by themselves, however, these results are not sufficient for complete characterization of the average behavior of such structures.

In this paper, we offer a general analysis of a *class of tries with adaptive multi-digit branching* in memoryless model, which, in various special cases, leads to expressions of average depths for all above mentioned implementations. We use these expressions to compare performance of these structures, and to predict properties of other possible implementations of tries in this class.

## 2    Definitions and Main Results

Consider a set $S = \{s_1, \ldots, s_n\}$ of $n$ distinct strings to be used for trie construction. For simplicity, we will assume that these strings contain symbols from a binary alphabet $\Sigma = \{0, 1\}$.

**Definition 1.** *A* multi-digit trie *with parameter* $r$ *(*$r \geqslant 1$*):* $T^{(r)}(S)$ *over* $S$ *is a data structure defined recursively as follows. If* $n = 0$*, the trie is an* empty node*. If* $n = 1$ *, the trie is an* external node *containing a pointer to a string in* $S$*. If* $n > 1$*, the trie is an* $r$*-digit* internal node *containing pointers to* $2^r$ *child tries:* $T^{(r)}(S_{0^r}), \ldots, T^{(r)}(S_{1^r})$*, constructed over sets of* suffixes *of strings from* $S$ *beginning with the corresponding* $r$*-digit words* $S_v = \{u_j \mid v\, u_j = s_i \in S\}$*,* $v \in \Sigma^r$*.*

In the simplest case, when $r = 1$, this structure turns in to a regular *binary trie* (see Fig. 1.a). When $r$ is fixed, this structure is a $2^r$-ary trie, which uses $r$-bits units of input data for branching. When $r$ is variable, we say that this trie belongs to a class of *adaptive multi-digit tries*. Below we define several important implementations of such data structures.

**Definition 2.** *An* LC-trie $T^{LC}(S)$ *over* $S$ *is an adaptive multi-digit trie, in which parameters* $r$ *are selected to reach the first levels at which there is at least one external or empty node:* $r = \min\{s : \sum_{v \in \Sigma^s} 1\{|S_v| \leqslant 1\} \geqslant 1\}$*.*

**Definition 3.** *An* $\varepsilon$-sparse LC-trie $T^{\varepsilon\text{-}LC}(S)$ *over* $S$ *is an adaptive multi-digit trie, in which parameters* $r$ *are selected to reach the deepest levels at which the ratio of the number of empty nodes to the total number of nodes is not greater than* $\varepsilon$*:* $r = \max\{s : \frac{1}{2^s}\sum_{v \in \Sigma^s} 1\{S_v = \emptyset\} \leqslant \varepsilon\}$*.*

**Definition 4.** *A logarithmic trie* $T^{\lg}(S)$ *over* $S$ *is an adaptive multi-digit trie, in which parameters* $r$ *for nodes with* $n$ *incoming strings are calculated using* $r = \lceil \log_2 n \rceil$.

Examples of the above defined types of tries are provided in Fig.1. Observe, that all input strings $s_1, \ldots, s_n$ inserted in a trie can be uniquely identified by the paths from the root node to the corresponding external nodes. The sum of lengths of these paths $C(T) = \sum_{i=1}^{n} |s_i|$ is called an *external path length* of a trie $T$, and the value $D(T) = C(T)/n$ – an *average depth* of this trie.

In order to study the average behavior of tries we will assume that input strings $S$ are generated by a binary *memoryless* (or *Bernoulli*) source [7]. In this model, symbols of the input alphabet $\Sigma = \{0, 1\}$ occur independently of one another with probabilities $p$ and $q = 1 - p$ correspondingly. If $p = q = 1/2$, such source is called *symmetric*, otherwise it is *asymmetric* (or *biased*).

Using this model, we can now define the quantity of our main interest:

$$D_n := E\{D(T)\} = \frac{E\{C(T)\}}{n}, \tag{1}$$

where the expectations are taken over all possible tries over $n$ strings when parameters of the memoryless source ($p$ and $q$) are fixed. Average depths of $LC$-, $\varepsilon$-sparse, and logarithmic tries over $n$ strings will be denoted by $D_n^{LC}$, $D_n^{\varepsilon-LC}$, and $D_n^{\lg}$ correspondingly.

In order to consolidate analysis of these (and possibly many other) implementations of adaptive multi-digit tries, we will assume, that in memoryless model, the numbers of digits $r_n$ assigned to their internal nodes with $n$ incoming strings have the following convergence (with $n \to \infty$):

$$r_n \to \frac{\log n}{\eta} \quad (pr.) \tag{2}$$

where $\eta$ is an algorithm-specific constant.

For example, it is well known, that convergence (2) takes place for LC-tries [21] (see also [9, 2]). In this case, the constant $\eta$ becomes:

$$\eta^{LC} = h_{-\infty}, \tag{3}$$

where $h_{-\infty} = -\log \min(p, q)$ is a special case of a Rényi's entropy [28]. In a case of $\varepsilon$-sparse LC-tries, an extension of analysis [25] suggests that: $r_n \to \frac{\log n}{\eta^{\varepsilon-LC}} + f(\varepsilon)\sqrt{\log n}$ (pr.), where $f(.)$ is a monotonic function, such that $f(1/2) = 0$, and

$$\eta^{\varepsilon-LC} = h_g, \tag{4}$$

where $h_g = -\log(\sqrt{pq})$ is another constant depending on the probabilities of the source. It is clear, that our model (2) is sufficient to describe $\varepsilon$-sparse LC-tries with $\varepsilon = 1/2$. Finally, the behavior of logarithmic tries can obviously be modelled by (2) with

$$\eta^{\lg} = \log 2. \tag{5}$$

Our main result for a class of adaptive multi-digit tries is formulated in the following theorem.

**Theorem 1.** *Consider a class of adaptive multi-digit tries, in which the numbers of digits $r_n$ processed by nodes with $n$ incoming strings, in binary memoryless model (with $n \to \infty$):*

$$r_n \to \frac{\log n}{\eta} \quad (pr.)$$

*where $\eta$ is a constant, such that: $h < \eta \leqslant h_{-\infty}$, where $h = -p \log p - q \log q$ is the Shannon's entropy of the source, $h_{-\infty} = -\log \min(p, q)$, and it is assumed that $p \neq q$.*

*Then, the average depth $D_n$ of such tries over $n$ strings is asymptotically (with $n \to \infty$):*

$$D_n = \frac{\log \log n}{-\log(1 - h/\eta)} (1 + o(1)). \quad (6)$$

## 3   Discussion

Using the result of Theorem 1 and the values of constants $\eta$ for each respective algorithm (3-5), we can now compare them. The results of such a comparison are presented in Fig.2. For completeness, we also show the behavior of the average depths of regular (binary) tries $D_n^{bin} = \frac{1}{h} \log n + O(1)$ (cf. [15, 8, 27]).

We first notice that when the source is nearly symmetric $p \to 1/2$:

$$\frac{D_n^{LC}}{\log \log n} \to 0, \quad \frac{D_n^{\varepsilon-LC}}{\log \log n} \to 0, \quad \frac{D_n^{lg}}{\log \log n} \to 0,$$

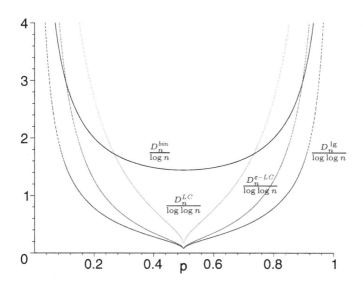

**Fig. 2.** Comparison of tries in binary memoryless model with $\Pr(0) = p$. $D_n^{bin}$, $D_n^{LC}$, $D_n^{\varepsilon-LC}$, and $D_n^{lg}$ denote average depths of binary, LC-, "sparse" LC-, and logarithmic tries correspondingly

which suggests that in a symmetric case, the average depths of these structures should have a much smaller order. We already know that this is true for LC-tries (whose depth in symmetric case is only $O(\log^* n)$ [1, 10]), but now we can predict such an effect for any adaptive trie with $\eta \to h$ ($p \to 1/2$).

We next observe that behavior of LC-tries and their "sparse" variants is not very much different. Thus, by plugging their respective constants, it can be shown that (with $n \to \infty$):

$$2 \leqslant \frac{D_n^{LC}}{D_n^{\varepsilon-LC}} \leqslant \max_{p \in [0,1]} \frac{\log(1 - h_g/h)}{\log(1 - h_{-\infty}/h)} \approx 2.165367...,$$

which suggests that sparse LC-tries should be approximately twice faster than LC-tries, and that, contrary to the intuition, this ratio cannot be influenced much by increasing $\varepsilon$.

We also observe that both sparse and dense LC-tries are much more sensitive to the asymmetry of the source than regular tries. Thus, it can be seen that with $p \to 0$:

$$\frac{D_n^{LC}/\log\log n}{D_n^{bin}/\log n} \to \infty, \quad \frac{D_n^{\varepsilon-LC}/\log\log n}{D_n^{bin}/\log n} \to \infty.$$

At the same time, the sensitivity of logarithmic tries on the asymmetry of the source remains similar to one of regular tries. Thus, it can be shown that with $p \to 0$:

$$\frac{D_n^{lg}/\log\log n}{D_n^{bin}/\log n} \to \log 2.$$

As obvious, logarithmic tries are the fastest among implementations that we have considered so far. Using the fact that the number of branches in their nodes is approximately equal to the number of passing strings, we can conjecture that the amount of space required by logarithmic tries is $O(n \log \log n)$, which is slightly larger than $O(n)$ space used by regular tries and LC-tries [15, 27, 2].

We conclude by pointing out that by using even larger nodes, e.g. with $h < \eta < \log 2$ in our model (2), it is possible to design tries that are faster than logarithmic tries. However, the amount of space required by such tries becomes $\Omega\left(n^{\frac{\log 2}{\eta}}\right)$, which in a practical world, might be too much of a price to be paid for a relatively small (in this case, limited to a constant-factor) improvement in speed.

## 4   Analysis

We start with deriving recurrent expressions for external path lengths of tries with $r_n$-digit nodes.

**Lemma 1.** *Parameters $C_n$ (average external path length of adaptive multi-digit tries over $n$ strings) in a binary memoryless model satisfy:*

$$C_n = n + \sum_{k=2}^{n} \binom{n}{k} \sum_{s=0}^{r_n} \binom{r_n}{s} \left(p^s q^{r_n-s}\right)^k \left(1 - p^s q^{r_n-s}\right)^{n-k} C_k;\tag{7}$$

$$C_0 = C_1 = 0.$$

*Proof.* Consider an $r_n$-digit node processing $n$ strings. Assuming that each of its $2^{r_n}$ branches have probabilities $p_1, \ldots, p_{2^{r_n}}$, and using the standard technique for enumeration of $C_n$ in tries [15–6.3-3], we can write:

$$C_n = n + \sum_{k_1+\ldots+k_{2^{r_n}}=n} \binom{n}{k_1,\ldots,k_{2^{r_n}}} p_1^{k_1} \ldots p_{2^{r_n}}^{k_{2^{r_n}}} \left(C_1 + \ldots + C_{2^{r_n}}\right),$$

$$= n + \sum_{k=0}^{n} \binom{n}{k} \left(p_1^k (1-p_1)^{n-k} + \ldots + p_{2^{r_n}}^k (1-p_{2^{r_n}})^{n-k}\right) C_k.\tag{8}$$

Recall now that our strings are generated by a binary memoryless source with probabilities $p$, and $q = 1 - p$. This means that:

$$p_i = p^{s_i} q^{r_n-s_i},\tag{9}$$

where $s_i$ is the number of occurrences of symbol 0 in a string leading to a branch $i$ $(1 \leqslant i \leqslant 2^{r_n})$. Combining (8) and (9), we arrive at the expression (7) claimed by the lemma. □

In order to find a solution of (7) we will use the following, very simple technique. We already know, that for a class of our tries $D_n = O(\log \log n)$, hence we can say that $C_n = \xi n \log \log n$ and plug it in (7). Ultimately, this will give us upper and lower bounds for the parameter $\xi$ such that the recurrence (7) holds. If these bounds are tight, then we have successfully deduced the constant factor in the $O(\log \log n)$ term.

We will need the following intermediate results. For simplicity, here and below we use natural logarithms.

**Lemma 2.** *Consider a sum:*

$$f(n, \theta, \lambda) = \sum_{k=2}^{n} \binom{n}{k} \theta^k (1-\theta)^{n-k} k \ln(\lambda + \ln k),\tag{10}$$

*where $\theta \in (0,1)$, and $\lambda > 1$ are some constants. Then, there exists $0 < \zeta < \infty$, such that for any $n \geqslant 2$:*

$$n\theta \ln(\lambda + \ln(n\theta)) - \zeta \leqslant f(n,\theta,\lambda) \leqslant n\theta \ln\left(\lambda + \ln\left(1 - \theta + n\theta\right)\right).\tag{11}$$

*Proof.* We start with a representation:

$$f(n, \theta, \lambda) = \sum_{k=1}^{n} \binom{n}{k} \theta^k (1-\theta)^{n-k} k \ln(\lambda + \ln k) - n\theta(1-\theta)^{n-1} \ln \lambda$$

where the last term can be easily bounded by:

$$n\theta(1-\theta)^{n-1}\ln\lambda \leqslant \frac{\theta e^{-1}}{(\theta-1)\ln(1-\theta)}\ln\lambda =: \zeta.$$

Next, by Jensen's inequality for $x\ln(\lambda+\ln x)$:

$$\sum_{k=1}^{n}\binom{n}{k}\theta^k(1-\theta)^{n-k}k\ln(\lambda+\ln k)$$
$$\geqslant \left(\sum_{k=1}^{n}\binom{n}{k}\theta^k(1-\theta)^{n-k}k\right)\ln\left(\lambda+\ln\left(\sum_{k=1}^{n}\binom{n}{k}\theta^k(1-\theta)^{n-k}k\right)\right)$$
$$= n\theta\ln(\lambda+\ln(n\theta)).$$

where convexity for $k \geqslant 1$ is assured by picking $\lambda > 1$.

To obtain an upper bound we use Jensen's inequality for $-\ln(\lambda+\ln(1+x))$:

$$\sum_{k=1}^{n}\binom{n}{k}\theta^k(1-\theta)^{n-k}k\ln(\lambda+\ln k)$$
$$= n\theta\sum_{k=0}^{n-1}\binom{n-1}{k}\theta^k(1-\theta)^{n-1-k}\ln(\lambda+\ln(1+k))$$
$$\leqslant n\theta\ln\left(\lambda+\ln\left(1+\sum_{k=0}^{n-1}\binom{n-1}{k}\theta^k(1-\theta)^{n-1-k}k\right)\right)$$
$$= n\theta\ln(\lambda+\ln(1-\theta+n\theta)).$$    □

**Lemma 3.** *Consider a sum:*

$$g(n,\theta,\alpha,\beta) = \sum_{k=0}^{n}\binom{n}{k}\theta^k(1-\theta)^{n-k}\ln(\alpha+\beta k),\qquad(12)$$

*where $\theta \in (0,1)$, $\alpha,\beta > 0$, and $\alpha > \beta$. Then, for any $n \geqslant 1$:*

$$\ln(\alpha-\beta(1-\theta)+\beta\theta n) \leqslant g(n,\theta,\alpha,\beta) \leqslant \ln(\alpha+\beta\theta n).\qquad(13)$$

*Proof.* We use the same technique as in the previous Lemma. By Jensen's inequality for $-\ln(\alpha+\beta x)$:

$$g(n,\theta,\alpha,\beta) \leqslant \ln\left(\alpha+\beta\sum_{k=0}^{n}\binom{n}{k}\theta^k(1-\theta)^{n-k}k\right) = \ln(\alpha+\beta\theta n).$$

The lower bound follows from Jensen's inequality for $x \ln(\alpha - \beta + \beta x)$:

$$g(n, \theta, \alpha, \beta) = \frac{1}{\theta(n+1)} \sum_{k=1}^{n+1} \binom{n+1}{k} \theta^k (1-\theta)^{n+1-k} k \ln(\alpha - \beta + \beta k)$$

$$\geq \frac{1}{\theta(n+1)} \left( \sum_{k=1}^{n+1} \binom{n+1}{k} \theta^k (1-\theta)^{n+1-k} k \right) \times$$

$$\times \ln \left( \alpha - \beta + \beta \sum_{k=1}^{n+1} \binom{n+1}{k} \theta^k (1-\theta)^{n+1-k} k \right)$$

$$= \ln(\alpha - \beta(1-\theta) + \beta\theta n).$$

It is clear, that convexity and continuity in both cases is assured when $\alpha > \beta > 0$.

$\square$

We are now prepared to solve our recurrence (7). For simplicity we assume that $p > 1/2$. Let $C_n = \xi n \ln(\lambda + \ln n)$, where $\lambda > 1$ is a constant. Then, according to Lemma 2:

$$C_n = n + \sum_{s=0}^{r_n} \binom{r_n}{s} \sum_{k=2}^{n} \binom{n}{k} (p^s q^{r_n-s})^k (1 - p^s q^{r_n-s})^{n-k} \xi k \ln(\lambda + \ln k)$$

$$\leq n + n\xi \sum_{s=0}^{r_n} \binom{r_n}{s} p^s q^{r_n-s} \ln\left(\lambda + \ln\left(n p^s q^{r_n-s} + 1 - p^s q^{r_n-s}\right)\right)$$

$$= n + n\xi \sum_{s=0}^{r_n} \binom{r_n}{s} p^s q^{r_n-s} \ln\left(\lambda + \ln(n p^s q^{r_n-s}) + \ln\left(1 + \frac{1}{n p^s q^{r_n-s}} - \frac{1}{n}\right)\right).$$

Next, by our assumed property (2), we can pick $\varepsilon > 0$, such that the probability that

$$\left| r_n - \frac{\ln n}{\eta} \right| \leq \varepsilon. \tag{14}$$

holds true is 1 with $n \to \infty$. If we further assume that $\eta \leq -\ln q$ [2], then

$$n p^s q^{r_n-s} \geq n q^{r_n} \geq n q^{\frac{\ln n}{\eta} + \varepsilon} = n^{1 - \frac{\ln q}{\eta}} q^\varepsilon \geq q^\varepsilon.$$

and consequently:

$$\ln\left(1 + \frac{1}{n p^s q^{r_n-s}} - \frac{1}{n}\right) \leq \ln\left(1 + q^{-\varepsilon} - \frac{1}{n}\right) < \ln(1 + q^{-\varepsilon}) = O(\varepsilon).$$

---

[2] A case when $\eta = -\ln q = h_{-\infty}$ corresponds to LC-tries. Smaller $\eta$ correspond to tries with larger nodes.

By incorporating this bound and using Lemma 3:

$$C_n \leqslant n + n\xi \sum_{s=0}^{r_n} \binom{r_n}{s} p^s q^{r_n-s} \ln \left(\lambda + \ln \left(n\, p^s q^{r_n-s}\right) + \ln \left(1 + q^{-\varepsilon} - 1/n\right)\right)$$

$$= n + n\xi \sum_{s=0}^{r_n} \binom{r_n}{s} p^s q^{r_n-s} \ln \left(\lambda + \ln n + r_n \ln q + s \ln(p/q) + \ln \left(1 + q^{-\varepsilon} - 1/n\right)\right)$$

$$\leqslant n + n\xi \ln \left(\lambda + \ln n - h\, r_n + \ln \left(1 + q^{-\varepsilon} - 1/n\right)\right),$$

where $h = -p \ln p - q \ln q$ is the entropy. Now, by applying (14), we have:

$$C_n \leqslant n + n\xi \ln \left(\lambda + \ln n \left(1 - \frac{h}{\eta}\right) + h\varepsilon + \ln \left(1 + q^{-\varepsilon} - 1/n\right)\right),$$

and by plugging $C_n = \xi n \ln(\lambda + \ln n)$ in the left side of the above inequality, we finally obtain:

$$\xi \leqslant \frac{1}{-\ln\left(1 - \frac{h}{\eta} + \frac{\lambda + h\varepsilon + \ln(1 + q^{-\varepsilon} - 1/n)}{\ln n}\right) + \ln\left(1 + \frac{\lambda}{\ln n}\right)}$$

$$= \frac{1}{-\ln\left(1 - \frac{h}{\eta}\right)} \left(1 + O\left(\frac{\varepsilon}{\ln n}\right)\right). \tag{15}$$

The procedure for finding a lower bound is very similar:

$$C_n = n + \sum_{s=0}^{r_n} \binom{r_n}{s} \sum_{k=2}^{n} \binom{n}{k} \left(p^s q^{r_n-s}\right)^k \left(1 - p^s q^{r_n-s}\right)^{n-k} \xi k \ln(\lambda + \ln k)$$

$$\geqslant n + n\xi \sum_{s=0}^{r_n} \binom{r_n}{s} p^s q^{r_n-s} \log \left(\lambda + \ln \left(n\, p^s q^{r_n-s}\right)\right) - \zeta$$

$$= n + n\xi \sum_{s=0}^{r_n} \binom{r_n}{s} p^s q^{r_n-s} \ln \left(\lambda + \ln n + r_n \ln q + s \ln(p/q)\right) - \zeta$$

$$\geqslant n + n\xi \ln \left(\lambda + \ln n - h\, r_n - q \ln(p/q)\right) - \zeta,$$

$$\geqslant n + n\xi \ln \left(\lambda + \ln n \left(1 - \frac{h}{\eta}\right) - h\varepsilon - q \ln(p/q)\right) - \zeta,$$

which (after plugging $C_n = \xi n \ln(\lambda + \ln n)$ in the right side) leads to the following inequality:

$$\xi \geqslant \frac{1 - \zeta/n}{-\ln\left(1 - \frac{h}{\eta} + \frac{\lambda - h\varepsilon - q \ln(p/q)}{\ln n}\right) + \ln\left(1 + \frac{\lambda}{\ln n}\right)}$$

$$= \frac{1}{-\ln\left(1 - \frac{h}{\eta}\right)} \left(1 + O\left(\frac{\varepsilon}{\ln n}\right)\right). \tag{16}$$

By combining our bounds (15) and (16) and taking into account the fact that for any $\varepsilon > 0$, the probability that they both hold true is approaching 1 with $n \to \infty$, we can conclude that:

$$\xi \to \frac{1}{-\ln\left(1 - \frac{h}{\eta}\right)}\left(1 + o(1)\right)$$

in probability.

# References

1. A. Andersson and S. Nilsson, Improved Behaviour of Tries by Adaptive Branching, *Information Processing Letters* **46** (1993) 295–300.
2. A. Andersson and S. Nilsson, Faster Searching in Tries and Quadtries – An Analysis of Level Compression. *Proc.* 2$^{\text{nd}}$ *Annual European Symp. on Algorithms* (1994) 82–93.
3. R. E. de la Briandeis, File searching using variable length keys. *Proc. Western Joint Computer Conference.* **15** (AFIPS Press, 1959).
4. J. Clement, P. Flajolet, and B. Vallée, The analysis of hybrid trie structures. *Proc. Annual ACM-SIAM Symp. on Discrete Algorithms.* (San Francisco, CA, 1998) 531–539.
5. Clement, J., Flajolet, P., and Vallée, B. (2001) Dynamic sources in information theory: A general analysis of trie structures. *Algorithmica* **29** (1/2) 307–369.
6. E. G. Coffman Jr. and J. Eve, File Structures Using Hashing Functions, *Comm. ACM*, 13 (7) (1970) 427–436.
7. T. M. Cover and J. M. Thomas, *Elements of Information Theory.* (John Wiley & Sons, New York, 1991).
8. L. Devroye, A Note on the Average Depths in Tries, *SIAM J. Computing* **28** (1982) 367–371.
9. L. Devroye, A Note on the Probabilistic Analysis of PATRICIA Tries, *Rand. Structures & Algorithms* **3** (1992) 203–214.
10. L. Devroye, Analysis of Random LC Tries, *Rand. Structures & Algorithms* **19** (3-4) (2001) 359–375.
11. L. Devroye and W. Szpankowski, Probabilistic Behavior of Asymmetric LC-Tries, *Rand. Structures & Algorithms* – submitted.
12. R. Fagin, J. Nievergelt, N. Pipinger, and H. R. Strong, Extendible Hashing – A Fast Access Method for Dynamic Files, *ACM Trans. Database Syst.*, 4 (3) (1979) 315–344.
13. P. Flajolet and R. Sedgewick, Digital Search Trees Revisited, *SIAM J. Computing* **15** (1986) 748–767.
14. E. Fredkin, Trie Memory, *Comm. ACM* **3** (1960) 490–500.
15. D. E. Knuth, *The Art of Computer Programming. Sorting and Searching. Vol. 3.* (Addison-Wesley, Reading MA, 1973).
16. D. A. Morrison, PATRICIA – Practical Algorithm To Retrieve Information Coded in Alphanumeric, *J. ACM*, 15 (4) (1968) 514–534.
17. S. Nilsson and G. Karlsson, Fast IP look-up for Internet routers. *Proc. IFIP* 4$^{\text{th}}$ *International Conference on Broadband Communication* (1998) 11–22.
18. S. Nilsson and G. Karlsson, IP-address look-up using LC-tries, *IEEE J. Selected Areas in Communication* **17** (6) (1999) 1083–1092.

19. S. Nilsson and M. Tikkanen, Implementing a Dynamic Compressed Trie, *Proc.* $2^{nd}$ *Workshop on Algorithm Engineering* (Saarbruecken, Germany, 1998) 25–36.
20. S. Nilsson and M. Tikkanen, An experimental study of compression methods for dynamic tries, *Algorithmica* **33** (1) (2002) 19–33.
21. B. Pittel, Asymptotic Growth of a Class of Random Trees, *Annals of Probability* **18** (1985) 414–427.
22. Yu. A. Reznik, Some Results on Tries with Adaptive Branching, *Theoretical Computer Science* **289** (2) (2002) 1009–1026.
23. Yu. A. Reznik, On Time/Space Efficiency of Tries with Adaptive Multi-Digit Branching, *Cybernetics and Systems Analysis* **39** (1) (2003) 32–46.
24. Yu. A. Reznik, On the Average Depth of Asymmetric LC-tries, *Information Processing Letters* – submitted.
25. Yu. A. Reznik, On the Average Density and Selectivity of Nodes in Multi-Digit Tries, *Proc.* $7^{th}$ *Workshop on Algorithm Engineering and Experiments and* $2^{nd}$ *Workshop on Analytic Algorithmics and Combinatorics (ALENEX/ANALCO 2005)* (SIAM, 2005).
26. E. H. Sussenguth, Jr. (1963) Use of Tree Structures for Processing Files, *Comm. ACM*, 6 (5) 272–279.
27. W. Szpankowski, Some results on V-ary asymmetric tries, *J. Algorithms* **9** (1988) 224–244.
28. W. Szpankowski, *Average Case Analysis of Algorithms on Sequences* (John Wiley & Sons, New York, 2001).

# Balanced Aspect Ratio Trees Revisited

Amitabh Chaudhary[1] and Michael T. Goodrich[2]

[1] Department of Computer Science & Engineering,
University of Notre Dame, Notre Dame IN 46556, USA
`Amitabh.Chaudhary.1@nd.edu`
[2] Department of Computer Science,
Bren School of Information & Computer Sciences,
University of California, Irvine CA 92697, USA
`goodrich@acm.org`

**Abstract.** Spatial databases support a variety of geometric queries on point data such as range searches, nearest neighbor searches, etc. Balanced Aspect Ratio (BAR) trees are hierarchical space decomposition structures that are general-purpose and space-efficient, and, in addition, enjoy a worst case performance poly-logarithmic in the number of points for approximate queries. They maintain limits on their depth, as well as on the aspect ratio (intuitively, how skinny the regions can be). BAR trees were initially developed for 2 dimensional spaces and a fixed set of partitioning planes, and then extended to $d$ dimensional spaces and more general partitioning planes. Here we revisit 2 dimensional spaces and show that, for any given set of 3 partitioning planes, it is not only possible to construct such trees, it is also possible to derive a simple closed-form upper bound on the aspect ratio. This bound, and the resulting algorithm, are much simpler than what is known for general BAR trees. We call the resulting BAR trees Parameterized BAR trees and empirically evaluate them for different partitioning planes. Our experiments show that our theoretical bound converges to the empirically obtained values in the lower ranges, and also make a case for using evenly oriented partitioning planes.

## 1 Introduction

Spatial databases for scientific applications need efficient data structures to solve a variety of geometric queries. Consider, e.g., the Sloan Digital Sky Survey (SDSS) [16, 17], a scientific application with which we have direct experience. It stores light intensities for over a 100 million celestial objects as points on a two-dimensional sphere, and needs support for geometric queries like the nearest neighbor queries, proximity queries, and general range queries (not just axis orthogonal). Similar needs arise in geographical information systems.

There are many access methods based on the hierarchical space decomposition data structures that are useful in solving geometric queries on point data. Quad trees and $k$-d trees are widely popular examples (e.g., see Samet [18, 19]). In these data structures two properties, *depth* and *aspect ratio*, play a crucial role

F. Dehne, A. López-Ortiz, and J.-R. Sack (Eds.): WADS 2005, LNCS 3608, pp. 73–85, 2005.
© Springer-Verlag Berlin Heidelberg 2005

in determining their efficiency in solving queries. The depth of a tree characterizes the number of nodes that have to be visited to find regions of interest. The
aspect ratio of a tree (intuitively, how "skinny" a region can be) characterizes the
number of wasteful nodes that can be in any particular region of interest. Most
queries require both of these values to be small for them to be solved efficiently.

Unfortunately, both quad trees and $k$-d trees optimize one of these properties
at the expense of the other. Quad trees produce regions with optimal aspect
ratios, but they can have terrible depth. $K$-d trees, on the other hand, have
optimal (logarithmic) depth, but they often produce lots of long-and-skinny
regions, which slow query times.

Balanced Aspect Ratio (BAR) trees are hierarchical space decomposition
data structures that have logarithmic depth and bounded aspect ratios. As a
result they have a worst case performance poly-logarithmic in the number of
points for approximate queries such as the approximate nearest neighbor, approximate farthest neighbor, approximate range query, etc. They were initially
developed for 2 dimensional spaces [12] for a particular fixed set $(0, \pi/4, \pi/2)$
of partitioning planes. Then [10], they were extended to $d$ dimensional spaces,
and the partitioning planes could be chosen flexibly as long as certain conditions were met. These conditions when applied to small dimensional spaces give
bounds that are known to be very loose. For instance, in $d$ dimensional spaces, as
long as $d$ of the partitioning planes are axis orthogonal, the aspect ratio (we give
a precise definition later) is bounded by $50\sqrt{d}+55$. In 2 dimensional spaces, the
aspect ratio is bounded by a number very close to 6. In this paper, we are interested in developing simpler conditions for choosing partitioning planes flexibly in
BAR trees for small dimensions. This will allow us to derive tighter bounds for
the aspect ratio, and discover the best set of partitioning planes for a particular
application.

## 1.1   Related Prior Work

In this subsection, we briefly review some known general-purpose hierarchical
spatial decomposition trees for a set of points, $S$.

**The Binary Space Partitioning (BSP) Trees.** The BSP tree [14, 13] is a recursive subdivision of space into regions by means of half-planes. Each node $u$ in
the tree represents a convex region and the points from $S$ lying in it. Initially, the
root node $r$ of the tree represents a bounding region of the point set $S$. At each
node $u$, an associated line partitions the region of $u$, $R_u$, into two disjoint regions
$R_l$ and $R_r$. The node $u$ then has two child nodes $l$ and $r$ representing $R_l$ and $R_r$
respectively. If the number of points from $S$ in $R_u$ is less than some constant,
$u$ is not partitioned and becomes a leaf. These structures satisfy our condition
of being general-purpose, but without further restricting how the cutting lines
are chosen, these structures are inefficient. Thus, much work has been done on
methods for specializing BSP trees to be more efficient, which we review next.hfil

**The $k$-d Tree.** This structure was introduced by Bentley [3, 4, 5] and has been
extensively studied. It is a special class of the BSP tree: the partitioning line

is orthogonal to one of the axes and such that it divides the set of points at the node in half by cardinality. This guarantees that the depth of the tree is $O(\log n)$. So point location queries, which take time proportional to the depth, can be answered efficiently. But, since there are no guarantees on the aspect ratio of the regions produced, with some exceptions [9], the running times of queries can nevertheless be poor.

**The Quadtree.** The quadtree (e.g., see [18, 19]) is another special class of the BSP tree. The point set $S$ is initially bounded by a square, and the partitions are such that a square region is divided into four smaller squares of equal area. (This notion can be extended to $d$ dimensional space, giving rise to a structure called the *octree*.) The aspect ratios of regions in quadtrees is bounded by a constant, but these trees can have unbounded depth. So even basic point location queries can take unbounded time. If the point set is uniformly distributed, however, then the depth is bounded, and in those situations quadtrees perform well for some geometric queries.

**Balanced Box Decomposition Trees.** In [1, 2], Arya *et al.* describe a relative of a binary-space partitioning tree called the *Balanced Box Decomposition* (BBD) tree. This structure is based on the fair-split tree of Callahan and Kosaraju [6, 7] and is defined such that its depth is $O(\log n)$ and all of the associated regions have low combinatorial complexity and bounded aspect ratio. Arya *et al.* show how BBD trees guarantee excellent performance in approximating general range queries and nearest-neighbor queries. (Approximate queries are like the regular exact versions, except they allow an error. See [10] for formal definitions.) The aspect ratio bound on each region allows them to bound the number of nodes visited during various approximate query searches by limiting the number of nodes that can be packed inside a query region. However, since these trees rely on using hole cuts during construction, they produce non-convex regions and thus are not true BSP trees. This is also a drawback with respect to several applications in computer graphics and graph drawing, where convexity of the partitioned regions is desirable (e.g., see [12, 15]).

**Balanced Aspect Ratio Trees.** Duncan *et al.* [12, 11] introduced the *Balanced Aspect Ratio* (BAR) trees. These are similar to $k$-d trees in 2-dimensional space, except that instead of allowing only axis-orthogonal partitions, they also allow a third partition orthogonal to a vector at a $\pi/4$ angle to the axes. This extra cut allows them to find partitions that not only divide the point set in a region evenly, but also ensure that the child regions have good (bounded) aspect ratio. In [10], Duncan extended BAR trees to $d$ dimensions. He showed that if a certain set of conditions is satisfied, BAR trees with bounded aspect ratio can be constructed. He also proved bounds on the running time of approximate queries. The $(1+\varepsilon)$-nearest neighbor query and the $(1-\varepsilon)$-farthest neighbor query can be answered in $O(\log n + (1/\varepsilon)\log(1/\varepsilon))$ time, where $n$ is the number of points. The $\varepsilon$-range query and the $\varepsilon$-proximity query can be answered in $O(\log n + (1/\varepsilon) + k)$ time, where $k$ is the size of the output.

## 1.2    Our Contributions

In this paper, we introduce the Parameterized Balanced Aspect Ratio (PBAR) trees in 2 dimensions, which take any three vectors as the partitioning planes. They enjoy all the advantages of BAR trees: general pupose, space efficient, logarithmic depth, bounded aspect ratio, poly-logarithmic worst case bounds for approximate versions of spatial queries. In addition, they have a bound on the aspect ratio which is a simple closed-form function of the given partitioning planes. The proofs used are significantly different from those for earlier BAR trees: they use the advantages of 2 dimensional spaces and yet work of any given set of partitioning vectors.

Our motivation for introducing PBAR trees comes from our experience with the Sloan Digital Sky Survey (SDSS) [16, 17]. Because objects in the SDSS are indexed by their positions on the night sky, the data can be viewed at a first level of indexing as two-dimensional points on a sphere. To allow for efficient access, astronomers overlay a quasi-uniform triangular "grid" on this sphere, to reduce the curvature of each "leaf" triangle to be "almost" planar and to reduce the number of points in each such triangle to a few hundred thousand. The difficulty is that when these leaf triangles are mapped to a projection plane to allow for fast queries via a secondary data structure, there are many different side angles that must be dealt with. A data structure like PBAR trees can conveniently use the given angles of a bounding triangle as its possible partitioning directions. Without this convenience, we would get poorly-shaped regions near the boundaries of these triangles.

The bounds on the running times for approximate queries, in [10], depend only on the fact that both BBD and BAR trees have $O(\log n)$ depth and ensure a constant bound on the aspect ratio of all their regions. These two conditions are satisfied by PBAR trees as well. So the same bounds hold for PBAR trees as well. We present empirical results for the $(1 + \varepsilon)$-nearest neighbor query using PBAR trees with various different partitioning planes using artificial data as well as real datasets from the SDSS. Our experiments also indicate that our bound is tight in some respects.

In the next section we give the foundations and definition for PBAR trees. In Section 3 we describe the algorithm for constructing PBAR trees and prove its correctness. In the last section we present our empirical results. We include details for the pseudo-code and proofs of correctness in an optional appendix in this extended abstract.

## 2    Parameterizing BAR Trees

PBAR trees are for point data in 2-dimensional space. The *distance* $\delta(p, q)$ between two points $p = (p_1, p_2)$ and $q = (q_1, q_2)$ is $\sqrt{(p_1 - q_1)^2 + (p_2 - q_2)^2}$. Extending this notion, the distance between two sets of points $P$ and $Q$ is

$$\delta(P, Q) = \min_{p \in P, q \in Q} \delta(p, q).$$

$S$ is the set of $n$ points given as input. The size $|R|$ of region $R$ is the number of points from $S$ in $R$.

*Partitioning Vectors.* We use vectors from $\mathbb{R}^2$ to specify partitioning directions. Note that partitioning vectors $l$ and $-l$ are equivalent. The *angle* $\theta_{lm}$ between two partitioning vectors $l$ and $m$ is the angle from $l$ to $m$ in the counterclockwise direction, except that we take into account that $m$ and $-m$ are equivalent. In the context of trigonometric functions, we shall prefer to use, for example, the short $\sin(lm)$ instead of $\sin(\theta_{lm})$.

To construct a PBAR tree, we use the 3 partitioning vectors from the given set $V = \{\lambda, \mu, \nu\}$. We make all partitions by taking a region $R$ and dividing it into two subregions, $R_1$ and $R_2$, with a line $c'_l$, called a *cut*, orthogonal to some $l \in V$. A cut orthogonal to $l$ is also called an *$l$-cut*. Note that if $R$ is convex, both $R_1$ and $R_2$ are convex too. We divide the set of points in $R$, call it $S$, between $R_1$ and $R_2$ in the natural fashion. For points in $S$ that are on $c_l$, we assign each of them to either $R_1$ or $R_2$ as per convenience.

Let the sequence $(\lambda, \mu, \nu)$ be in the counterclockwise order. All the 3 sequences in the set $\mathcal{P}(V) = \{(\lambda, \mu, \nu), (\mu, \nu, \lambda), (\nu, \lambda, \mu)\}$ are equivalent for our purpose. So, often, we shall speak in terms of the general $(l, m, n)$, where $(l, m, n) \in \mathcal{P}(V)$.

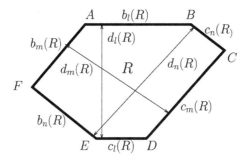

**Fig. 1.** The names used for the sides and the diameters

*Canonical Regions and Canonical Aspect Ratios.* We assume that the given set of points $S$ has an initial convex bounding region with sides that are orthogonal to $\lambda, \mu$, or $\nu$. Since in constructing PBAR trees we make all partitions with lines orthogonal to these 3 partitioning vectors, the regions we construct are always hexagons with sides orthogonal to $\lambda, \mu$, or $\nu$. Some sides may be degenerate, that is, of length 0. We call these hexagonal regions *canonical regions*. See Figure 1. In a canonical region $R$, $b_l(R)$ and $c_l(R)$ are the two unique opposing sides orthogonal to the partitioning vector $l$. The *diameter* $d_l(R)$ of $R$ with respect to the partitioning vector $l$ is the distance $\delta(b_l(R), c_l(R))$. The *maximum diameter* of $R$ is $d_{\max}(R) = \max_{l \in V} d_l(R)$, and the *minimum diameter* of $R$ is $d_{\min}(R) = \min_{l \in V} d_l(R)$. When the region is understood from the context we drop the argument in the above notations and use, for example, $b_l$ instead of $b_l(R)$. A *canonical trapezoidal region* is of special interest, and is a canonical region that is quadrilateral and has exactly one pair of parallel sides.

The *canonical aspect ratio* $\mathsf{casp}(R)$ of canonical region $R$ is the ratio of $d_{\max}(R)$ to $d_{\min}(R)$. In this paper, we use the terms aspect ratio and canonical aspect ratio synonymously

*PBAR Trees, One-Cuts, and Two-Cuts.* Given a *balancing factor* $\alpha$, $R$ is $\alpha$-*balanced* or has a *balanced aspect ratio* if $\mathsf{casp}(R) \leq \alpha$. $R$ is *critically balanced* if $\mathsf{casp}(R) = \alpha$. A cut $c_l$ orthogonal to $l \in V$ that divides an $\alpha$-balanced region $R$ into $R_1$ and $R_2$ is *feasible* if both $R_1$ and $R_2$ are $\alpha$-balanced.

Given a set $S$ of $n$ points in 2-dimensional space, a set of 3 partitioning vectors $V = \{\lambda, \mu, \nu\}$, a balancing factor $\alpha$, $\alpha \geq 1$, and a reduction factor $\beta$, $0.5 \leq \beta < 1$, a *Parameterized Balanced Aspect Ratio* tree $T$ is a BSP tree on $S$ such that

1. All partitions are made with cuts orthogonal to the vectors in $V$;
2. The canonical aspect ratio of each region is at most $\alpha$;
3. The number of points in each leaf cell of $T$ is a constant with respect to $n$.
4. The depth of $T$ is $O(\log_{1/\beta} n)$.

Given a balancing factor $\alpha$ and reduction factor $\beta$, an $\alpha$-balanced region $R$ is *one-cuttable* if there is a cut $c$, called a *one-cut*, orthogonal to a vector in $V$ that divides $R$ into two canonical subregions $R_1$ and $R_2$ such that

1. $c$ is feasible;
2. $|R_1| \leq \beta|R|$ and $|R_2| \leq \beta|R|$.

(Note that if there is a continuum of feasible cuts that cover the entire region $R$, then $R$ is one-cuttable, as at least one of these cuts will satisfy 2 above.) A region $R$ is *two-cuttable* if there is a cut $c$, called a *two-cut*, orthogonal to a vector in $V$ that divides $R$ into two canonical subregions $R_1$ and $R_2$ such that

1. $c$ is feasible;
2. $|R_1| \leq \beta|R|$;
3. $|R_2| \leq \beta|R|$ or $R_2$ is one-cuttable.

*Shield Regions.* Let $R$ be an $\alpha$-balanced canonical region and let $x_l$ be a side of $R$, $x \in \{b, c\}$, $l \in V$. Now sweep a cut $x'_l$ starting from the side opposite to $x_l$ toward $x_l$. Let $P$ be the subregion formed between $x_l$ and $x'_l$. In the beginning, $\mathsf{casp}(P) \leq \alpha$. Sweep $x'_l$ toward $x_l$ and stop when $P$ is critically balanced. $P$ is called the *shield region* $\mathsf{shield}_{x_l}(R)$ of $R$ with respect to $x_l$. $x'_l$ is the *cut for* $\mathsf{shield}_{x_l}(R)$. $R$ has two shield regions for each $l \in V$, $\mathsf{shield}_{b_l}(R)$ and $\mathsf{shield}_{c_l}(R)$. Note that $R$ has a feasible $l$-cut if and only if $\mathsf{shield}_{b_l}(R) \cap \mathsf{shield}_{c_l}(R) = \emptyset$.

For a given $l \in V$, the *maximal shield region* $\mathsf{maxshield}_l(R)$ of $R$ with respect to $l$ is one among $\mathsf{shield}_{b_l}(R)$ and $\mathsf{shield}_{c_l}(R)$ that has the maximum size. (Remember, the size of a region is the number of points in it.) Note that $R$ has a one-cut orthogonal to $l$ only when $|\mathsf{maxshield}_l(R)| \leq \beta|R|$.

## 3   The PBAR Tree Algorithm

In this section we present the PBAR tree algorithm that, given a set of partitioning vectors $V = \{\lambda, \mu, \nu\}$, a reduction factor $\beta$, and a balancing factor $\alpha$, constructs a PBAR tree on any set $S$ of $n$ points in 2-dimensional space; as long as $0.5 \leq \beta < 1$, and $\alpha$ is at least

$$f(V) = \frac{4.38}{\sin(\theta_{\min})\sin(\lambda\mu)\sin(\mu\nu)\sin(\nu\lambda)},$$

where $\theta_{\min}$ is the minimum among the angles $(\lambda\mu), (\mu\nu), (\nu\lambda)$.

The PBAR tree algorithm takes an initial $\alpha$-balanced canonical region $R$ that bounds $S$ and recursively subdivides it by first searching for a one-cut, and if no such cut exists, by searching for a two-cut. For details see [8].

The algorithm for searching for a one-cut, OneCut, considers each partitioning vector in turn. For a partitioning vector $l$, a one-cut orthogonal to it exists if and only if the shield regions with respect to $l$ don't overlap and the maximal shield region contains at most $\beta|R|$ points. Details are in [8].

The algorithm for searching for a two-cut, TwoCut, considers very few cuts as potential two-cuts. Only cuts for the maximal shield regions for the 3 partitioning vectors are considered as potential two-cuts. This is sufficient as long as $\alpha \geq f(V)$ — that this is true is our main result and we prove it in the rest of the section. Details for TwoCut are in [8].

**Theorem 1 (Main Result).** *Given a set $S$ of $n$ points in 2-dimensional space, a set of 3 partitioning vectors $V = \{\lambda, \mu, \nu\}$, a balancing factor $\alpha$, $\alpha \geq f(V)$, and a reduction factor $\beta$, $0.5 \leq \beta < 1$, the PBAR tree algorithm constructs a PBAR tree on $S$ in $O(n \log n)$ time.*

We first prove some preliminary lemmas. For all of these we shall assume that $0.5 \leq \beta < 1$ and $\alpha \geq f(V)$.

**Lemma 1.** *Given a set of partitioning vectors $V$, a balancing factor $\alpha$, and a reduction factor $\beta$, if every $\alpha$-balanced region $R$ is two-cuttable, then a PBAR tree can be constructed for every set $S$ of $n$ points.*

*Proof.* Start with any initial $\alpha$-balanced canonical region that bounds $S$. Since this region is two-cuttable, divide it into a maximum of 3 $\alpha$-balanced subregions such that each contains less than $\beta n$ points. Repeat this process for each of the resulting subregions until each of the final leaf regions has at most a constant number of points. The process, along any path of subregions, cannot be repeated more than $O(\log_{1/\beta} n)$ times.

**Lemma 2.** *A canonical region $R$ that is a triangle is always $\alpha$-balanced.*

Due to lack of space, proofs for the lemmas are in the full version of the paper[8].

**Lemma 3.** *Let $(l, m, n) \in \mathcal{P}(V)$. Let $R$ be an $\alpha$-balanced canonical region that is not critically balanced. If $P$ is a critically balanced subregion created by partitioning $R$ with an $l$-cut, then the minimum diameter of $P$ is $d_l(P)$.*

**Corollary 1.** *For a critically balanced canonical region $R$ that is a trapezoid, if $b_l$ and $c_l$ are the two parallel sides, then the minimum diameter of $R$ is $d_l$.*

**Lemma 4.** *Let $(l, m, n) \in \mathcal{P}(V)$, and let $R$ be an $\alpha$-balanced region that has no feasible $l$-cut. Let $S$ be the region formed by extending $R$ such that $b_n(S)$ is of length 0. If $P$ is $\mathsf{shield}_{c_m}(S)$, then $c_l(P) \leq c_l(R)$.*

**Lemma 5.** *Let $(l, m, n) \in \mathcal{P}(V)$. If an $\alpha$-balanced region $R$ has no feasible $l$-cut, then it has a feasible $m$-cut and a feasible $n$-cut.*

**Corollary 2.** *Let $(l, m, n) \in \mathcal{P}(V)$. If an $\alpha$-balanced region $R$ has no one-cut and no feasible $l$-cut, then the maximal shield regions with respect to $m$ and $n$ intersect.*

**Lemma 6.** *A critically balanced region $R$ that is a trapezoidal is one-cuttable.*

**Lemma 7.** *Let $(l, m, n) \in \mathcal{P}(V)$. If an $\alpha$-balanced region $R$ has no one-cut and no feasible $l$-cut, then $R$ has a two-cut.*

**Lemma 8.** *Let $(l, m, n) \in \mathcal{P}(V)$. If an $\alpha$-balanced region $R$ does not have a one-cut and yet there are feasible cuts along all 3 partitioning vectors $l$, $m$, and $n$, then $R$ has a two-cut.*

**Proof of Theorem 1.**     The PBAR tree algorithm recursively subdivides the initial bounding region by first searching for a one-cut, and if no such cut exists, by searching for a two-cut. By Lemma 1, if it always succeeds in finding a one-cut or a two-cut, it constructs a PBAR tree. It is easy to see that when the algorithm does not find a one-cut, no such cut exists. In such a situation, the algorithm searches for a two-cut by checking if any of the 3 maximal shield regions are one-cuttable. The proofs for Lemmas 7 and 8 show that at least one of these shield regions is one-cuttable, and so the algorithm always succeeds in finding either a one-cut or a two-cut. For the time analysis see proof in [8].     □

# 4     Empirical Tests

In this section we present the preliminary empirical results we have obtained and analyze them. We look at measures like number of nodes created, the depth of the tree, and the number of leaves visited instead of the actual time or space requirements. This is because the time and space measures are dependent on the efficiency of the implementation, the load on the machine during testing, etc. The other measures are not as dependent on the kind of testing carried out. In addition, the number of nodes visited is the dominant term if the data structure is stored in external memory (as is the case in SDSS).

We took 2 data sets and varied the partitioning planes in small increments and for each we found the best aspect ratio that can be obtained. First, we present plots that summarize the results of this experiment. Later, we present detailed results for 6 different data sets in which we compare BAR trees with $(0, \pi/4, \pi/2)$ partitioning angles with 2 instances of PBAR trees.

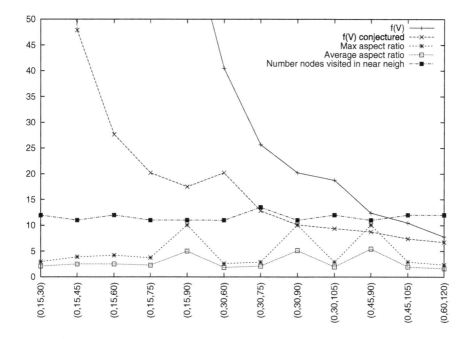

**Fig. 2.** Effects of varying the partitioning planes on SDSS data

## 4.1    Varying the Partitioning Planes

We varied the set of partitioning planes for a given dataset and found the best possible aspect ratio that can be obtained for that dataset. We plot this best empirical aspect ratio in Figures 2 and 3 (it is called the Maximum aspect ratio in the figures). We also plot, alongside, the bound $f(V)$ on the aspect ratio that we have proved. We had conjectured that $f(V)$ can be tightened by removing the $\sin(\theta_{\min})$ term to obtain $f(V) = 4.38/(\sin(\lambda\mu)\sin(\mu\nu)\sin(\nu\lambda))$. We plot the conjectured bound as well. We also plot the average aspect ratio of the nodes in tree.

The bound and the conjectured bound both decrease dramatically as the planes become evenly oriented. But the empirically obtained values do not follow their lead, though there is a reasonable amount of variation in the best (maximum) aspect ratio possible. The value of $f(V)$ converges towards the empirical value as it reduces, which indicates that it is possibly tight for the evenly oriented planes. The same is true for the conjectured bound; but, if you look closely at Figure 3 for the orientation (0, 45, 90) the conjectured value is actually lower than the best aspect ratio obtained through experiments. This indicates that the conjecture is wrong. We also plot the number of nodes visited during the nearest neighbor searches (the details for these searches are described in the next section) for the various planes. There is very slight variation in this, or in the average aspect ratio with the change in plane orientations.

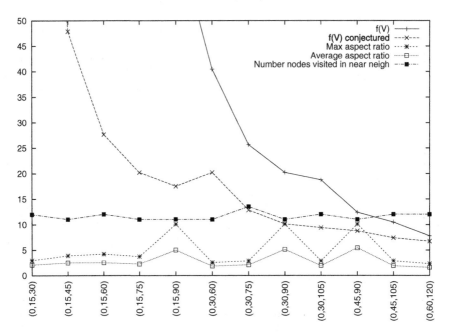

**Fig. 3.** Effects of varying the partitioning planes on data created uniformly at random along a circle

## 4.2    Comparing BAR Trees and PBAR Trees

To compare the performance of BAR trees (with mostly axis orthogonal planes: $0, \pi/4, \pi/2$) and PBAR trees we constructed PBAR trees with three different sets of partitioning vectors $V$. In the first set, $V$ is such that $\theta_{\lambda\mu} = \pi/4$ and $\theta_{\lambda\nu} = \pi/2$. In the second set, $V$ is such that $\theta_{\lambda\mu} = \pi/3$ and $\theta_{\lambda\nu} = 2\pi/3$, and in the third set, $V$ is such that $\theta_{\lambda\mu} = \pi/6$ and $\theta_{\lambda\nu} = \pi/2$. Note that the partitioning vectors in the first set are that used by BAR trees. The PBAR trees constructed in this case closely mimic BAR trees and the performance is representative of the performance of BAR trees. In the second case, the partitioning vectors are more evenly oriented, while in the third case they are less evenly oriented, than the first case. We refer to the former case as the BAR tree results and the latter two cases as the PBAR-even tree and PBAR-uneven tree results respectively. In all cases $\alpha$ is 20, $\beta$ is 0.6, and the maximum points in any leaf $k$ is 5. A 100 $(1 + \varepsilon)$-nearest neighbor queries were solved using each tree. For each data set, a point $q$ is first chosen uniformly at random from among the data points. This is the first query point. Then a random increment is chosen and repeatedly added to $q$ to get 99 other query points. $\varepsilon$ for the queries is always 0.001. The results for BAR tree are in Figure 4, for PBAR-even tree are in Figure 5, and for PBAR-uneven tree are in Figure 6. Of the 6 data sets data sets, 4 were chosen rather arbitrarily, and the last two are real data from the Sloan Digital Sky Survey(SDSS). The data sets are described in [8].

| Data Set | Nodes in tree | Depth of tree | Avg. casp(·) of regions | Nodes visited during query | Leaves visited during query |
|---|---|---|---|---|---|
| Set 1 | 4615 | 12 | 9.70 | 57.2 | 11.74 |
| Set 2 | 4719 | 13 | 8.78 | 73.64 | 13.97 |
| Set 3 | 4709 | 12 | 9.08 | 148.51 | 30.97 |
| Set 4 | 4631 | 12 | 7.84 | 16.47 | 2.49 |
| Set 5 | 4749 | 12 | 8.98 | 13.49 | 1.60 |
| Set 6 | 2079 | 11 | 9.13 | 21.84 | 3.68 |

**Fig. 4.** Results for $(1 + \varepsilon)$-nearest neighbor queries on BAR trees. (Number of nodes and leaves visited during query are averaged over a 100 queries.)

| Data Set | Nodes in tree | Depth of tree | Avg. casp(·) of regions | Nodes visited during query | Leaves visited during query |
|---|---|---|---|---|---|
| Set 1 | 4647 | 12 | 9.93 | 21.5 | 3.66 |
| Set 2 | 4791 | 12 | 8.78 | 22.61 | 4.38 |
| Set 3 | 4741 | 12 | 9.36 | 13.7 | 1.48 |
| Set 4 | 4641 | 12 | 8.14 | 15.59 | 2.23 |
| Set 5 | 4673 | 12 | 8.41 | 32.44 | 5.91 |
| Set 6 | 2067 | 11 | 8.77 | 21.19 | 3.49 |

**Fig. 5.** Results for $(1 + \varepsilon)$-nearest neighbor queries on PBAR-even trees; $V$ such that $\theta_{\lambda\mu} = \pi/3$ and $\theta_{\lambda\nu} = 2\pi/3$. (Number of nodes and leaves visited during query are averaged over a 100 queries.)

| Data Set | Nodes in tree | Depth of tree | Avg. casp(·) of regions | Nodes visited during query | Leaves visited during query |
|---|---|---|---|---|---|
| Set 1 | 4635 | 12 | 9.64 | 58.04 | 12.03 |
| Set 2 | 4697 | 13 | 8.58 | 76.32 | 14.91 |
| Set 3 | 4655 | 12 | 8.86 | 152.81 | 31.62 |
| Set 4 | 4629 | 12 | 7.44 | 15.96 | 2.37 |
| Set 5 | 4683 | 12 | 8.57 | 13.48 | 1.6 |
| Set 6 | 2077 | 11 | 8.82 | 21.21 | 3.51 |

**Fig. 6.** Results for $(1 + \varepsilon)$-nearest neighbor queries on PBAR-uneven trees; $V$ such that $\theta_{\lambda\mu} = \pi/6$ and $\theta_{\lambda\nu} = \pi/2$. (Number of nodes and leaves visited during query are averaged over a 100 queries.)

The number of nodes in the tree are about the same, for the first 5 data sets, irrespective of $V$. This is expected as the number of data points and the maximum size of a leaf are the same in all cases. The depth of the trees are about the same too, irrespective of the data set. This, again, is expected as the $\beta$ values are the same, and we don't except too many regions that require two-cuts. Set 6 has far fewer points and so has much fewer nodes. Surprisingly, the average values of the canonical aspect ratio are not very different in the three trees for the different data sets. Neither is one of the trees always better than the other. Such is not the case for number of nodes and number of leaves visited during query processing. PBAR-even trees almost always visit fewer nodes and fewer leaves and in one particular case the difference with both BAR and PBAR-uneven trees is a factor of 10. That PBAR-even trees perform better at approximate nearest neighbor searches may be expected given the theoretical results, it is surprising that this should be the case when the canonical aspect ratio values are about the same. For Set 5, which is real data set from the SDSS, however, PBAR-even trees are not the best. For this set, both BAR trees and PBAR-uneven trees perform better, with PBAR-uneven slightly ahead of BAR. For the other real data, Set 6, again PBAR-even is the best.

In conclusion, our experiments show that the flexibility of PBAR trees can help in increasing the efficiency of approximate nearest neighbor searches.

## 5     Conclusion and Future Work

In this paper we revisited BAR trees in 2 dimensional spaces and developed the Parameterized Balanced Aspect Ratio (PBAR) trees. These allow any given set of 3 partitioning planes and yet retain all the advantages of BAR trees — general purpose data structures, space efficient, logarithmic depth, bounded aspect ratio, and poly-logarithmic approximate query processing. These are the first known "BAR-type" trees in which the aspect ratio can be bounded by a simple closed-form function (it depends on the orientation of the partitioning planes). We conducted empirical tests that show that in many instances the evenly oriented partitioning planes are better than the mostly axis orthogonal planes that have been mostly studied prior to this. In addition, our experiments indicate that our bound is tight in some respects: it converges to empirical values for evenly oriented planes, and that a natural modification to tighten it (our conjecture that $\sin(\theta_{\min})$ factor can be removed) is wrong.

Having bounds on the aspect ratio can be useful in ways other than solving queries faster. For example, PBAR trees can be used to efficiently compute the density of a region around a given point. This can be useful in detecting density based outliers. We want to explore this and other possible applications of PBAR trees in spatial data mining.

## Acknowledgment

The authors will like to thank Breno de Medeiros for many helpful suggestions, including helping tighten the bounds in Lemma 7 and Christian Duncan for helpful discussions.

# References

1. S. Arya, D. M. Mount, N. S. Netanyahu, R. Silverman, and A. Wu. An optimal algorithm for approximate nearest neighbor searching. In *Proc. 5th ACM-SIAM Sympos. Discrete Algorithms*, pages 573–582, 1994.
2. Sunil Arya and David M. Mount. Approximate range searching. In *Proc. 11th Annu. ACM Sympos. Comput. Geom.*, pages 172–181, 1995.
3. J. L. Bentley. Multidimensional binary search trees used for associative searching. *Commun. ACM*, 18(9):509–517, September 1975.
4. J. L. Bentley. Multidimensional binary search trees in database applications. *IEEE Trans. Softw. Eng.*, SE-5:333–340, 1979.
5. J. L. Bentley. $K$-d trees for semidynamic point sets. In *Proc. 6th Annu. ACM Sympos. Comput. Geom.*, pages 187–197, 1990.
6. P. B. Callahan and S. R. Kosaraju. A decomposition of multidimensional point sets with applications to $k$-nearest-neighbors and $n$-body potential fields. *J. ACM*, 42:67–90, 1995.
7. Paul B. Callahan and S. Rao Kosaraju. Algorithms for dynamic closest-pair and $n$-body potential fields. In *Proc. 6th ACM-SIAM Sympos. Discrete Algorithms*, pages 263–272, 1995.
8. A. Chaudhary and M.T. Goodrich. Balanced aspect ratio trees revisited. http://www.cse.nd.edu/~achaudha/research. Full version of the paper.
9. M. Dickerson, C. A. Duncan, and M. T. Goodrich. K-D trees are better when cut on the longest side. In *Proc. 8th European Symp. on Algorithms*, volume 1879 of *Lecture Notes Comput. Sci.*, pages 179–190. Springer-Verlag, 2000.
10. C. Duncan. *Balanced Aspect Ratio Trees*. PhD thesis, The Johns Hopkins University, Baltimore, Maryland, Sep 1999.
11. C. A. Duncan, M. T. Goodrich, and S. Kobourov. Balanced aspect ratio trees: combining the advantages of k-d trees and octrees. In *Proc. 10th Annu. ACM-SIAM Sympos. Discrete Alg.*, pages 300–309, 1999.
12. C. A. Duncan, M. T. Goodrich, and S. G. Kobourov. Balanced aspect ratio trees and their use for drawing very large graphs. In *Graph Drawing*, Lecture Notes in Computer Science, pages 111–124. Springer-Verlag, 1998.
13. H. Fuchs, G. D. Abrams, and E. D. Grant. Near real-time shaded display of rigid objects. *Comput. Graph.*, 17(3):65–72, 1983. Proc. SIGGRAPH '83.
14. H. Fuchs, Z. M. Kedem, and B. Naylor. On visible surface generation by a priori tree structures. *Comput. Graph.*, 14(3):124–133, 1980. Proc. SIGGRAPH '80.
15. David Luebke and Carl Erikson. View-dependent simplification of arbitrary polygonal environments. In Turner Whitted, editor, *SIGGRAPH 97 Conference Proceedings*, Annual Conference Series, pages 199–208. ACM SIGGRAPH, Addison Wesley, August 1997. ISBN 0-89791-896-7.
16. Robert Lupton, F. Miller Maley, and Neal Young. Sloan digital sky survey. http://www.sdss.org/sdss.html.
17. Robert Lupton, F. Miller Maley, and Neal Young. Data collection for the Sloan Digital Sky Survey—A network-flow heuristic. *Journal of Algorithms*, 27(2):339–356, 1998.
18. H. Samet. *Spatial Data Structures: Quadtrees, Octrees, and Other Hierarchical Methods*. Addison-Wesley, Reading, MA, 1989.
19. H. Samet. *Applications of Spatial Data Structures: Computer Graphics, Image Processing, and GIS*. Addison-Wesley, Reading, MA, 1990.

# Improved Combinatorial Group Testing
# for Real-World Problem Sizes

David Eppstein, Michael T. Goodrich, and Daniel S. Hirschberg

Dept. of Computer Science,
Univ. of California, Irvine, CA 92697-3425 USA
{eppstein, goodrich, dan}(at)ics.uci.edu

**Abstract.** We study practically efficient methods for performing combinatorial group testing. We present efficient non-adaptive and two-stage combinatorial group testing algorithms, which identify the at most $d$ items out of a given set of $n$ items that are defective, using fewer tests for all practical set sizes. For example, our two-stage algorithm matches the information theoretic lower bound for the number of tests in a combinatorial group testing regimen.

**Keywords:** combinatorial group testing, Chinese remaindering, Bloom filters.

## 1    Introduction

The problem of combinatorial group testing dates back to World War II, for the problem of determining which in a group of $n$ blood samples contain the syphilis antigen (hence, are contaminated). Formally, in combinatorial group testing, we are given a set of $n$ items, at most $d$ of which are defective (or contaminated), and we are interested in identifying exactly which of the $n$ items are defective. In addition, items can be "sampled" and these samples can be "mixed" together, so tests for contamination can be applied to arbitrary subsets of these items. The result of a test may be positive, indicating that at least one of the items of that subset is defective, or negative, indicating that all items in that subset are good. Example applications that fit this framework include:

- *Screening blood samples for diseases.* In this application, items are blood samples and tests are disease detections done on mixtures taken from selected samples.
- *Screening vaccines for contamination.* In this case, items are vaccines and tests are cultures done on mixtures of samples taken from selected vaccines.
- *Clone libraries for a DNA sequence.* Here, the items are DNA subsequences (called *clones*) and tests are done on pools of clones to determine which clones contain a particular DNA sequence (called a *probe*) [8].
- *Data forensics.* In this case, items are documents and the tests are applications of one-way hash functions with known expected values applied to selected collections of documents.

The primary goal of a testing algorithm is to identify all defective items using as few tests as possible. That is, we wish to minimize the following function:

- $t(n, d)$: The number of tests needed to identify up to $d$ defectives among $n$ items.

F. Dehne, A. López-Ortiz, and J.-R. Sack (Eds.): WADS 2005, LNCS 3608, pp. 86–98, 2005.

This minimization may be subject to possibly additional constraints, as well. For example, we may wish to identify all the defective items in a single (*non-adaptive*) round of testing, we may wish to do this in two (*partially-adaptive*) rounds, or we may wish to perform the tests sequentially one after the other in a *fully adaptive* fashion.

In this paper we are interested in efficient solutions to combinatorial group testing problems for realistic problem sizes, which could be applied to solve the motivating examples given above. That is, we wish solutions that minimize $t(n, d)$ for practical values of $n$ and $d$ as well as asymptotically. Because of the inherent delays that are built into fully adaptive, sequential solutions, we are interested only in solutions that can be completed in one or two rounds. Moreover, we desire solutions that are efficient not only in terms of the total number of tests performed, but also for the following measures:

- $A(n, t)$: The *analysis* time needed to determine which items are defective.
- $S(n, d)$: The *sampling* rate—the maximum number of tests any item may be included in.

An analysis algorithm is said to be *efficient* if $A(n, t)$ is $O(tn)$, where $n$ is the number of items and $t$ is the number of tests conducted. It is *time-optimal* if $A(n, t)$ is $O(t)$. Likewise, we desire efficient sampling rates for our algorithms; that is, we desire that $S(n, d)$ be $O(t(n, d)/d)$. Moreover, we are interested in this paper in solutions that improve previous results, either asymptotically or by constant factors, for realistic problem sizes. We do not define such "realistic" problem sizes formally, but we may wish to consider as unrealistic a problem that is larger than the total memory capacity (in bytes) of all CDs and DVDs in the world ($< 10^{25}$), the number of atomic particles in the earth ($< 10^{50}$), or the number of atomic particles in the universe ($< 10^{80}$).

*Viewing Testing Regimens as Matrices.* A single round in a combinatorial group testing algorithm consists of a test regimen and an analysis algorithm (which, in a non-adaptive (one-stage) algorithm, must identify all the defectives). The test regimen can be modeled by a $t \times n$ Boolean matrix, $M$. Each of the $n$ columns of $M$ corresponds to one of the $n$ items. Each of the $t$ rows of $M$ represents a test of items whose corresponding column has a 1-entry in that row. All tests are conducted before the results of any test is made available. The analysis algorithm uses the results of the $t$ tests to determine which of the $n$ items are defective.

As described by Du and Hwang [5](p. 133), the matrix $M$ is *d-disjunct* if the Boolean sum of any $d$ columns does not contain any other column. In the analysis of a *d-disjunct* testing algorithm, items included in a test with negative outcome can be identified as pure. Using a $d$-disjunct matrix enables the conclusion that if there are $d$ or fewer items that cannot be identified as pure in this manner then all those items must be defective and there are no other defective items. If more than $d$ items remain then at least $d + 1$ of them are defective. Thus, using a $d$-disjunct matrix enables an efficient analysis algorithm, with $A(n, t)$ being $O(tn)$.

$M$ is *d-separable* ($\bar{d}$-separable) if the Boolean sums of $d$ (up to $d$) columns are all distinct. The $\bar{d}$-separable property implies that each selection of up to $d$ defective items induces a different set of tests with positive outcomes. Thus, it is possible to identify which are the up to $d$ defective items by checking, for each possible selection, whether its induced positive test set is exactly the obtained positive outcomes. However, it might not be possible to detect that there are more than $d$ defective items. This analysis algorithm takes time $\Theta(n^d)$ or requires a large table mapping $t$-subsets to $d$-subsets.

Generally, $\overline{d}$-separable matrices can be constructed with fewer rows than can $d$-disjunct matrices having the same number of columns. Although the analysis algorithm described above for $d$-separable matrices is not efficient, some $\overline{d}$-separable matrices that are not $d$-disjunct have an efficient analysis algorithm.

*Previous Related Work.* Combinatorial group testing is a rich research area with many applications to many other areas, including communications, cryptography, and networking [3]. For an excellent discussion of this topic, the reader is referred to the book by Du and Hwang [5]. For general $d$, Du and Hwang [5](p. 149) describe a slight modification of the analysis of a construction due to Hwang and Sós [9] that results in a $t \times n$ $d$-disjunct matrix, with $n \geq (2/3)3^{t/16d^2}$, and so $t \leq 16d^2(1 + \log_3 2 + (\log_3 2) \lg n)$. For two-stage testing, Debonis *et al.* [4] provide a scheme that achieves a number of tests within a factor of $7.54(1 + o(1))$ of the information-theoretic lower bound of $d \log(n/d)$. For $d = 2$, Kautz and Singleton [10] construct a 2-disjunct matrix with $t = 3^{q+1}$ and $n = 3^{2^q}$, for any positive integer $q$. Macula and Reuter [11] describe a $\overline{2}$-separable matrix and a time-optimal analysis algorithm with $t = (q^2 + 3q)/2$ and $n = 2^q - 1$, for any positive integer $q$. For $d = 3$, Du and Hwang [5](p. 159) describe the construction of a $\overline{3}$-separable matrix (but do not describe the analysis algorithm) with $t = 4\binom{3q}{2} = 18q^2 - 6q$ and $n = 2^q - 1$, for any positive integer $q$.

*Our Results.* In this paper, we consider problems of identifying defectives using non-adaptive or two-stage protocols with efficient analysis algorithms. We present several such algorithms that require fewer tests than do previous algorithms for practical-sized sets, although we omit the proofs of some supporting lemmas in this paper, due to space constraints. Our general case algorithm, which is based on a method we call the Chinese Remainder Sieve, improves the construction of Hwang and Sós [9] for all values of $d$ for real-world problem instances as well as for $d \geq n^{1/5}$ and $n \geq e^{10}$. Our two-stage algorithm achieves a bound for $t(n,d)$ that is within a factor of $4(1 + o(1))$ of the information-theoretic lower bound. This bound improves the bound achieved by Debonis *et al.* [4] by almost a factor of 2. Likewise, our algorithm for $d = 2$ improves on the number of tests required for all real-world problem sizes and is time-optimal (that is, with $A(n,t) \in O(t)$). Our algorithm for $d = 3$ is the first known time-optimal testing algorithm for that $d$-value. Moreover, our algorithms all have efficient sampling rates.

## 2   The Chinese Remainder Sieve

In this section, we present a solution to the problem for determining which items are defective when we know that there are at most $d < n$ defectives. Using a simple number-theoretic method, which we call the *Chinese Remainder Sieve* method, we describe the construction of a $d$-disjunct matrix with $t = O(d^2 \log^2 n/(\log d + \log \log n))$. As we will show, our bound is superior to that of the method of Hwang and Sós [9], for all realistic instances of the combinatorial group testing problem.

Suppose we are given $n$ items, numbered $0, 1, \ldots, n-1$, such that at most $d <$ $n$ are defective. Let $\{p_1^{e_1}, p_2^{e_2}, \ldots, p_k^{e_k}\}$ be a sequence of powers of distinct primes, multiplying to at least $n^d$. That is, $\prod_j p_j^{e_j} \geq n^d$. We construct a $t \times n$ matrix $M$ as the vertical concatenation of $k$ submatrices, $M_1, M_2, \ldots, M_k$. Each submatrix $M_j$ is a $t_j \times n$ testing matrix, where $t_j = p_j^{e_j}$; hence, $t = \sum_{j=1}^{k} p_j^{e_j}$. We form each row of $M_j$ by associating it with a non-negative value $x$ less than $p_j^{e_j}$. Specifically, for each $x$, $0 \leq x < p_j^{e_j}$, form a test in $M_j$ consisting of the item indices (in the range $0, 1, \ldots, n-1$) that equal $x \pmod{p_j^{e_j}}$. For example, if $x = 2$ and $p_j^{e_j} = 3^2$, then the row for $x$ in $M_j$ has a 1 only in columns 2, 11, 20, and so on.

The following lemma shows that the test matrix $M$ is $d$-disjunct.

**Lemma 1.** *If there are at most $d$ defective items, and all tests in $M$ are positive for $i$, then $i$ is defective.*

*Proof.* If all $k$ tests for $i$ (one for each prime power $p_j^{e_j}$) are positive, then there exists at least one defective item. With each positive test that includes $i$ (that is, it has a 1 in column $i$), let $p_j^{e_j}$ be the modulus used for this test, and associate with $j$ a defective index $i_j$ that was included in that test (choosing $i_j$ arbitrarily in case test $j$ includes multiple defective indices). For any defective index $i'$, let $P_{i'} = \prod_{j \text{ s.t. } i_j = i'} p_j^{e_j}$. That is, $P_{i'}$ is the product of all the prime powers such that $i'$ caused a positive test that included $i$ for that prime power. Since there are $k$ tests that are positive for $i$, each $p_j^{e_j}$ appears in exactly one of these products, $P_{i'}$. So $\prod P_{i'} = \prod p_j^{e_j} \geq n^d$. Moreover, there are at most $d$ products, $P_{i'}$. Therefore, $\max_{i'} P_{i'} \geq (n^d)^{1/d} = n$; hence, there exists at least one defective index $i'$ for which $P_{i'} \geq n$. By construction, $i'$ is congruent to the same values to which $i$ is congruent, modulo each of the prime powers in $P_{i'}$. By the Chinese Remainder Theorem, the solution to these common congruences is unique modulo the least common multiple of these prime powers, which is $P_{i'}$ itself. Therefore, $i$ is equal to $i'$ modulo a number that is at least $n$, so $i = i'$; hence, $i$ is defective.

The important role of the Chinese Remainder Theorem in the proof of the above lemma gives rise to our name for this construction—the Chinese Remainder Sieve.

*Analysis.* As mentioned above, the total number of tests, $t(n, d)$, constructed in the Chinese Remainder Sieve is $\sum_{j=1}^{k} p_j^{e_j}$, where $\prod p_j^{e_j} \geq n^d$. If we let each $e_j = 1$, we can simplify our analysis to note that $t(n, d) = \sum_{j=1}^{k} p_j$, where $p_j$ denotes the $j$-th prime number and $k$ is chosen so that $\prod_{j=1}^{k} p_j \geq n^d$. To produce a closed-form upper bound for $t(n, d)$, we make use of the prime counting function, $\pi(x)$, which is the number of primes less than or equal to $x$. We also use the well-known *Chebyshev function*, $\theta(x) = \sum_{j=1}^{\pi(x)} \ln p_j$. In addition, we make use of the following (less well-known) prime summation function, $\sigma(x) = \sum_{j=1}^{\pi(x)} p_j$. Using these functions, we bound the number of tests in the Chinese Remainder Sieve method as $t(n, d) \leq \sigma(x)$, where $x$ is chosen so that $\theta(x) \geq d \ln n$, since $\ln \prod_{p_j \leq x} p_j = \theta(x)$. For the Chebyshev function, it can be shown [1] that $\theta(x) \geq x/2$ for $x > 4$ and that $\theta(x) \sim x$ for large $x$. So if we let $x = \lceil 2d \ln n \rceil$, then $\theta(x) \geq d \ln n$. Thus, we can bound the number of tests in our method as $t(n, d) \leq \sigma(\lceil 2d \ln n \rceil)$. To further bound $t(n, d)$, we use the following lemma, which may be of mild independent interest.

**Lemma 2.** *For integer $x \geq 2$,*

$$\sigma(x) < \frac{x^2}{2 \ln x} \left( 1 + \frac{1.2762}{\ln x} \right).$$

*Proof.* Let $n = \pi(x)$. Dusart [6, 7] shows that, for $n \geq 799$, $(1/n) \sum_{j=1}^{n} p_j < p_n/2$; that is, the average of the first $n$ primes is half the value of the $n$th prime. Thus,

$$\sigma(x) = \sum_{j=1}^{\pi(x)} p_j < \frac{\pi(x)}{2} p_n \leq \frac{\pi(x)}{2} x,$$

for integer $x \geq 6131$ (the 799th prime). Dusart [6, 7] also shows that

$$\pi(x) < \frac{x}{\ln x} \left( 1 + \frac{1.2762}{\ln x} \right),$$

for $x \geq 2$. Therefore, for integer $x \geq 6131$,

$$\sigma(x) < \frac{x^2}{\ln x} \left( 1 + \frac{1.2762}{\ln x} \right).$$

In addition, we have verified by an exhaustive computer search that this inequality also holds for all integers $2 \leq x < 6131$. This completes the proof.

Thus, we can characterize the Chinese Remainder Sieve method as follows.

**Theorem 1.** *Given a set of $n$ items, at most $d$ of which are defective, the Chinese Remainder Sieve method can identify the defective items using a number of tests*

$$t(n, d) < \frac{\lceil 2d \ln n \rceil^2}{2 \ln \lceil 2d \ln n \rceil} \left( 1 + \frac{1.2762}{\ln \lceil 2d \ln n \rceil} \right).$$

By calculating the exact numbers of tests required by the Chinese Remainder Sieve method for particular parameter values and comparing these numbers to the claimed bounds for Hwang and Sós [9], we see that our algorithm is an improvement when:

- $d = 2$ and $n \leq 10^{57}$ • $d = 3$ and $n \leq 10^{66}$
- $d = 4$ and $n \leq 10^{70}$ • $d = 5$ and $n \leq 10^{74}$
- $d = 6$ and $n \leq 10^{77}$ • $d \geq 7$ and $n \leq 10^{80}$.

Of course, these are the most likely cases for any expected actual instance of the combinatorial group testing problem. In addition, our analysis shows that our method is superior to the claimed bounds of Hwang and Sós [9] for $d \geq n^{1/5}$ and $n \geq e^{10}$. Less precisely, we can say that $t(n, d)$ is $O(d^2 \log^2 n/(\log d + \log \log n))$, that $S(n, d)$ is $O(d \log n/(\log d + \log \log n))$, and $A(n, t)$ is $O(tn)$, which is $O(d^2 n \log^2 n/(\log d + \log \log n))$.

## 3   A Two-Stage Rake-and-Winnow Protocol

In this section, we present a randomized construction for two-stage group testing. This two-stage method uses a number of tests within a constant factor of the information-theoretic lower bound. It improves previous upper bounds [4] by almost a factor of 2. In addition, it has an efficient sampling rate, with $S(n, d)$ being only $O(\log(n/d))$. All the constant factors "hiding" behind the big-ohs in these bounds are small.

*Preliminaries.* One of the important tools we use in our analysis is the following lemma for bounding the tail of a certain distribution. It is a form of Chernoff bound [12].

**Lemma 3.** *Let $X$ be the sum of $n$ independent indicator random variables, such that $X = \sum_{i=1}^{n} X_i$, where each $X_i = 1$ with probability $p_i$, for $i = 1, 2, \ldots, n$. If $E[X] = \sum_{i=1}^{n} p_i \leq \hat{\mu} < 1$, then, for any integer $k > 0$, $\Pr(X \geq k) \leq (e\hat{\mu}/k)^k$.*

*Proof.* Let $\mu = E[X]$ be the actual expected value of $X$. Then, by a well-known Chernoff bound [12], for any $\delta > 0$,

$$\Pr[X \geq (1+\delta)\mu] \leq \left[ \frac{e^{\delta}}{(1+\delta)^{1+\delta}} \right]^{\mu}.$$

(The bound in [12] is for strict inequality, but the same bound holds for nonstrict inequality.) We are interested in the case when $(1+\delta)\mu = k$, that is, when $1 + \delta = k/\mu$. Observing that $\delta < 1 + \delta$, we can therefore deduce that

$$\Pr(X \geq k) \leq \left[ \frac{e^{k/\mu}}{(k/\mu)^{k/\mu}} \right]^{\mu} = \frac{e^k}{(k/\mu)^k} = \left( \frac{e\mu}{k} \right)^k.$$

Finally, noting that $\mu \leq \hat{\mu}$, $\Pr(X \geq k) \leq (e\hat{\mu}/k)^k$.

**Lemma 4.** *If $d < n$, then $\binom{n}{d} < (en/d)^d$.*

*Identifying Defective Items in Two Stages.* As with our Chinese Remainder Sieve method, our randomized combinatorial group testing construction is based on the use of a Boolean matrix $M$ where columns correspond to items and rows correspond to tests, so that if $M[i, j] = 1$, then item $j$ is included in test $j$. Let $C$ denote the set of columns of $M$. Given a set $D$ of $d$ columns in $M$, and a specific column $j \in C - D$, we say that $j$ is *distinguishable* from $D$ if there is a row $i$ of $M$ such that $M[i, j] = 1$ but $i$ contains a 0 in each of the columns in $D$. Such a property is useful in the context of group testing, for the set $D$ could correspond to the defective items and if a column $j$ is distinguishable from the set $D$, then there would be a test in our regimen that would determine that the item corresponding to column $j$ is not defective.

An alternate and equivalent definition [5](p. 165) for a matrix $M$ to be *d-disjunct* is if, for any $d$-sized subset $D$ of $C$, each column in $C - D$ is distinguishable from $D$. Such a matrix determines a powerful group testing regimen, but, unfortunately, building such a matrix requires $M$ to have $\Omega(d^2 \log n / \log d)$ rows, by a result of Ruszinkó [13](see also [5], p. 139). The best known constructions have $\Theta(d^2 \log(n/d))$

rows [5], which is a factor of $d$ greater than information-theoretic lower bound, which is $\Omega(d \log(n/d))$.

Instead of trying to use a matrix $M$ to determine all the defectives immediately, we will settle for a weaker property for $M$, which nevertheless is still powerful enough to define a good group testing regimen. We say that $M$ is $(d, k)$-*resolvable* if, for any $d$-sized subset $D$ of $C$, there are fewer than $k$ columns in $C - D$ that are not distinguishable from $D$. Such a matrix defines a powerful group testing regimen, for defining tests according to the rows of a $d$-resolvable matrix allows us to restrict the set of defective items to a group $D'$ of smaller than $d + k$ size. Given this set, we can then perform an additional round of individual tests on all the items in $D'$. This two-stage approach is sometimes called the trivial two-stage algorithm; we refer to this two-stage algorithm as the *rake-and-winnow* approach.

Thus, a $(d, k)$-resolvable matrix determines a powerful group testing regimen. Of course, a matrix is $d$-disjunct if and only if it is $(d, 1)$-resolvable. Unfortunately, as mentioned above, constructing a $(d, 1)$-resolvable matrix requires that the number of rows (which correspond to tests) be significantly greater than the information theoretical lower bound. Nevertheless, if we are willing to use a $(d, k)$-resolvable matrix, for a reasonably small value of $k$, we can come within a constant factor of the information theoretical lower bound.

Our construction of a $(d, k)$-resolvable matrix is based on a simple, randomized *sample-injection* strategy, which itself is based on the approach popularized by the Bloom filter [2]. This novel approach also allows us to provide a strong worst-case bound for the sample rate, $S(n, d)$, of our method. Given a parameter $t$, which is a multiple of $d$ that will be set in the analysis, we construct a $2t \times n$ matrix $M$ in a column-wise fashion. For each column $j$ of $M$, we choose $t/d$ rows at random and we set the values of these entries to 1. The other entries in column $j$ are set to 0. In other words, we "inject" the sample $j$ into each of the $t/d$ random tests we pick for the corresponding column (since rows of $M$ correspond to tests and the columns correspond to samples). Note, then, that for any set of $d$ defective samples, there are at most $t$ tests that will have positive outcomes and, therefore, at least $t$ tests that will have negative outcomes. The columns that correspond to samples that are distinguishable from the defectives ones can be immediately identified. The remaining issue, then, is to determine the value of $t$ needed so that, for a given value of $k$, $M$ is a $(d, k)$-resolvable matrix with high probability.

Let $D$ be a fixed set of $d$ defectives samples. For each (column) item $i$ in $C - D$, let $X_i$ denote the indicator random variable that is 1 if $i$ is falsely identified as a positive sample by $M$ (that is, $i$ is not included in the set of (negative) items distinguished from those in $D$), and is 0 otherwise. Observe that the $X_i$'s are independent, since $X_i$ depends only on whether the choice of rows we picked for column $i$ collide with the at most $t$ rows of $M$ that we picked for the columns corresponding to items in $D$. Furthermore, this observation implies that any $X_i$ is 1 (a false positive) with probability at most $2^{-t/d}$. Therefore, the expected value of $X$, $E[X]$, is at most $\hat{\mu} = n/2^{t/d}$. This fact allows us to apply Lemma 3 to bound the probability that $M$ does not satisfy the $(d, k)$-resolvable property for this particular choice, $D$, of $d$ defective samples. In particular,

$$\Pr(X \geq k) \leq \left(\frac{e\hat{\mu}}{k}\right)^k = \frac{\left(\frac{en}{k}\right)^k}{2^{(t/d)k}}.$$

Note that this bound immediately implies that if $k = 1$ and $t \geq d(e+1) \log n$, then $M$ will be completely $(d, 1)$-resolvable with high probability $(1 - 1/n)$ for any particular set of defective items, $D$.

We are interested, however, in a bound implying that for *any* subset $D$ of $d$ defectives (of which there are $\binom{n}{d} < (en/d)^d$, by Lemma 4), our matrix $M$ is $(d, k)$-resolvable with high probability, that is, probability at least $1 - 1/n$. That is, we are interested in the value of $t$ such that the above probability bound is $(en/d)^{-d}/n$. From the above probability bound, therefore, we are interested in a value of $t$ such that

$$2^{(t/d)k} \geq \left(\frac{en}{d}\right)^d \left(\frac{en}{k}\right)^k n.$$

This bound will hold whenever $t \geq (d^2/k) \log(en/d) + d \log(en/k) + (d/k) \log n$. Thus, we have the following.

**Theorem 2.** *If $t \geq (d^2/k) \log(en/d) + d \log(en/k) + (d/k) \log n$, then a $2t \times n$ random matrix $M$ constructed by sample-injection is $(d, k)$-resolvable with high probability, that is, with probability at least $1 - 1/n$.*

As mentioned above, a productive way of using the sample-injection construction is to build a $(d, k)$-resolvable matrix $M$ for a reasonably small value of $k$. We can then use this matrix as the first round in a two-round rake-and-winnow testing strategy, where the second round simply involves our individual testing of the at most $d + k$ samples left as potential positive samples from the first round.

**Corollary 1.** *If $t \geq 2d \log(en/d) + \log n$, then the $2t \times n$ random matrix $M$ constructed by sample-injection is $(d, d)$-resolvable with high probability.*

This corollary implies that we can construct a rake-and-winnow algorithm where the first stage involves performing $O(d \log(n/d))$ tests, which is within a (small) constant factor of the information theoretic lower bound, and the second round involves individually testing at most $2d$ samples.

## 4    Improved Bounds for Small $d$ Values

In this section, we consider efficient algorithms for the special cases when $d = 2$ and $d = 3$. We present time-optimal algorithms for these cases; that is, with $A(n, t)$ being $O(t)$. Our algorithm for $d = 3$ is the first known such algorithm.

*Finding up to Two Defectives.* Consider the problem of determining which items are defective when we know that there are at most two defectives. We describe a $\overline{2}$-separable matrix and a time-optimal analysis algorithm with $t = (q^2 + 5q)/2$ and $n = 3^q$, for any positive integer $q$.

Let the number of items be $n = 3^q$, and let the item indices be expressed in radix 3. Index $X = X_{q-1} \cdots X_0$, where each digit $X_p \in \{0, 1, 2\}$.

Hereafter, $X$ ranges over the item index numbers $\{0, \ldots n - 1\}$, $p$ ranges over the radix positions $\{0, \ldots q - 1\}$, and $v$ ranges over the digit values $\{0, 1, 2\}$.

For our construction, matrix $M$ is partitioned into submatrices $B$ and $C$. Matrix $B$ is the submatrix of $M$ consisting of its first $3q$ rows. Row $\langle p, v \rangle$ of $B$ is associated with radix position $p$ and value $v$. $B[\langle p, v \rangle, X] = 1$ iff $X_p = v$.

Matrix $C$ is the submatrix of $M$ consisting of its last $\binom{q}{2}$ rows. Row $\langle p, p' \rangle$ of $C$ is associated with distinct radix positions $p$ and $p'$, where $p < p'$. $C[\langle p, p' \rangle, X] = 1$ iff $X_p = X_{p'}$.

Let $test_B(p, v)$ be the result (1 for positive, 0 for negative) of the test of items having a 1-entry in row $\langle p, v \rangle$ in $B$. Similarly, let $test_C(p, p')$ be the result of testing row $\langle p, p' \rangle$ in $C$. Let $test1(p)$ be the number of different values held by defectives in radix position $p$. $test1(p)$ can be computed by $test_B(p, 0) + test_B(p, 1) + test_B(p, 2)$.

The analysis algorithm is shown in the Appendix in Figure 1.

It is easy to determine how many defective items are present. There are no defective items when $test1(0) = 0$. There is only one defective item when $test1(p) = 1$ for all $p$, since if there were two defective items then there must be at least one position $p$ in which their indices differ and $test1(p)$ would then have value 2. The one defective item has index $D = D_{q-1} \cdots D_0$, where digit $D_p$ is the value $v$ for which $test_B(p, v) = 1$.

Otherwise, there must be 2 defective items, $D = D_{q-1} \cdots D_0$ and $E = E_{q-1} \cdots E_0$. We iteratively determine the values of the digits of indices $D$ and $E$.

For radix positions in which defective items exist for only one value of that digit, both $D$ and $E$ must have that value for that digit. For each other radix position, two distinct values for that digit occur in the defective items.

The first radix position in which $D$ and $E$ differ is recorded in the variable $p^*$ and the value of that digit in $D$ (respectively, $E$) is recorded in $v_1^*$ (respectively, $v_2^*$).

For any subsequent position $p$ in which $D$ and $E$ differ, the digit values of the defectives in that position are $v_a$ and $v_b$, which are two distinct values from $\{0, 1, 2\}$, as are $v_1^*$ and $v_2^*$, and therefore there must be at least one value in common between $\{v_a, v_b\}$ and $\{v_1^*, v_2^*\}$.

Let a common value be $v_a$ and, without loss of generality, let $v_a = v_1^*$.

**Lemma 5.** *The digit assignment for $p$ is $D_p = v_a$ and $E_p = v_b$ iff $test_C(p^*, p) = 1$.*

We have determined the values of defectives D and E for all positions – those where they are the same and those where they differ. For each position, only a constant amount of work is required to determine the assignment of digit values. Therefore, we have proven the following theorem.

**Theorem 3.** *A $\overline{2}$-separable matrix that has a time-optimal analysis algorithm can be constructed with $t = (q^2 + 5q)/2$ and $n = 3^q$, for any positive integer $q$.*

*Comparison of the Number of Tests Required for $d = 2$ Method.* For all $n \leq 3^{63}$, our $d = 2$ algorithm uses the smallest number of tests. For higher values of $n \leq 3^{130}$, the Kautz/Singleton and our $d = 2$ and general (Chinese Remainder Sieve) algorithms alternate being dominant. For all $n \geq 3^{131}$, the Hwang/Sós algorithm uses the fewest tests.

*Finding up to Three Defectives.* Consider the problem of determining which items are defective when we know that there are at most three defectives. We describe a $\overline{3}$-separable matrix and a time-optimal analysis algorithm with $t = 2q^2 - 2q$ and $n = 2^q$, for any positive integer $q$.

Let the number of items be $n = 2^q$, and let the item indices be expressed in radix 2. Index $X = X_{q-1} \cdots X_0$, where each digit $X_p \in \{0,1\}$.

Hereafter, $X$ ranges over the item index numbers $\{0, \ldots n - 1\}$, $p$ ranges over the radix positions $\{0, \ldots q - 1\}$, and $v$ ranges over the digit values $\{0,1\}$.

Matrix $M$ has $2q^2 - 2q$ rows. Row $\langle p, p', v, v' \rangle$ of $M$ is associated with distinct radix positions $p$ and $p'$, where $p < p'$, and with values $v$ and $v'$, each of which is in $\{0,1\}$. $M[\langle p, p', v, v' \rangle, X] = 1$ iff $X_p = v$ and $X_{p'} = v'$.

Let $test_M(p, p', v, v')$ be the result (1 for positive, 0 for negative) of testing items having a 1-entry in row $\langle p, p', v, v' \rangle$ in $M$. For $p' > p$, define $test_M(p', p, v', v) = test_M(p, p', v, v')$.

The following three functions can be computed in terms of $test_M$.

- $test_B(p, v)$ has value 1 (0) if there are (not) any defectives having value $v$ in radix position $p$. Hence, $test_B(0, v) = 0$ if $test_M(0, 1, v, 0) + test_M(0, 1, v, 1) = 0$, and 1 otherwise. For $p > 0$, $test_B(p, v) = 0$ if $test_M(p, 0, v, 0) + test_M(p, 0, v, 1) = 0$, and 1 otherwise.
- $test1(p)$ is the number of different binary values held by defectives in radix position $p$. Thus, $test1(p) = test_B(p, 0) + test_B(p, 1)$.
- $test2(p, p')$ is the number of different ordered pairs of binary values held by defectives in the designated ordered pair of radix positions. Therefore, $test2(p, p') = test_M(p, p', 0, 0) + test_M(p, p', 0, 1) + test_M(p, p', 1, 0) + test_M(p, p', 1, 1)$.

The analysis algorithm is shown in the Appendix in Figure 1.

We determine the number of defective items and the value of their digits. There are no defective items when $test1(0) = 0$. At each radix position $p$ in which $test1(p) = 1$, all defective items have the same value of that digit. If all defectives agree on all digit values, then there is only one defective. Otherwise there are at least two defectives, and we need to consider how to assign digit values for only the set of positions $P$ in which there is at least one defective having each of the two possible binary digit values.

**Lemma 6.** *There are only two defectives if and only if, for $p, p' \in P$, $test2(p, p') = 2$.*

Accordingly, if there is no pair of positions for which $test2$ has value 3, we can conclude that there are only two defectives. Otherwise, there are positions $p_1, p_2$ for which $test2(p_1, p_2) = 3$, and one of the four combinations of two binary values will not appear. Let that missing combination be $v_1, v_2$. Thus, while position $p_1$ uniquely identifies one defective, say $D$, as the only defective having value $v_1$ at that position, position $p_2$ uniquely identifies one of the other defectives, say $E$, as having value $v_2$.

**Lemma 7.** *If the position $p^*$ uniquely identifies the defective $X$ to have value $v^*$, then the value of the defective $X$ at any other position $p$ will be that value $v$ such that $test_M(p^*, p, v^*, v) = 1$.*

**if** $test1(0) = 0$ **then**
    **return** there are no defective items
$p^* \leftarrow -1$
**for** $p \leftarrow 0$ **to** $q - 1$ **do**
    **if** $test1(p) = 1$ **then**
        Let $D_p$ and $E_p$ be the (same)
        value $v$ such that $test_B(p, v) = 1$
    **else** // $test1(p)$ has value 2
        Let $v_1, v_2$ be the two values
        of $v$ such that $test_B(p, v) = 1$
        **if** $p^* < 0$ **then**
            $p^* \leftarrow p$
            $v_1^* \leftarrow D_p \leftarrow v_1$
            $v_2^* \leftarrow E_p \leftarrow v_2$
        **else**
            **if** $test_C(p^*, p) = 1$
            **and** ( $v_1^* = v_1$ **or** $v_2^* = v_2$ ) **then**
                $D_p \leftarrow v_1$
                $E_p \leftarrow v_2$
            **else**
                $D_p \leftarrow v_2$
                $E_p \leftarrow v_1$
**if** $p^* < 0$ **then**
    **return** one defective, $D$
**else**
    **return** two defectives, $D$ and $E$

(a)

**if** $test1(0) = 0$ **then**
    **return** there are no defective items
$P \leftarrow \emptyset$
**for** $p \leftarrow 0$ **to** $q - 1$ **do**
    **if** $test1(p) = 1$ **then**
        Let $D_p$, $E_p$, and $F_p$ be the (same)
        value $v$ such that $test_B(p, v) = 1$
    **else** $P \leftarrow P \cup \{p\}$
**if** $P = \emptyset$ **then return** there is one defective item $D$
**if** $test2(p_1, p_2) = 2$ for all $p_1, p_2 \in P$ **then**
    $p^* \leftarrow -1$
    **for** $p \in P$ **do**
        **if** $p^* < 0$ **then**
            $p^* \leftarrow p$
            $v^* \leftarrow D_p \leftarrow 0$
        **else if** $test_M(p^*, p, v^*, 0) = 1$ **then**
            $D_p \leftarrow 0$
         **else** $D_p \leftarrow 1$
        $E_p \leftarrow 1 - D_p$
    **return** there are two defective items $D, E$
**else**
    Let $p_1, p_2$ be positions s.t. $test2(p_1, p_2) = 3$
    Let $v_1, v_2$ be values s.t. $test_M(p_1, p_2, v_1, v_2) = 0$
    $D_{p_1} \leftarrow v_1$
    $F_{p_1} \leftarrow E_{p_1} \leftarrow 1 - v_1$
    $E_{p_2} \leftarrow v_2$
    $F_{p_2} \leftarrow D_{p_2} \leftarrow 1 - v_2$
    **for** $p \in P - \{p_1, p_2\}$ **do**
        **if** $test_M(p_1, p, v_1, 0) = 1$ **then**
            $D_p \leftarrow 0$
         **else** $D_p \leftarrow 1$
        **if** $test_M(p_2, p, v_2, 0) = 1$ **then**
            $E_p \leftarrow 0$
         **else** $E_p \leftarrow 1$
        $v \leftarrow E_p$
        **if** $test_M(p_1, p, 1 - v_1, 1 - v) = 1$ **then**
            $F_p \leftarrow 1 - v$
         **else** $F_p \leftarrow v$
    **return** there are three defective items $D, E$, and $F$

(b)

**Fig. 1.** Analysis algorithms. (a) for up to 2 defectives; (b) for up to 3 defectives

Since we have positions that uniquely identify $D$ and $E$, we can determine the values of all their other digits and the only remaining problem is to determine the values of the digits of defective $F$.

Since position $p_1$ uniquely identifies $D$, we know that $F_{p_1} = \overline{v}_1$. For any other position $p$, after determining that $E_p = v$, we note that if $test_M(p_1, p, \overline{v}_1, \overline{v}) = 1$ then there must be at least one defective, $X$, for which $X_{p_1} = \overline{v}_1$ and $X_p = \overline{v}$. Defective $D$ is ruled out since $D_{p_1} = v_1$, and defective $E$ is ruled out since $E_p = v$. Therefore, it must be that $F_p = \overline{v}$. Otherwise, if that $test_M = 0$ then $F_p = v$, since $F_p = \overline{v}$ would have caused $test_M = 1$.

We have determined the values of defectives D, E and F for all positions. For each position, only a constant amount of work is required to determine the assignment of digit values. Therefore, we have proven the following theorem.

**Theorem 4.** *A $\overline{3}$-separable matrix that has a time-optimal analysis algorithm can be constructed with $t = 2q^2 - 2q$ and $n = 2^q$, for any positive integer $q$.*

*Comparison of the Number of Tests Required for $d = 3$ Method.* The general $d$ algorithm due to Hwang and Sós [9] requires fewer tests than does the algorithm for $d = 3$ suggested by Du and Hwang [5]. For $n < 10^{10}$, our ($d = 3$) algorithm requires even fewer tests and our general (Chinese Remainder Sieve) algorithm fewest. However, asymptotically Hwang/Sós uses the fewest tests. We note that, unlike these other efficient algorithms, our ($d = 3$) algorithm is time-optimal.

***Acknowledgments.*** We would like to thank George Lueker and Dennis Shasha for several helpful discussions related to the topics of this paper. This work was supported in part by NSF Grants CCR-0312760, CCR-0311720, CCR-0225642, and CCR-0098068.

# References

1. E. Bach and J. Shallit. *Algorithmic Number Theory, Vol. 1: Efficient Algorithms.* MIT Press, Cambridge, MA, 1996.
2. B. H. Bloom. Space/time trade-offs in hash coding with allowable errors. *Commun. ACM,* 13:422–426, 1970.
3. Colbourn, Dinitz, and Stinson. Applications of combinatorial designs to communications, cryptography, and networking. In *Surveys in Combinatorics, 1993, Walker (Ed.), London Mathematical Society Lecture Note Series 187.* Cambridge University Press, 1999.
4. A. DeBonis, L. Gasieniec, and U. Vaccaro. Generalized framework for selectors with applications in optimal group testing. In *Proceedings of 30th International Colloquium on Automata, Languages and Programming (ICALP'03),* pages 81–96. Springer, 2003.
5. D.-Z. Du and F. K. Hwang. *Combinatorial Group Testing and Its Applications, 2nd ed.* World Scientific, 2000.
6. P. Dusart. Encadrements effectifs des functions de Chebyshev: (sharper bounds for $\phi$, $\theta$, $\pi$, $p_k$). Report, Laboratoire d'Arithmétique, de Calcul formel et d'Optimisation, 1998. Rapport no. 1998-06, http://www.unilim.fr/laco/rapports/1998/R1998_06.pdf.
7. P. Dusart. The $k$th prime is greater than $k(\ln k + \ln \ln k - 1)$ for $k \geq 2$. *Math. Comp.,* 68(225):411–415, 1999.
8. M. Farach, S. Kannan, E. Knill, and S. Muthukrishnan. Group testing problems with sequences in experimental molecular biology. In *SEQUENCES,* page 357. IEEE Press, 1997.
9. F. K. Hwang and V. T. Sós. Non-adaptive hypergeometric group testing. *Studia Scient. Math. Hungarica,* 22:257–263, 1987.

10. W. H. Kautz and R. C. Singleton. Nonrandom binary superimposed codes. *IEEE Trans. Inf. Th.*, 10:363–377, 1964.

11. A. J. Macula and G. R. Reuter. Simplified searching for two defects. *J. Stat. Plan. Inf.*, 66:77–82, 1998.

12. R. Motwani and P. Raghavan. *Randomized Algorithms*. Cambridge University Press, New York, NY, 1995.

13. M. Ruszinkó. On the upper bound of the size of the $r$-cover-free families. *J. Combin. Th. Ser. A*, 66:302–310, 1994.

# Parameterized Counting Algorithms for General Graph Covering Problems*

Naomi Nishimura[1], Prabhakar Ragde[1], and Dimitrios M. Thilikos[2]

[1] School of Computer Science, University of Waterloo,
Waterloo, Ontario, Canada, N2L 3G1
[2] Departament de Llenguatges i Sistemes Informàtics,
Universitat Politècnica de Catalunya, Campus Nord,
Desp. $\Omega$-228, c/Jordi Girona Salgado,
1-3. E-08034, Barcelona, Spain

**Abstract.** We examine the general problem of covering graphs by graphs: given a graph $G$, a collection $\mathcal{P}$ of graphs each on at most $p$ vertices, and an integer $r$, is there a collection $\mathcal{C}$ of subgraphs of $G$, each belonging to $\mathcal{P}$, such that the removal of the graphs in $\mathcal{C}$ from $G$ creates a graph none of whose components have more than $r$ vertices? We can also require that the graphs in $\mathcal{C}$ be disjoint (forming a "matching"). This framework generalizes vertex cover, edge dominating set, and minimal maximum matching. In this paper, we examine the parameterized complexity of the counting version of the above general problem. In particular, we show how to count the solutions of size at most $k$ of the covering and matching problems in time $O(n \cdot r(pk+r) + 2^{f(k,p,r)})$, where $n$ is the number of vertices in $G$ and $f$ is a simple polynomial. In order to achieve the additive relation between the polynomial and the non-polynomial parts of the time complexity of our algorithms, we use the compactor technique, the counting analogue of kernelization for parameterized decision problems.

## 1 Introduction

Parameterized algorithms offer an approach to solving NP-hard problems through the observation that many such problems come with one or more natural parameters which may be small in practice, and so algorithms that are polynomial in the input size but exponential in the parameters may be of practical use. The considerable literature on parameterized complexity provides both algorithms for certain problems (e.g. vertex cover, where the parameter $k$ is the size of the cover) and evidence (in the form of completeness results) that other problems (e.g. clique) do not have efficient parameterized algorithms.

One common technique in designing parameterized algorithms is to find a problem kernel. This consists of reducing an instance of a problem to a smaller

* The two first authors were supported by the Natural Sciences and Engineering Research Council of Canada (NSERC). The third author was supported by the Spanish CICYT project TIN-2004-07925 (GRAMMARS).

instance of size dependent only on the parameters, such that the smaller instance has a solution if and only if the original instance does. Inefficient algorithms (e.g. brute force search) can then be used on the kernel. While this approach is appealing, it may be difficult to find a kernel for a given problem.

Our focus in this paper is on counting the number of solutions constrained by the parameters (e.g. the number of vertex covers of size at most $k$), as first considered by Flum and Grohe [FG04]. Fernau [Fer02] defines fixed-parameter enumerability, and considers the two problems of enumerating all solutions (producing each one, as opposed to counting the total), and of enumerating all optimal solutions. But for many problems (e.g. vertex cover) enumerating all solutions is not fixed-parameter enumerable, as there are too many solutions. This naturally suggests our approach of counting the solutions without enumerating them.

We consider a different sort of kernel-like structure from that used in traditional parameterized algorithms, one specialized for counting. Such a kernel comes with a function mapping a solution in the original instance to one in the kernel, in a fashion that allows us to compute the size of each preimage. That way, we reduce the problem of counting the solutions of the original problem to the problem of enumerating the solutions in the kernel. We count solutions in the original instance by using a (possibly inefficient) algorithm to enumerate solutions in the kernel and summing the sizes of preimages. This method was used by Díaz, Serna, and Thilikos [DST04a, DST04b] in the context of colouring problems; here we apply it to covering and matching problems. This method is a departure from previous work on parameterized counting [Fer02, Dam04, AR02, DST04b]. Our goal is to obtain running times with an additive relation between the part that is polynomial in the input size and the part that is possibly exponential in the parameters.

Our problems are defined by a graph $G$, a collection $\mathcal{P}$ of graphs each on at most $p$ vertices, an integer $r$, and a parameter $k$. We wish to "cover" $G$ by $k$ graphs chosen from $\mathcal{P}$, such that the connected subgraphs left uncovered have no more than $r$ vertices each. That is, we ask whether or not there is a collection $\mathcal{C}$ of subgraphs of $G$ (with $|\mathcal{C}| \leq k$), each belonging to $\mathcal{P}$, such that the removal of the graphs in $\mathcal{C}$ from $G$ creates a graph none of whose components have more than $r$ vertices. In this formulation, we allow the graphs in $\mathcal{C}$ to overlap, forming a "covering". Another variation of the problem requires that the graphs in $\mathcal{C}$ be disjoint, forming a "matching".

This framework generalizes vertex cover, edge dominating set, and minimal maximum matching. For vertex cover, $\mathcal{P}$ contains only the graph with one vertex, and $r = 1$; for edge dominating set, $\mathcal{P}$ contains only the graph with two connected vertices, and $r = 1$; for minimal maximum matching, we add the constraint that the graphs in $\mathcal{C}$ must be disjoint. Interestingly, minimum maximal maximal matching and edge dominating set are polynomially equivalent as decision problems, but not as counting problems. In this paper, we show how to count the number of solutions of the general covering and matching problems problems in time $O(n \cdot r(pk + r) + 2^{O(pkr(pk+r))})$ where $n$ is the number of vertices in $G$.

## 2    Basic Definitions

All graphs in this paper are undirected, loopless and without multiple edges. For each graph $G$ considered, we will denote as $V(G)$ and $E(G)$ its vertex and edge set, respectively. Given a set $S \subseteq V(G)$ we define $N_G(S)$ as the set of all vertices $v \in V(G) - S$ such that $v$ has a neighbour in $S$. For a set of graphs $\mathcal{C}$, we denote as $\mathbf{V}(\mathcal{C})$ the set of all vertices of the graphs in $\mathcal{C}$, i.e. $\mathbf{V}(\mathcal{C}) = \bigcup_{G \in \mathcal{C}} V(G)$.

For $p$ a fixed constant and $\mathcal{P}$ a fixed set of graphs of no more than $p$ vertices, we define the following parameterized problems:

| $(k,r)$-MINIMUM COVERING BY GRAPHS IN $\mathcal{P}$ $((k,r)$-MCG-$(\mathcal{P}))$ |
|---|
| *Input:* A graph $G$, a collection of graphs $\mathcal{P}$, and an integer $r$. |
| *Parameter:* A non-negative integer $k$. |
| *Question:* Does $G$ contain a collection $\mathcal{C}$ of $k$ subgraphs each isomorphic to some graph in $\mathcal{P}$ and such that $G[V(G) - \mathbf{V}(\mathcal{C}))]$ has no component of size more than $r$? |

If in the above problem we demand that the graphs in $\mathcal{C}$ be pairwise disjoint (i.e. no vertices in common) then we define the $(k,r)$-MINIMUM MAXIMAL MATCHING BY GRAPHS IN $\mathcal{P}$ $((k,r)$-MMM-$(\mathcal{P}))$.

We denote by $\mathbf{MCG}_k(G)$ the set of solutions of the $(k,r)$-MCG-$(\mathcal{P})$ problem when the input is $G$ and the parameter is $k$. Similarly, we define $\mathbf{MMM}_k(G)$ and notice that $\mathbf{MCG}_k(G) \subseteq \mathbf{MMM}_k(G)$. Also we define $\mathrm{mcg}_k(G) = |\mathbf{MCG}_k(G)|$ and $\mathrm{mmm}_k(G) = |\mathbf{MMM}_k(G)|$.

In what follows we will give a parameterized counting algorithm for each of the above problems. In particular, we will give two algorithms that output $\mathrm{mcg}_k(G)$ and $\mathrm{mmm}_k(G)$, respectively, in $O(n \cdot r(pk+r) + 2^{O(pkr(pk+r))})$.

A basic tool for our algorithms is the notion of $(a,b)$-*central set*. In particular, we say that a subset $P$ of $V(G)$ is an $(a,b)$-*central set* (or *central set* for short) if $|P| \leq a$ and each connected component of $G[V(G) - P]$ has at most $b$ vertices. Notice that a vertex cover of size at most $k$ is a $(k,1)$-central set and vice versa. For convenience, we refer to the connected components of $G[V(G) - P]$ as *satellite graphs*. Of particular interest are those satellite graphs that have neighbours in $P$; we will call these *dependent* satellite graphs and all others *independent* satellite graphs.

To form our counting algorithms, we use the notion of *compactor enumeration* as introduced by Díaz, Serna, and Thilikos [DST04a, DST04b]. The idea is to find a particular kind of kernel for a parameterized problem such that any solution of the problem can be mapped to a solution within the kernel. If we can enumerate all the solutions within the kernel, and for each one, compute (in a reasonable amount of time) the number of preimages of general solutions mapping to it, we can count the number of general solutions of the problem. More formally, a compactor $\mathsf{Cmp}(\Pi, k)$ for a parameterized problem $\Pi$ with parameter $k$ and set of solutions $\mathsf{Sol}(\Pi, k)$ has the following properties: $|\mathsf{Cmp}(\Pi, k)|$ is a function that depends only on $k$; $\mathsf{Cmp}(\Pi, k)$ can be enumerated with an algorithm whose complexity depends only on $k$; there is a surjective function $m : \mathsf{Sol}(\Pi, k) \rightarrow$

$\mathsf{Cmp}(\Pi, k)$; and for any $C \in \mathsf{Cmp}(\Pi, k)$, $|m^{-1}(C)|$ can be computed within time $O(f(k)n^c)$.

## 3    Central Sets

The notion of a central set plays a key role in our algorithms, as a necessary condition for a nonzero number of solutions (Lemma 1) and as an important step towards forming a compactor.

**Lemma 1.** *If for a graph $G$ $\mathsf{mcg}_k(G) > 0$ or $\mathsf{mmm}_k(G) > 0$, then $G$ contains a $(pk, r)$-central set.*

*Proof.* We present the proof for the case $\mathsf{mcg}_k(G) > 0$, as the proof for the case $\mathsf{mmm}_k(G) > 0$ is identical. Since there is at least one solution to the problem, we let $\mathcal{C}$ be one such solution. By definition, the collection $\mathcal{C}$ consists of $k$ subgraphs each isomorphic to a graph in $\mathcal{P}$, and hence the total number of vertices in $\mathbf{V}(\mathcal{C})$ is at most $pk$. Moreover, again by the definition of the problem, $G[V(G) - \mathbf{V}(\mathcal{C}))]$ has no component of size more than $r$. This implies that $\mathcal{C}$ is a $(pk, r)$-central set, as claimed.

As central sets are used in our covering and matching algorithms, the complexity of finding an $(a, b)$-central set for a graph $G$ has an impact on the complexity of our counting algorithms. The problem of determining whether $G$ has an $(a, b)$-central set is NP-hard when $a$ and $b$ are both part of the input, as for $b = 1$ the problem is vertex cover. If $a$ is fixed, the brute-force checking of all $O(n^a)$ candidate solutions constitutes a polynomial-time algorithm. For the case in which $b$ is fixed, the problem can be shown to be NP-hard using a reduction from vertex cover in which $G$ is transformed into a graph $G'$ by attaching an $b$-clique to each vertex $v \in V(G)$: then $G$ has a vertex cover of size $a$ if and only if $G'$ has an $(a, b)$-central set. Our parameterized solution follows.

**Lemma 2.** *An $(a, b)$-central set of a graph $G$ can be found, if it exists, in time $O(n(a + b) + (ab(a + b - 1) + a)(b + 1)^a)$, where $n$ is the number of vertices in $G$; otherwise, in the same time bound it can be determined that $G$ has no $(a, b)$-central set.*

*Proof.* We present an algorithm, **FIND-CENTRAL-SET**$(a, b, G)$, that determines whether $G$ has an $(a, b)$-central set and, if so, returns one $(a, b)$-central set. We first observe that if a vertex $v$ of $G$ has degree greater than $a + b - 1$, it must be in the $(a, b)$-central set $C$, as otherwise the placement of any $a$ of its neighbours in $C$ would leave a graph of size at least $b + 1$ in $G[V(G) - C]$ (namely $v$ and its remaining neighbours), violating the definition of an $(a, b)$-central set. Consequently, if there are more than $a$ such high-degree vertices, since the size of the $(a, b)$-central set is at most $a$, we can conclude that we have a NO-instance, as indicated in Step 1 below.

---

**FIND-CENTRAL-SET**$(a, b, G)$

**1**. Let $A$ contain all vertices of $G$ that have degree greater than $a+b-1$. If $|A| > a$ then **return** NO.

**2**. Let $G' = G[V(G) - A]$ and $a' = a - |A|$. Let $G^*$ be the union of the connected components of $G'$ that have more than $b$ vertices. If $|V(G^*)| > a'b(a+b-1) + a'$ then **return** NO.

**3**. If **ST-CENTRAL-SET**$(a', b, G^*)$ returns NO, **return** NO and **return**; otherwise, let $C$ be the $(a', b)$-central set of $G^*$ that is returned.

**4**. **return** YES and $A \cup C$ as an $(a, b)$-central set of $G$.

---

We can then reduce the problem to that of finding an $(a', b)$-central set in a graph $G^*$ of degree at most $a + b - 1$ in Steps 2 and 3, where $a'$ is $a$ minus the number of high-degree vertices found in the previous paragraph. We observe that if $G' = G[V(G) - A]$ is a YES-instance, there can be at most $a'(a + b - 1)$ dependent satellite graphs, since each of the $a'$ vertices in the $(a', b)$-central set have at most $a + b - 1$ neighbours. As each dependent satellite graph has at most $b$ vertices and the central set has at most $a'$, the size of $G^*$ can be at most $a'b(a + b - 1) + a'$. Having obtained a graph of size dependent only on the parameters $a$ and $b$, it is now possible to obtain a solution using the search-tree based routine **ST-CENTRAL-SET**$(a, b, G)$ below. The routine consists of checking if all connected components are of size at most $b$ for the base case $a = 0$, and otherwise choosing $b + 1$ vertices that share a component and trying each as a possible member of the central set.

We first determine the running time of **ST-CENTRAL-SET**$(a', b, G^*)$, observing that the depth of the recursion will be at most $a'$. We can find connected components in time linear in the size of $G^*$, or in time $O(a'b(a + b - 1) + a')$. In Step 3, the routine is called $b + 1$ times, giving a total running time of $O((a'b(a + b - 1) + a')(b + 1)^{a'}) = O((ab(a + b - 1) + a)(b + 1)^{a})$.

---

**ST-CENTRAL-SET**$(a, b, G)$

**1**. If $a = 0$, check whether each connected component of $G$ has at most $b$ vertices; if so, **return** YES, and if not, **return** NO.

**2**. Let $K$ be any set of $b + 1$ vertices inducing a connected subgraph of $G$. If no such $K$ can be found, **return** YES.

**3**. For each $v \in K$, determine **ST-CENTRAL-SET**$(a - 1, b, G')$ where $G' = G[V(G) - \{v\}]$. If any answer is YES, **return** YES and the set of vertices removed in the sequence of calls leading to the YES answer. Otherwise, **return** NO.

---

The running time of **FIND-CENTRAL-SET**$(a, b, G)$ can be determined as follows. Step 1 requires checking at most $a + b$ neighbours of each of the $n$ nodes, in total time $O(n(a + b))$. Determining the connected components of $G'$ and counting the number of vertices in components of size more than $b$ can be completed in linear time, $O(n)$. Thus, using the result above for Step 3, we conclude that the total running time is $O(n(a + b) + (ab(a + b - 1) + a)(b + 1)^{a})$.

## 4   Forming a Compactor

The compactor for $\mathsf{mcg}_k(G)$ is based on the fact that our graph can be viewed as a central set surrounded by satellite graphs. We first group satellite graphs into equivalence classes and then prune the classes to form a reduced graph $G'$. In each counting algorithm, the number of solutions in $G$ is computed by determining the number of solutions in $G$ represented by each solution in $G'$.

To be able to substitute a satellite graph in an equivalence class by another graph in the same class, satellite graphs in the same equivalence class should be isomorphic to each other, have the same neighbourhood in the central set, and have the same attachments to those neighbours. More formally, we first define the graphs formed by the satellites and their neighbours, and then formally define the necessary property. For $H$ a subgraph of $G$, we denote as $\partial_G(H)$ the graph $G[V(H) \cup N_G(V(H))]$. Then, for $G$ a graph and $G_1$ and $G_2$ subgraphs of $G$, we say that $G_1$ and $G_2$ are *friends for* $G$ if the following conditions are satisfied.

1. $N_G(V(G_1)) = N_G(V(G_2))$,
2. There is an isomorphism $\phi$ from $\partial_G(G_1)$ to $\partial_G(G_2)$ where for each $v \in N_G(V(G_1)), \phi(v) = v$.

The counting algorithms proceed by finding a $(pk, r)$-central set, grouping satellites into equivalence classes, pruning the graph $G$ to form a graph $G'$ by reducing the size of each sufficiently large equivalence class, solving the problem on $G'$, and then counting the number of solutions to $G$ represented by the solutions found for $G'$. The pruned graph $G'$ plays the role of the compactor in the formalization in Section 2. Crucial to the algorithm is the formation of $G'$, identifying for each equivalence class $\mathcal{S}$ which graphs are to be retained ($\mathcal{R}_{\mathcal{S}}$) and how many have been omitted ($o_{\mathcal{S}}$).

**Lemma 3.** *Given a graph $G$ and a $(pk, r)$-central set $C$ of $G$, it is possible to determine the following, where $\mathcal{C}$ is the set of connected components of $G[V(G) - C]$, $\mathbf{S}$ is a partition of $\mathcal{C}$ such that any two graphs in the same part are friends, and $\mathbf{S}^*$ is the collection of sets in $\mathbf{S}$ with more than $pk + 1$ graphs:*

$\mathcal{R}_{\mathcal{S}}$ : *a set of $pk + 1$ graphs from $\mathcal{S}$.*
$o_{\mathcal{S}}$ : *the number of graphs that have been omitted from $\mathcal{S}$ to form $\mathcal{R}_{\mathcal{S}}$, namely $|\mathcal{S}| - pk - 1$.*
$G'$ : *the graph formed by removing graphs associated with each $\mathcal{S}$, namely $G' = G[V(G) - \bigcup_{\mathcal{S} \in \mathbf{S}^*} \mathbf{V}(\mathcal{S} - \mathcal{R}_{\mathcal{S}})]$.*

*for each $\mathcal{S} \in \mathbf{S}^*$ in time $O(nr(pk + r) + 2^{r(pk+r)})$, where $|\mathbf{S}^*| \in O(2^{r(pk+r)})$.*

*Proof.* The algorithm **CREATE-KERNEL-SETS**$(p, k, r, C, G)$ partitions satellites into equivalence classes based on an arbitrary ordering on the vertices in the central set $C$ and each satellite in $\mathcal{C}$ (Steps 1 and 2) and a bit vector used to indicate the edges that exist both in the satellite and between the satellite and the central set. In particular, each bit vector has one entry for each potential edge between vertices in the satellite (at most $\binom{r}{2}$ in total) and for each potential edge between a vertex in the satellite and a vertex in the $(pk, r)$-central set (at most $pkr$ in total). Satellites with the same bit vectors will be placed in the same equivalence class.

It is worth noting that the algorithm does not guarantee that friends are in the same part, only that graphs in the same part are friends. This is due to the fact that we are choosing arbitrary orderings of vertices in satellites; if different orderings are chosen, friends will appear in different parts of the partition. We settle for this finer partition in order to realize the goal of having the relation between the polynomial and the exponential parts of the running time be additive.

As we identify the equivalence classes to which satellites belong, we keep track of the number of satellites in each class, retaining the first $pk + 1$ in each class by marking the vertices for inclusion in $G'$ (Step 5). It then follows that $G'$ will consist of all marked vertices and $C$, $\mathcal{R}_S$ will be a set of $pk + 1$ retained satellites, and $o_S$ will be the number of satellites seen minus $pk + 1$ to indicate how many have been omitted.

---

**CREATE-KERNEL-SETS**$(p, k, r, C, G)$
**1.** Arbitrarily label the vertices in $C$ from 1 through $pk$.
**2.** For each component $D \in C$, arbitrarily label the vertices in $D$ from $pk + 1$ through $pk + |V(D)| \leq pk + r$.
**3.** Create $\sigma$ to map the integers 1 through $d = pkr + \binom{r}{2}$ to the pairs $(i, j)$ where $1 \leq i \leq pk + r$ and $pk + 1 \leq j \leq pk + r$.
**4.** Create an array $R$ of size $2^d$ with each entry storing an integer and a pointer.
**5.** For each component $D$ form a bit vector of size $d$, where entry $\ell$ is set to 1 if and only if $\sigma(\ell) = (i, j)$ such that there is an edge between the vertices with labels $i$ and $j$ in $D$ and (if $i \leq pk$) $C$. Using the value of the bit vector as an index to $R$, increment the entry in $R$; if the entry in $R$ is now at most $pk + 1$, add $D$ to the linked list at $R$ and mark all vertices in $D$.
**6.** Form $o_S$ by subtracting $pk + 1$ from each value in $R$ of size greater than $pk + 1$.
**7.** Create $G'$ by marking all vertices in $C$ and forming the subgraph of $G$ induced on the marked vertices.
**8.** Return as $\mathcal{R}_S$ all linked lists of entries of $R$ with values greater than $pk + 1$, all values $o_S$, and $G'$.

---

To see that the running time is as claimed, we observe that the labels in Steps 1 and 2 can be created in time $O(n)$ for $n$ the number of vertices in $G$, and in Step 3 $\sigma$ can be created in time $O(d) = O(r(pk + r))$. As there are at most $n$ components $D$ and each bit vector is of length $d$, bit vector formation and marking of vertices in Step 5 can be executed in $O(nd)$ time. As there are $2^d$ entries in $R$, Step 6 can be executed in time $O(2^d)$. Finally, since Step 7 will require at most $O(n)$ time, the running time of the algorithm is at most $O(nd + 2^d) = O(nr(pk + r) + 2^{r(pk+r)})$. We observe that $|\mathbf{S}^*| \leq 2^d$ and thus is in $O(2^{r(pk+r)})$.

## 5    Counting Coverings and Matchings

The following theorems present algorithms that make use of the compactor defined in the previous section.

**Theorem 1.** *The value* $\mathrm{mcg}_k(G)$ *can be determined in time*

$$O(n \cdot r(pk + r) + r^{pk}(pk + 1)^{pk}2^{pkr(pk+r)}(r(pk+1)2^{r(pk+r)} + 2^{p^2} \cdot (p+3)!)).$$

*Proof.* The algorithm below makes use of the earlier subroutines to find a central set (Step 1) and construct $G'$ by removing all but $pk + 1$ remaining satellites in each part of a partition (Step 3); it then finds the solution in $G'$ (Step 4), and counts the number of solutions in $G$ (Step 5).

---

**COMPUTE-mcg-k**$(p, r, G)$
1. Use **FIND-CENTRAL-SET**$(pk, r, G)$ to check whether $G$ contains a $(pk, r)$-central set. If the answer is NO then **return** 0.
2. Let $C$ be the $(pk, r)$-central set of $G$. Let $\mathcal{C}$ be the set of connected components of $G[V(G) - C]$. Recall that each graph in $\mathcal{C}$ has at most $r$ vertices.
3. Use **CREATE-KERNEL-SETS**$(p, k, r, C, G)$ to obtain remaining graphs $\mathcal{R}_S$, numbers of removed graphs $o_S$, and $G'$.
4. Compute **MCG**$_k(G')$ using brute force.
5. Compute and **return** the following number, for $j_{H,S} = |\{J \in \mathcal{R}_S \mid V(J) \cap V(H) \neq \emptyset\}|$:

$$\sum_{\mathcal{G} \in \mathbf{MCG}_k(G')} \prod_{S \in \mathbf{S}^*} \sum_{\mathcal{G}' \subseteq \mathcal{G}} \prod_{H \in \mathcal{G}'} \binom{j_{H,S} + o_S}{j_{H,S}}.$$

---

The correctness of Step 1 follows from Lemma 1, as any graph $G$ with a nonzero solution will have a $(pk, r)$-central set. In forming $G'$, we need to ensure that in any solution, there is at least one satellite with no vertex in the solution (in essence, representing all the pruned satellites). As any solution will be of at most $k$ graphs of at most $p$ vertices each, the entire solution will consist of at most $pk$ vertices; retaining $pk+1$ satellites will thus satisfy the needed condition.

The correctness of the algorithm depends on the counting in Step 5, summing over each solution $\mathcal{G}$ in the reduced graph the number of solutions in the original graph that are represented by $\mathcal{G}$. In particular, graphs involving remaining satellites (i.e., those in $\mathcal{R}_S$) in a particular equivalence class can be replaced by graphs involving satellites that were omitted from the class to form the reduced graph. To count all such possibilities for a particular solution, we consider one particular pruned equivalence class $S$, and observe that our total result will be formed by taking the product of all such values; this is because the effects of making such alterations are independent for each equivalence class.

For a fixed solution $\mathcal{G}$ and equivalence class $S$, we consider all possible ways of exchanging a subset $\mathcal{G}'$ of the collection of graphs forming the solution for satellites that have been omitted. This entails summing over all subsets $\mathcal{G}'$ of graphs in $\mathcal{G}$, and then for each graph $H$ in the subset $\mathcal{G}'$ counting all the ways of making exchanges. The graph $H$ may use more than one satellite in the set $\mathcal{R}_{\mathcal{S}}$, where a graph $J$ is used by $H$ precisely when it is a remaining graph (and hence in $\mathcal{R}_{\mathcal{S}}$) and includes at least one vertex of $H$ (and hence $V(J) \cap V(H)$). The number $j_{H,\mathcal{S}}$ of graphs used is thus $j_{H,\mathcal{S}} = |\{J \in \mathcal{R}_{\mathcal{S}} \mid V(J) \cap V(H) \neq \emptyset\}|$, and we need to count the number of ways to choose $j_{H,\mathcal{S}}$ such graphs out of the ones used plus the set $\mathcal{O}_{\mathcal{S}}$, in order to count the solutions in the general graph represented by this one.

To see that we are not overcounting or undercounting the number of solutions in the complete graph that are represented by a solution in the reduced graph, we first impose an arbitrary ordering on all graphs in each equivalence class and for each choose an isomorphism as defined in the term *friends*. We now observe that if there is more than one way to map the same part of $H$ to a particular satellite, each of the mappings entails a different solution in the reduced graph, and hence we count these different possibilities by our counting ways of swapping satellites for the other solutions.

To determining the running time of the algorithm, we first observe that due to Lemma 2, the running time of Step 1 is in $O(n(pk + r) + pkr(pk + r - 1) + pk)(r + 1)^{pk})$. Finding connected components in Step 2 will take linear time. The running time of Step 3, $O(nr(pk + r) + 2^{r(pk+r)})$, is a direct consequence of Lemma 3.

A brute-force approach for Step 4 will consist of trying all possible choices of a collection $\mathcal{C}$ of $k$ subgraphs of $G'$ such that each subgraph is isomorphic to a graph in $\mathcal{P}$ and such that the graph obtained by removing $\mathcal{C}$ has no component of size more than $r$. This can be accomplished by first choosing $k$ subsets of vertices of $V(G')$, each of size at most $p$, and then checking that each subset induces a graph that contains a graph $P \in \mathcal{P}$ as a subset, where each vertex in the subset is in $P$. More formally, we choose $k$ subsets $S_1, \ldots, S_k$ of vertices of $V(G')$ of size at most $p$; there are $O(m^p)$ choices for each subset, and $O(m^{pk})$ choices for the set of $k$ subsets, where $m = |V(G')|$ (for the matching problem, we ensure that the subsets are disjoint). Checking if a particular graph $P \in \mathcal{P}$ is a subgraph of the graph induced on a set $S_i$ can be accomplished by trying all possible mappings of vertices to vertices and then checking for all edges, or $O(p^2 p!)$ time in total. Finally, to check that no component is of size greater than $r$, we use a linear number of steps. In total, the number of steps will thus be $O(m^{pk}(m+|\mathcal{P}|p^2 p!)) = O(m^{pk}(m + |\mathcal{P}|(p + 3)!))$. We observe that since $G'$ contains the central set and at most $pk + 1$ satellites in each equivalence class, since there are $2^d = 2^{pkr+\binom{r}{2}}$ equivalence classes and each satellite has at most $r$ vertices, the size of $m$ is at most $pk + r(pk + 1)2^{pkr+\binom{r}{2}} = O(r(pk+1)2^{r(pk+r)})$. Thus Step 4 can be executed in time $O(r^{pk}(pk + 1)^{pk}2^{pkr(pk+r)}(r(pk + 1)2^{r(pk+r)} + |\mathcal{P}|(p + 3)!))$.

Finally, we determine the running time of Step 5. By the argument in the previous paragraph, the number of possible solutions in $\mathbf{MCG}_k(G')$ is in $O(m^{pk}) = O(r^{pk}(pk + 1)^{pk}2^{pkr(pk+r)})$. The size of $\mathbf{S}^*$ is no greater than the number of possible equivalence classes, which was shown in Lemma 3 to be in $O(2^{r(pk+r)})$.

There are $O(2^k)$ subsets of $\mathcal{G}$, as $|\mathcal{G}| \leq k$, and at most $k$ choices of $H$ for the same reason. The cost of computing the binomial coefficient is $pk + 1 = |\mathcal{R}_{\mathcal{S}}|$, assuming constant-time arithmetic operations. The total cost of Step 5 will thus be $O(r^{pk}(pk+1)^{pk}2^{pkr(pk+r)}2^{r(pk+r)}2^k k(pk+1))$.

The dominating steps are Steps 3 and 4; using the fact that $|\mathcal{P}| \leq 2^{\binom{p}{2}} \leq 2^{p^2}$, we obtain the claimed time bound of

$$O(n \cdot r(pk+r) + r^{pk}(pk+1)^{pk}2^{pkr(pk+r)}(r(pk+1)2^{r(pk+r)} + 2^{p^2} \cdot (p+3)!)).$$

We are able to count matchings by making a small modification in the algorithm of the previous theorem.

**Theorem 2.** *The value* $\mathsf{mmm}_k(G)$ *can be determined in time*

$$O(n \cdot r(pk+r) + r^{pk}(pk+1)^{pk}2^{pkr(pk+r)}(r(pk+1)2^{r(pk+r)} + 2^{p^2} \cdot (p+3)!)).$$

*Proof.* The proof of the theorem follows from the proof of Theorem 1 and the fact that we can compute $\mathsf{mmm}_k(G)$ by replacing the last two lines of the algorithm **COMPUTE-mcg-k**$(p, r, G)$ with the following lines.

---

**COMPUTE-mmm**$_k(p, r, G)$
4. Compute **MMM**$_k(G')$ using brute force.
5. Compute and **return** the following number:

$$\sum_{\mathcal{G} \in \mathsf{MCG}_k(G')} \prod_{S \in \mathbf{S}^*} \sum_{\mathcal{G}' \subseteq \mathcal{G}} \binom{o_{\mathcal{S}} + |\{J \in \mathcal{R}_{\mathcal{S}} : V(J) \cap \mathbf{V}(\mathcal{G}') \neq \emptyset\}|}{|\{J \in \mathcal{R}_{\mathcal{S}} : V(J) \cap \mathbf{V}(\mathcal{G}') \neq \emptyset\}|}.$$

---

Here we cannot replace each graph in the solution independently; instead we choose a subset $\mathcal{G}'$ to replace and then consider all ways of choosing the right number of satellites in each class either from the remaining or the omitted satellites. The analysis is very similar to that given in the proof of Theorem 1.

## 6    Conclusions

Our primary concern in developing the algorithms was to maintain an additive, rather than multiplicative, relationship between the polynomial (on $n$) and non-polynomial (on the parameters) parts of the running time. We also stress that our analysis for determining the super-polynomial part of the time complexity is a worst-case analysis, and the algorithm should be expected to run even faster in practice, at least for small values of $k$.

We observe that the algorithm **COMPUTE-mcg-k**$(p, r, G)$ can be used to determine the existence of a $(k, r)$-central set, as the problems are equivalent for $p = 1$. We can thus count the number of $(k, r)$-central sets in a graph in time $O(n \cdot r(k+r) + 2^{O(kr(k+r))})$.

We leave as an open problem the question of whether enumerating all optimal solutions of the covering and matching problems is fixed-parameter enumerable.

# References

[AR02]      V. Arvind and V. Raman. Approximation algorithms for some param-
            eterized counting problems. In *Electronic Colloquium on Computational
            Complexity*, 2002.

[Dam04]     P. Damaschke. Parameterized enumeration, transversals, and imperfect
            phylogeny reconstruction. In *Proc., 1st International Workshop on Param-
            eterized and Exact Computation (IWPEC 2004)*, pages 1–12, 2004.

[DST04a]    J. Díaz, M. Serna, and D. M. Thilikos. Fixed parameter algorithms for
            counting and deciding bounded restrictive list h-colorings. In *Proc., 12th
            Annual European Symposium on Algorithms (ESA 2004)*, pages 275–286,
            2004.

[DST04b]    J. Díaz, M. Serna, and D. M. Thilikos. Recent results on parameterized $h$-
            coloring. In Jarik Nesetril and P. Winkler (eds.) DIMACS Series in Discrete
            Mathematics and Theoretical Computer Science, *Morphisms and Statistical
            Physics*, volume 63, pages 65–86. Amer. Math. Soc., Providence, RI, 2004.

[Fer02]     H. Fernau. On parameterized enumeration. In *Proc., 8th Annual Inter-
            national Conference on Computing and Combinatorics (COCOON 2002)*,
            pages 564–573, 2002.

[FG04]      J. Flum and M. Grohe. The parameterized complexity of counting prob-
            lems. *SIAM J. Comput.*, 33(4):892–922 (electronic), 2004.

# Approximating the Online Set Multicover Problems via Randomized Winnowing

Piotr Berman[1,*] and Bhaskar DasGupta[2,**]

[1] Department of Computer Science and Engineering,
Pennsylvania State University, University Park, PA 16802
`berman@cse.psu.edu`
[2] Computer Science Department,
University of Illinois at Chicago, Chicago, IL 60607
`dasgupta@cs.uic.edu`

**Abstract.** In this paper, we consider the weighted online set $k$-multicover problem. In this problem, we have an universe $V$ of elements, a family $\mathcal{S}$ of subsets of $V$ with a positive real cost for every $S \in \mathcal{S}$, and a "coverage factor" (positive integer) $k$. A subset $\{i_0, i_1, \ldots\} \subseteq V$ of elements are presented online in an arbitrary order. When each element $i_p$ is presented, we are also told the collection of all (at least $k$) sets $\mathcal{S}_{i_p} \subseteq \mathcal{S}$ and their costs in which $i_p$ belongs and we need to select additional sets from $\mathcal{S}_{i_p}$ if necessary such that our collection of selected sets contains *at least $k$ sets* that contain the element $i_p$. The goal is to *minimize* the *total cost* of the selected sets[1]. In this paper, we describe a new randomized algorithm for the online multicover problem based on the randomized winnowing approach of [11]. This algorithm generalizes and improves some earlier results in [1]. We also discuss lower bounds on competitive ratios for *deterministic algorithms* for general $k$ based on the approaches in [1].

## 1   Introduction

In this paper, we consider the Weighted Online Set $k$-multicover problem (abbreviated as **WOSC$_k$**) defined as follows. We have an universe $V = \{1, 2, \ldots, n\}$ of elements, a family $\mathcal{S}$ of subsets of $U$ with a cost (positive real number) $c_S$ for every $S \in \mathcal{S}$, and a "coverage factor" (positive integer) $k$. A subset $\{i_0, i_1, \ldots\} \subseteq V$ of elements are presented in an arbitrary order. When each element $i_p$ is presented, we are also told the collection of all (at least $k$) sets $\mathcal{S}_{i_p} \subseteq \mathcal{S}$ in which $i_p$ belongs and we need to select additional sets from $\mathcal{S}_{i_p}$, if necessary, such that our

---

* Supported by NSF grant CCR-O208821.
** Supported in part by NSF grants CCR-0206795, CCR-0208749 and a CAREER grant IIS-0346973.
[1] Our algorithm and competitive ratio bounds can be extended to the case when a set can be selected at most a prespecified number of times instead of just once; we do not report these extensions for simplicity.

F. Dehne, A. López-Ortiz, and J.-R. Sack (Eds.): WADS 2005, LNCS 3608, pp. 110–121, 2005.
© Springer-Verlag Berlin Heidelberg 2005

collection of sets contains *at least* $k$ sets that contain the element $i_p$. The goal is to minimize the total cost of the selected sets. The special case of $k = 1$ will be simply denoted by **WOSC** (Weighted Online Set Cover). The unweighted versions of these problems, when the cost any set is one, will be denoted by **OSC**$_k$ or **OSC**.

The performance of any online algorithm can be measured by the *competitive ratio*, *i.e.*, the ratio of the total cost of the online algorithm to that of an optimal offline algorithm that knows the entire input in advance; for randomized algorithms, we measure the performance by the *expected* competitive ratio, *i.e.*, the ratio of the expected cost of the solution found by our algorithm to the optimum cost computed by an adversary that knows the entire input sequence and has no limits on computational power, but who is *not familiar* with our random choices.

The following notations will be used uniformly throughout the rest of the paper unless otherwise stated explicitly:

- $V$ is the universe of elements;
- $m = \max\limits_{i \in V} |\{S \in \mathcal{S} \mid i \in S\}|$ is the maximum *frequency*, *i.e.*, the maximum number of sets in which any element of $V$ belongs;
- $d = \max\limits_{S \in \mathcal{S}} |S|$ is the maximum set size;
- $k$ is the coverage factor.

*None of $m$, $d$ or $|V|$ is known to the online algorithm in advance.*

## 1.1   Motivations and Applications

There are several applications for investigating the online settings in **WOSC**$_k$. Below we mention two such applications:

**Client/Server Protocols [1]:** Such a situation is modeled by the problem **WOSC** in which there is a network of servers, clients arrive one-by-one in arbitrary order, and the each client can be served by a subset of the servers based on their geographical distance from the client. An extension to **WOSC**$_k$ handles the scenario in which a client must be attended to by at least a minimum number of servers for, say, reliability, robustness and improved response time. In addition, in our motivation, we want a distributed algorithm for the various servers, namely an algorithm in which each server locally decide about the requests without communicating with the other servers or knowing their actions (and, thus for example, not allowed to maintain a potential function based on a subset of the servers such as in [1]).

**Reverse Engineering of Gene/Protein Networks [2, 4, 6, 9, 10, 14, 15]:** We briefly explain this motivation here due to lack of space; the reader may consult the references for more details. This motivation concerns unraveling (or "reverse engineering") the web of interactions among the components of complex protein and genetic regulatory networks by observing global changes to derive interactions between individual nodes. In one such setup, one assumes that the time evolution of a vector of state variables $x(t) = (x_1(t), \ldots, x_n(t))$ is described by a system of differential equations:

$$\frac{\partial \boldsymbol{x}}{\partial t} = f(\boldsymbol{x}, \boldsymbol{p}) \equiv \begin{cases} \frac{\partial x_1}{\partial t} = f_1(x_1, \ldots, x_n, p_1, \ldots, p_m) \\ \frac{\partial x_2}{\partial t} = f_2(x_1, \ldots, x_n, p_1, \ldots, p_m) \\ \quad \vdots \\ \frac{\partial x_n}{\partial t} = f_n(x_1, \ldots, x_n, p_1, \ldots, p_m) \end{cases}$$

where $\boldsymbol{p} = (p_1, \ldots, p_m)$ is a vector of parameters, such as levels of hormones or of enzymes, whose half-lives are long compared to the rate at which the variables evolve and which can be manipulated but remain constant during any given experiment. The components $x_i(t)$ of the state vector represent quantities that can be in principle measured, such as levels of activity of selected proteins or transcription rates of certain genes. There is a reference value $\bar{p}$ of $\boldsymbol{p}$, which represents "wild type" (that is, normal) conditions, and a corresponding steady state $\bar{x}$ of $\boldsymbol{x}$, such that $f(\bar{x}, \bar{p}) = 0$. We are interested in obtaining information about the Jacobian of the vector field $f$ evaluated at $(\bar{x}, \bar{p})$, or at least about the signs of the derivatives $\partial f_i / \partial x_j(\bar{x}, \bar{p})$. For example, if $\partial f_i / \partial x_j > 0$, this means that $x_j$ has a positive (catalytic) effect upon the rate of formation of $x_i$. The critical assumption is that, while we may not know the form of $f$, we often do know that *certain parameters $p_j$ do not directly affect certain variables $x_i$*. This amounts to *a priori* biological knowledge of specificity of enzymes and similar data. Consequently, an "online" experimental protocol to achieve the above goal, that gives rise to the problems **WOSC$_k$** and **OSC$_k$** is as follows:

- Change one parameter, say $p_k$.
- Measure the resulting steady state vector $\boldsymbol{x} = \xi(p)$. Experimentally, this may for instance mean that the concentration of a certain chemical represented by $p_k$ is kept are a slightly altered level, compared to the default value $\bar{p}_k$; then, the system is allowed to relax to steady state, after which the complete state $\boldsymbol{x}$ is measured, for example by means of a suitable biological reporting mechanism, such as a microarray used to measure the expression profile of the variables $x_i$.
- For each of the possible $m$ experiments, in which a given $p_j$ is perturbed, we may estimate the $n$ "sensitivities"

$$b_{ij} = \frac{\partial \xi_i}{\partial p_j}(\bar{p}) \approx \frac{1}{\bar{p}_j - p_j} \left( \xi_i(\bar{p} + p_j e_j) - \xi_i(\bar{p}) \right)$$

for $i = 1, \ldots, n$ (where $e_j \in \mathbb{R}^m$ is the $j$th canonical basis vector).

From these data, via some linear-algebraic reductions and depending on whether each experiment has the same or different cost, one can arrive at the problems **WOSC$_k$** and **OSC$_k$** with "large" $k$, *e.g.*, when $k \approx |V|$.

## 1.2   Summary of Prior Work

Offline versions **SC$_1$** and **SC$_k$** of the problems **WOSC$_k$** and **OSC$_k$**, in which all the $|V|$ elements are presented at the same time, have been well studied in the literature. Assuming $NP \not\subseteq DTIME(n^{\log \log n})$, the $\mathbf{SC}_1$ problem cannot be approximated to within a factor of $(1 - \varepsilon) \ln |V|$ for any constant $0 < \varepsilon < 1$ in polynomial time [7]; a slightly weaker lower bound under the more standard complexity-theoretic assumption of P$\neq$NP was obtained by Raz and Safra [13] who showed that there is a constant $c$ such that it is NP-hard to approximate the $\mathbf{SC}_1$ problem to within a factor of $c \ln |V|$. An instance of the $\mathbf{SC}_k$ problem can be $(1 + \ln d)$-approximated in $O(|V| \cdot |\mathcal{S}| \cdot k)$ time by a simple greedy heuristic that, at every step, selects a new set that covers the maximum number of those elements that has not been covered at least $k$ times yet [8, 16]; these results was recently improved upon in [4] who provided a randomized approximation algorithm with an expected performance ratio that about $\ln(d/k)$ when $d/k$ is at least about $\mathbf{e}^2 \approx 7.39$, and for smaller values of $d/k$ it decreases towards 1 as a linear function of $\sqrt{d/k}$.

Regarding previous results for the online versions, the authors in [1] considered the $\mathbf{WOSC}$ problem and provided both a deterministic algorithm with a competitive ratio of $O(\log m \log |V|)$ and an almost matching lower bound of $\Omega \left( \frac{\log |\mathcal{S}| \log |V|}{\log \log |\mathcal{S}| + \log \log |V|} \right)$ on the competitive ratio for any deterministic algorithm for almost all values[2] of $|V|$ and $|\mathcal{S}|$. The authors in [3] provided an efficient randomized online approximation algorithm and a corresponding matching lower bound (for any randomized algorithm) for a different version of the online set-cover problem in which one is allowed to pick at most $k$ sets for a given $k$ and the goal is to maximize the number of presented elements for which at least one set containing them was selected on or before the element was presented. To the best of our knowledge, there are no prior non-trivial results for either $\mathbf{WOSC}_k$ or $\mathbf{OSC}_k$ for general $k > 1$.

### 1.3   Summary of Our Results and Techniques

Let $r(m, d, k)$ denote the competitive ratio of any online algorithm for $\mathbf{WOSC}_k$ as a function of $m$, $d$ and $k$. In this paper, we describe a new randomized algorithm for the online multicover problem based on the randomized winnowing approach of [11]. Our main contributions are then as follows:

- We first provide an uniform analysis of our algorithm for all cases of the online set multicover problems. As a corollary of our analysis, we observe the following.
  - For $\mathbf{OSC}$, $\mathbf{WOSC}$ and $\mathbf{WOSC}_k$ our randomized algorithm has $E[r(m, d, k)]$ equal to $\log_2 m \ln d$ plus small lower order terms. While the authors in [1] did obtain a deterministic algorithm for $\mathbf{OSC}$ with $O(\log m \log |V|)$ competitive ratio, the advantages of our approach are more uniform algorithm with simpler analysis, as well as better constant factors and usage of the maximum set size $d$ rather than the larger universe size $|V|$ in the competitive ratio bound. Unlike the approach in [1],

---

[2] To be precise, when $\log_2 |V| \leq |\mathcal{S}| \leq \mathbf{e}^{|V|^{\frac{1}{2} - \delta}}$ for any fixed $\delta > 0$; we will refer to similar bounds as "almost all values" of these parameters in the sequel.

our algorithm does not need to maintain a global potential function over a subcollection of sets.

- For (the unweighted version) $\mathbf{OSC}_k$ for general $k$ the expected competitive ratio $E\left[r(m, d, k)\right]$ decreases logarithmically with increasing $k$ with a value of roughly $5 \log_2 m$ in the limit for all sufficiently large $k$.

— We next provide an improved analysis of $E\left[r(m, d, 1)\right]$ for $\mathbf{OSC}$ with better constants.

— We next provide an improved analysis of $E\left[r(m, d, k)\right]$ for $\mathbf{OSC}_k$ with better constants and asymptotic limit for large $k$. The case of large $k$ is important for its application in reverse engineering of biological networks as outlined in Section 1.1. More precisely, we show that $E\left[r(m, d, 1)\right]$ is at most

$$\left(\tfrac{1}{2} + \log_2 m\right) \cdot \left(2 \ln \tfrac{d}{k} + 3.4\right) + 1 + 2 \log_2 m, \text{ if } k \leq (2\mathbf{e}) \cdot d$$
$$1 + 2 \log_2 m, \qquad\qquad\qquad\qquad\qquad\qquad \text{otherwise}$$

— Finally, we discuss lower bounds on competitive ratios for *deterministic algorithms* for $\mathbf{OSC}_k$ and $\mathbf{WOSC}_k$ general $k$ using the approaches in [1]. The lower bounds obtained are $\Omega\left(\max\left\{1, \frac{\log \frac{|S|}{k} \log \frac{|V|}{k}}{\log\log \frac{|S|}{k} + \log\log \frac{|V|}{k}}\right\}\right)$ for $\mathbf{OSC}_k$ and $\Omega\left(\frac{\log |S| \log |V|}{\log\log |S| + \log\log |V|}\right)$ for $\mathbf{WOSC}_k$ for almost all values of the parameters.

*All proofs omitted due to space limitations will appear in the full version of the paper.*

## 2   A Generic Randomized Winnowing Algorithm

We first describe a generic randomized winnowing algorithm **A-Universal** below in Fig. 1. The winnowing algorithm has two scaling factors: a multiplicative scaling factor $\frac{\mu}{c_S}$ that depends on the particular set $S$ containing $i$ and another additive scaling factor $|\mathcal{S}_i|^{-1}$ that depends on the number of sets that contain $i$. These scaling factors quantify the appropriate level of "promotion" in the winnowing approach. In the next few sections, we will analyze the above algorithm for the various online set-multicover problems. The following notations will be used uniformly throughout the analysis:

- $\mathcal{J} \subseteq V$ be the set of elements received in a run of the algorithm.
- $\mathcal{T}^*$ be an optimum solution.

### 2.1   Probabilistic Preliminaries

For the analysis of Algorithm **A-Universal**, we will use the following combinatorial and probabilistic facts and results.

**Fact 1.** *If $f$ is a non-negative integer random function, then $E\left[f\right] = \sum_{i=1}^{\infty} Pr\left[f \geq i\right]$.*

```
// definition //
D1    for (i ∈ V)
D2        S_i ← {s ∈ S : i ∈ S}
```

```
// initialization //
I1    T ← ∅                  // T is our collection of selected sets //
I2    for (S ∈ S)
I3        ap[S] ← 0          // accumulated probability of each set //
```

```
// after receiving an element i //
A1    deficit ← k − |S_i ∩ T|        // k is the coverage factor //
A2    if deficit = 0                 // we need deficit more sets for i //
A3        finish the processing of i
A4    A ← ∅
A5    repeat deficit times
A6        S ←least cost set from S_i − T − A
A7        insert S to A
A8    μ ← c_S                        // μ is the cost of the last set added to A //
A9    for (S ∈ S_i − T)
A10       p[S] ← min { μ/c_S (ap[S] + |S_i|^{−1}), 1 }  // probability for this step //
A11       ap[S] ← ap[S] + p[S]                          // accumulated probability //
A12       with probability p[S]
A13           insert S to T                             // randomized selection //
A14   deficit ← k − |S_i ∩ T|
A15   repeat deficit times                              // greedy selection //
A16       insert a least cost set from S_i − T to T
```

**Fig. 1.** Algorithm **A-Universal**

**Fact 2.** *The function* $f(x) = xe^{-x}$ *is maximized for* $x = 1$.

The subsequent lemmas deal with $N$ independent 0-1 random functions $\tau_1, \ldots, \tau_N$ called trials with event$\{\tau_i = 1\}$ is the *success* of trial number $i$ and $s = \sum_{i=1}^{N} \tau_i$ is the number of successful trials. Let $x_i = Pr[\tau_i = 1] = E[\tau_i]$ and $X = \sum_{i=1}^{N} x_i = E[s]$.

**Lemma 3.** *If* $0 \le \alpha \le X/2$ *then* $Pr[s \le \alpha] < e^{-X} X^\alpha / \alpha!$.

## 3    An Uniform Analysis of Algorithm A-Universal

In this section, we present an uniform analysis of Algorithm **A-Universal** that applies to all versions of the online set multicover problems, *i.e.*, **OSC**, **OSC_k**, **WOSC** and **WOSC_k**. Abusing notations slightly, define $c(S') = \sum_{S \in S'} c_S$ for any subcollection of sets $S' \subseteq S$. Our bound on the competitive ratio will be influenced by the parameter $\kappa$ defined as: $\kappa = \min_{i \in J \ \& \ S \in S_i \cap T^*} \left\{ \dfrac{c(S_i \cap T^*)}{c_S} \right\}$. It is easy to check that $\kappa = \begin{cases} 1 \text{ for } \mathbf{OSC} \text{ and } \mathbf{WOSC} \\ k \text{ for } \mathbf{OSC}_k \\ \geq 1 \text{ for } \mathbf{WOSC}_k \end{cases}$ . The main result proved

in this section is the following theorem.

**Theorem 1.** *The expected competitive ratio* $E\left[r(m, d, k)\right]$ *of Algorithm* **A-Universal** *is at most*

$$\max\left\{1 + 5(\log_2(m+1)+1),\ 1 + (1 + \log_2(m+1))\left(2 + \ln\left(\frac{d}{\kappa(\log_2(m+1)+1)}\right)\right)\right\}$$

**Corollary 4**

(a) *For* **OSC**, **WOSC** *and* **WOSC**$_k$, *setting* $\kappa = 1$ *we obtain* $E\left[r(m, d, k)\right]$ *to be at most* $\log_2 m \ln d$ *plus lower order terms.*
(b) *For* **OSC**$_k$, *setting* $\kappa = k$, *we obtain* $E\left[r(m, d, k)\right]$ *to be at most*

$$\max\left\{6 + 5\log_2(m+1),\ 1 + (1 + \log_2(m+1))\left(2 + \ln\left(\frac{d}{k\log_2(m+1)}\right)\right)\right\}.$$

In the next few subsections we prove the above theorem.

### 3.1    The Overall Scheme

We first roughly describe the overall scheme of our analysis. The average cost of a run of **A-Universal** is the sum of average costs that are incurred when elements $i \in \mathcal{J}$ are received. We will account for these costs by dividing these costs into three parts $\text{cost}_1 + \sum_{i \in \mathcal{J}} \text{cost}_2^i + \sum_{i \in \mathcal{J}} \text{cost}_3^i$ where:

$\text{cost}_1 \leq c(\mathcal{T}^*)$ upper bounds the *total* cost incurred by the algorithm for selecting sets in $\mathcal{T} \cap \mathcal{T}^*$.
$\text{cost}_2^i$ is the cost of selecting sets from $\mathcal{S}_i - \mathcal{T}^*$ in line A13 for each $i \in \mathcal{J}$.
$\text{cost}_3^i$ is the cost of selecting sets from $\mathcal{S}_i - \mathcal{T}^*$ in line A16 for each $i \in \mathcal{J}$.

We will use the accounting scheme to count these costs by creating the following three types of accounts:

$\quad account(\mathcal{T}^*);$
$\quad account(S)\quad$ for each set $S \in \mathcal{T}^* - \mathcal{T}$;
$\quad account(i)\quad$ for each received element $i \in \mathcal{J}$.

$\text{cost}_1$ obviously adds at most 1 to the average competitive ratio; we will charge this cost to $account(\mathcal{T}^*)$. The other two kinds of costs, namely $\text{cost}_2^i + \text{cost}_3^i$ for each $i$, will be distributed to the remaining two accounts. Let $L(m)$ be a function of $m$ satisfying $L(m) \leq 1 + \log_2(m+1)$ and let $D = \frac{d}{\kappa(\log_2(m+1)+1)}$. The distribution of charges to these two accounts will satisfy the following:

- $\sum_{i \in \mathcal{J}} account(i) \leq L(m) \cdot c(\mathcal{T}^*)$. This claim in turn will be satisfied by:
  - dividing the optimal cost $c(\mathcal{T}^*)$ into pieces $c_i(\mathcal{T}^*)$ for each $i \in \mathcal{J}$ such that $\sum_{i \in \mathcal{J}} c_i(\mathcal{T}^*) \leq c(\mathcal{T}^*)$; and
  - showing that, for each $i \in \mathcal{J}$, $account(i) \leq L(m) \cdot c_i(\mathcal{T}^*)$.
- $\sum_{S \in \mathcal{T}^*} account(S) \leq L(m) \cdot \max\{4, \ln D + 1\} \cdot c(\mathcal{T}^*)$.

This will obviously prove an expected competitive ratio of at most the maximum of $1 + 5(\log_2(m+1) + 1)$ and $1 + (\log_2(m+1) + 1)(2 + \ln D)$, as promised.

We will perform our analysis from the point of view of each received element $i \in \mathcal{J}$. To define and analyze the charges we will define several quantities:

| | |
|---|---|
| $\mu(i)$ | the value of $\mu$ calculated in line A8 after receiving $i$ |
| $\xi(i)$ | the sum of $\alpha p[S]$'s over $S \in \mathcal{S}_i - T^*$ at the time when $i$ is received |
| $a(i)$ | $\lvert \mathcal{T} \cap \mathcal{S}_i - T^* \rvert$ at the time when $i$ is received |
| $\Lambda(S)$ | $\log_2(m \cdot \alpha p[S] + 1)$ for each $S \in \mathcal{S}$; it changes during the execution of **A-Universal** |

Finally, let $\Delta(X)$ denote the amount of change (increase or decrease) of a quantity $X$ when an element $i$ is processed.

## 3.2   The Role of $\Lambda(S)$

Our goal is to ensure that $\sum_{S \in T^* - T} account(S)$ is bounded by at most $\max\{4, \ln D + 1\}$ times $\sum_{S \in T^*} c_S \Lambda(S)$. For a $S \in \mathcal{S}_i \cap T^* - T$ corresponding to the case when element $i \in \mathcal{J}$ is processed, we will do this by ensuring that $\Delta(account(S))$, the change in $account(S)$, is at most a *suitable* multiple of $\Delta(c_S \Lambda(S))$. Roughly, we will partition the sets in $T^* - T$ into the so-called "heavy" and "light" sets that we will define later and show that

- for a light set, $\Delta(account(S))$ will be at most $\Delta(c_S \Lambda(S))$, and
- for a heavy set $\Delta(account(S))$ will be at most $\max\{4, \ln D + 1\} \Delta(c_S \Lambda(S))$.

The general approach to prove that $\Delta(account(S))$ is at least some multiple of $\Delta(c_S \Lambda(S))$ will generally involve two steps:

- $\Delta(c_S \Lambda(S)) \geq \min\{c_S, \mu(i)\}$;
- $\Delta(account(S))$ is at most a multiple of $\min\{c_S, \mu(i)\}$.

Of course, such an approach makes sense only if we can prove an upper bound on $E[\Lambda(S)]$. As a first attempt, the following lemma seems useful.

**Lemma 5.** $E[\Lambda(S)] \leq log_2(m + 1)$.

How does $\Lambda(S)$ increase when **A-Universal** handles its element $i$? A preliminary glance at tha algorithm suggests the following. First we calculate $\mu$ in line A8, then we calculate $p[S]$ in line A10 to be at least $\frac{\mu(i)}{c_S} \frac{1}{m}(m \cdot \alpha p[S] + 1)$, then we increase $\alpha p[S]$ by $p[S]$, thus we increase $m \cdot \alpha p[S] + 1$ by a factor of at least $1 + \frac{\mu(i)}{c_S}$. Therefore $\log_2(m \cdot \alpha p[S] + 1)$ seems to increase by at least $\log_2(1 + \frac{\mu(i)}{c_S})$.

However, some corrections may need to be made to the upper bound of $Ave\Lambda(S)$ in Lemma 5 to ensure that $\log_2(m \cdot \alpha p[S] + 1)$ increases by at least $\log_2(1 + \frac{\mu(i)}{c_S})$ for the very last time $p[S]$ and consequently $\alpha p[S]$ is updated. The reason for this is that in line $A10$ of algorithm AUn we calculate $p[S] \leftarrow \min\{\frac{\mu}{c_S}(\alpha p[S] + \lvert \mathcal{S}_i \rvert^{-1}), 1\}$ instead of calculating just $p[S] \leftarrow \frac{\mu}{c_S}(\alpha p[S] + \lvert \mathcal{S}_i \rvert^{-1})$ and it may be the case that $\frac{\mu}{c_S}(\alpha p[S] + \lvert \mathcal{S}_i \rvert^{-1}) > 1$. Note that for each

$S$ such a problem may occur only once and for the last increment since if we calculate $p[S] = 1$ then $S$ is surely inserted to $\mathcal{T}$. Thus, the very last increment of $\Lambda(S) = \log_2(m \cdot \alpha p[S] + 1)$ may be smaller than $\log_2(1 + \frac{\mu(i)}{c_S})$ (and, consequently, the very last increment of $c_S\Lambda(S)$ may be smaller than $c_S \log_2(1 + \frac{\mu(i)}{c_S})$). Intead of separately arguing for this case repeatedly at various places, we handle this by extending the upper bound for $E[\Lambda(S)]$ in Lemma 5 so that we can consider this last increment of $c_S \log_2(m \cdot \alpha p[S] + 1)$ also to be at least $c_S \log_2(1 + \frac{\mu(i)}{c_S})$. We omit the details here, but to summarize, we can alter the definition of $\Lambda(S)$ so that for $S \in \mathcal{S}_i \cap \mathcal{T}^* - \mathcal{T}$

- if $c_S \geq \mu(i)$, $\Delta(\Lambda(S)) \geq \log_2(1 + \frac{\mu}{c_S})$;
- if $c_S \leq \mu(i)$, $\Delta(\Lambda(S)) \geq 1$;
- the expected final value of $\Lambda(S)$ is $L(m) < 1 + \log_2(m + 1)$.

Now we are able to prove the following lemma.

**Lemma 6.** *If* $S \in \mathcal{S}_i \cap \mathcal{T}^* - \mathcal{T}$ *then* $\Delta(c_S\Lambda(S)) \geq \min\{c_S, \mu(i)\}$.

### 3.3    Definition of Light/Heavy Sets and Charges to Light Sets

When an element $i$ is received, we will make charges to $account(S)$ for $S \in \mathcal{S}_i \cap \mathcal{T}^* - \mathcal{T}$. Note that these are accounts of *at least* $deficit + a(i)$ many sets. We number these sets as $S(1), S(2), \dots$ in nondecreasing order of their costs with. We will define the *last* $a(i) + 1$ sets in this ordering as *heavy* and the rest as *light*.

Consider the sets inserted to $\mathcal{A}$ in lines A5-7, say $A(1), \dots, A(deficit)$. We pessimistically assume that except for its last — and most costly — element, $\mathcal{A}$ is inserted to $T$ in line A16. We charge the cost of that to the accounts of light sets — these sets will not receive any other charges. More specifically, we charge $c_{A(j)}$ to $account(S(j))$. Because $c_{A(j)} \leq \min\{c_{S(j)}, \mu(i)\}$, this charge is not larger than $\Delta(c_{S(j)}\Lambda(S(j)))$ by Lemma 6.

### 3.4    Charges to $Account(i)$

The sum of charges to accounts of heavy set and $account(i)$ can be estimated as $\mu(i)\xi(i) + 2\mu(i)$, where the part $\mu(i)\xi(i) + \mu(i)$ refers to line A13 and the remaining part $\mu(i)$ refers to the cost of line A16 that is not attributed to the accounts of light sets. *To simplify our calculations, we rescale the costs of sets so* $\mu(i) = 1$ *and thus* $c_S \geq 1$ *for each heavy set* $S$ *and the sum of charges to accounts of heavy set and* $account(i)$ *is simply* $\xi(i) + 2$.

We associate with $i$ a piece $c_i(\mathcal{T}^*)$ of the optimum cost $c(\mathcal{T}^*)$:

$$c_i(\mathcal{T}^*) = \sum_{S \in \mathcal{S}_i \cap \mathcal{T}^*} c_S/|S| \leq \frac{1}{d}c(\mathcal{S}_i \cap \mathcal{T}^*) \leq \frac{\kappa}{d}\mu(i) = \kappa/d.$$

It is then easy to verify that $\sum_{i \in \mathcal{J}} c_i(\mathcal{T}^*) \leq \sum_{i \in \mathcal{J}} \frac{1}{d}c(\mathcal{S}_i \cap \mathcal{T}^*) \leq c(\mathcal{T} \cap \mathcal{T}^*) \leq c(\mathcal{T}^*)$. As explained in the overview of this approach, we will charge $account(i)$

in such a way that on average it receives $D^{-1} = L(m)\kappa/d$. We will define a random events $E(i,a)$ so that the probability of event $E(i,a)$ is a function of the form $p(\xi(i),a)$ and when such an event happens, we charge $account(i)$ with some $f(\xi(i),a)$. We will show in the next subsection that the event $E(i,a)$ can be appropriately defined such that the expected sum of charges is *sufficiently small*, i.e., that $\sum_a p(X,a)f(X,a) < D^{-1}$.

### 3.5    Charges to Heavy Sets

Let $\psi = \max\{1, \ln D - 1\}$. Suppose that we charge each heavy set $S$ with an amount of $\psi$ of $\xi$ plus the two additional amounts, for a total of $\max\{3, \ln D + 1\}$. Then, $\Delta(c_S\Lambda(S)) \geq \min\{1, c_S\} \geq 1$ and the maximum charge is within a factor $\max\{3, \ln D + 1\}$ of $\Delta(c_S\Lambda(S))$.

If $\psi(a(i)+1) \geq \xi(i)$ we have no problem because we charge $a(i)+1$ accounts, each with at most $\psi$. Otherwise we need to charge $account(i)$ with $\xi(i) - \psi(a(i) + 1)$. We describe this case using the following events: $E(i,a)$ means that $a(i) \leq a$.

Let us identify $E(i,a)$ with a zero-one random funtion, and $charge(i, \psi, \ell, x)$ is the formula for the charge to $account(i)$ assuming we use $\psi$, $\ell\psi \leq X \leq (\ell+1)\psi$ and $\xi(i) = x$. If $E(i, \ell - 1)$ happens, we have to charge $account(i)$ with $x - \ell\psi$; if $E(i, \ell - 2)$ happens than $E(i, a - 1)$ also happens, so we charged $x - \ell\psi$, but we need to charge $account(i)$ with another $\psi$. One can see that for each $a \leq \ell - 2$, if $E(i,a)$ happens we charge $account(i)$ with $\psi$. One can see that

$$charge(i, \psi, \ell, x) = E(i, \ell - 1)(\xi(i) - \ell\psi) + \psi \sum_{j=0}^{\ell-2} E(i, \psi, j).$$

Let $C(\psi, \ell, x)$ be the estimate of $E\left[charge(i, \psi, \ell, x)\right]$ that uses Lemma 3:

$$C(\psi, \ell, x) = e^{-x}\left(\frac{x^{\ell-1}}{(k-1)!}(x - \ell\psi) + \psi \sum_{j=0}^{\ell-2} \frac{x^j}{j!}\right).$$

**Lemma 7.** *If $\psi \geq 2$, $x \geq 1$ and $\ell = \lfloor x/\psi \rfloor$ then $C(\psi, \ell, x) \leq e^{-(\psi+1)}$.*

As a result of the above lemma, setting $\psi = \max\{2, \ln D - 1\}$ we conclude that the average charge to $account(i)$ is at most $D^{-1}$.

## 4    Improved Analysis of Algorithm A-Universal for Unweighted Cases

In this section, we provide improved analysis of the expected competitive ratios of Algorithm **A-Universal** or its minor variation for the unweighted cases of the online set multicover problems. These improvements pertain to providing improved constants in the bound for $E\left[r(m, d, k)\right]$.

## 4.1    Improved Performance Bounds for OSC

**Theorem 2.** $E\left[r(m,d,1)\right] \leq \begin{cases} \log_2 m \ln d, & \text{if } m > 15 \\ \left(\frac{1}{2} + \log_2 m\right)(1 + \ln d), & \text{otherwise} \end{cases}$

## 4.2    Improved Performance Bounds for OSC$_k$

Note that for **OSC$_k$** we substitute $\mu = c_S = 1$ in the psuedocode of Algorithm **A-Universal** and that $deficit \in \{0, 1, 2, \ldots, k\}$. For improved analysis, we change Algorithm **A-Universal**slightly, namely, line A10 (with $\mu = c_S = 1$)

A10    $p[S] \leftarrow \min\left\{\left(\alpha p[S] + |\mathcal{S}_i|^{-1}\right), 1\right\}$ // probability for this step //

is changed to

A10'    $p[S] \leftarrow \min\left\{\left(\alpha p[S] + deficit \cdot |\mathcal{S}_i|^{-1}\right), 1\right\}$ // probability for this step //

**Theorem 3.** *With the above modification of Algorithm* **A-Universal**,

$$E\left[r(m,d,k)\right] \leq \begin{cases} \left(\frac{1}{2} + \log_2 m\right) \cdot \left(2\ln\frac{d}{k} + 3.4\right) + 1 + 2\log_2 m & \text{if } k \leq (2e) \cdot d \\ 1 + 2\log_2 m & \text{otherwise} \end{cases}$$

## 4.3    Lower Bounds on Competitive Ratios for OSC$_k$ and WOSC$_k$

**Lemma 1.** *For any $k$, there exists an instance of* **OSC$_k$** *and* **WOSC$_k$** *for almost all values of $|V|$ and $|\mathcal{S}|$ such that any deterministic algorithm must have a competitive ratio of* $\Omega\left(\max\left\{1, \frac{\log\frac{|\mathcal{S}|}{k}\log\frac{|V|}{k}}{\log\log\frac{|\mathcal{S}|}{k} + \log\log\frac{|V|}{k}}\right\}\right)$ *for* **OSC$_k$** *and* $\Omega\left(\frac{\log|\mathcal{S}|\log|V|}{\log\log|\mathcal{S}| + \log\log|V|}\right)$ *for* **WOSC$_k$**.

**Acknowledgments.** The second author wishes to thank the organizers of the Online Algorithms 2004 Workshop (OLA-2004) in Denmark for invitation which provided motivatations to look at these cover problems.

# References

1. N. Alon, B. Awerbuch, Y. Azar, N. Buchbinder, and J. Naor. *The online set cover problem*, 35th annual ACM Symposium on the Theory of Computing, pp. 100-105, 2003.
2. M. Andrec, B.N. Kholodenko, R.M. Levy, and E.D. Sontag. *Inference of signaling and gene regulatory networks by steady-state perturbation experiments: Structure and accuracy*, J. Theoretical Biology, in press.
3. B. Awerbuch, Y. Azar, A. Fiat and T. Leighton. *Making Commitments in the Face of Uncertainty: How to Pick a Winner Almost Every Time*, 28th annual ACM Symposium on the Theory of Computing, pp. 519-530, 1996.

4. P. Berman, B. DasGupta and E. Sontag. *Randomized Approximation Algorithms for Set Multicover Problems with Applications to Reverse Engineering of Protein and Gene Networks*, to appear in the special issue of Discrete Applied Mathematics on computational biology (the conference version in 7th International Workshop on Approximation Algorithms for Combinatorial Optimization Problems, LNCS 3122, K. Jansen, S. Khanna, J. D. P. Rolim and D. Ron (editors), Springer Verlag, pp. 39-50, August 2004).
5. H. Chernoff. *A measure of asymptotic efficiency of tests of a hypothesis based on the sum of observations*, Annals of Mathematical Statistics, 23: 493–509, 1952.
6. E.J. Crampin, S. Schnell, and P.E. McSharry. *Mathematical and computational techniques to deduce complex biochemical reaction mechanisms*, Progress in Biophysics & Molecular Biology, 86, pp. 77-112, 2004.
7. U. Feige. *A threshold for approximating set cover*, Journal of the ACM, Vol. 45, 1998, pp. 634-652.
8. D. S. Johnson. *Approximation Algorithms for Combinatorial Problems*, Journal of Computer and Systems Sciences, Vol. 9, 1974, pp. 256-278.
9. B. N. Kholodenko, A. Kiyatkin, F. Bruggeman, E.D. Sontag, H. Westerhoff, and J. Hoek. *Untangling the wires: a novel strategy to trace functional interactions in signaling and gene networks*, Proceedings of the National Academy of Sciences USA 99, pp. 12841-12846, 2002.
10. B. N. Kholodenko and E.D. Sontag. *Determination of functional network structure from local parameter dependence data*, arXiv physics/0205003, May 2002.
11. N. Littlestone. *Learning Quickly When Irrelevant Attributes Abound: A New Linear-Threshold Algorithm*, Machine Learning, 2, pp. 285-318, 1988.
12. R. Motwani and P. Raghavan. *Randomized Algorithms*, Cambridge University Press, New York, NY, 1995.
13. R. Raz and S. Safra. *A sub-constant error-probability low-degre test and sub-constant error-probability PCP characterization of NP* , proceedings of the 29th Annual ACM Symposium on Theory of Computing, pp. 475-484, 1997.
14. E.D. Sontag, A. Kiyatkin, and B.N. Kholodenko. *Inferring dynamic architecture of cellular networks using time series of gene expression, protein and metabolite data*, Bioinformatics 20, pp. 1877-1886, 2004.
15. J. Stark, R. Callard and M. Hubank. *From the top down: towards a predictive biology of signaling networks*, Trends Biotechnol. 21, pp. 290-293, 2003.
16. V. Vazirani. *Approximation Algorithms*, Springer-Verlag, July 2001.

# Max-stretch Reduction for Tree Spanners

Kazuo Iwama[1], Andrzej Lingas[2], and Masaki Okita[1]

[1] School of Informatics, Kyoto University, Kyoto 606-8501, Japan
{iwama, okita}@kuis.kyoto-u.ac.jp
[2] Department of Computer Science, Lund University, 221 00 Lund, Sweden
andrzej@cs.lth.se

**Abstract.** A tree $t$-spanner $T$ of a graph $G$ is a spanning tree of $G$ whose max-stretch is $t$, i.e., the distance between any two vertices in $T$ is at most $t$ times their distance in $G$. If $G$ has a tree $t$-spanner but not a tree $(t-1)$-spanner, then $G$ is said to have max-stretch of $t$. In this paper, we study the Max-Stretch Reduction Problem: for an unweighted graph $G = (V, E)$, find a set of edges not in $E$ originally whose insertion into $G$ can decrease the max-stretch of $G$. Our results are as follows: (i) For a ring graph, we give a linear-time algorithm which inserts $k$ edges improving the max-stretch optimally. (ii) For a grid graph, we give a nearly optimal max-stretch reduction algorithm which preserves the structure of the grid. (iii) In the general case, we show that it is $\mathcal{NP}$-hard to decide, for a given graph $G$ and its spanning tree of max-stretch $t$, whether or not one-edge insertion can decrease the max-stretch to $t-1$. (iv) Finally, we show that the max-stretch of an arbitrary graph on $n$ vertices can be reduced to $s' \geq 2$ by inserting $O(n/s')$ edges, which can be determined in linear time, and observe that this number of edges is optimal up to a constant.

## 1 Introduction

If a communication network is a *tree*, then we can enjoy countless merits. For example, we can design optimal routing schemes with succinct routing tables [1, 9, 17, 15, 16], there are several labeling schemes which allow us to deduce useful information such as distance and adjacency only from vertex labels [12], efficient broadcasting [3], and many more (see, e.g., [6]). This naturally leads us to the idea that we first construct a spanning tree for the (complicated) network we encounter and then do our tasks on this tree. Since the spanning tree covers all the vertices, we can do virtually everything in theory.

One drawback of this idea is that the distance between two vertices in the original network is not preserved in the spanning tree. For example, although two vertices are connected directly by an edge on the original network, they may be far apart on its spanning tree; a packet is forced to make a long detour if we use the spanning tree for routing. The notion of *tree $t$-spanners* was thus introduced to relax this demerit. A tree $t$-spanner is a spanning tree such that the distance between any two vertices on the tree is at most $t$ times their distance on the

F. Dehne, A. López-Ortiz, and J.-R. Sack (Eds.): WADS 2005, LNCS 3608, pp. 122–133, 2005.

original graph. The value $t$ is usually called the *max-stretch*. Thus our main goal is to find, given a graph $G$, its spanning tree whose max-stretch is as small as possible.

As easily expected, however, this problem is not so easy. For many cases it is $\mathcal{NP}$-hard [6] to achieve an optimal max-stretch, and the best approximation factor known is $O(\log n)$ at present [7]. Even more seriously for application purposes, many graphs do not admit a small max-stretch; for such graphs we can never attain our goal. Fortunately, it is also true that in many cases the max-stretch can be greatly improved by adding a reasonable number of new edges. This rather general approach, i.e., adding new edges to improve graph parameters, has always been popular in the field of graph algorithms; a typical example is to increase a connectivity of a graph for which there is a huge literature (see e.g., [11]). In this paper we investigate how to add edges to improve (reduce) the max-stretch of given graphs. This, little surprisingly, has not been discussed in the literature but for a related problem on hotlink assignments in [4] and a related problem for geometric graphs in [10] very recently.

**Our Contribution.** Our problem is called the *Max-Stretch Reduction* problem, which is to find, given a (connected, unweighted) graph $G$, a set of edges which are not in $G$ originally and whose insertion to $G$ reduces its max-stretch. Since complete graphs has a max-stretch of two, max-stretch reduction is obviously possible if we insert many edges. Thus our goal is to find as few such edges as possible to decrease max-stretch as much as possible. Our results include: (i) For a ring graph, we give a linear-time algorithm which finds, given an integer $k$, $k$ edges whose insertion improves the max-stretch optimally. (ii) For a grid graph, we give a nearly optimal max-stretch reduction algorithm which preserves the structure of the grid. (iii) For a general graph, we assume that its spanning tree of max-stretch $t$ is also given. Even so, we can show that it is $\mathcal{NP}$-complete to decide whether or not a spanning tree of max-stretch at most $t - 1$ can be obtained by adding a single edge. (iv) Furthermore, we can demonstrate an infinite family of graphs with $n$ vertices and max-stretch $s$ which require $\Omega(n/s)$ edge insertions to decrease the max-stretch at least by one. On the positive side, we can show that the max-stretch of an arbitrary graph on $n$ vertices can be reduced to $s' \geq 2$ by inserting $O(n/s')$ edges and that these edges can be determined in time $O(n)$.

**Previous Work.** A spanning subgraph (which may not be a tree) with max-stretch $t$ was first studied in [14]. Finding sparse spanning subgraphs with max-stretch $t$ is intractable in general [5], but for some restricted cases, there are polynomial time algorithms [2]. For the spanning tree case, the problem is called the *Minimum Max-Stretch Spanning Tree (MMST)* problem. Its $\mathcal{NP}$-hardness was established in [6]. Several tractability results for restricted problems are found in [8]. Emek and Peleg first obtained a nontrivial approximation algorithm for MMST, whose approximation factor is $O(\log n)$ [7]. Their paper also gives a very nice exposition of the history of this problem. For applications of MMSTs, see [6].

## 2    Preliminaries

In this paper, a graph $G(V, E)$ is unweighted and connected. We always use $n$ for $|V|$ and $m$ for $|E|$. Let $dist_G(u, v)$ denote the *distance* between vertices $u$ and $v$ in $G$. For a tree $T$, let $path_T(u, v)$ denote the *path* between vertices $u$ and $v$ in $T$.

A *t-spanner* of a graph $G$ is a spanning subgraph $G'$ in which the distance between every pair of vertices is at most $t$ times their distance in $G$. The value $t$ is called the max-stretch of $G'$. If a graph $G$ has a spanning tree of max-stretch $t$, but not of max-stretch $t - 1$, then it is said that the max-stretch of $G$ is $t$. For a graph $G$, its spanning tree $T$ and an edge $e \in E$, $e = \{u, v\}$, it is said that $e$ has a stretch of $t$ if $dist_T(u, v) = t$. If $e$ is included in $T$, then its stretch is obviously one. Following lemma is useful when the graph is unweighted [13].

**Lemma 1.** *For an unweighted graph $G = (V, E)$, the subgraph $G' = (V, E')$ is a t-spanner of $G$ iff for every $\{u, \ v\} \in E$, $dist_{G'}(u, v) \leq t$.* □

Thus, the max-stretch of $T$ is equal to the maximal stretch of edges. The *Minimum Max-Stretch Spanning Tree* (MMST) problem is to find the spanning tree $T$ that minimizes its max-stretch. The problem was shown to be $\mathcal{NP}$-hard in [6]. For a given graph $G$ and a positive integer $k$, the *Max-Stretch Reduction* problem is to find a set $S$ of $k$ edges which are not included in $G$ originally such that $G' = (V, E \cup S)$ has a max-stretch which is less than that of $G$.

Let $E^+$ denote the set of edges which are added to $G$ to obtain the spanning tree of reduced max-stretch. As shown later, we can assume without loss of generality that $T$ includes all the edges in $E^+$. $E^-$ denotes the set of edges in $E$ which are not included in $T$. An edge $\bar{e}$ in $E^-$ is called a *non-tree edge* and sometimes the cycle consisting of such an edge $\bar{e}$ and edges in $T$ is called a *fundamental cycle*. Conversely, a fundamental cycle $\bar{c}$ includes a unique non-tree edge $\bar{e}$.

## 3    Edge Insertion for Ring Graphs

Let $G = (V, E)$ be a ring graph on $n$ vertices, i.e., a simple cycle on $n$ vertices. We may assume w.l.o.g that $G$ is embedded in the plane and has vertices $v_1, v_2, ..., v_n$, and edges $\{v_1, v_2\}, ..., \{v_{n-1}, v_n\}, \{v_n, v_0\}$ in the clockwise direction. The max-stretch of $G$ is obviously $n - 1$, since its spanning tree must have $n - 1$ edges out of the $n$ ones. This max-stretch is quite large, but our approach really works well for this graph, namely the max-stretch is reduced approximately by a factor of $1/k$ by adding $k$ edges. The idea of our approach is simply to improve the diameter of the graph by adding few edges, and then take a shortest spanning tree of the augmented graph. Our algorithm, called `InsertRing(G)` is as follows:

1. Let $\alpha = \lceil \frac{n-2}{k+1} \rceil + 1$, select any integer $\beta$ such that $0 \leq \beta \leq \lceil \frac{n-2}{k+1} \rceil$, and let $\gamma = \alpha - \beta - 1$.
2. Select an arbitrary vertex $u^1$ on the ring and then select $v^1$ which is $\alpha$ apart from $u^1$ in the clockwise direction.

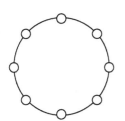 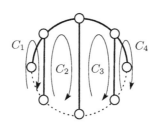 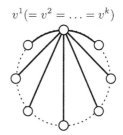

**Fig. 1.** Ring graph($n = 8$)      **Fig. 2.** MMST and its fun-      **Fig. 3.** Bad case of $\beta = 0$
damental cycles

3. Set $E^+ = \{\{u^1, v^1\}\}$. For $i = 2$ to $k$,
   (a) select vertex $v^i$ which is $\beta$ apart from $v^{i-1}$ in the clockwise direction,
   (b) select vertex $u^i$ which is $\gamma$ apart from $u^{i-1}$ to in the anti-clockwise direction, and
   (c) insert a new edge $\{u^i, v^i\}$ and let $E^+ = E^+ \cup \{\{u^i, v^i\}\}$.
4. Let $E^- = \phi$. For $i = 1$ to $k - 1$, select an arbitrary edge on the path from $u^i$ to $u^{i+1}$ and add it to $E^-$. Similarly select two edges, one on the path of the ring from $v^1$ to $u^1$ and the other on the path from $v^k$ to $u^k$, and add them to $E^-$. Let $T = (E \cup E^+) - E^-$.

**Remark.** Recall that $\beta$ can be any integer between 0 and $\lceil\frac{n-2}{k+1}\rceil$. If we select $\beta = 0$ or $\lceil\frac{n-2}{k+1}\rceil$, then all the inserted edges in $E^+$ share the single vertex $v^1 (= v^2 = \ldots = v^k)$ (See Fig. 3) or $u^1$. For the routing purposes, this might not be too good since much load is concentrated to this single vertex. For the remaining values of $\beta$ the maximum degree of the resulting tree spanner is just three.

**Theorem 1.** *The spanning tree $T$ obtained by* InsertRing(G) *is an MMST of $G' = (V, E \cup E^+)$ whose max-stretch is $\lceil\frac{n-2}{k+1}\rceil + 1$. These $k$ edge insertions performed by* InsertRing(G) *are optimal since for any graph $G^k$ resulting from inserting $k$ new edges into $G$ its max-stretch is at least $\lceil\frac{n-2}{k+1}\rceil + 1$.* InsertRing(G) *runs in linear time.*

*Proof.* Obviously $T$ is an MMST. Consider $k + 1$ cycles, $C_1, C_2, \ldots, C_{k+1}$, in $G'$ as shown in Fig.2. One can see that the length of $C_1 = \alpha + 1$. The length of $C_2, \ldots, C_k$ is all the same, which is $\beta + \gamma + 2 = \alpha + 1$. Finally the length of $C_{k+1}$ is $n - \{\alpha + 1 + (k - 1)(\beta + \gamma)\} + 2 = n - k(\alpha - 1) \leq \alpha + 1$. Note that the length of all those $k + 1$ cycles is at most $\alpha + 1$. Each cycle includes exactly one (deleted) edge in $E^-$, so the max-stretch is at most $\alpha = \lceil\frac{n-2}{k+1}\rceil + 1$.

For the optimality, we first show the following observation about the relation between the added edges and the fundamental cycles: let $X$ be any MMST for $G^k$. Without loss of generality, we can assume $X$ includes all the inserted edges (otherwise, we can achieve the same stretch with $k - 1$ or less inserted edges). Let $e$ be an edge in $X$. By removing this $e$ from $X$, $X$ is decomposed into two trees $X_l$ and $X_r$ (one of them might be a single vertex). Now we give a label

$l$ ($r$, respectively.) to all the vertices connected by $X_l$ ($X_r$, respectively.). Since $X$ is a spanning tree, every vertex has a label $l$ or $r$ and not both. Since the original graph $G = (V, E)$ is a ring, there are at least two edges, say $e_1$ and $e_2$, in $E$ whose two endpoints are labeled differently. If the originally selected edge $e$ is an inserted edge, i.e., in $E^+$, then both of these $e_1$ and $e_2$ must be removed edges, i.e., in $E^-$ and the two fundamental cycles for $e_1$ and $e_2$ must go through $e$. If is $e$ in $E \cap X$, one of $e_1$ and $e_2$, say $e_1$, can be the original $e$. Then the fundamental cycle for $e_2$ must go through $e$. As a result: if $e$ is in $E^+$ (in $E \cap X$, respectively.), then $e$ is included in at least two (one, respectively.) fundamental cycle(s).

Now consider the total length of fundamental cycles. As shown above, fundamental cycles go through each edge in $E \cap X$ (their number is $n - (k+1)$) at least once and each edge in $E^+$ (their number is $k$) at least twice, which implies that the total length is at least $n - (k+1) + 2k = n + k - 1$. Since the number of non-tree edges is $k + 1$, the average length of each fundamental cycle is $\frac{n+k-1}{k+1}$, i.e., $\frac{n-2}{k+1} + 1$. Therefore, the tree spanner, constructed by any $k$-insertion algorithm, has at least one fundamental cycle whose length is greater than or equal to $\frac{n-2}{k+1} + 1$. This gives the lower bound of the max-stretch by the $k$-insertion algorithm for ring graphs.

Clearly, `InsertRing(G)` can be implemented in linear time.

## 4   Edge Insertion for Grid Graphs

In this section, we assume that $g = \sqrt{n}$ is an integer and $G$ is a $g \times g$ grid graph. Vertices are denoted as $v_j^i$ for each $i, j \in \{1, \ldots, g\}$ and there exist edges $\{v_j^i, v_j^{i+1}\}$ and $\{v_j^i, v_{j+1}^i\}$ for every $i, j \in \{1, \ldots, g-1\}$. Suppose that we can insert $k$ edges into the grid graph $G$. As in the previous section, we follow the idea of reducing the diameter of $G$ as much as possible by such $k$ edge insertions. One way of doing so is illustrated in Fig.4. Namely, the whole graph is divided into $\sqrt{k+1} \times \sqrt{k+1}$ small grids, each has a size (= the number of vertices) of $\frac{g}{\sqrt{k+1}} \times \frac{g}{\sqrt{k+1}}$. (When $k = 3$, it is $2 \times 2 = 4$ small grids as shown in Fig. 4).

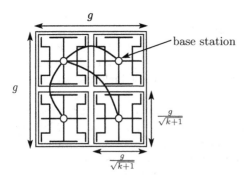

**Fig. 4.** MMST of grid graph

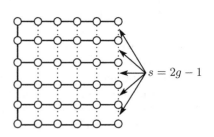

**Fig. 5.** Comb MMST

Then we select a single vertex called a *base station* for each small grid, and insert $k-1$ edges between the base station of the upper-left corner and all other $k-1$ base stations. To construct an MMST $T$, we include all the inserted edges into $T$. Furthermore we construct a shortest-path tree from the base station in each small grid and include all those edges constituting the tree into $T$. By a simple calculation, this construction achieves a stretch factor of $2\lceil\sqrt{\frac{n}{k+1}}\rceil + 4$ $(k \geq 3)$, which is (almost) optimal as shown later. Thus the construction is simple and efficient, but the resulting MMST no longer preserves the structure of the grid.

Another method is to apply the grid structure recursively. We first fix some scheme for constructing an MMST for a grid. For simplicity, we consider the *comb* MMST as shown in Fig.5. This scheme gives us a stretch factor of $2g - 1$, which is almost twice as bad as more sophisticated schemes like the shortest-path scheme (constructing a shortest-path tree from the center of the mesh). However, the following argument does not change at all for such better schemes. (We can choose any scheme if its radius is not too large. For example, the snake MMST has the same stretch as comb, but its radius is $\Omega(g^2)$, which is too large for our purpose.) Now suppose again that we insert $k$ edges. Then we divide the whole grid into $\sqrt{k+1} \times \sqrt{k+1}$ small grids and select a base station for each small grid, as before. Now we construct a comb MMST for each small grid and for the set of base stations as shown in Fig.6. The latter MMST is called an MMST at a higher level. It is not hard to see that the resulting MMST achieves a stretch factor of $2\sqrt{k+1} - 1 + 3\frac{g}{\sqrt{k+1}} + 2 \leq O(\sqrt{\frac{n}{k+1}})$, if $k$ is at most $g = \sqrt{n}$.

If $k$ is larger than $g$, then the stretch for the higher level MMST becomes larger than the stretch for the small grid, by which the total stretch increases as $k$ grows. This can be avoided by providing more higher-level MMSTs. For example, suppose that $n^{1/2} \leq k \leq n^{2/3}$. Then, we first construct $\sqrt{k+1} \times \sqrt{k+1}$ small grids as before. Note that the size of each small grid is $\sqrt{\frac{n}{k+1}} \times \sqrt{\frac{n}{k+1}}$. To make a balance with this size, we then construct a middle-sized grid each of which consists of $\sqrt{\frac{n}{k+1}} \times \sqrt{\frac{n}{k+1}}$ small grids (= base stations). Finally we construct the highest level MMST for these middle-sized grids whose size (= the number of base stations for middle-sized grids) is at most $\sqrt{\frac{n}{k+1}} \times \sqrt{\frac{n}{k+1}}$. One can easily

**higher level MMST**

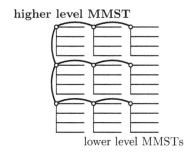

lower level MMSTs

**Fig. 6.** Recursive construction of MMST

**highest level MMST**

**Fig. 7.** Example for $n = 9^3 (g = 27)$ and $k = 80$

see that the stretch factor is $O(\sqrt{\frac{n}{k+1}})$. The case that $n = 9^3 (g = 27)$ and $k = 80$ is illustrated in Fig.7. Although details are omitted, one can extend this method to work for furthermore larger $k$ by introducing more levels.

**Theorem 2.** *The recursive insertion achieves the max-stretch of* $O(\sqrt{\frac{n}{k+1}})$ *if* $k \leq n^d$ *for some constant* $0 < d < 1$*, and this is within an* $O(\sqrt{k})$ *factor from the optimum.*

*Proof.* The upper bound follows from straightforward calculations.

For the lower bound, let $E'$ be an optimal set of $k$ edges to insert, and let $T'$ be the MMST of the augmented graph. By removing $E'$ from $T'$, we obtain $k + 1$ subtrees $T'_0, \ldots, T'_k$ of $T'$. Note that each of these subtrees is embedded in the grid without crossings. For $i = 0, \ldots, k$, let $border(T'_i)$ be the set of vertices in $T'_i$ that are incident to non-tree edges having their other endpoint in another tree $T'_j$. By the planar embedding in the grid, we can observe that (i) $|border(T'_i)| = \Omega(\sqrt{|T'_i|})$ holds for $i = 0, \ldots, k$. Furthermore, by a straightforward fraction argument, (ii) there must exist a subtree $T'_l$ such that the proportion between $|T'_l|$ and the number of vertices in $T'_l$ adjacent to at least one of the $k$ inserted edges is $\Omega(\frac{n}{k+1})$. By (i), $|border(T'_l)| = \Omega(\sqrt{\frac{nq}{k+1}})$ holds where $q$ is the number of vertices in $T'_l$ incident to the inserted edges. Now, cluster the vertices in $border(T'_l)$ into $q$ clusters according to their minimum distance to a vertex in $T'_l$ adjacent to an inserted edge. It follows that there is a cluster, say $C$, for which $|border(C)| = \Omega(\sqrt{\frac{n}{q(k+1)}})$ holds. By a simple geometric case analysis, there is a vertex $v$ in $C$ such that (a) $v$ is incident to a non-tree edge going to another subtree $T'_j$, (b) the minimum distance between $v$ and a vertex in $T'_l$ incident to an inserted edge is $\Omega(\sqrt{\frac{n}{q(k+1)}})$. This yields the $\Omega(\sqrt{\frac{n}{k(k+1)}})$ bound on the max-stretch.

## 5    Intractability for General Graphs

### 5.1    NP-Completeness of Stretch Reducibility

As mentioned before, the MMST problem itself was shown to be $\mathcal{NP}$-hard in [6], where the authors gave a graph $G(s)$ for which to construct an MMST with stretch $s (\geq 5)$ is easy but to construct one with stretch $s - 1$ is hard. Using this fact, we can show that the most basic version of our current problem is also intractable even if an optimal MMST for the original graph is given. We consider the following problem called MSR1 (*Max-Stretch Reduction by adding one edge*):

*Instance:* A graph $G$ and its MMST $T$ of max-stretch $s$.
*Question:* Is it possible to achieve a max-stretch of less than $s$ by adding one edge?

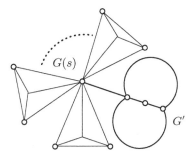

**Fig. 8.** Graph for $\mathcal{NP}$-hard

**Theorem 3.** *MSR1 is $\mathcal{NP}$-hard.*

*Proof.* Let $G(s)$ be the above mentioned graph in [6], which consists of the center vertex and petal-like subgraphs around it as illustrated in Fig.8. The idea is to add a graph $G'$, which consists of two rings of length $s+1$ with two edges shared, to $G(s)$ as shown in Fig.8.

Consider an MMST for the whole graph $G$ obtained by removing a single edge from each ring combined with an (easy) MMST of stretch $s$ for the portion of $G(s)$. Since the max-stretch of the single ring is at least $s$, this MMST, denoted by $T$, of stretch $s$ for $G$ is optimal. Thus we are given the graph $G$ and its (optimal) MMST $T$, and try to reduce its max-stretch by adding one edge.

Now consider inserting one edge $e$ to $G$ such that $e$ shortcuts the two edges shared by the two rings. By doing this, we can reduce the stretch of the $G'$ portion by one, namely from $s$ to $s-1$. Furthermore, one can see that this is the only way of getting such a max-stretch reduction. (Especially, we cannot obtain any stretch reduction of the $G'$ portion by adding an edge between $G$ and $G'$.) This means that whether or not we can reduce the stretch of the whole graph $G$ by inserting one edge is exactly depend on whether or not we can reduce the stretch of the $G(s)$ part *without adding any edge*. The latter is hard as mentioned before.                                                                                     □

### 5.2    Approximation Factors

The natural optimization version of MSR1 is the problem to obtain, given a graph $G = (V, E)$, a pair of vertices $\{u, v\}(\notin E)$ such that the insertion of the edge $\{u, v\}$ reduces max-stretch as much as possible. Approximability of this problem appears to be interesting. We observe the following simple upper and lower bounds on the approximability of this problem.

**Theorem 4.** *The optimization version of MSR1 admits a 2-approximation and does not admit a 1.25 approximation unless $P = NP$.*

*Proof.* First we show that one-edge insertion cannot reduce max-stretch to less than one half, which is enough to claim the upper bound of 2.0. Let $G'$ be the graph which is obtained from the original graph $G$ by adding an (optimal) edge

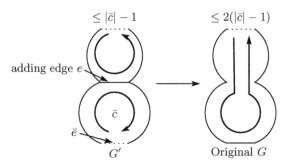

**Fig. 9.** Stretch change by adding edge

$e$, and $T'$ be its corresponding MMST. We shall show that the max-stretch of the original graph is at most twice that of $G'$. Let $\bar{c}$ be a longest fundamental cycle including $e$ with respect to $T'$, and $\bar{e}$ be its non-tree edge. If $|\bar{c}| = p$, then the max-stretch of $G'$ is obviously at least $p - 1$. Now construct an MMST $T$ of the original graph $G$ by setting $T = (T' - e) \cup \bar{e}$. Then a fundamental cycle which goes through $e$ in $G'$ must detour in $G$ as shown in Fig.9. However, since $\bar{c}$ is the longest fundamental cycle including $e$, the length of the detoured cycle is obviously less than twice $|\bar{c}|$.

To show the lower bound, we can use the graph $G$ described in the above theorem. Suppose that $s = 5$. Then if we have an algorithm whose approximation factor is less than 1.25, we can obtain an MMST of max-stretch 4 iff $G(5)$ has an MMST of max-stretch 4 in polynomial time. This contradicts the intractability result in [6].

# 6    General Algorithms for Stretch Reduction

Even though the optimal one-edge insertion to arbitrary graphs is $\mathcal{NP}$-hard, we can still find some sufficient conditions in the case of general graphs. We discuss a multi-edge insertion algorithm to reduce the max-stretch of the graph under the assumption that the MMST of the graph is given.

**Theorem 5.** *For any sufficiently large $n$ and $s \geq 3$, where $n$ is divisible by $s$, one can construct a graph $G$ such that the stretch of its MMST tree is $s$ and $G$ has to be augmented by $\Omega(n/s)$ edges in order to reduce the stretch $s$ at least by one.*

*Proof.* Let $k = n/s$. The graph $G$ is simply composed of $k$ simple cycles on $s+1$ vertices so that the $i$th cycle shares a single vertex with the $(i \bmod k) + 1$th for $i = 1, 2, ..., k$. In any spanning tree of $G$, there is at least one cycle in which one edge is not included and all others are included in the spanning tree. (Otherwise, the spanner would not be a connected graph, a contradiction.) Therefore it must have stretch at least $s$. To reduce the stretch, conversely, at least two edges of each cycle must not be included in the spanning tree. In the original graph, there are $k$ cycles and $k$ edges are not already included in the spanning tree and so at least $k = n/s$ new edges must be inserted.

We shall show that the following procedure not only provides an upper bound on the number of edge augmentations matching the lower bound given in Theorem 5 but also can yield a stretch reduction by a constant fraction.

**procedure** *Cut-stretch(T, s')*

> *Input:* A spanning tree $T$ of a graph $G$ with max-stretch at most $s$ and some even integer $s'$, where $s > s' \geq 2$.
> *Output:* A graph $G'$ resulting from inserting $O(n/s')$ edges into $G$ and its spanning tree $T'$ having max-stretch at most $s'$.

**begin**

1. $G' \leftarrow G$; $T' \leftarrow T$;
2. Root $T'$ at some vertex $r$;
3. **while** there is a leaf in $T'$ which has a $s'/2 - 1$ distant ancestor in $T'$ − $\{r$ and its neighbors$\}$ **then** select such a leaf and "hang" its $s'/2 - 1$ distant ancestor on $r$, i.e., connect it with the root of $T'$ by inserting a new edge;
4. Insert the edges of $T'$ not appearing in $G$ into $G'$, and output the resulting graph $G'$ and the tree $T'$;

**end**.

**Lemma 2.** *The number of iterations of the while block in Cut-stretch(T, s') is at most $2n/s'$.*

*Proof.* When a vertex $v$ is to be hanged on the root of $T'$, neither $v$ nor any of its nearest $s'/2 - 1$ descendants could have been hanged on the root before or can be hanged on the root afterwards. Hence, with each iteration of the while block a distinct set of at least $s'/2$ vertices of $G$ can be associated uniquely.

**Lemma 3.** *Cut-stretch(T, s') inserts at most $2n/s'$ edges in $G$ and outputs a spanning tree of the resulting graph having max-stretch at most $s'$. It runs in time $O(n)$.*

*Proof.* The first part follows from Lemma 2. When there is no leaf in $T'$ satisfying the condition of the while instruction then $T'$ has to have height at most $s'/2$. Otherwise, there would be a path from the root to a leaf of $T'$ having length at least $s'/2 + 1$ and the leaf would satisfy this condition, a contradiction. Hence, any vertex is at most $s'$ distant from another vertex in $T'$ and $T'$ has max-stretch at most $s'$.

Each iteration of the while block in Cut-stretch(T, s') can be performed in time $O(s)$. The leaf selections can be implemented by using a dynamic list of relevant leaves given with their depth which takes $O(n)$ preprocessing time and has processing time proportional to the number of iterations of the while block. Hence, Lemma 2 yields the $O(n)$ running time of this algorithm.

By Lemma 3, we obtain our main result in this section.

**Theorem 6.** *For any graph with $n$ vertices and its spanning tree of max-stretch $s > 2$, and even integer $s'$ satisfying $s > s' \geq 2$, $O(n/s')$ edge augmentations are sufficient to reduce the stretch $s$ to $s'$. The edge augmentations can be found in time $O(n)$.*

**Corollary 1.** *For any graph with $n$ vertices and its spanning tree of max-stretch $s > 2$, $O(n/s)$ edge augmentations are sufficient to reduce the stretch $s$ to $2s/3$. The edge augmentations can be found in time $O(n)$.*

*Proof.* Set $s' = 2\lfloor s/3 \rfloor$ in Theorem 6 and observe that $n/s' = O(n/s)$.

## 7 Final Remark

It seems that our max-stretch reduction algorithm for grid graphs is optimal up to a constant. We are currently working on getting rid off the $\frac{1}{\sqrt{k}}$ factor in our corresponding lower bound (see Theorem 2).

## Acknowledgements

The authors are very grateful to unknown referees for several valuable comments and suggestions on improving an earlier version of this paper.

## References

1. I. Abraham, C. Gavoille, and D. Malkhi. Routing with Improved Communication-Space Trade-Off. In *Proceedings of the 18th International Symposium on Distributed Computing (DISC)*, pages 305–319, Amsterdam, October 2004.
2. I. Althöffer, G. Das, D. Dobkin, D. Josepth, and J. Soares. On sparse spanners of weighted graphs. *Discrete and Computational Geometry*, 9:81–100, 1993.
3. B. Awerbuch, A. Baratz, and David Peleg. Efficient Broadcast and Light-Weight Spanners. Technical Report CS92-22, 1992. The Weizmann Institute of Science, Rehovot, Israel.
4. P. Bose, J. Czyzowicz, L. Gasieniec, E. Kranakis, D. Krizanc, A. Pelc, and M.V. Martin. Strategies for hotlink assignments. In *Proceedings of the Eleventh Annual International Symposium on Algorithms and Computation (ISAAC 2000)*, Lecture Notes in Computer Science, pages 23–34. Springer-Verlag, 2000.
5. L. Cai. Np-Completeness of Minimum Spanner Problems. *Discrete Applied Mathematics*, 48:187–194, 1994.
6. L. Cai and D. Corneil. Tree Spanners. *SIAM Journal on Discrete Mathematics*, 8(3):359–387, 1995.
7. Y. Emek and D. Peleg. Approximating Minimum Max-Stretch Spanning Trees on Unweighted Graphs. In *Proceedings of the 15th Symposium on Discrete Algorithms (SODA)*, January 2004.

8. S. P. Fekete and J. Kremer. Tree spanners in planar graphs. *Discrete Applicated Mathematics*, 108:85–103, 2001. Extended abstract version appears in the Proceedings of the 24th International Annual Workshop on Graph-Theoretic Concepts in Computer Science (WG98).
9. P. Fraigniaud and C. Gavoille. Routing in Trees. In *Proceedings of the 28th International Colloquium on Automata, Languages and Programming (ICALP)*, volume LNCS 2076, pages 757–772, July 2001.
10. P. Giannopoulos M. Farshi and J. Gudmundsson. Finding the best shortcut in a geometric network. In *Proceedings of 21st annual ACM Symposium on Computational Geometry*, June 2005.
11. H. Nagamochi, T. Shiraki, and T. Ibaraki. Computing Edge-Connectivity Augmentation Function. In *Proceedings of the 8th Annual ACM-SIAM Symposium on Discrete Algorithms (SODA)*, pages 649–658, 1997.
12. D. Peleg. Proximity-Preserving Labeling Schemes and Their Applications. In *Proceedings of the 25th International Workshop on Graph-Theoretic Concepts in Computer Science*, pages 30–41, June 1999. Ascona , Switzerland.
13. D. Peleg. *Distributed Computing: A Locality-Sensitive Approach*. SIAM monographs on Discrete Mathematics and Applications, 2000.
14. D. Peleg and J. Ullman. An Optimal Syncronizer for the Hypercube. In *Proceedings of the 6th Annual ACM Symposium on Principles of Distributed Computing (PODC)*, pages 77–85, 1987. Vancouver.
15. D. Peleg and E. Upfal. A Tradeoff between Space and Efficiency for Routing Tables. In *Proceedings of the 20th ACM Symposium on Theory of Computing (STOC)*, pages 43–52, May 1988.
16. M. Thorup and U. Zwick. Compact Routing Schemes. In *Proceedings of the 13th Annual ACM Symposium on Parallel Algorithms and Architectures (SPAA)*, pages 1–10, May 2001.
17. J. van Leeuwen and R. B. Tan. Interval Routing. *The Computer Journal*, 30: 298–307, 1987.

# Succinct Representation of Triangulations with a Boundary[*]

L. Castelli Aleardi[1,2], Olivier Devillers[2], and Gilles Schaeffer[1]

[1] Ecole Polytechnique, Palaiseau, France
{amturing, schaeffe}@lix.polytechnique.fr
[2] Inria Sophia-Antipolis, Projet Geometrica, France
olivier.devillers@sophia.inria.fr

**Abstract.** We consider the problem of designing succinct geometric data structures while maintaining efficient navigation operations. A data structure is said succinct if the asymptotic amount of space it uses matches the entropy of the class of structures represented.

For the case of planar triangulations with a boundary we propose a succinct representation of the combinatorial information that improves to 2.175 bits per triangle the asymptotic amount of space required and that supports the navigation between adjacent triangles in constant time (as well as other standard operations). For triangulations with $m$ faces of a surface with genus $g$, our representation requires asymptotically an extra amount of $36(g-1) \lg m$ bits (which is negligible as long as $g \ll m/\lg m$).

## 1 Introduction

The problem of representing compactly the connectivity information of a two-dimensional triangulation has been largely addressed for compression purpose [2]. Indeed for a triangulation with $m$ triangles and $n$ vertices, the usual description of the incidence relations between faces, vertices and edges involves $6m + n$ pointers (each triangle knows its neighbors and its incident vertices, and each vertex has a reference to an incident triangle). In practice, this connectivity information uses $32 \times 7 = 224$ bits/triangle, or in theory $7 \log m$ bits/triangle (as for a triangle mesh it holds $n < m$), that is much more than the cost of point coordinates [4]. The enumeration of all different structures that the connectivity can assume shows that for the case of planar triangulations (with degree 3 faces) an encoding requires asymptotically 1.62 bits/triangle (or 3.24 bits/vertex, see [12] for a recent optimal encoding). Similarly, 2.175 bits per triangle are needed to code triangulations when a larger boundary of arbitrary size is allowed (in [1] the entropy of this class of triangulations is computed). In this paper, our purpose is not to compress the data for storage or network transmission, but to design a compact representation that can be used in main memory and supports navigation queries. Since we care for coming down to the entropy bound, this work pertains to the algorithmics of *succinct* data structures, as discussed below.

---

[*] This work has been supported by the French "ACI Masses de données" program, via the Geocomp project, http://www.lix.polytechnique.fr/~schaeffe/GeoComp/.

F. Dehne, A. López-Ortiz, and J.-R. Sack (Eds.): WADS 2005, LNCS 3608, pp. 134–145, 2005.
© Springer-Verlag Berlin Heidelberg 2005

**Contribution.** At a conceptual level, our contribution is to show that two-dimensional geometric objects have nice local neighborhood relations that allow to apply a direct hierarchical approach to represent them succinctly, without even using separators or canonical orders. More precisely, given a triangulation of $m$ triangles, we propose a structure that uses $2.175m + O\left(m\frac{\lg \lg m}{\lg m}\right)$ bits and supports access from a triangle to its neighbors in $O(1)$ worst case time.

This storage is asymptotically optimal for the class of planar triangulations with a boundary. To our knowledge this is the first optimal representation supporting queries in constant time for geometric data structures.

Our approach extends directly to the more general case of triangulations with $m$ triangles of a surface with genus $g$. In this case, the structure uses $2.175m + 36(g - 1)\lg m + O\left(m\frac{\lg \lg m}{\lg m} + g \lg \lg m\right)$ bits, which remains asymptotically optimal for $g = o(m/\lg m)$. For $g = \Theta(m)$, we still have an explicit dominant term, which is of the same order as the cost of a pointer-based representation. Finally, our approach allows to take advantage of low diversity in the local structure: for instance, when applied to the class of triangulations with vertex degree at most 10, our construction automatically adjusts to the corresponding entropy.

### Related Work on Compact Representations of Graphs.
A first approach to design better geometric data structures is, as done in [9], to look from the programming point of view for practical solutions that improve by a constant factor on usual representations [4]. From a theoretical point of view however, standard representations are intrinsically non optimal since they use global pointers across the structure: $\Theta(m \log m)$ bits are needed to index $m$ triangles.

The seminal work of Jacobson [8] showed that it is possible to represent planar graphs with $O(n)$ bits, allowing adjacency queries in $O(\lg n)$ time. The result is based on a compact representation for balanced parenthesis systems, and on the four page decomposition of planar graphs. This two step approach was pushed further by improving on the representation of parenthesis systems or by using alternative graph encodings. Munro and Raman [10] achieve $O(1)$ time for adjacency between vertices and degree queries: for planar triangulations with $e$ edges and $n$ vertices only $2e + 8n$ bits are asymptotically required, that is, in terms of the number $m$ of faces, between $7m$ and $12m$ bits depending on the boundary size. The best result for triangulations is due to Chuang et al. [6] $2e + n$ bits (which is equivalent to $3.5m$ for triangulations with a triangular boundary), with slightly different navigation primitives than ours. For the general case of planar graphs, Chiang at al. further extended and improved this result [5].

Although this literature is mainly focused on the asymptotic behaviors, it has been demonstrated by Blandford et al. [3] that $O(n)$ data structures can be competitive in practice: using graph separators and local labellings, they propose a compact representation for separable graphs that supports adjacency and degree queries in constant time and saves space already for middle size graphs. The design of compact data structures ultimately relies on partitioning into small regions, inside which local labels can be used, and on describing efficiently inter-region relations. A compact data structure is called *succinct* when its space requirement matches asymptotically the entropy bound at first order. For the previous approach to yield a succinct data structure, the partitioning must be done without increase of entropy, and the inter-region relations must be described within sub-linear space. To our knowledge, this was done successfully only for simpler structures like bit vectors, dictionaries or trees [13, 10, 11].

**Overview of our Structure.** As opposed to the previous approaches for triangulations, we apply the partitioning process directly to the triangulation, following a three level scheme similar to what was done for trees in [11].

The initial triangulation of $m$ triangles is divided in pieces (*small triangulations*) having $\Theta(\lg^2 m)$ triangles, and each small triangulation is then divided into *planar* sub-triangulations (*tiny triangulations*) of size $\Theta(\lg m)$. Any such subdivision is acceptable for our approach. We produce one in linear time using a tree partitioning algorithm on a spanning tree of the dual graph.

Then we construct a three level structure. The first level is a graph linking the $\Theta\left(\frac{m}{\lg^2 m}\right)$ small triangulations. This graph is classically represented with pointers of size $O(\lg m)$, and in view of its number of nodes and edges, its storage requires $o(m)$ bits. The second level consists in a graph linking the tiny triangulations, or more precisely a map, since the relative order of neighbors around a tiny triangulation matters here. The nodes of this map are grouped according to the small triangulation they belong to. This allows to use local pointers of size $O(\lg \lg m)$ to store adjacencies between tiny triangulations. The combinatorial information of a tiny triangulation is not explicitly stored at this second level, we just store a pointer to the third level: the catalog of all possible tiny triangulations. The whole size of all these pointers to the third level can be proved to be 2.175 bits per triangle and all other informations are sub-linear.

As such, the structure would not describe completely the initial triangulation: the combinatorics of combining tiny triangulations into a big triangulation is more involved than, *e.g.*, the combinatorics of combining subtrees into a big tree as in [11]. The second level must therefore be enriched with a coloring of the vertices on the boundary of tiny triangulations according to the number of tiny triangulations they belongs to. This coloring describes how the boundary edges of a tiny triangulation are distributed between its neighbors (which are given by the second level map). Like its combinatorial description, the coloring of a tiny triangulation is encoded through a pointer to a catalog of all possible border colorings. The subtle point is that the total size of these pointers is sub-linear, even though the total length of the borders themselves can be linear (recall no assumption is made on the quality of the decomposition in tiny triangulations). The space requirement is dictated by the cost of the pointers to tiny pieces: since these are planar triangulations with boundary the representation is succinct for this class. On the other hand, restraining the catalog to a subclass (like triangulations with degree at most 10) immediately cuts the pointer sizes and reduces the cost to the associated entropy.

The construction of our representation can be performed in linear time and a complete analysis is provided in [1].

## 2    Preliminaries

**Model of Computation.** As in previous works about succinct representations of binary trees, our model of computation is a RAM machine with $O(1)$ time access and arithmetic operation on words of size $\log_2 m$. Any element in a memory word can be accessed in constant time, once we are given a pointer to a word and an integer index inside. The machine word size matches the problem size, in the sense that a word is large enough to store the input problem size. We use $\lg m$ to denote $\lceil \log_2(1 + m) \rceil$. From now on, when we speak about the time complexity of an algorithm we refer to the number of elementary operations on words of size $\lg m$, and about storage we refer to the size of an object in term of bits.

**Notations and Vocabulary.** The initial triangulation is denoted $T$ and its size $m$ (from now on the size of any triangulation is its number of triangles). When the triangulation $T$ is not planar, we denote by $g$ its genus. The *small* triangulations, of size between $\frac{1}{3} \lg^2 m$ and $\lg^2 m$, are denoted $ST_i$. Finally the *tiny* triangulations, of size between $\frac{1}{12} \lg m$ and $\frac{1}{4} \lg m$, are denoted $TT_j$.

All tiny triangulations shall be *planar* triangulations with *one* boundary cycle. As subtriangulations of $T$, these tiny triangulations will share their boundary edges. More precisely a boundary edge can be shared by two different tiny triangulations or can also appear twice on the boundary of one tiny triangulation. We call *multiple vertices* those vertices that are incident to at least 3 boundary edges (generically they are shared by more than two tiny triangulations, but self-intersections of boundaries also create multiple vertices). A *side* of a tiny triangulation $TT_j$ is a sequence of consecutive boundary edges between two multiple vertices: edges of a same side are shared by $TT_j$ with a same tiny triangulation $TT_{j'}$ (possibly with $j' = j$). The boundary of a tiny triangulation is divided in this way in a cyclic sequence of sides, called the *coloring* of the boundary. As just seen, this coloring is induced by the distinction multiple/normal vertices.

The exhaustive set of all possible tiny triangulations with at most $\frac{1}{4} \lg m$ triangles is stored in a structure denoted $A$ while the set of all colorings of a boundary with less than $\frac{1}{4} \lg m$ vertices is stored in a structure called $B$. The adjacencies between the small triangulations $ST_i$ are stored in a graph denoted $F$, those between tiny triangulations $TT_j$ in a graph $G$. The part of $G$ corresponding to pieces of $ST_i$ is denoted $G_i$. To be more precise $G$ must be a *map*: at each node the set of incident arcs (one for each side) is stored in an array, sorted to reflect the circular arrangement of the sides of the triangulation. For $F$ we could content with a graph structure, but it is convenient, as discussed in Appendix A, to construct both $F$ and $G$ by the same simplification process: in particular, although $F$ and $G$ can have loops (corresponding to boundary self intersections) and multiple arcs (corresponding to two subtriangulations sharing different sides), their number of edges is linearly bounded in the genus and number of vertices because they are constructed with all faces of degree at least 3.

For the sake of clarity, from now on we will use the word *arcs* and *nodes* to refer to edges and vertices of the maps $F$, $G$ and $G_i$, and keep the word edges and vertices only for the edges and vertices of $T$ and of the subtriangulations.

**Operations on the Triangulation.** The following primitive operations are supported by our representation in $O(1)$ time.

- $Triangle(v)$: returns a triangle incident to vertex $v$;
- $Index(\triangle, v)$: returns the index of vertex $v$ in triangle $\triangle$;
- $Neighbor(\triangle, v)$: returns the triangle adjacent to $\triangle$ opposite to vertex $v$ of $\triangle$;
- $Vertex(\triangle, i)$: returns the vertex of $\triangle$ of index $i$.

With marginal modifications, the structure could also allow for other local operations, like degree queries or adjacency between vertices in constant time.

## 3   Exhaustive List of All Tiny Triangulations

All possible triangulations having $i$ triangles ($i \leq \frac{1}{4} \lg m$) are generated and their explicit representations are stored in a collection $A$ of tables $A_i$. A reference to a tiny triangulation in $A_i$ costs asymptotically $2.175i$ bits because there are at most $2^{2.175i}$ triangulations with $i$ triangles (for more details refer to [1]).

In the rest of this section we describe the organization of the structure (see also Figure 1) and we analyze the storage. The construction of the structure can be done in sub-linear time (see [1]).

### Description of the Representation

- $A$ is a table of size $\frac{1}{4} \lg m$, in which the $i$th element is a pointer to Table $A_i$.
- $A_i$ is a table containing all possible triangulations having exactly $i$ triangles. The $j$th element is a pointer to an explicit representation $A_{i,j}^{explicit}$ of the triangulation $A_{i,j}$.
- $A_{i,j}^{explicit}$ contains at least two fields:

  — $A_{i,j}^{explicit}.vertices$ is the table of the vertices of $A_{i,j}$. Each vertex just contains the index of an incident triangle in Table $A_{i,j}^{explicit}.triangles$. By convention, the boundary vertices appear first in that table, and are stored in the counterclockwise order of the boundary of $A_{i,j}$. For boundary vertices, the incident triangle stored is required to be the one incident to next edge on the boundary.

  — $A_{i,j}^{explicit}.triangles$ is the table of the triangles of $A_{i,j}$. Each triangle contains the indices of its vertices in $A_{i,j}^{explicit}.vertices$ and of its neighbors in $A_{i,j}^{explicit}.triangles$. Triangles on the boundary have *null* neighbors.

**Storage Analysis.** *The storage of Table $A$, and of all the information associated with Tables $A_i$ requires asymptotically $O(m^{0.55})$ bits.*

- $A$ is a table of size $\frac{1}{4} \lg m$ of pointers of size $\lg m$ and thus costs $O(\lg^2 m)$.
- $A_i$ is a table of at most $2^{2.175i}$ pointers on $\lg m$ bits, thus the storage of $A_i$ requires less than $O(2^{2.175i} \lg m)$ bits.
- The explicit representation $A_{i,j}^{explicit}$:

  — $A_{i,j}^{explicit}.triangles$ (resp. $A_{i,j}^{explicit}.vertices$) is a table of size less than $i \leq \lg m$ (resp. less than $i + 2 \leq \lg m + 2$). Each element consists in several indices of value less than $i$, thus representable on $\lg \lg m$ bits.

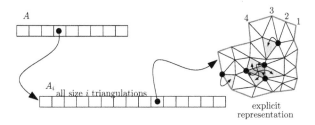

**Fig. 1.** Storage of all tiny triangulations

Thus the size of one $A_{i,j}^{explicit}$ is $O(\lg m \lg \lg m)$ bits and the total size of the $A_{i,j}^{explicit}$ indexed in Table $A_i$ is less than $O(2^{2.175i} \lg m \lg \lg m)$ bits.

Finally the storage requirement for the whole structure $A$ is obtained by summing over $i$, which yields $O(2^{2.175\frac{1}{4}\lg m} \lg m \lg \lg m) = O(m^{0.55})$.

## 4    Boundary Descriptions

As already explained, we need to distinguish some vertices on the boundary of each tiny triangulation. This will be done with the help of a structure essentially equivalent to a bit vector supporting rank and select operations in constant time. This problem was addressed very much in detail in the literature and compact solutions have been proposed (see [14], [13], [7] and ref. therein). Since the bit vectors we use have size at most $\frac{1}{4}\lg m$, we can content with a simple explicit encoding of all bit-vectors of size $p$ and weight $q$ in a collection $B$ of tables $B_{pq}$. Then $B_{pq}$ contains $\binom{p}{q}$ elements and a reference to one entry of $B_{pq}$ has size $\lg\binom{p}{q} \le \min(q\lg p, p)$ bits (observe that the size of a reference is at most $\frac{1}{4}\lg m$, which allows to index in tables $B_{pq}$ in $O(1)$ time). In the rest of this section we provide the description and analysis of the structure.

**Description of the Representation**
- $B$ is a bi-dimensional array of size $\frac{1}{4}\lg m \times \frac{1}{4}\lg m$: each entry $B(p,q)$ is a pointer to Table $B_{pq}$.
- $B_{pq}$ is a table containing for the $k$th bit-vector of size $p$ and weight $q$ a pointer to a structure $B_{pqk}^{RS}$ allowing Rank/Select in constant time.
- $B_{pqk}^{RS}$ is a table of length $p$ with two fields storing the precomputed result for $Rank_1$ and $Select_1$:
  — $B_{pqk}^{RS}(i).rank$ is the number of '1's that precede the $i$-th bit.
  — $B_{pqk}^{RS}(i).select$ is the position of the $i$-th '1' in the vector.

**Storage Analysis.** *The storage of Table $B$, and of all the information associated with Tables $B_{pqk}$ requires asymptotically $O(m^{\frac{1}{4}} \lg m \lg \lg m)$ bits.*

- $B$ is a table of $(\frac{1}{4}\lg m)^2$ pointers of size $\lg m$, its size is $O(\lg^3 m)$ bits.

- $B_{pq}$ is a table containing $\binom{p}{q}$ pointers of size $O(\lg m)$.
- $B_{pqk}^{RS}(i).rank$, $B_{pqk}^{RS}(i).select$ are all integers less than $\frac{1}{4}\lg m$ and then representable on $\lg\lg m$ bits. The size of $B_{pqk}^{RS}$ is $O(\lg m \lg\lg m)$.

The total amount of space required for storing all the bit-vectors of size (and weight) less than $\frac{1}{4}\lg m$ is then $\sum_{p,q}\binom{p}{q}O(\lg m \lg\lg m) = \left(\sum_p 2^q\right)O(\lg m \lg\lg m)$, which is bounded by $2^{\frac{1}{4}\lg m+1}O(\lg m \lg\lg m) = O(m^{\frac{1}{4}}\lg m \lg\lg m))$.

## 5    Map of Tiny Triangulations

The main triangulation is split into small triangulations which are themselves split into tiny triangulations. In this section we describe the map $G$ that stores the incidences between tiny triangulations. The memory for this map is organized by gathering nodes of $G$ that correspond to tiny triangulations that are part of the same small triangulation $\mathcal{ST}_i$ in a sub-map $G_i$. The purpose of this partition is to allow for the use of local pointers of small size for references inside a given sub-map $G_i$.

The map $G$ may have multiple arcs or loops but all its faces have degree $\geq 3$. Each arc of $G$ between $\mathcal{TT}_j$ and $\mathcal{TT}_{j'}$ corresponds to a side shared by $\mathcal{TT}_j$ and $\mathcal{TT}_{j'}$.

**Description of the Representation.** The memory dedicated to $G$ is organized in a sequence of variable size zones, each dedicated to a $G_i$. The memory requirements are analyzed afterward.

In the zone for $G_i$, for each node $G_{i,j}$ corresponding to a tiny triangulation $\mathcal{TT}_{i,j}$, we have the following informations:

— $G_{i,j}^t$ is the number of triangles in $\mathcal{TT}_{i,j}$.
— $G_{i,j}^b$ is the size of the boundary of $\mathcal{TT}_{i,j}$.
— $G_{i,j}^A$ is the index of the explicit representation of $\mathcal{TT}_{i,j}$ in Table $A_{G_{i,j}^t}$.
— $G_{i,j}^s$ is the degree of the node $G_{i,j}$ (it is also the number of sides of $\mathcal{TT}_{i,j}$)
— $G_{i,j}^B$ is the index in Table $B_{G_{i,j}^b, G_{i,j}^s}$ of a bit vector of size $G_{i,j}^b$ and weight $G_{i,j}^s$. (This bit vector encodes the way the boundary of $\mathcal{TT}_{i,j}$ splits into sides: the $i$th bit is 0 if the $i$th vertex on the boundary of $\mathcal{TT}_{i,j}$ is inside a side, or 1 if this is a multiple vertex that separates two sides)

— Each of the $G_{i,j}^s$ arcs of $G_i$ that are incident to $G_{i,j}$ is described by some additional information (beware that loops appear twice). Assume that the $k$th such arc connects $G_{i,j}$ to a neighbor $G_{i',j'}$ in $G$, then we store:

— $G_{i,j,k}^{address}$ the relative address of the first bit concerning the node of the neighbor in the memory zone associated to its small triangulation $G_{i'}$.

— $G_{i,j,k}^{back}$ the index $k'$ of the side corresponding to the current arc in the numbering of sides at the opposite node $G_{i',j'}$.

— $G_{i,j,k}^{small}$ the index of the small triangulation $G_{i'}$ in the table of the neighbors of $G_i$ in the main map $F$ (if $i'=i$ then this index is set to 0).

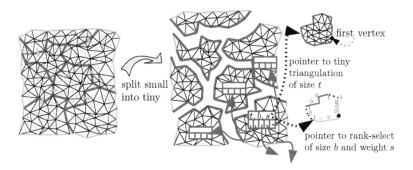

**Fig. 2.** This Figure shows the decomposition of a small triangulation into tiny triangulations and the map $G_i$ that describes their adjacency relations

**Storage Analysis.** *The storage of map $G$ requires asymptotically* $2.175m + O(g \lg \lg m) + O\left(m \frac{\lg \lg m}{\lg m}\right)$ *bits.*

For each node:

- $G_{i,j}^t$, $G_{i,j}^b$ and $G_{i,j}^s$ are less than $\frac{1}{4} \lg m$: each is stored in $\lg \lg m$ bits.
- $G_{i,j}^A$ is an index in Table $A_{G_{i,j}^t}$ stored in $2.175 G_{i,j}^t$ bits (see Section 3)
- $G_{i,j}^B$ is an index in $B_{G_{i,j}^b, G_{i,j}^s}$ stored in $G_{i,j}^s \lg G_{i,j}^b$ bits (see Section 4)
- The number of tiny triangulations neighboring $G_{i,j}$ is $G_{i,j}^s < \frac{1}{4} \lg m$. We have for each:
    - the pointers $G_{i,j,k}^{address}$ are stored in $K \lg \lg m$ bits ($K$ chosen below).
    - $G_{i,j,k}^{back}$ is less than $\frac{1}{4} \lg m$ and thus can be stored in $\lg \lg m$ bits.
    - $G_{i,j,k}^{small}$ requires $2 \lg \lg m$ bits of storage: indeed a small triangulation

has at most $\lg^2 m$ triangles, hence at most $\lg^2 m$ edges on its boundary, thus the table of the neighbors of $G_i$ in $F$ has less than $\lg^2 m$ entries.

Since each arc appears on at most two nodes, the cost per arc can be evaluated independently as $2(K+3) \lg \lg m$ bits per arc. It then remains for node $G_{i,j}$ of $G_i$ a cost of $3 \lg \lg m + 2.175 G_{i,j}^t + G_{i,j}^s \lg G_{i,j}^b$.

The number of nodes is at most $12 \lg m$ and the number of arcs (including arcs directed to other $G_{i'}$) is bounded by the number of edges of $\mathcal{T}$ incident to triangles of $\mathcal{ST}_i$, that is by $\lg^2 m$.

The cost for $G_i$ is thus $C_i \le 2(K+3) \lg^2 m \lg \lg m + 12 \lg m (3 \lg \lg m + 2.175 \frac{1}{4} \lg m + \frac{1}{4} \lg m \lg \lg m)$. Taking $K = 5$, we have $\lg C_i < K \lg \lg m$ for all $m \ge 2$, which validates our hypothesis for the storage of $G_{i,j,k}^{address}$.

The overall cost for the complete map $G$ is obtained by summing over $i, j$:

$$\sum_i \sum_j \left(3 \lg \lg m + 2.175 G_{i,j}^t + G_{i,j}^s (\lg G_{i,j}^b) + G_{i,j}^s \cdot 8 \lg \lg m\right)$$

$$\le 2.175 \sum_{i,j} G_{i,j}^t + 9 \lg \lg m \sum_{i,j} G_{i,j}^s + 3 \lg \lg m \cdot 12 \frac{m}{\lg m}.$$

The sum over $G_{i,j}^t$ is the total number of triangles, *i.e.* $m$. The sum over $G_{i,j}^s$ is the sum of the degrees of the nodes of the map $G$, or equivalently, twice its number of arcs.

Since $G$ has only faces of degree $\geq 3$, its number $a$ of arcs linearly bounds its number $f$ of faces: $2a = \sum_f d(f) \geq 3f$. Euler's formula can then be written (with $n$ for the number of nodes and $g$ for the genus of $G$ which is also the genus of $T$):

$$3(a + 2) = 3n + 3f + 6g \qquad \Leftrightarrow \qquad a \leq 3n + 6(g - 1).$$

Finally the number $n$ of nodes of $G$ is bounded by $12m/\lg m$, so that the cost of representing $G$ is

$$C = 2.175m + 9\lg\lg m \cdot 2\left(3 \cdot 12\frac{m}{\lg m} + 6(g - 1)\right) + 3\lg\lg m \cdot 12\frac{m}{\lg m},$$

and the lemma follows. Observe also that the bound $g \leq \frac{1}{2}m + 1$ yields $\lg C \leq 2\lg m + 8$ for all $m$ which will be used in the next section.

## 6    Graph of Small Triangulations

The last data structure needed is a graph $F$ that describes the adjacency relations between small triangulations. The circular arrangement of neighbors is not used here so do not need a map structure as for $G$. However, $F$ is obtained by construction as a map and it is convenient for the storage analysis to observe that, as a map, $F$ has a genus smaller or equal to the genus of $G$ and contains no faces of degree less than 3. We adopt here an explicit pointer based representation.

**Description of the Representation.** We store for each node of $F$ its degree, a link to the corresponding part of $G$ and the list of its neighbors. More precisely, for a node $F_i$ corresponding to a small triangulation $ST_i$:

- $F_i^s$ is the degree of node $F_i$ in the map $F$ (it corresponds to the number of small triangulations adjacent to $ST_i$);
- $F_i^G$ is a pointer to the sub-map $G_i$ of $G$ that is associated to the small triangulation $ST_i$.
- A table of pointers to neighbors: $F_{i,k}^{address}$ is the address of the $k$th neighbor of $F_i$ in $F$.

**Storage Analysis.** *The graph $F$ uses $36(g - 1)\lg m + O\left(\frac{m}{\lg m}\right)$ bits.*

Recall that a small triangulation contains between $\frac{1}{3}\lg^2 m$ and $\lg^2 m$ triangles, thus map $F$ has at most $3m/\lg^2 m$ nodes.

- $F_i^s$ is less than $\lg^2 m$, and thus representable on $2\lg\lg m$ bits.
- the address of $G_i$ is a pointer of size bounded by $2\lg m + 8$.
- the pointers $F_{i,k}^{address}$ are stored on $K'\lg m$ bits each ($K'$ chosen below).

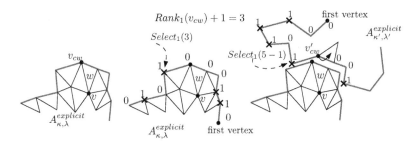

**Fig. 3.** Going to the neighbor

Summing on all the small triangulations we obtain that the bit size of $F$ is $(2\lg m + 8) \cdot 3m/\lg^2 m + K' \lg m \sum_i F_i^s$. The sum of the $F_i^s$ is the sum of the degrees of nodes of $F$, which is also twice its number of arcs.

In analogy with what was done for the map $G$, the number of arcs of $F$ can be bounded more precisely by three times its number of nodes, which is less than $3m/\lg^2 m$, plus six times the genus minus one of $F$, which is bounded by the genus of $T$. Using the bound $g < \frac{1}{2}m + 1$ on the genus, the value $K' = 3$ is seen to satisfy the constraints for $m \geq 5$. Finally the total bit cost for $F$ is thus: $36(g-1)\lg m + O(m/\lg m)$.

## 7 Navigation

**Triangle and Vertex Representations.** In our structure, a triangle $t$ is represented by a triple $(F_i, a, w)$ where $F_i$ is a node of $F$ such that $t \in \mathcal{ST}_i$, $a$ is the address of $G_{ij}$ in the memory zone of $G_i$ such that $t \in \mathcal{TT}_{ij}$ and $w$ is the index of the triangle corresponding to $t$ in $A_{\kappa,\lambda}^{explicit}$ where $A_{\kappa,\lambda}$ is the triangulation to which $G_{ij}$ points.

Similarly a vertex is represented by a triple $(F_i, a, v)$. Observe that, as such, the representation of a vertex is not unique since, as opposed to triangles, vertices may be shared by several tiny triangulations. As sketched in Section 8, upon adding a negligible amount of information in the map $G$, a unique representation could be defined if needed (*e.g.* to test adjacency of vertices in constant time, or to attach data to vertices). However this is not needed for the four operations we have chosen to describe.

**Operations on the Triangulation.** Given a triangle $(F_i, a, w)$ or a vertex $(F_i, a, v)$ the operations *Triangle*, *Index* and *Vertex* are implemented directly by performing the operation in the explicit representation $A_{\kappa,\lambda}^{explicit}$.

The difficulty is with $Neighbor((F_i, a, w), (F_i, a, v))$.

- Check if the corresponding neighbor $w'$ of $w$ exists in the explicit representation $A_{\kappa,\lambda}$: if it does return $(F_i, a, w')$.

Otherwise, the neighbor must be found across the boundary of the current tiny triangulation:

— Find in $A^{explicit}_{\kappa,\lambda}$ the vertex $v_{cw}$ following $v$ in clockwise order around $w$.

— Compute $l = Rank_1(v_{cw}) + 1$ in the bit vector associated to $G_{ij}$: it says that we are on the $l$th side of $TT_{ij}$ ($l = 3$ in Figure 3);

— Compute $l' = Select_1(l) - v_{cw}$: it says that $v_{cw}$ is $l'$th vertex before the end of the side; ($l' = 1$ in Figure 3); recall that $Select_1(l)$ gives the position of the last vertex of the current $l$th side.

— Let $x = G^{address}_{i,j,l}$, $y = G^{back}_{i,j,l}$ and $z = G^{small}_{i,j,l}$.

— If $z > 0$ then we must change also small triangulation: let $G_{i'}$ be the sub-map of $G$ pointed at by the $z$th neighbor $F_{i'}$ of $F_i$ in $F$.

Otherwise (that is, $z = 0$) let $G_{i'}$ be equal to $G_i$.

— Let $G_{i',j'}$ be the node of $G$ at address $x$ in the memory zone of $G_{i'}$ and $A^{explicit}_{\kappa',\lambda'}$ the tiny triangulation it points at ($y = 5$ in Figure 3, the $y$th side of $G_{i',j'}$ matches the $l$th side of $G_{i,j}$).

— Let $v'_{cw} = Select_1(y - 1) + l'$ in the bit vector associated to $G_{i',j'}$: then $v'_{cw}$ in $A^{explicit}_{\kappa',\lambda'}$ matches $v_{cw}$ in $A^{explicit}_{\kappa,\lambda}$.

— Let $w'$ be the triangle pointed at by $v'_{cw}$ in $A^{explicit}_{\kappa',\lambda'}$.

— Return triangle $(F_{i'}, x, w')$.

# 8    Concluding Remarks

*Unique representation for vertices* A vertex on boundary of a tiny triangulation has several representations $(F_i, a, v)$. To show how to test that two such representations correspond to the same vertex of $T$ in constant time, let us distinguish three types of ambiguous vertices: vertices incident to only two boundary edges, multiple vertices incident to at most two small triangulations, and multiple vertices incident to at least three small triangulations. Identity can already be tested for the first type. For the $O(n/\lg n)$ vertices of the second type, a $\lg \lg n$ labelling (local to each $T_i$) can be used to describe the multiple vertices on the boundary of $TT_{ij}$ in an ordered array at each $G_{ij}$, and with the boundary description this allows to test identity. Finally upon listing in a table $F^{vertex}_i$ the vertices of the third type appearing in each $ST_i$, $O(\lg \lg n)$ indices to this table can be added in $G_{ij}$ to allow for the test. The extra storage is negligible at first order.

*Attaching information.* The proposed structure represents only the connectivity of the triangulation. One may want to attach information to vertices or triangles, such as vertices coordinates (or colors...). This should be done by adding the information to nodes of $G$. For instance one can add to $G_{i,j}$ a table $G^{coordinate}_{i,j}$ describing the coordinates of the vertices of $TT_{i,j}$. This Table contains the coordinates of all the internal vertices of $TT_{i,j}$ and a selection of its boundary vertices, so that vertices lying on the side between two tiny triangulations are stored only once. To retrieve vertices shared by several tiny triangulations, one uses the above unique representation. Basic compression on these coordinates can be obtained by giving them in a frame local to $G_{i,j}$.

*Practical implementation.* The result here is mainly theoretical: if $m$ is one billion, $\frac{1}{4} \lg m$ is only 7. However the value $\frac{1}{4}$ is chosen to ensure that the table $A$ can be constructed in sub-linear time: looking at the actual number of triangulations with $p$ faces for small $p$, one can check that constructing the table of all tiny triangulations up to size 13 is actually be feasible. In particular Table A can be computed once and for all and stored. We intend to implement and test a simplified version of this work, by gathering triangles in small groups of 3 to 5 triangles and making a map of these groups.

# References

1. L. Castelli Aleardi, O. Devillers, and G. Schaeffer.   Compact representation of triangulations.   Technical report, RR-5433 INRIA, 2004.   available at `http://www.inria.fr/rrrt/rr-5433.html`.
2. P. Alliez and C. Gotsman. Recent advances in compression of 3d meshes. In N.A. Dodgson, M.S. Floater, and M.A. Sabin, editors, *Advances in Multiresolution for Geometric Modelling*, pages 3–26. Springer-Verlag, 2005.
3. D. Blanford, G. Blelloch, and I. Kash. Compact representations of separable graphs. In *Proc. of the Annual ACM-SIAM Symp. on Discrete Algorithms*, pages 342–351, 2003.
4. J.-D. Boissonnat, O. Devillers, S. Pion, M. Teillaud, and M. Yvinec. Triangulations in CGAL. *Comput. Geom. Theory Appl.*, 22:5–19, 2002.
5. Y.-T. Chiang, C.-C. Lin, and H.-I. Lu. Orderly spanning trees with applications to graph encoding and graph drawing. *SODA*, pages 506–515, 2001.
6. R.C.-N Chuang, A. Garg, X. He, M.-Y. Kao, and H.-I. Lu. Compact encodings of planar graphs via canonical orderings and multiple parentheses. *Automata, Laguages and Programming*, pages 118–129, 1998.
7. D. R. Clark and J. I. Munro. Efficient suffix trees on secondary storage. In *SODA*, pages 383–391, 1996.
8. G. Jacobson. Space efficients static trees and graphs. In *Proceedings of the IEEE Symposium on Foundations of Computerb Science (FOCS)*, pages 549–554, 1989.
9. M. Kallmann and D. Thalmann. Star-vertices: a compact representation for planar meshes with adjacency information. *Journal of Graphics Tools*, 6:7–18, 2002.
10. J. I. Munro and V. Raman. Succint representation of balanced parantheses and static trees. *SIAM J. on Computing*, 31:762–776, 2001.
11. J. I. Munro, V. Raman, and A. J. Storm. Representing dynamic binary trees succintly. In *SODA*, pages 529–536, 2001.
12. D. Poulalhon and G. Schaeffer. Optimal coding and sampling of triangulations. In *Proc. Intern. Colloqium ICALP'03*, pages 1080–1094, 2003.
13. R. Raman, V. Raman, and S.S. Rao. Succint indexable dictionaries with application to encoding k-ary trees and multisets. In *SODA*, pages 233–242, 2002.
14. V. Raman and S.S. Rao. Static dictionaries supporting rank. In *ISAAC*, pages 18–26, 1999.

# Line-Segment Intersection Made In-Place

Jan Vahrenhold

Westfälische Wilhelms-Universität Münster,
Institut für Informatik, 48149 Münster, Germany
`jan@math.uni-muenster.de`

**Abstract.** We present a space-efficient algorithm for reporting all $k$ intersections induced by a set of $n$ line segments in the place. Our algorithm is an in-place variant of Balaban's algorithm and runs in $\mathcal{O}(n \log_2^2 n + k)$ time using $\mathcal{O}(1)$ extra words of memory over and above the space used for the input to the algorithm.

## 1 Introduction

Researchers have studied space-efficient algorithms since the early 70's. Examples include merging, (multiset) sorting, and partitioning problems; see [8, 9, 11]. Brönnimann *et al.* [5] were the first to consider space-efficient geometric algorithms and showed how to compute the convex hull of a planar set of $n$ points in $\mathcal{O}(n \log_2 h)$ time using $\mathcal{O}(1)$ extra space, where $h$ denotes the size of the output. Recently, Brönnimann *et al.* [4] developed some space-efficient data structures and used them to solve a number of geometric problems such as convex hull, Delaunay triangulation and nearest neighbor queries. Bose *et al.* [3] developed a general framework for geometric divide-and-conquer algorithmus and derived space-efficient algorithms for the nearest neighbor, bichromatic nearest neighbor, and orthogonal line segment intersection problems, and Chen and Chan [7] presented an algorithm for the general line segment intersection problem: to report all $k$ intersections induced by a set of $n$ line segments in the plane.

*The Model.* The goal is to design algorithms that use very little extra space over and above the space used for the input to the algorithm. The input is assumed to be stored in an array A of size $n$, thereby allowing random access. We assume that a constant size memory can hold a constant number of words. Each word can hold one pointer, or an $\mathcal{O}(\log_2 n)$ bit integer, and a constant number of words can hold one element of the input array. The extra memory used by an algorithm is measured in terms of the number of extra words. In certain cases, the output may be much larger than the size of the input. For example, given a set of $n$ line segments, the number $k$ of intersections may be as large as $\Omega(n^2)$. We consider the output memory to be write-only space that is usable for output but cannot be used as extra storage space by the algorithm. This model has been used by Chen and Chan [7] for variable size output, space-efficient algorithms and accurately models algorithms that have output streams with write-only buffer

F. Dehne, A. López-Ortiz, and J.-R. Sack (Eds.): WADS 2005, LNCS 3608, pp. 146–157, 2005.

space. In the space-efficient model, an algorithm is said to work *in-place* iff it uses $\mathcal{O}(1)$ extra words of memory.

*Related Work.* There is a large number of algorithms for the line segment intersection problem that are not in-place, and we refer the reader to the recent survey by Mount [12]. In the space-efficient model of computation, Bose *et al.* [3] have presented an optimal in-place algorithm for the restricted setting when the input consists of only horizontal and vertical segments. Their algorithm runs in $\mathcal{O}(n \log_2 n + k)$ time and uses $\mathcal{O}(1)$ words of extra memory. Chen and Chan [7] modified the well-known algorithm of Bentley and Ottmann [2] and obtained a space-efficient algorithm that runs in $\mathcal{O}((n + k) \log_2^2 n)$ time and uses $\mathcal{O}(\log_2^2 n)$ extra words of memory.[1] We will improve these bounds to $\mathcal{O}(n \log_2^2 n + k)$ time and $\mathcal{O}(1)$ extra space thus making the algorithm in-place and establishing an optimal linear dependency on the number $k$ of intersections reported.

## 2    The Algorithm

Our algorithm is an in-place version of the optimal $\mathcal{O}(n \log_2 n + k)$ algorithm proposed by Balaban [1]. Balaban obtained this complexity by first developing an intermediate algorithm with running time $\mathcal{O}(n \log_2^2 n + k)$ and then applying the well-known concept of *fractional cascading* [6]. As fractional cascading relies on explicitly maintained copies of certain elements, this concept can only applied with $\mathcal{O}(n)$ extra space which is prohibitive for an in-place algorithm. Thus, we build upon the (suboptimal) intermediate algorithm.

### 2.1    Divide-and-Conquer and the Recursion Tree

Balaban's intermediate algorithm is a clever combination of plane-sweeping and divide-and-conquer; the plane is subdivided into two vertical strips each containing the same number of segment endpoints, and each strip is (recursively) processed from left to right—see Figure 1. While doing so, the algorithm maintains the following invariants:

**Invariant 1:** Prior to processing a strip, all segments crossing the left strip boundary are vertically ordered at the $x$-coordinate of the left strip boundary.

**Invariant 2:** During the sweep over a strip, all intersections inside the strip are reported.

**Invariant 3:** After having processed a strip, the segments crossing the right strip boundary are rearranged such that they are vertically ordered at the $x$-coordinate of the right strip boundary.

The base of recursion is the case when a set $\mathcal{L}$ of line segments spans a vertical strip $\langle b, e \rangle := [b, e] \times \mathbb{R}$ that does not contain any endpoint of a segment.

---

[1] If the model is changed such that the input can be destroyed, the bounds can be improved to $\mathcal{O}((n + k) \log_2 n)$ time and $\mathcal{O}(1)$ extra space.

Invariant 1 implies that this set of segments is sorted according to $<_b$, the vertical order at $x$-coordinate $b$.

Balaban explains his algorithm based upon the intuition that the recursive calls of a divide-and-conquer algorithm can be modelled as a recursion tree where each node is assigned the subproblem to be solved in the corresponding recursive call. The recursion starts at the root node of the recursion tree $\mathcal{T}$, and hence the algorithm can been said to process the nodes (and hence the strips) along an Euler tour of $\mathcal{T}$. A closer look at the algorithm will reveal that, during the execution of the algorithm, some of the intersections detected while processing a strip corresponding to a node $v \in \mathcal{T}$ are found while $v$ is being visited for the first time whereas some of these intersections are found while $v$ is being visited for the last time. This in turn implies that the algorithm follows a divide-and-conquer strategy similar to the one described in Algorithm 1:

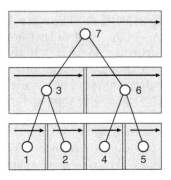

**Fig. 1.** Processing the recursion tree. Numbers indicate the order in which the strips are *finished*

---

**Algorithm 1.** RECURSIVE($A, b, e$): Recursive divide-and-conquer [3]

1: **if** $e - b \leq s$ where $s$ is the size of the recursion base. **then**
2:    BASE-CODE($A, b, e$)                    {Code for solving small instances}
3: **else**
4:    PRE-CODE($A, b, e$)                    {Setup Subproblem 1 in $A[b, \ldots, \lfloor e/2 \rfloor - 1]$}
5:    RECURSIVE($A, b, \lfloor e/2 \rfloor$)                    {First recursive call}
6:    MID-CODE($A, b, e$)                    {Setup Subproblem 2 in $A[\lfloor e/2 \rfloor, \ldots, e - 1]$}
7:    RECURSIVE($A, \lfloor e/2 \rfloor, e$)                    {Second recursive call}
8:    POST-CODE($A, b, e$)                    {Merge Subproblems 1 and 2 in $A[b, \ldots, e - 1]$}

---

This algorithm operates on an array $A[0, \ldots, n - 1]$ and makes calls to 4 subroutines: BASE-CODE is used to solve small instances, PRE-CODE is executed before any recursive calls, MID-CODE is executed after the first recursive call but before the second, and POST-CODE is executed after the second recursive call. In our previous work, we have shown that this general template can be realized in-place [3]. In the following subsections, we will demonstrate how both the subroutines and the partitioning of the segments to be processed can be realized using only $\mathcal{O}(1)$ extra space.

## 2.2    The Base of Recursion

As mentioned above, the base of recursion is the case when a set $\mathcal{L}$ of line segments spans a vertical strip $\langle b, e \rangle := [b, e] \times \mathbb{R}$ that does not contain any endpoint of a segment. By Invariant 1, the set $\mathcal{L}$ of segments is sorted according to $<_b$, the vertical order at $x$-coordinate $b$.

**Algorithm 2.** The algorithm $\text{SPLIT}_{b,e}(\mathcal{L}, \mathcal{Q}, \mathcal{L}')$ [1]

**Require:** $\mathcal{L} = (s_1, \ldots, s_m)$ is ordered by $<_b$.
**Ensure:** $\mathcal{L}'$ and $\mathcal{Q}$ are ordered by $<_b$; $\mathcal{Q}$ is complete relative to $\langle b, e \rangle$.
1:   $\mathcal{Q} := \emptyset$; $\mathcal{L}' := \emptyset$;
2:   **for** $j = 1$ to $m$ **do**
3:      **if** $s_j$ spans $\langle b, e \rangle$ and does not intersect the last segment of $\mathcal{Q}$ within $\langle b, e \rangle$ **then**
4:         $\mathcal{Q} \leftarrow s_j$.
5:      **else**
6:         $\mathcal{L}' \leftarrow s_j$.

Algorithm 2 partitions $\mathcal{L}$ into two sets $\mathcal{Q}$ and $\mathcal{L}' = \mathcal{L} \setminus \mathcal{Q}$ such that both sets are sorted according to $<_b$, that there are no intersections induced by the segments in $\mathcal{Q}$ ($\mathcal{Q}$ is called a *staircase*), and that $\mathcal{Q}$ is maximal: that is, *complete relative to* $\langle b, e \rangle$.

The correctness of the algorithms depends on the invariant that both the staircase $\mathcal{Q}$ and the remaining subset $\mathcal{L}'$ remain ordered by $<_b$. This condition cannot be enforced with a linear-time in-place algorithm as the only known such algorithm for stable partitioning [11] is a variant of $\{0, 1\}$-sorting. This implies that the algorithm has to be able to decide for any given element whether it should belong to $\mathcal{Q}$ or $\mathcal{L}'$—independent of whether the algorithm has seen any other element before and independent of the processing order of the elements. As constructing a staircase has to be done incrementally, using this non-incremental stable in-place partitioning is not feasible.

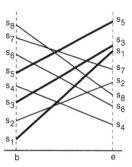

**Fig. 2.** Base of recursion. Fat lines indicate a maximal staircase

Let us for the moment, however, assume that such an in-place partitioning algorithm exists, and let us see how it can be used as a subroutine. Algorithm 3 recursively uses SPLIT to partition a set $\mathcal{L}$ of segments spanning $\langle b, e \rangle$ and sorted by $<_b$ such that the set $\text{Int}_{b,e}(\mathcal{L})$ of intersections induced by $\mathcal{L}$ and falling into the strip $\langle b, e \rangle$ can be found easily using a synchronized scan over the staircase $\mathcal{Q}$ and the set $\mathcal{L}'$, both of which are ordered by $<_b$. As a side effect, Algorithm 3 reorders the segments in $\mathcal{L}$ such that they are sorted according to $<_e$. This implies that, in the process of sweeping the plane, $\mathcal{L}$ can be used as the input for processing an adjacent strip $\langle b', e' \rangle$, i.e. a strip $\langle b', e' \rangle$ for which $b' = e$.

The running time of Algorithm 3 is linear in the number of segments in $\mathcal{L}$ and the number of intersections reported. To see this, note that Steps 1, 3, and 7 run in time linear in $|\mathcal{L}|$ and that Step 5 runs in time linear in $|\mathcal{L}|$ plus the number of intersections reported. For the recursive calls, observe that a segment is *not* assigned to a staircase (during the executing of Algorithm 2) hence being processed in a recursive call iff there exists at least one intersection with a staircase. The effects of the recursive calls to SPLIT are reverted by the repeated calls to MERGE (Line 7 of Algorithm 3). This operation is a linear-time operation as both $\mathcal{Q}$ and $\mathcal{R}'$ are ordered by $<_e$, and using the algorithm by Geffert *et al.* [9], it can also be performed in-place.

---

**Algorithm 3.** The algorithm $\text{SEARCHINSTRIP}_{b,e}(\mathcal{L}, \mathcal{R})$ [1]

---

**Require:** $\mathcal{L} = (s_1, \ldots, s_k)$ is ordered by $<_b$; each $s_i$ spans $\langle b, e \rangle$.
**Ensure:** $\mathcal{R} = (s_{\pi(1)}, \ldots, s_{\pi(k)})$ is ordered by $<_e$.
1: $\text{SPLIT}_{b,e}(\mathcal{L}, \mathcal{Q}, \mathcal{L}')$.          {Partition $\mathcal{L}$ into (a staircase) $\mathcal{Q}$ and $\mathcal{L}' = \mathcal{L} \setminus \mathcal{Q}$.}
2: **if** $\mathcal{L}' = \emptyset$ **then**
3:     $\mathcal{R} := \mathcal{Q}$.     {Base of recursion: No intersections within $\mathcal{L}$; $\mathcal{R}$ is ordered by $<_b$.}
4: **else**
5:     Find all intersections of $\mathcal{Q}$ and $\mathcal{L}'$ inside $\langle b, e \rangle$ using a synchronized scan.
6:     $\text{SEARCHINSTRIP}_{b,e}(\mathcal{L}', \mathcal{R}')$.          {Recursively find intersections within $\mathcal{L}'$.}
7:     $\mathcal{R} := \text{MERGE}_e(\mathcal{Q}, \mathcal{R}')$.          {Both $\mathcal{Q}$ and $\mathcal{R}'$ are ordered by $<_e$.}

---

We have noted [3] that, due to the use of a recursion stack, algorithms that use recursion generally are not in-place. However, if we assume the existence of an in-place partitioning algorithm $\text{INPLACESPLIT}(\mathtt{A}, b, e, \ell_b, \ell_e)$ that partitions $\mathcal{L} := \mathtt{A}[\ell_b, \ldots, \ell_e - 1]$ into $\mathcal{L}' := \mathtt{A}[\ell_b, \ldots, \ell_c - 1]$ and $\mathcal{Q} := \mathtt{A}[\ell_c, \ldots, \ell_e - 1]$ and returns the split index $\ell_c$, Algorithm 3, can be made in-place using this subroutine inside a simple **repeat-until**-loop (see Algorithm 4):

---

**Algorithm 4.** Algorithm $\text{INPLACESEARCHINSTRIP}(\mathtt{A}, b, e, \ell_b, \ell_e)$

---

**Require:** $\mathtt{A}[\ell_b, \ldots, \ell_e - 1]$ is ordered by $<_b$; each $s_i$ spans $\langle b, e \rangle$.
**Ensure:** $\mathtt{A}[\ell_b, \ldots, \ell_e - 1]$ is ordered by $<_e$.
1: Let $\ell_c := \ell_e$.
2: **repeat**
3:     Let $\ell := \ell_c$.
4:     $\ell_c := \text{INPLACESPLIT}(\mathtt{A}, b, e, \ell_b, \ell)$     {$\mathcal{L}' = \mathtt{A}[\ell_b, \ldots, \ell_c - 1]$; $\mathcal{Q} = \mathtt{A}[\ell_c, \ldots, \ell - 1]$}
5:     **if** $\ell_c > \ell_b$ **then**
6:         Find all intersections of $\mathcal{Q}$ and $\mathcal{L}'$ inside $\langle b, e \rangle$ using a synchronized scan.
7: **until** $\ell_c = \ell_b$
8: Repeatedly identify and merge staircases into $\mathcal{R} = \mathtt{A}[\ell_b, \ldots, \ell_e - 1]$.

---

Because Algorithm $\text{INPLACESPLIT}$ partitions the set $\mathcal{L}$ (that is the subarray $\mathtt{A}[\ell_b, \ldots, \ell_e - 1]$) such that the elements of the non-staircase set $\mathcal{L}'$ always appear in the front of the subarray, Algorithm 4 only needs to maintain the following pointers: one pointer to the beginning and the end of the original set $\mathcal{L}$, say $\ell_b$ and $\ell_e$, and one pointer $\ell$ to the end of the *current* set $\mathcal{L}$. The final merging step is implemented as follows: starting from $\ell$ and advancing towards at most $\ell_e$, we find the index $\ell'$ denoting the end of the next staircase to be merged by exploiting the fact that each staircase is sorted according to $<_e$. We then merge the staircases $\mathtt{A}[\ell_b, \ldots, \ell - 1]$ and $\mathtt{A}[\ell, \ldots, \ell' - 1]$ according to $<_e$ using an in-place merging algorithm [9], update $\ell$ with the value of $\ell'$, and repeat the merging process as long as $\ell < \ell_e$. By the argument used above, the overall runtime for merging all staircases is linear in $|\mathcal{L}| + |\text{Int}_{b,e}(\mathcal{L})|$.

Let us now come back to the problem of finding a stable in-place partitioning of $\mathcal{L} = \mathcal{L}' + \mathcal{Q}$. In order to still obtain both sets in sorted order, we proceed as follows: we implement INPLACESPLIT such as to use the approach of Algorithm SORTEDSUBSETSELECTION [3] to stably move the non-stairs, i.e. the set $\mathcal{L}'$ to the front of $\mathcal{L}$. We can do so incrementally, as we only need to keep track of the (position of the) topmost stair in order to decide whether the next segment in question can be added to the staircase or not. We then sort the segments in $\mathcal{Q}$ using an in-place sorting algorithm, e.g. *heapsort* [8].

**Lemma 1.** *Algorithm* INPLACESPLIT, *when invoked at a node $v$ of the recursion tree $\mathcal{T}$, runs in time* $\mathcal{O}(|\mathcal{L}|+|\mathrm{Int}_{b,e}(\mathcal{L})|+H_v)$ *where* $\sum_{v \in \mathcal{T}} H_v \in \mathcal{O}(n \log_2 n)$.

*Proof.* Selecting the non-stairs can be done in-place in linear time using SORTEDSUBSETSELECTION, and each segment not added to a staircase and thus considered in another pass can be charged to (at least) one intersection with the topmost stair. Each segment appears in a staircase exactly once, so the overall running time of sorting all staircases is in $\mathcal{O}(n \log_2 n)$.

### 2.3    The "Divide" and "Conquer" Phases

The main concept of Balaban's algorithm is to report a pair $(s,t)$ of intersecting segment at the highest node $v$ in the recursion tree where one of the segments, say $s$, is part of the staircase $\mathcal{Q}_v$ spanning the strip $\langle b, e \rangle$ assigned to $v$ and where the intersection point lies within $\langle b, e \rangle$. The other segment $t$ cannot be part of the staircase at $v$ because segments in the same staircase do not intersect. There are three possible situations: (1) $t$ crosses the left boundary of $\langle b, e \rangle$, (2) $t$ lies completely within $\langle b, e \rangle$, or (3) $t$ crosses the right boundary of $\langle b, e \rangle$.[2]

Invariant 1 implies that, upon entering a node $v$, all segments intersecting the left boundary of the strip $\langle b, e \rangle$ are available in the form of an ordered set $\mathcal{L}_v$ that is sorted according to $<_b$. Similarly, Invariant 3 requires the existence of an ordered set $\mathcal{R}_v$ (which, in the parameter list of Balaban's algorithm (Algorithm 5) is a reference parameter to be modified by the algorithm) that contains the segments crossing the right strip boundary—again in sorted order. The unordered set $\mathcal{I}_v$ contains all segments that lie completely within the strip $\langle b, e \rangle$. Handling Situations (1)–(3) then consists of computing $\mathrm{Int}_{b,e}(\mathcal{Q}_v, \mathcal{L}_v)$, $\mathrm{Int}_{b,e}(\mathcal{Q}_v, \mathcal{I}_v)$, and $\mathrm{Int}_{b,e}(\mathcal{Q}_v, \mathcal{R}_v)$, the sets of intersections inside $\langle b, e \rangle$ and induced by segments in the staircase $\mathcal{Q}_v$ and in the sets $\mathcal{L}_v$, $\mathcal{I}_v$, and $\mathcal{R}_v$, respectively. The intersections inside $\langle b, e \rangle$ that do not involve any $s \in \mathcal{Q}_v$ are found recursively.

To obtain a logarithmic depth of recursion, Balaban subdivides the set of segments that are not part of the staircase at the current node in such a way that the same number of *endpoints* is processed in each of the recursive call. Under the simplifying assumption that the $x$-coordinates of the segments are the

---

[2] There might be segments appearing both in Situation (1) and Situation (3); we can detect (and skip) those segments when handling Situation (3) because these segments are exactly the segments crossing both strip boundaries.

integers $[1\ldots 2n]$,[3] this corresponds to subdividing with respect to the median $c := \lfloor (b+e)/2 \rfloor$, and the Balaban's algorithm can be stated as follows ($\text{LSON}(v)$ and $\text{RSON}(v)$ denote the left and right child of $v$, respectively):

---

**Algorithm 5.** The algorithm $\text{TREESEARCH}(\mathcal{L}_v, \mathcal{I}_v, b, e, \mathcal{R}_v)$ [1]

---

1: **if** $e - b = 1$ **then**
2:     $\text{SEARCHINSTRIP}_{b,e}(\mathcal{L}_v, \mathcal{R}_v)$;
3: **else**
4:     $\text{SPLIT}_{b,e}(\mathcal{L}_v, \mathcal{Q}_v, \mathcal{L}_{\text{LSON}(v)})$;                              {Compute staircase.}
5:     Compute $\text{Int}_{b,e}(\mathcal{Q}_v, \mathcal{L}_{\text{LSON}(v)})$.                        {Handle Situation (1).}
6:     $c := \lfloor (b+e)/2 \rfloor$;
7:     Construct $\mathcal{I}_{\text{LSON}(v)}$ and $\mathcal{I}_{\text{RSON}(v)}$ from $\mathcal{I}_v$;
8:     $\text{TREESEARCH}(\mathcal{L}_{\text{LSON}(v)}, \mathcal{I}_{\text{LSON}(v)}, b, c, \mathcal{R}_{\text{LSON}(v)})$;
9:     Construct $\mathcal{L}_{\text{RSON}(v)}$ from $\mathcal{R}_{\text{LSON}(v)}$ by insertion/deletion;
10:    $\text{TREESEARCH}(\mathcal{L}_{\text{RSON}(v)}, \mathcal{I}_{\text{RSON}(v)}, c, e, \mathcal{R}_{\text{RSON}(v)})$;
11:    Compute $\text{Int}_{b,e}(\mathcal{Q}_v, \mathcal{R}_{\text{RSON}(v)})$.                     {Handle Situation (3).}
12:    Compute $\text{Loc}(\mathcal{Q}_v, \{s\})$ for each $s \in \mathcal{I}_v$.
13:    Compute $\text{Int}(\mathcal{Q}_v, \mathcal{I}_v)$ based upon $\text{Loc}(\mathcal{Q}_v, \mathcal{I}_v)$.     {Handle Situation (2).}
14:    $\mathcal{R}_v := \text{MERGE}_e(\mathcal{Q}_v, \mathcal{R}_{\text{RSON}(v)})$;          {Establish Invariant (3).}

---

There are several issues that complicate making this algorithm in-place: First of all, like in any recursive algorithm that has to be transformed into an in-place algorithm, one has to keep track of the subarrays processed in each recursive call. It is not feasible to keep the start and end indices on a stack as this would result in using $\Omega(\log_2 n)$ extra words of memory. The second issue to be resolved is how to partition the data prior to "going into recursion". Whereas algorithms working on point data can easily subdivide the data based upon, say, the $x$-coordinate by first sorting and then halving the point set, subdividing a set of segments such that the same number of endpoints appear on each side of the dividing line, seems impossible to do without splitting or copying the segments. Both splitting and copying, however, is infeasible in an in-place setting.

To guarantee both the correctness of the algorithm and the property that is uses only $\mathcal{O}(1)$ extra space, we will require the following invariants to be established at each invocation $\text{INPLACETREESEARCH}(\mathtt{A}, b, e, \ell_b, \ell_e)$:

**Invariant A:** All segments that cross the left boundary of $\langle b, e \rangle$ are stored in sorted $<_b$ order at the front of $\mathtt{A}[\ell_b, \ldots, \ell_e - 1]$ (see Invariant (1)).
**Invariant B:** $\mathtt{A}[\ell_b, \ldots, \ell_e - 1]$ contains all segments in $\mathtt{A}$ that have at least one endpoint inside $\langle b, e \rangle$.

Additionally, we will require the following invariants to be established whenever we return from a call to $\text{INPLACETREESEARCH}(\mathtt{A}, b, e, \ell_b, \ell_e)$:

---

[3] This assumption is impossible to make in an in-place setting as one would need an extra lookup-table for translating the integer $i$ to the $x$-coordinate with rank $i$.

**Invariant C:** The strip boundaries $\langle b', e' \rangle$ of the "parent strip" are known.

**Invariant D:** There exists an integer $i \in \{0, \ldots \ell_b - \ell_e\}$ such that all segments of $\mathtt{A}[\ell_b, \ell_e - 1]$ that do not cross the right strip boundary are stored in $\mathtt{A}[\ell_b, \ldots, \ell_b + i - 1]$ and that all other segments are stored in $\mathtt{A}[\ell_b + i, \ldots, \ell_e - 1]$ sorted according to $<_e$ (see Invariant (3)).

Establishing Invariant (C) in-place is one of the most crucial steps of the algorithm. We will establish this invariant as follows: Prior to "going into recursion", we select the segments $q_b$ and $q_e$ whose endpoints define the strip boundary and move it (using a linear number of swaps) to the front of the staircase $\mathcal{Q}_v$. When moving these segments, however, it is important to keep in mind that they might be part of the staircase (Fig. 3 (a)), part of $\mathcal{L}_v$ and/or $\mathcal{R}_v$ (Fig. 3 (b)), identical (Fig. 3 (c)), or not intersecting the interior of $\langle b, e \rangle$ at all (Fig. 3 (d)), and that $q_b$ and $q_e$ need to be handled accordingly when looking for intersections.

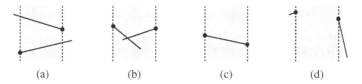

(a)          (b)          (c)          (d)

**Fig. 3.** Some of the configurations of segments whose endpoints define $\langle b, e \rangle$

Any combination of these configurations is possible, but as the overall number of combinations is constant, a constant number of bits is sufficient to encode the specific combination. Thus a "configuration" stack $\mathcal{C}$ of $\mathcal{O}(\log_2 n)$ bits, i.e. using $\mathcal{O}(1)$ extra space, can be used to store the information necessary to recover the subset(s) into which $q_b$ and $q_e$ have to be reinserted when returning from the "recursive" calls.

We first use Algorithm INPLACESPLIT to compute the staircase $\mathcal{Q}_v$ and interchange the subarrays containing $\mathcal{L}'$ and $\mathcal{Q}_v{}^4$ such that the subarray looks as follows:

We also maintain a "staircase" stack $\mathcal{S}$ of depth $\mathcal{O}(\log_2 n)$ to indicate whether $\mathcal{Q}_v$ contains zero, one, or more segments in addition to $(q_b \cup q_e) \cap \mathcal{Q}_v$. This information can be encoded using $\mathcal{O}(1)$ bits per entry, i.e. using $\mathcal{O}(1)$ extra space in total. We then establish Invariant (A) by shifting $\mathcal{Q}_v$ in front of $q_b$, and prior to going into the "left recursion" we also prepare for establishing Invariant (C):

---

[4] Interchanging two blocks $\mathtt{A}[x_0, \ldots, x_1 - 1]$ and $\mathtt{A}[x_1, \ldots, x_2 - 1]$ can be done in-place and in linear time by first using swaps to revert the order of the elements in each of the blocks separately and by then reverting the order of $\mathtt{A}[x_0, \ldots, x_2 - 1]$ again using swaps. Katajainen and Pasanen [11] attribute this to "computer folklore".

We determine the segment $q_c$ whose endpoint induces the right boundary of the left subslab (see Section 2.4 for the details of how to do this in-place) and shift the segments $q_e$, $q_b$, and $q_c$ in front of $\mathcal{L}_{\text{LSON}(v)} := \mathcal{L}'$. We then establish Invariant (B) by moving all elements in $\mathcal{I}_{\text{LSON}(v)} \cup \mathcal{R}_{\text{LSON}(v)}$ to immediately behind $\mathcal{L}_{\text{LSON}}(v)$ using simple swaps (we use $\mathcal{N}_v$ to denote the set of segments not moved). We also update $\ell_b$ to point to the first element in $\mathcal{L}_{\text{LSON}(v)}$ and update $\ell_e$ to point to the first segment not in $\mathcal{I}_{\text{LSON}(v)} \cup \mathcal{R}_{\text{LSON}(v)}$:

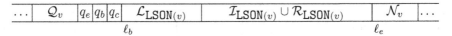

By Invariant (D), we know that upon returning from the "left recursive" call, the array has the following form ($\mathcal{O}_{\text{LSON}(v)}$ denotes the segments whose right endpoint lies inside the left subslab):

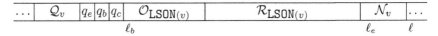

We can recover $q_e$, hence establishing Invariant (C), by simply looking at the at most three entries in front of $\mathtt{A}[\ell_b]$ (depending on the configuration encoded by the topmost element of the configuration stack $\mathcal{C}$). As doing this we have also recovered the old value of $e$, the index $\ell$ which corresponds to the old value of $\ell_e$ prior to going into the "left recursion" can be recovered by scanning forward from $\ell_e$ until we find the first segment not intersecting $\langle b, e \rangle$ (or reach the end of the array). We interchange $\mathcal{R}_{\text{LSON}(v)} \cup \mathcal{N}_v$ and $\mathcal{O}_{\text{LSON}(v)}$. Note that $\mathcal{R}_{\text{LSON}(v)} = \mathcal{L}_{\text{RSON}(v)}$ in our setting as these sets only differ by the segment $q_c$ which is stored separately. Also, relative to the subslab at $\text{RSON}(v)$, $\mathcal{N}_v = \mathcal{I}_{\text{RSON}(v)} \cup \mathcal{R}_{\text{RSON}(v)}$. We update $\ell_e$ to point to the first element in $\mathcal{O}_v$, and shift $q_e$, $q_b$, and $q_c$.

| $\cdots$ | $\mathcal{Q}_v$ | $q_b$ | $q_c$ | $q_e$ | $\mathcal{L}_{\text{RSON}(v)}$ | $\mathcal{I}_{\text{RSON}(v)} \cup \mathcal{R}_{\text{RSON}(v)}$ | $\mathcal{O}_{\text{LSON}(v)}$ | $\cdots$ |

(with $\ell_b$ below $q_b q_c q_e$, $\ell_e$ below $\mathcal{I}_{\text{RSON}(v)} \cup \mathcal{R}_{\text{RSON}(v)}$, and $\ell$ at the right)

By Invariant (D), we know that upon returning from the "right recursive" call, the array has the following form:

| $\cdots$ | $\mathcal{Q}_v$ | $q_b$ | $q_c$ | $q_e$ | $\mathcal{O}_{\text{RSON}(v)}$ | $\mathcal{R}_{\text{RSON}(v)}$ | $\mathcal{O}_{\text{LSON}(v)}$ | $\cdots$ |

(with $\ell_b$ below $q_b q_c q_e$, $\ell_e$ below $\mathcal{R}_{\text{RSON}(v)}$, and $\ell$ at the right)

Again, we recover the values of $b$ and $e$, and find the index $\ell$ by scanning forward from $\ell_e$. Depending of whether $q_c$ crosses the right boundary of $\langle b, e \rangle$ or not, we insert $q_c$ into $\mathcal{R}_{\text{RSON}(v)}$ or into $\mathcal{O}_{\text{RSON}(v)}$. Scanning backward from $\ell_b$ and using the information on top of the staircase stack $\mathcal{S}$ as well as the fact that the segments in $\mathcal{Q}_v$ span $\langle b, e \rangle$, are non-intersecting, and are ordered by $<_b$, we determine the start of the subarray in which $\mathcal{Q}_v$ is stored. We then interchange the blocks such that $\mathcal{O}_{\text{LSON}(v)}$ and $\mathcal{O}_{\text{RSON}(v)}$ as well as $\mathcal{Q}_v$ and $\mathcal{R}_{\text{RSON}(v)}$ appear next to each other. Finally, we use an in-place merging algorithm [9] to construct $\mathcal{R}_v (= \mathcal{R}_{\text{RSON}(v)}) \cup \mathcal{Q}_v$, thus establishing Invariant (D). Note that all interchanging, shifting, and scanning done so far takes time linear in $|\mathcal{L}_v \cup \mathcal{I}_v \cup \mathcal{R}_v|$.

As all invariants can be established for the base case of the recursion, we conclude that the invariants can be established for each "recursive call", and thus we have established the correctness of the following algorithm:

---

**Algorithm 6.** Algorithm INPLACETREESEARCH($A, b, e, \ell_b, \ell_e$)

---

1: **if** $\langle b, e \rangle$ does not contain any endpoint of a segment $s \in A[\ell_b, \ldots, \ell_e - 1]$ **then**
2:     INPLACESEARCHINSTRIP($A, b, e, \ell_b, \ell_e$);
3: **else**
4:     Let $\ell_l$ be the index of the first segment in $A[\ell_b, \ldots, \ell_e - 1]$ that does not cross the left strip boundary.                   $\{ \mathcal{L}_v = A[\ell_b, \ldots, \ell_l - 1] \}$
5:     $\ell_c := $ INPLACESPLIT($A, b, e, \ell_b, \ell_l$). $\{ \mathcal{L}' = A[\ell_b, \ldots, \ell_c - 1]; \mathcal{Q}_v = A[\ell_c, \ldots, \ell_l - 1] \}$
6:     Stably exchange the subarrays $A[\ell_b, \ldots, \ell_c - 1]$ and $A[\ell_c, \ldots, \ell_l - 1]$.
7:     Compute $\mathrm{Int}_{b,e}(\mathcal{Q}_v, \mathcal{L}')$.                    {Handle Situation (1).}
8:     **for** each segment $s \in A[\ell_l, \ldots, \ell_e - 1]$ than lies inside $\langle b, e \rangle$ **do**
9:         Using binary search, locate the lower endpoint of $s$ w.r.t. the stairs of $\mathcal{Q}_v$ and compute $\mathrm{Int}_{b,e}(\mathcal{Q}_v, \{s\})$.                    {Handle Situation (2).}
10:     Find the index of the median of the endpoints inside the current strip. Let $c$ be the $x$-coordinate of this endpoint.
11:     Establish Invariants (A) and (B), update $\ell_b$ and $\ell_e$.
12:     INPLACETREESEARCH($A, b, c, \ell_b, \ell_e$);
13:     Recover the old values of $e$ and $\ell_e$. Establish Invariants (A) and (B), update $\ell_b$ and $\ell_e$.
14:     INPLACETREESEARCH($A, c, e, \ell_b, \ell_e$);
15:     Recover the old values of $b$ and $\ell_e$.
16:     Compute $\mathrm{Int}_{b,e}(\mathcal{Q}_v, \mathcal{R}')$.         {Handle Situation (3); check for duplicates.}
17:     $\mathcal{R}_v := $ MERGE$_e(\mathcal{Q}_v, \mathcal{R}_{\mathrm{RSON}(v)})$; Establish Invariant (D).

---

Due to space constraints, the above description does not explicitly contains code for simulating the two "recursive" calls to INPLACETREESEARCH, since we have shown previously [3] that it is possible to handle these calls using a stack of $\mathcal{O}(\log_2 n)$ bits, that is using $\mathcal{O}(1)$ extra space. To do so we need to be able to retrieve the subset to work with upon returning from a recursive call using only $\mathcal{O}(1)$ extra space, and this is guaranteed by Invariant (C). We also did not include the code for handling the segments $q_b$, $q_e$, and $q_c$. We need, however, to fill in the details of how to select the endpoint with median $x$-coordinate.

## 2.4    Selecting the Median In-Place

When selecting the median of the endpoints in line 10 of Algorithm 6, we have to do so while maintaining the set $\mathcal{L}'$ in sorted order.

To make sure that the overall cost of median-finding does not depend on the number $k$ of intersections reported by the algorithm, we make sure to only process segments not spanning $\langle b, e \rangle$. Doing so, we can guarantee that each segment participates in $\mathcal{O}(\log_2 n)$ invocations of median-finding, namely in $\mathcal{O}(1)$ such invocations on each level of the recursion tree. To make the algorithm reflect this, we use SORTEDSUBSETSELECTION to stably select the segments of $\mathcal{L}'$ spanning $\langle b, e \rangle$. The segments that have at least one endpoint in $\langle b, e \rangle$ are then stored consecutively in $A[\ell_b + i, \ldots, \ell_e - 1]$ (for some $i \in \{0, \ldots, \ell_e - \ell_b\}$).

**Lemma 2.** *Given $m$ segments and a strip $\langle b, e \rangle$, the $k$-th endpoint in sorted order inside $\langle b, e \rangle$ can be found in-place in $\mathcal{O}(m \log_2 m)$ time.*

*Proof.* We simulate a plane-sweep over the set of segments and maintain the current $x$-coordinate $\xi$ as well as the number $o$ of endpoints inside $\langle b, e \rangle$ that have already be swept over. The segments are maintained in-place in a heap-based priority queue $\mathcal{H}$, the priority of $s$ being the smallest $x$-coordinate of $s$'s endpoints that still is at least $\xi$. When deleting the minimal element $s$ from $\mathcal{H}$ we increment $o$ iff $\xi \in [b, e]$ and re-insert $s$ iff the $x$-coordinate of its right endpoint is larger than $\xi$. If $o = k$, we report $s$ and $\xi$, else we continue. As there are at most $2m$ priority queue operations, the algorithm runs in time $\mathcal{O}(m \log_2 m)$.

After we have found the median using the algorithm implied by Lemma 2, we need to restore $\mathcal{L}'$ in sorted $<_b$ order. To this end, we then select the elements from $A[\ell_b + i, \ldots, \ell_e - 1]$ that cross the left strip boundary, sort them in-place by $<_b$, and then merge them in-place with the segments in $A[\ell_b, \ldots, \ell_b + i - 1]$.

**Lemma 3.** *The global cost incurred by median-finding is $\mathcal{O}(n \log_2^2 n)$.*

*Proof.* The median-finding algorithm considers only those segments that have at least one endpoint in the current strip. Hence, on each level of recursion, each segment is considered at most twice, so we can charge each segment $s$ $\mathcal{O}(\log_2 n)$ cost per level for median-finding (see Lemma 2). We charge $s$ an additional $\mathcal{O}(\log_2 n)$ cost per level for the at most one sorting step it participates in (when restoring $\mathcal{L}'$). As all other operations require only linear time per level, the global cost incurred by median-finding is $\mathcal{O}(n \log_2^2 n)$ as claimed.

## 2.5    Analysis of the Running Time

For the main part of the analysis, Balaban's results carry over. Using the notation $\mathcal{S}_v = \mathcal{L}_v \cup \mathcal{I}_v \cup \mathcal{R}_v$, the following theorem holds for the recursion tree $\mathcal{T}$:

**Theorem 1 (Theorem 2 in [1]).** $\sum_{v \in \mathcal{T}} |\mathcal{S}_v| \leq n \lceil 4 \log_2 n + 5 \rceil + 2k$.

To make the algorithm in-place, we had to resort to some algorithmic techniques not captured in Balaban's analysis. The global extra cost for making the algorithm SPLIT in-place is $\mathcal{O}(n \log_2 n)$ (see Lemma 1). From Theorem 1 follows that the overall extra cost for establishing the invariants is $\mathcal{O}(n \log_2 n + k)$ as all operations performed at a node $v \in \mathcal{T}$ take time linear in $|\mathcal{S}_v|$. Finally, we had to realize the median-finding in-place and restoring the original order of the elements. By Lemma 3, the overall cost for this is in $\mathcal{O}(n \log_2^2 n)$. The last component of the analysis is the for-loop in Line 8 of Algorithm 4: Each iteration of this loop takes $\mathcal{O}(\log_2 |\mathcal{Q}_v|) \subseteq \mathcal{O}(\log_2 n)$ time, and each of the $n$ segments can be part of $\mathcal{I}_w$ for $\mathcal{O}(\log_2 n)$ nodes $w \in \mathcal{T}$. Combining this with Balaban's original analysis, we obtain the main result of this paper:

**Theorem 2.** *All $k$ intersections induced by a set of $n$ segments in the plane can be computed in $\mathcal{O}(n \log_2^2 n + k)$ time using $\mathcal{O}(1)$ extra words of memory.*

We conclude with the obvious open problem: Is it possible to compute all $k$ intersections induced by a set of $n$ segments in the plane in-place *and* in optimal time $\mathcal{O}(n \log_2 n + k)$?

# References

1. I. J. Balaban. An optimal algorithm for finding segments [*sic!*] intersections. In *Proceedings of the Eleventh Annual Symposium on Computational Geometry*, pages 211–219, New York, 1995. ACM Press.
2. J. L. Bentley and T. A. Ottmann. Algorithms for reporting and counting geometric intersections. *IEEE Transactions on Computers*, C-28(9):643–647, Sept. 1979.
3. P. Bose, A. Maheshwari, P. Morin, J. Morrison, M. Smid, and J. Vahrenhold. Space-efficient geometric divide-and-conquer algorithms. *Computational Geometry: Theory & Applications*, 2005. To appear, accepted November 2004. An extend abstract appeared in *Proceedings of the 20th European Workshop on Computational Geometry*, pages 65–68, 2004.
4. H. Brönnimann, T. M.-Y. Chan, and E. Y. Chen. Towards in-place geometric algorithms. In *Proceedings of the Twentieth Annual Symposium on Computational Geometry*, pages 239–246. ACM Press, 2004.
5. H. Brönnimann, J. Iacono, J. Katajainen, P. Morin, J. Morrison, and G. T. Toussaint. Optimal in-place planar convex hull algorithms. *Theoretical Computer Science*, 321(1):25–40, June 2004. An extended abstract appeared in the *Proceedings of the Fifth Latin American Symposium on Theoretical Informatics* (2002), pages 494–507.
6. B. M. Chazelle and L. J. Guibas. Fractional cascading: I. A data structuring technique. *Algorithmica*, 1(2):133–162, 1986.
7. E. Y. Chen and T. M.-Y. Chan. A space-efficient algorithm for line segment intersection. In *Proceedings of the 15th Canadian Conference on Computational Geometry*, pages 68–71, 2003.
8. R. W. Floyd. Algorithm 245: Treesort. *Communications of the ACM*, 7(12):701, Dec. 1964.
9. V. Geffert, J. Katajainen, and T. Pasanen. Asymptotically efficient in-place merging. *Theoretical Computer Science*, 237(1–2):159–181, Apr. 2000.
10. J. E. Goodman and J. O'Rourke, editors. *Handbook of Discrete and Computational Geometry*. Discrete Mathematics and its Applications. CRC Press, Boca Raton, FL, second edition, 2004.
11. J. Katajainen and T. Pasanen. Stable minimum space partitioning in linear time. *BIT*, 32:580–585, 1992.
12. D. M. Mount. Geometric intersection. In Goodman and O'Rourke [10], chapter 38, pages 857–876.

# Improved Fixed-Parameter Algorithms for Two Feedback Set Problems

Jiong Guo[1,*], Jens Gramm[2,**], Falk Hüffner[1,*],
Rolf Niedermeier[1], and Sebastian Wernicke[1,***]

[1] Institut für Informatik, Friedrich-Schiller-Universität Jena,
Ernst-Abbe-Platz 2, D-07743 Jena, Germany
{guo, hueffner, niedermr, wernicke}@minet.uni-jena.de
[2] Wilhelm-Schickard-Institut für Informatik, Universität Tübingen,
Sand 13, D-72076 Tübingen, Germany
gramm@informatik.uni-tuebingen.de

**Abstract.** Settling a ten years open question, we show that the NP-complete FEEDBACK VERTEX SET problem is deterministically solvable in $O(c^k \cdot m)$ time, where $m$ denotes the number of graph edges, $k$ denotes the size of the feedback vertex set searched for, and $c$ is a constant. As a second result, we present a fixed-parameter algorithm for the NP-complete EDGE BIPARTIZATION problem with runtime $O(2^k \cdot m^2)$.

## 1   Introduction

In feedback set problems the task is, given a graph $G$ and a collection $C$ of cycles in $G$, to find a minimum size set of vertices or edges that meets all cycles in $C$. We refer to Festa, Pardalos, and Resende [9] for a 1999 survey. In this work we restrict our attention to undirected and unweighted graphs, giving significantly improved exact algorithms for two NP-complete feedback set problems.

- FEEDBACK VERTEX SET (FVS): Here, the task is to find a minimum cardinality set of *vertices* that meets *all* cycles in the graph.
- EDGE BIPARTIZATION: Here, the task is to find a minimum cardinality set of *edges* that meets *all odd-length* cycles in the graph.[1]

Concerning the FVS problem, it is known that an optimal solution can be approximated to a factor of 2 in polynomial time [1]. FVS is MaxSNP-hard [15]

---

* Supported by the Deutsche Forschungsgemeinschaft, Emmy Noether research group PIAF (fixed-parameter algorithms), NI 369/4.
** Supported by the Deutsche Forschungsgemeinschaft, project OPAL (optimal solutions for hard problems in computational biology), NI 369/2.
*** Supported by the Deutsche Telekom Stiftung and the Studienstiftung des deutschen Volkes. Main work done while the author was with TU München.
[1] That is, the deletion of those edges would make the graph bipartite.

F. Dehne, A. López-Ortiz, and J.-R. Sack (Eds.): WADS 2005, LNCS 3608, pp. 158–168, 2005.
© Springer-Verlag Berlin Heidelberg 2005

(hence, there is no hope for polynomial-time approximation schemes). A question of similar importance as approximability is to ask how fast one can find an *optimal* feedback vertex set. There is a very simple randomized algorithm due to Becker et al. [3] which solves the FVS problem in $O(c \cdot 4^k \cdot kn)$ time by finding a feedback vertex set of size $k$ with probability at least $1 - (1 - 4^{-k})^{c4^k}$ for an arbitrary constant $c$. Note that this means that by choosing an appropriate value for $c$, one can achieve any constant error probability independent of $k$. As to deterministic algorithms, Bodlaender [4] and Downey and Fellows [6] were the first to show that the problem is fixed-parameter tractable. An exact algorithm with runtime $O((2k + 1)^k \cdot n^2)$ was described by Downey and Fellows [7]. In 2002, Raman, Saurabh, and Subramanian [20] made a significant step forward by proving the upper bound $O(\max\{12^k, (4 \log k)^k\} \cdot n^\omega)$ ($n^\omega$ denotes the time to multiply two $n \times n$ matrices). Recently, this bound was slightly improved to $O((2 \log k + 2 \log \log k + 18)^k \cdot n^2)$ by Kanj, Pelsmajer, and Schaefer [14] using results from extremal graph theory. Lastly, Raman et al. [21] published an algorithm running in $O((12 \log k / \log \log k + 6)^k \cdot n^\omega)$ time.

The central question left open for more than ten years is whether there is an $O(c^k \cdot n^{O(1)})$ time algorithm for FVS for some constant $c$. We settle this open problem by giving an $O(c^k \cdot mn)$ time algorithm. Independently, this result was also shown by Dehne et al. [5], proving the constant $c \approx 10.6$. Surprisingly, although both studies were performed completely independent of each other, the developed algorithms turn out to be quite similar. The advantage of the result by Dehne et al. is a better upper bound on the constant $c$, whereas our advantage seems to be a more compact, easier accessible presentation of the algorithm. Since it seems hard to bring the constant $c$ close to the constant 4 achieved by Becker et al., the described deterministic algorithms for FVS are of more theoretical interest.

Compared with Dehne et al. our algorithm also shows that FVS can be solved deterministically in *linear* time for constant $k$, a property which also holds for the randomized algorithm. Hence, with our corresponding $O(c^k \cdot m)$ algorithm we can conclude that FVS is "*linear-time* fixed-parameter tractable." Very recently, Fiorini et al. [10] showed, by significant technical expenditure, the analogous result concerning the GRAPH BIPARTIZATION problem (which is basically the same problem as EDGE BIPARTIZATION, only deleting vertices instead of edges) restricted to planar graphs.

We now turn our attention to the EDGE BIPARTIZATION problem. This problem is known to be MaxSNP-hard [18] and can be approximated to a factor of $O(\log n)$ in polynomial time [11]. It has applications in genome sequence assembly [19] and VLSI chip design [13]. In a recent breakthrough paper, Reed, Smith, and Vetta [22] proved that the GRAPH BIPARTIZATION problem is solvable in $O(4^k \cdot kmn)$ time, where $k$ denotes the number of vertices to be deleted for making the graph bipartite. (Actually, it is straightforward to observe that the exponential factor $4^k$ can be lowered to $3^k$ by a more careful analysis of the algorithm [12].) Since there is a "parameter-preserving" reduction from EDGE BIPARTIZATION to GRAPH BIPARTIZATION [23], one can use the algorithm by

Reed et al. to directly obtain a runtime of $O(3^k \cdot k^3 m^2 n)$ for EDGE BIPARTIZA-TION, $k$ denoting the size of the set of edges to be deleted. In this work our main concern is to shrink the combinatorial explosion and the polynomial complexity related with the fixed-parameter tractability of EDGE BIPARTIZATION. We achieve an algorithm running in $O(2^k \cdot m^2)$ time. This shows that we can save a cubic-time factor $k^3$ as well as a linear-time factor $n$, and that we can shrink the combinatorial explosion from $3^k$ to $2^k$.

## 2  Preliminaries and Previous Work

This work considers undirected graphs $G = (V, E)$ with $n := |V|$ and $m := |E|$. Given a set $E' \subseteq E$ of edges, $V(E')$ denotes the set $\bigcup_{\{u,v\}\in E'}\{u, v\}$ of endpoints. We use $G[X]$ to denote the subgraph of $G$ induced by the vertices $X \subseteq V$. For a set of edges $E' \subseteq E$, we write $G\backslash E'$ for the graph $(V, E\backslash E')$. For $u \in V$, we use $N(u)$ to denote the neighbor set $\{v \in V \mid \{u, v\} \in E\}$. With a *side* of a bipartite graph $G$, we mean one of the two classes of an arbitrary but fixed two-coloring of $G$.

The two problems we study are formally defined as follows:

FEEDBACK VERTEX SET (FVS)
Given an undirected graph $G = (V, E)$ and a nonnegative integer $k$, find a subset $V' \subseteq V$ of vertices with $|V'| \leq k$ such that each cycle in $G$ contains at least one vertex from $V'$. (The removal of all vertices in $V'$ from $G$ therefore results in a forest.)

EDGE BIPARTIZATION
Given an undirected graph $G = (V, E)$ and a nonnegative integer $k$, find a subset $E' \subseteq E$ of edges with $|E'| \leq k$ such that each odd-length cycle in $G$ contains at least one edge from $E'$. (The removal of all edges in $E'$ from $G$ therefore results in a bipartite graph.)

We investigate FVS and EDGE BIPARTIZATION in the context of parameterized complexity [7, 17] (see [8, 16] for surveys). A parameterized problem is *fixed-parameter tractable* if it can be solved in $f(k)\cdot n^{O(1)}$ time, where $f$ is a computable function solely depending on the parameter $k$, not on the input size $n$.

To the best of our knowledge, Reed et al. [22] were the first to make the following simple observation: To show that a minimization problem is fixed-parameter tractable with respect to the size of the solution $k$, it suffices to give a fixed-parameter algorithm which, given a size-$(k + 1)$ solution, proves that there is no size-$k$ solution or constructs one. Starting with a trivial instance and inductively applying this compression step a linear number of rounds to larger instances, one obtains the fixed-parameter tractability of the problem. This method is called *iterative compression*. The main challenge of applying it lies in showing that there is a "fixed-parameter compression algorithm." It is this hard part where Reed et al. achieved a breakthrough concerning GRAPH BIPARTIZATION. The compression step, however, is highly problem-specific and no universal standard techniques are known.

## 3  Algorithm for Feedback Vertex Set

In this section we show that FEEDBACK VERTEX SET can be solved in $O(c^k \cdot m)$ time for a constant $c$ by presenting an algorithm based on iterative compression. The following lemma provides the compression step.

**Lemma 1.** *Given a graph $G$ and a size-$(k+1)$ feedback vertex set (fvs) $X$ for $G$, we can decide in $O(c^k \cdot m)$ time for some constant $c$ whether there exists a size-$k$ fvs $X'$ for $G$ and if so provide one.*

*Proof.* Consider the smaller fvs $X'$ as a modification of the larger fvs $X$. The smaller fvs retains some vertices $Y \subseteq X$ while the other vertices $S := X \setminus Y$ are replaced with $|S| - 1$ new vertices from $V \setminus X$. The idea is to try by brute force all $2^{|X|}$ partitions of $X$ into such sets $Y$ and $S$. In each case, we then have significant information about a possible smaller fvs $X'$—it contains $Y$, but not $S$—and it turns out that there is only a "small" set $V'$ of candidate vertices to draw from in order to complete $Y$ to $X'$. More precisely, we later show in Lemma 4 that the size of $V'$ is bounded by $14 \cdot |S|$ and that, given $S$, we can compute $V'$ in $O(m)$ time. Since $|S| \leq k + 1$, $|V'|$ thus only depends on the problem parameter $k$ and not on the input size. We again use brute force and consider each of the at most $\binom{14 \cdot |S|}{|S|-1}$ possible choices of vertices from $V'$ that can be added to $Y$ to form $X'$. The test whether a choice of vertices from $V'$ together with $Y$ forms an fvs can be easily done in $O(m)$ time. We can now bound the overall runtime $T$, where the index $i$ corresponds to a partition of $X$ into $Y$ and $S$ with $|Y| = i$ and $|S| = |X| - i$:

$$T = O\left(\sum_{i=0}^{k} \binom{|X|}{i} \cdot \left(O(m) + \binom{14 \cdot (|X| - i)}{|X| - i - 1} \cdot O(m)\right)\right).$$

With Stirling's inequality, we arrive at the lemma's claim with $c \approx 37.7$.[2]     □

**Theorem 2.** FEEDBACK VERTEX SET *can be solved in $O(c^k \cdot mn)$ time for a constant $c$.*

*Proof.* Given as input a graph $G$ with vertex set $\{v_1, \ldots, v_n\}$, we can apply iterative compression to solve FEEDBACK VERTEX SET for $G$ by iteratively considering the subgraphs $G_i := G[\{v_1, \ldots, v_i\}]$. For $i = 1$, the optimal fvs is empty. For $i > 1$, assume that an optimal fvs $X_i$ for $G_i$ is known. Obviously, $X_i \cup \{v_{i+1}\}$ is an fvs for $G_{i+1}$. Using Lemma 1, we can in $O(c^k \cdot m)$ time either determine that $X_i \cup \{v_{i+1}\}$ is an optimal fvs for $G_{i+1}$, or, if not, compute an optimal fvs for $G_{i+1}$. For $i = n$, we thus have computed an optimal fvs for $G$ in $O(c^k \cdot mn)$ time.     □

---

[2] The value of $c$ can be significantly improved by a more careful analysis in Lemma 4. Indeed, Dehne et al. [5] achieve $c \approx 10.6$.

Theorem 2 shows that FVS is fixed-parameter tractable with the combinatorial explosion bounded from above by $c^k$ for some constant $c$. Next, we show that FVS is also *linear-time* fixed-parameter tractable (with the combinatorial explosion bounded by $c^k$ for a larger constant $c$). The result of Fiorini et al. [10], accepting a much worse combinatorial explosion compared to [22], is to show the analogous result for GRAPH BIPARTIZATION restricted to planar graphs.

**Theorem 3.** FEEDBACK VERTEX SET *can be solved in* $O(c^k \cdot m)$ *time for a constant* $c$.

*Proof.* We first calculate in $O(m)$ time a factor-4 approximation as described by Bar-Yehuda et al. [2]. This gives us the precondition for Lemma 1 with $|X| = 4k$ instead of $|X| = k + 1$. Now, we can employ the same techniques as in the proof of Lemma 1 to obtain the desired runtime: we examine $2^{4k}$ partitions $S \,\dot\cup\, Y$ of $X$, and—by applying the arguments from Lemma 4—for each there is some constant $c'$ such that the number of candidate vertices is bounded from above by $c' \cdot |S|$. In summary, there is some constant $c$ such that the runtime of the compression step is bounded from above by $O(c^k \cdot m)$. Since one of the $2^{4k}$ partitions must lead to the optimal solution of size $k$, we need only one compression step to obtain an optimal solution, which proves the claimed runtime bound.  □

Note that any improvement of the approximation factor of a linear-time approximation algorithm for FEEDBACK VERTEX SET below 4 will immediately improve the runtime of the exact algorithm described in Theorem 3.

It remains to show the size bound of the "candidate vertices set" $V'$ for fixed partition $Y$ and $S$ of a size-$(k+1)$ fvs $X$. To this end, we make use of two simple data reduction rules.

**Lemma 4.** *Given a graph* $G = (V, E)$, *a size-$(k+1)$ fvs $X$ for $G$, and a partition of $X$ into two sets $Y$ and $S$. Let $X'$ denote a size-$k$ fvs for $G$ with $X' \cap X = Y$ and $X' \cap S = \emptyset$. In $O(m)$ time, we can either decide that no such $X'$ exists or compute a subset $V'$ of $V \setminus X$ with $|V'| < 14 \cdot |S|$ such that there exists an $X'$ as desired consisting of $|S| - 1$ vertices from $V'$ and all vertices from $Y$.*

*Proof.* The idea of the proof is to use a well-known data reduction technique for FVS to get rid of degree-1 and degree-2 vertices and to show that if the resulting instance is too large as compared to the part $S$ (whose vertices we are not allowed to add to $X'$), then there exists no set $X'$ as desired.

First, check that $S$ does not induce a cycle; otherwise, no $X'$ with $X' \cap S = \emptyset$ can be an fvs for $G$. Then, remove in $G$ all vertices from $Y$ as they are determined to be in $X'$. Finally, apply a standard data reduction to the vertices in $V \setminus X$ (the vertices in $S$ remain unmodified): remove degree-1 vertices and successively bypass any degree-2 vertex by a new edge between its neighbors (thereby removing the bypassed degree-2 vertex). There are two exceptions to note: One exception is that we do not bypass a degree-2 vertex which has two neighbors in $S$. The other exception is the way to deal with parallel edges. If we create two parallel edges between two vertices during the data reduction process—these two edges form a length-two cycle—, then exactly one of the two endpoints of these edges has to be in $S$ since $S$ is an fvs of $G[V \setminus Y]$ and $G[S]$ contains no cycle. Thus, we have to delete the other endpoint and add it to $X'$ since we are not allowed to add vertices from $S$ to $X'$. Given an appropriate graph data structure, all of the above steps can be accomplished in $O(m)$ time. Proofs for the correctness and the time bound of the data reduction technique are basically straightforward and omitted here.

In the following we use $G' = (V' \cup S, E')$ with $V' \subseteq V \setminus X$ to denote the graph resulting after exhaustive application of the data reduction described above; note that none of the vertices in $S$ have been removed during the data reduction process. In order to prove that $|V'| < 14 \cdot |S|$, we partition $V'$ into three subsets, each of which will have a provable size bound linearly depending on $|S|$ (the partition is illustrated in Fig. 1):

$$A := \{v \in V' \mid |N(v) \cap S| \geq 2\},$$
$$B := \{v \in V' \setminus A \mid |N(v) \cap V'| \geq 3\},$$
$$C := V' \setminus (A \cup B).$$

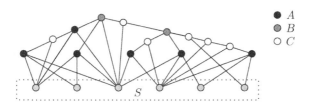

●  $A$
◑  $B$
○  $C$

**Fig. 1.** Partition of the vertices in $V'$ into three disjoint subsets $A$, $B$, and $C$

To bound the number of vertices in $A$, consider the bipartite subgraph $G_A = (A \cup S, E_A)$ of $G' = (V' \cup S, E')$ with $E_A := (A \times S) \cap E'$. Observe that if there are more than $|S| - 1$ vertices in $A$, then there is a cycle in $G_A$: If $G_A$ is acyclic, then $G_A$ is a forest, and, thus, $|E_A| \leq |S| + |A| - 1$. Moreover, since each vertex in $A$ has at least two incident edges in $G_A$, $|E_A| \geq 2|A|$, which implies that $|A| \leq |S| - 1$ if $G_A$ is acyclic. It follows directly that if $|A| \geq 2|S|$, it is impossible to delete at most $|S|$ vertices from $A$ such that $G'[A \cup S]$ is acyclic.

To bound the number of vertices in $B$, observe that $G'[V']$ is a forest. Furthermore, all leaves of the trees in $G'[V']$ are from $A$ since $G'$ is reduced with respect to the above data reduction rules. By the definition of $B$, each vertex in $B$ has at least three vertices in $V'$ as neighbors. Thus, there cannot be more vertices in $B$ than in $A$, and therefore $|B| < 2|S|$.

Finally, consider the vertices in $C$. By the definitions of $A$ and $B$, and since $G$ is reduced, each vertex in $C$ has degree two in $G'[V']$ and exactly one neighbor in $S$. Hence, graph $G'[C]$ is a forest consisting of paths and isolated vertices. We now separately bound the number of isolated vertices and those participating in paths.

Each of the isolated vertices in $G'[C]$ connects two vertices from $A \cup B$ in $G'[V']$. Since $G'[V']$ is acyclic, the number of isolated vertices in $G'[C]$ cannot exceed $|A \cup B| - 1 < 4|S|$. The total number of vertices participating in paths in $G'[C]$ can be bounded as follows: Consider the subgraph $G'[C \cup S]$. Each edge in $G'[C]$ creates a path between two vertices in $S$, that is, if $|E(G'[C])| \geq |S|$, then there exists a cycle in $G'[C \cup S]$. By an analogous argument to the one that bounded the size of $A$ (and considering that removing a vertex from $G'[C]$ destroys at most two edges), the total number of edges in $G'[C]$ may thus not exceed $|S| + 2|S|$, bounding the total number of vertices participating in paths in $G'[C]$ by $6|S|$.

Altogether, $|V'| = |A| + |B| + |C| < 2|S| + 2|S| + (4|S| + 6|S|) = 14|S|$.  □

## 4    Algorithm for Edge Bipartization

In this section we present a new algorithm for EDGE BIPARTIZATION which is based on iterative compression and runs in $O(2^k \cdot m^2)$ time. The algorithm is structurally similar to the $O(4^k \cdot kmn)$ time algorithm for GRAPH BIPARTIZATION given by Reed et al. [22]: Their compression routine starts by enumerating all partitions of the known solution into two parts, one containing vertices to keep in the solution and one containing the vertices to exchange. This is followed by a second step that tries to find a compressed bipartization set under this constraint. Our algorithm for EDGE BIPARTIZATION does not need the first step by enforcing that the smaller solution is disjoint from the known one, thereby gaining a factor of $O(2^k)$ in the runtime.

We note that a similar runtime of $O(2^k \cdot |G|^{O(1)})$ for EDGE BIPARTIZATION can be achieved by first reducing the input instance to GRAPH BIPARTIZATION [23], and exploiting a solution disjointness property analogously to the presented algorithm. This, however, involves several nontrivial modifications to the algorithm of Reed et al., whereas we give a self-contained presentation here. Moreover, our proof reveals details about the structure of EDGE BIPARTIZATION that might be of independent interest.

The following lemma provides some central insight into the structure of a minimal edge bipartization set. (Note that in this section, we always use the notion of paths in which every vertex is allowed to occur at most once.)

**Lemma 5.** *Given a graph $G = (V, E)$ with a minimal edge bipartization set $X$ for $G$, the following two properties hold:*

1. *For every odd-length cycle $C$ in $G$, $|E(C) \cap X|$ is odd.*
2. *For every even-length cycle $C$ in $G$, $|E(C) \cap X|$ is even.*

*Proof.* For each edge $e = \{u, v\} \in X$, note that $u$ and $v$ are on the same side of the bipartite graph $G \setminus X$, since otherwise we do not need $e$ to be in $X$ and $X$ would not be minimal. Consider a cycle $C$ in $G$. The edges in $E(C) \setminus X$ are all between the two sides of $G \setminus X$, while the edges in $E(C) \cap X$ are between vertices of the same side as argued above. In order for $C$ to be a cycle, however, this implies that $|E(C) \setminus X|$ is even. Since $|E(C)| = |E(C) \setminus X| + |E(C) \cap X|$, we conclude that $|E(C)|$ and $|E(C) \cap X|$ have the same parity.    □

When subdividing all edges in a graph $G$ that are contained in an edge bipartization set $X$ for $G$ by two vertices, we can assume without loss of generality that an edge bipartization set smaller than $X$ is disjoint from $X$. This input transformation is formalized in the following definition.

**Definition 6.** *For a graph $G = (V, E)$ with minimal edge bipartization $X$, let the corresponding* edge-extension graph *$\tilde{G} := (\tilde{V}, \tilde{E})$ be given by*

$$\tilde{V} := V \cup \{u_e, v_e \mid e \in X\} \text{ and}$$
$$\tilde{E} := (E \setminus X) \cup \{\{u, u_e\}, \{u_e, v_e\}, \{v_e, v\} \mid e = \{u, v\} \in X\}.$$

*Let $\tilde{X} := \{\{u_e, v_e\} \mid e \in X\}$. A mapping $\Phi : V(\tilde{X}) \rightarrow \{A, B\}$ is called* valid 2-partition *of $V(\tilde{X})$ if for each $\{u_e, v_e\} \in \tilde{X}$, either $\Phi(u_e) = A$ and $\Phi(v_e) = B$ or $\Phi(u_e) = B$ and $\Phi(v_e) = A$.*

An illustration of edge-extension graphs is given in Fig. 2. It is easy to see that $\tilde{G}$ has an edge bipartization with $k$ edges if and only if $G$ has an edge bipartization with $k$ edges. Observe that, hence, the set $\tilde{X}$ as defined above constitutes a minimal edge bipartization for $\tilde{G}$.

$G:$ 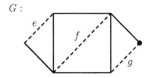    $\tilde{G}:$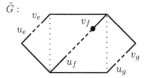

**Fig. 2.** Left: Graph $G$ with a minimal edge bipartization marked by dashed lines. Right: Edge-extension graph $\tilde{G}$ of $G$ with the corresponding edge bipartization $\tilde{X}$ marked by dashed lines. The mapping $\Phi$ which maps $\Phi(u_e) = \Phi(u_f) = \Phi(u_g) = A$, and $\Phi(v_e) = \Phi(v_f) = \Phi(v_g) = B$ is a valid 2-partition of $V(\tilde{X})$. Note that when choosing this valid 2-partition $\Phi$, then the dotted edges are an edge cut between the $A$-vertices and the $B$-vertices in $\tilde{G} \setminus \tilde{X}$. Therefore, the dotted edges are an edge bipartization for the graph on the left (Lemma 7)

**Lemma 7.** *Consider an edge-extension graph $G = (V, E)$ and a minimal edge bipartization $X$ for $G$. For a set of edges $Y \subseteq E$ with $X \cap Y = \emptyset$, the following are equivalent:*

*(1) $Y$ is an edge bipartization for $G$.*
*(2) There is a valid 2-partition $\Phi$ of $V(X)$ such that $Y$ is an edge cut between $A_\Phi := \Phi^{-1}(A)$ and $B_\Phi := \Phi^{-1}(B)$ in $G \setminus X$ (see Fig. 2).*

*Proof.* (2) $\Rightarrow$ (1): Consider any odd-length cycle $C$ in $G$. We show that $E(C) \cap Y \neq \emptyset$. Let $s := |E(C) \cap X|$. By Property 1 in Lemma 5, $s$ is odd. Without loss of generality, we assume that $E(C) \cap X = \{\{u_0, v_0\}, \{u_1, v_1\}, \dots, \{u_{s-1}, v_{s-1}\}\}$ with vertices $v_i$ and $u_{(i+1) \bmod s}$ being connected by a path in $C \setminus X$. Since $\Phi$ is a valid 2-partition, we have $\Phi(u_i) \neq \Phi(v_i)$ for all $0 \leq i < s$. With $s$ being odd, this implies that there is a pair $v_i, u_{(i+1) \bmod s}$ such that $\Phi(v_i) \neq \Phi(u_{(i+1) \bmod s})$. Since the removal of $Y$ destroys all paths in $G \setminus X$ between $A_\Phi$ and $B_\Phi$, we obtain that $E(C) \cap Y \neq \emptyset$.

$(1) \Rightarrow (2)$: Let $C_X : V \to \{A, B\}$ be a two-coloring of the bipartite graph $G \setminus X$ and $C_Y : V \to \{A, B\}$ a two-coloring of the bipartite graph $G \setminus Y$. Define

$$\Phi : V \to \{A, B\}, \quad v \mapsto \begin{cases} A & \text{if } C_X(v) = C_Y(v) \\ B & \text{otherwise.} \end{cases}$$

We show that $\Phi|_{V(X)}$ (that is, $\Phi$ with domain restricted to $V(X)$) is a valid 2-partition with the desired property.

First we show that $\Phi|_{V(X)}$ is a valid 2-partition. Consider an edge $\{u, v\} \in X$. There must be at least one even path in $G \setminus X$ from $u$ to $v$, or $\{u, v\}$ would be redundant; therefore $C_X(u) = C_X(v)$. In $G \setminus Y$, the vertices $u$ and $v$ are connected by an edge, and therefore $C_Y(u) \neq C_Y(v)$. It follows that $\Phi(u) \neq \Phi(v)$.

Since both $C_X$ and $C_Y$ change in value when going from a vertex to its neighbor in $G \setminus (X \cup Y)$, the value of $\Phi$ is constant along any path in $G \setminus (X \cup Y)$. Therefore, there can be no path from any $u \in A_\Phi$ to any $v \in B_\Phi$ in $G \setminus (X \cup Y)$, that is, $Y$ is an edge cut between $A_\Phi$ and $B_\Phi$ in $G \setminus X$. $\qquad\square$

**Theorem 8.** EDGE BIPARTIZATION *can be solved in* $O(2^k \cdot m^2)$ *time.*

*Proof.* Through Lemma 7 we obtain the compression step that, from a given minimal edge bipartization $X$, computes a smaller edge bipartization $Y$ in $O(2^k \cdot km)$ time or proves that no such $Y$ exists: We enumerate all $2^k$ valid 2-partitions $\Phi$ of $V(X)$ and determine a minimum edge cut between $A_\Phi$ and $B_\Phi$ until we find an edge cut $Y$ of size $k - 1$ (see Fig. 2). Note that the condition of Lemma 7 that $Y \cap X = \emptyset$ does not restrict generality: Since $G$ is an edge extension graph (Definition 6), we can replace each edge in $Y \cap X$ by one of its two adjacent edges in $G$. Each of the minimum cut problems can individually be solved in $O(km)$ time with the Ford-Fulkerson method that finds and augments a flow augmenting path $k$ times. By Lemma 7, $Y$ is an edge bipartization; furthermore, if no such $Y$ is found, we know that $|X|$ is minimum.

Given as input a graph $G$ with edge set $\{e_1, \dots, e_m\}$, we can apply iterative compression to solve EDGE BIPARTIZATION for $G$ by iteratively considering the graphs $G_i$ containing edges $\{e_1, \dots, e_i\}$, for $i = 1, \dots, m$. For $i = 1$, the optimal edge bipartization is empty. For $i > 1$, assume that an optimal edge bipartization $X_{i-1}$ with $|X_{i-1}| \leq k$ for $G_{i-1}$ is known. If $X_{i-1}$ is not an edge bipartization for $G_i$, then we consider the set $X_{i-1} \cup \{e_i\}$, which obviously is a minimal edge bipartization for $G_i$. Using Lemma 7, we can in $O(2^{k'} \cdot k'i)$ time (where $k' := |X_{i-1} \cup \{e_i\}| \leq k+1$) either determine that $X_{i-1} \cup \{e_i\}$ is an optimal edge bipartization for $G_i$ or, if not, compute an optimal edge bipartization $X_i$ for $G_i$. This process can be stopped if $|X_i| > k$.

Summing over all iterations, we have an algorithm that computes an optimal edge bipartization for $G$ in $O(\sum_{i=1}^{m} 2^k \cdot ki) = O(2^k \cdot km^2)$ time.

With the same technique used by Hüffner [12] to improve the runtime of the iterative compression algorithm for GRAPH BIPARTIZATION, the runtime here can also be improved to $O(2^k \cdot m^2)$. For this, one uses a Gray code to enumerate the valid 2-partitions in such a way that consecutive 2-partitions differ in only one element. For each of these (but the first one), one can then solve the flow problem by a constant number of augmentation operations on the previous flow network in $O(m)$ time.                                                              $\square$

## 5   Conclusion

We present significantly improved results on the fixed-parameter tractability of FEEDBACK VERTEX SET and EDGE BIPARTIZATION. To our belief, the iterative compression strategy due to Reed et al. employed in this work will become an important tool in the design of efficient fixed-parameter algorithms.

We succeeded in proving that FVS is even solvable in *linear* time for constant parameter value $k$. Employing a completely different technique, a similar result could very recently be shown for GRAPH BIPARTIZATION restricted to planar graphs (where the problem remains NP-complete) [10]. For general GRAPH BIPARTIZATION as well as for EDGE BIPARTIZATION, this remains open.

Finally, it remains a long-standing open problem whether FEEDBACK VERTEX SET on *directed* graphs is fixed-parameter tractable. The answer to this question would mean a significant breakthrough in the field.

## Acknowledgement

The authors would like to thank the anonymous referees of WADS 2005 for helpful and inspiring comments.

## References

1. V. Bafna, P. Berman, and T. Fujito. A 2-approximation algorithm for the undirected feedback vertex set problem. *SIAM Journal on Discrete Mathematics*, 3(2):289–297, 1999.
2. R. Bar-Yehuda, D. Geiger, J. Naor, and R. M. Roth. Approximation algorithms for the feedback vertex set problem with applications to constraint satisfaction and Bayesian inference. *SIAM Journal on Computing*, 27(4):942–959, 1998.
3. A. Becker, R. Bar-Yehuda, and D. Geiger. Randomized algorithms for the Loop Cutset problem. *Journal of Artificial Intelligence Research*, 12:219–234, 2000.
4. H. L. Bodlaender. On disjoint cycles. *International Journal of Foundations of Computer Science*, 5:59–68, 1994.
5. F. Dehne, M. Fellows, M. Langston, F. Rosamond, and K. Stevens. An $\mathcal{O}^*(2^{O(k)})$ FPT algorithm for the undirected feedback vertex set problem. In *Proc. 11th COCOON*, LNCS. Springer, Aug. 2005.

6. R. G. Downey and M. R. Fellows. Fixed-parameter tractability and completeness. *Congressus Numerantium*, 87:161–187, 1992.
7. R. G. Downey and M. R. Fellows. *Parameterized Complexity*. Springer, 1999.
8. M. R. Fellows. New directions and new challenges in algorithm design and complexity, parameterized. In *Proc. 8th WADS*, volume 2748 of *LNCS*, pages 505–520. Springer, 2003.
9. P. Festa, P. M. Pardalos, and M. G. C. Resende. Feedback set problems. In D. Z. Du and P. M. Pardalos, editors, *Handbook of Combinatorial Optimization, Vol. A*, pages 209–258. Kluwer, 1999.
10. S. Fiorini, N. Hardy, B. Reed, and A. Vetta. Planar graph bipartization in linear time. In *Proc. 2nd GRACO*, Electronic Notes in Discrete Mathematics, 2005.
11. N. Garg, V. V. Vazirani, and M. Yannakakis. Approximate max-flow min-(multi)cut theorems and their applications. *SIAM Journal on Computing*, 25(2):235–251, 1996.
12. F. Hüffner. Algorithm engineering for optimal graph bipartization. In *Proc. 4th WEA*, volume 3503 of *LNCS*, pages 240–252. Springer, 2005.
13. A. B. Kahng, S. Vaya, and A. Zelikovsky. New graph bipartizations for double-exposure, bright field alternating phase-shift mask layout. In *Proc. Asia and South Pacific Design Automation Conference*, pages 133–138, 2001.
14. I. Kanj, M. Pelsmajer, and M. Schaefer. Parameterized algorithms for feedback vertex set. In *Proc. 1st IWPEC*, volume 3162 of *LNCS*, pages 235–247. Springer, 2004.
15. C. Lund and M. Yannakakis. The approximation of maximum subgraph problems. In *Proc. 20th ICALP*, volume 700 of *LNCS*, pages 40–51. Springer, 1993.
16. R. Niedermeier. Ubiquitous parameterization—invitation to fixed-parameter algorithms. In *Proc. 29th MFCS*, volume 3153 of *LNCS*, pages 84–103. Springer, 2004.
17. R. Niedermeier. *Invitation to Fixed-Parameter Algorithms*. Oxford University Press, forthcoming, 2005.
18. C. H. Papadimitriou and M. Yannakakis. Optimization, approximation, and complexity classes. *Journal of Computer and System Sciences*, 43:425–440, 1991.
19. M. Pop, D. S. Kosack, and S. L. Salzberg. Hierarchical scaffolding with Bambus. *Genome Research*, 14:149–159, 2004.
20. V. Raman, S. Saurabh, and C. R. Subramanian. Faster fixed parameter tractable algorithms for undirected feedback vertex set. In *Proc. 13th ISAAC*, volume 2518 of *LNCS*, pages 241–248. Springer, 2002.
21. V. Raman, S. Saurabh, and C. R. Subramanian. Faster algorithms for feedback vertex set. In *Proc. 2nd GRACO*, Electronic Notes in Discrete Mathematics, 2005.
22. B. Reed, K. Smith, and A. Vetta. Finding odd cycle transversals. *Operations Research Letters*, 32:299–301, 2004.
23. S. Wernicke. On the algorithmic tractability of single nucleotide polymorphism (SNP) analysis and related problems. Diplomarbeit, WSI für Informatik, Universität Tübingen, 2003.

# Communication-Aware Processor Allocation for Supercomputers[*]

Michael A. Bender[1], David P. Bunde[2], Erik D. Demaine[3], Sándor P. Fekete[4],
Vitus J. Leung[5], Henk Meijer[6], and Cynthia A. Phillips[5]

[1] Department of Computer Science, SUNY Stony Brook,
Stony Brook, NY 11794-4400, USA
bender@cs.sunysb.edu
[2] Department of Computer Science, University of Illinois,
Urbana, IL 61801, USA
bunde@uiuc.edu
[3] MIT Computer Science and Artificial Intelligence Laboratory,
Cambridge, MA 02139, USA
edemaine@mit.edu
[4] Dept. of Mathematical Optimization,
Braunschweig University of Technology,
38106 Braunschweig, Germany
s.fekete@tu-bs.de
[5] Discrete Algorithms & Math Department, Sandia National Laboratories,
Albuquerque, NM 87185-1110, USA
{vjleung, caphill}@sandia.gov
[6] Dept. of Computing and Information Science, Queen's University,
Kingston, Ontario, K7L 3N6, Canada
henk@cs.queensu.ca

**Abstract.** We give processor-allocation algorithms for grid architectures, where the objective is to select processors from a set of available processors to minimize the average number of communication hops.

The associated clustering problem is as follows: Given $n$ points in $\Re^d$, find a size-$k$ subset with minimum average pairwise $L_1$ distance. We present a natural approximation algorithm and show that it is a $\frac{7}{4}$-approximation for 2D grids. In $d$ dimensions, the approximation guarantee is $2 - \frac{1}{2d}$, which is tight. We also give a polynomial-time approximation scheme (PTAS) for constant dimension $d$ and report on experimental results.

## 1 Introduction

We give processor-allocation algorithms for grid architectures. Our objective is to select processors to run a job from a set of available processors so that the average number of communication hops between processors assigned to the job is minimized. Our problem is restated as follows: given a set $P$ of $n$ points in $\Re^d$, find a subset $S$ of $k$ points with minimum average pairwise $L_1$ distance.

---

[*] Extended Abstract. A full version is available as [5].

F. Dehne, A. López-Ortiz, and J.-R. Sack (Eds.): WADS 2005, LNCS 3608, pp. 169–181, 2005.

*Motivation: Processor Allocation in Supercomputers.* Our algorithmic work is motivated by a problem in the operation of supercomputers. The supercomputer for which we targeted our simulations and experiments is called Computational Plant or Cplant [7, 25], a commodity-based supercomputer developed at Sandia National Laboratories. In Cplant, a scheduler selects the next job to run based on priority. The allocator then independently places the job on a set of processors which exclusively run that job to completion. Security constraints forbid migration, preemption, or multitasking. To obtain maximum throughput in a network-limited computing system, the processors allocated to a single job should be physically near each other. This placement reduces communication costs and avoids bandwidth contention caused by overlapping jobs. Experiments have shown that processor allocation affects throughput on a range of architectures [3,17,20,21,23]. Several papers suggest that minimizing the *average number of communication hops* is an appropriate metric for job placement [20, 21, 16]. Experiments with a communication test suite demonstrate that this metric correlates with a job's completion time [17].

Early processor-allocation algorithms allocate only convex sets of processors to each job [18, 9, 29, 6]. For such allocations, each job's communication can be routed entirely within processors assigned to that job, so jobs contend only with themselves. But requiring convex allocations reduces the achievable system utilization to levels unacceptable for a government-audited system [15, 26].

Recent work [19,22,8,17,26] allows discontiguous allocation of processors but tries to cluster them and minimize contention with previously allocated jobs. Mache, Lo, and Windisch [22] propose the MC algorithm for grid architectures: For each free processor, algorithm MC evaluates the quality of an allocation centered on that processor. It counts the number of free processors within a submesh of the requested size centered on the given processor and within "shells" of processors around this submesh. The cost of the allocation is the sum of the shell numbers in which free processors occur; see Figure 1 reproduced from [22]. MC chooses the allocation with lowest cost. Since users of Cplant do not request processors in a particular shape, in this paper, we consider MC1x1, a variant in which shell 0 is $1 \times 1$ and subsequent shells grow in the same way as in MC.

Until recently, processor allocation on the Cplant system was *not* based on the locations of the free processors. The allocator simply verified that enough processors were free before dispatching a job. The current allocator uses space-filling curves and 1D bin-packing techniques based upon work of Leung et al. [17].

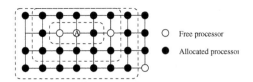

**Fig. 1.** Illustration of MC: Shells around processor $A$ for a $3 \times 1$ request

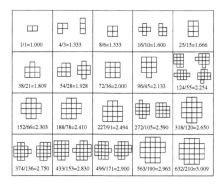
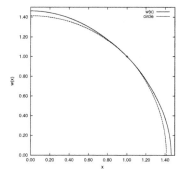

**Fig. 2.** (Left) Optimal unconstrained clusters for small values of $k$; numbers shown are the average $L_1$ distances, with truncated decimal values. (Right) Plot from [4] of a quarter of the optimal limiting boundary curve; the dotted line is a circle

*Related Algorithmic Work.* Krumke et al. [16] consider a generalization of our problem on arbitrary topologies for several measures of locality, motivated by allocation on the CM5. They prove it is NP-hard to approximate average pairwise distance in general, but give a 2-approximation for distances obeying the triangle inequality.

A natural special case of the allocation problem is the *unconstrained* problem, in the absence of occupied processors: For any number $k$, find $k$ grid points minimizing average pairwise $L_1$ distance. For moderate values of $k$, these sets can be found by exhaustive search; see Figure 2. The resulting shapes appear to approximate some "ideal" rounded shape, with better and better approximation for growing $k$. Karp et al. [14] and Bender et al. [4] study the exact nature of this shape. Surprisingly, the resulting convex curve can only be described by a differential equation; the closed-form solution is unknown. The complexity of this special case remains open, but its mathematical difficulty emphasizes the hardness of obtaining good solutions for the general constrained problem.

In reconfigurable computing on field-programmable gate arrays (FPGAs), varying processor sizes give rise to a generalization of our problem: place a set of rectangular modules on a grid to minimize the overall weighted sum of $L_1$ distances between modules. Ahmadinia et al. [1] give an optimal $\Theta(n \log n)$ algorithm for finding an optimal feasible location for a module given a set of $n$ existing modules. At this point, no results are known for the general off-line problem (place $n$ modules simultaneously) or for on-line versions.

Another related problem is *min-sum k-clustering*: separate a graph into $k$ clusters to minimize the sum of distances between nodes in the same cluster. For general graphs, Sahni and Gonzalez [24] show it is NP-hard to approximate this problem to within any constant factor for $k \geq 3$. In a metric space, Guttmann-Beck and Hassin [12] give a 2-approximation, Indyk [13] gives a PTAS for $k = 2$, and Bartel et al. [2] give an $O((1/\epsilon) \log^{1+\epsilon} n)$-approximation for general $k$.

Fekete and Meijer [11] consider the problem of *maximizing* the average $L_1$ distance. They give a PTAS for this *dispersion* problem in $\Re^d$ for constant $d$, and show that an optimal set of any fixed size can be found in $O(n)$ time.

*Our Results.* We develop algorithms for minimizing the average $L_1$ distance between allocated processors in a mesh supercomputer. A greedy heuristic we analyze called MM and a 3D version of MC1x1 have been implemented on Cplant. In particular, we give the following results:

- We prove that MM is a $\frac{7}{4}$-approximation algorithm for $2D$ grids, reducing the previous best factor of 2 [16], and we show that this analysis is tight.
- We present a simple generalization to general $d$-dimensional space with fixed $d$ and prove that the algorithm gives a $2 - \frac{1}{2d}$-approximation algorithm, which is tight.
- We give an efficient polynomial-time approximation scheme (PTAS) for points in $\Re^d$ for constant $d$.
- Using simulations, we compare the allocation performance of our algorithm to that of other algorithms. As a byproduct, we get insight on how to place a stream of jobs in an online setting.

In addition, we have a number of other results whose details are omitted due to space constraints: We have a linear-time exact algorithm for the 1D case based on dynamic programming. We prove that the $d$-dimensional version of MC1x1 has approximation factor at most $d$ times that of MM. We have an algorithm to solve the 2-dimensional case for $k = 3$ in time $O(n \log n)$.

## 2    Manhattan Median Algorithm for Two-Dimensional Point Sets

### 2.1    Median-Based Algorithms

Given a set $S$ of $k$ points in the plane, a point that minimizes the total $L_1$ distance to these points is called an $(L_1)$ median. Given the nature of $L_1$ distances, this is a point whose $x$-coordinate (resp. $y$-coordinate) is the median of the $x$ (resp. $y$) values of the given point set. We can always pick a median whose coordinates are from the coordinates in $S$. There is a unique median if $k$ is odd; if $k$ is even, possible median coordinates may form intervals.

The natural greedy algorithm for our clustering problem is as follows:

---

Consider the $O(n^2)$ intersection points of the horizontal and vertical lines through the points in $P$. For each of these points $p$ do:

1. Take the $k$ points closest to $p$ (using the $L_1$ metric), breaking ties arbitrarily.
2. Compute the total pairwise distance between all $k$ points.

Return the set of $k$ points with smallest total pairwise distance.

---

We call this strategy MM, for **M**anhattan **M**edian. We prove that MM is a $\frac{7}{4}$-approximation on 2D meshes. (Note that Krumke et al. [16] call this algorithm Gen-Alg and show it is a 2-approximation in arbitrary metric spaces.)

## 2.2 Analysis of the Algorithm

For $S \subseteq P$, let $|S|$ denote the sum of $L_1$ distances between points in $S$. For a point $p$ in the plane, we use $p_x$ and $p_y$ to denote its $x$- and $y$-coordinates respectively.

**Lemma 1.** *MM is not better than a 7/4 approximation.*

*Proof.* For a class of examples establishing the lower bound, consider the situation shown in Figure 3. For any $\epsilon > 0$, it has clusters of $k/2$ points at $(0,0)$ and $(1,0)$. In addition, it has clusters of $k/8$ points at $(0, \pm(1-\epsilon))$, $(1, \pm(1-\epsilon))$, $(2-\epsilon, 0)$, and $(-1+\epsilon, 0)$. The best choices of median are $(0,0)$ and $(1,0)$, which yield a total distance of $7k^2(1-\Theta(\epsilon))/16$. The optimal solution is the points at $(0,0)$ and $(1,0)$, which yield a total distance of $k^2/4$.  □

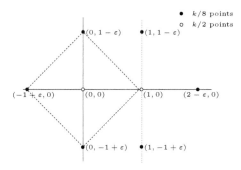

**Fig. 3.** A class of examples where MM yields a ratio of 7/4

Now we show that 7/4 is indeed the worst-case bound. We focus on possible worst-case arrangements and use local optimality to restrict the possible arrangements until the claim follows.

Let $OPT$ be a subset of $P$ of size $k$ for which $|OPT|$ is minimum. Without loss of generality assume that the origin is a median point of $OPT$. This means that the number of points of $OPT$ with positive or negative $x$- or $y$-coordinates is at most $k/2$. Let $MM$ be the set of $k$ points closest to the origin. (Since this is one candidate solution for the algorithm, its sum of pairwise distances is at least as high as that of the solution returned by the algorithm.)

Without loss of generality, assume that the largest distance of a point in $MM$ to the origin is 1, so $MM$ lies in the unit circle $C$. We say that points are either inside $C$, on $C$, or outside $C$. All points of $P$ inside $C$ are in $MM$ and at least some points on $C$ are in $MM$. If there are more than $k$ points on and inside $C$, we select all points inside $C$ plus those points on $C$ maximizing $|MM|$.

Clearly $1 \leq |MM|/|OPT|$. Let $\rho_k$ be the supremum of $|MM|/|OPT|$ over all input configurations $P$. By assuming that ties are broken badly, we can assume that there is a configuration $S \subseteq P$ for which $|MM|/|OPT| = \rho_k$:

**Lemma 2.** *For any $n$ and $k$, there are point sets $P$ with $|P| = n$ for which $|MM|/|OPT|$ attains the value $\rho_k$.*

*Proof.* The set of arrangements of $n$ points in the unit circle $C$ is a compact set in 2$d$-dimensional space. By our assumption on breaking ties, $|MM|/|OPT|$ is upper semi-continuous, so it attains a maximum.    □

For $k \leq 8n/11$ we show $|MM|$ is at most 7/4 times larger than $|OPT|$.

**Lemma 3.** *For $k \leq 8n/11$ we have $\rho_k \leq 7/4$.*

*Sketch of Proof.* We assume that we have a point set $P$ for which $\rho_k$ is equal to 7/4. We can assume without loss of generality that $P = MM \cup OPT$. If there is a point $p \in P$ that does not lie in a corner of $C$ or on the origin, we look at all points that lie on the axis-parallel rectangle through $p$ with corners on $C$. We move these points simultaneously, in such a way that they stay on an axis-parallel rectangle with corners on $C$. This move changes $|MM|$ by some small amount $\delta_a$ and $|OPT|$ by some amount $\delta_o$. However if we move all points in the opposite direction $|MM|$ and $|OPT|$ change by $-\delta_a$ and $-\delta_o$ respectively. So if $\delta_a/\delta_o \neq \rho_k$, one of these two moves increases $|MM|/|OPT|$, which is impossible. If $\delta_a/\delta_o = \rho_k$ we keep moving the points in the same direction until there is a combinatorial change in $P$. We can then repeat this argument until all points of $P$ lie on a corner of $C$ or on the origin.

It is now not too hard to show that the ratio $MM/OPT$ is maximal if there are $k/2$ points at the origin, $k/2$ points in one corner of $C$ and $k/8$ points at each of the other three corners. So we have $|MM|/|OPT| = 7/4$. Notice that $n$ has to be at least $11k/8$ for this value to be obtained.    □

For larger values of $k$ it can be shown that $\rho_k$ decreases, so we summarize:

**Theorem 1.** *MM is a 7/4-approximation algorithm for minimizing the sum of pairwise $L_1$ distances in a 2D mesh.*

## 3  PTAS for Two Dimensions

Let $w(S,T)$ be the sum of all the distances from points in $S$ to points in $T$. Let $w_x(S,T)$ and $w_y(S,T)$ be the sum of $x$- and $y$- distances from points in $S$ to points in $T$, respectively. So $w(S,T) = w_x(S,T) + w_y(S,T)$. Let $w(S) = w(S,S)$, $w_x(S) = w_x(S,S)$, and $w_y(S) = w_y(S,S)$. We call $w(S)$ the *weight* of $S$.

Let $S = \{s_0, s_1, \ldots, s_{k-1}\}$ be a minimum-weight subset of $P$, where $k$ is an integer greater than 1. We label the $x$- and $y$-coordinates of a point $s \in S$ by some $(x_a, y_b)$ with $0 \leq a < k$ and $0 \leq b < k$ such that $x_0 \leq x_1 \leq \ldots \leq x_{k-1}$ and $y_0 \leq y_1 \leq \ldots \leq y_{k-1}$. (Note that in general, $a \neq b$ for a point $s = (x_a, y_b)$.) We can derive the following equations: $w_x(S) = (k-1)(x_{k-1} - x_0) + (k-3)(x_{k-2} - x_1) + \ldots$ and $w_y(S) = (k-1)(y_{k-1} - y_0) + (k-3)(y_{k-2} - y_1) + \ldots$ We show that there is a polynomial-time approximation scheme (PTAS), i.e., for any fixed positive $m = 1/\varepsilon$, there is a polynomial approximation algorithm that finds a solution within $(1 + \varepsilon)$ of the optimum.

The basic idea is similar to the one used by Fekete and Meijer [11] to select a set of points maximizing the overall distance: We find (by enumeration) a subdivision of an optimal solution into $m \times m$ rectangular cells $C_{ij}$, each containing a

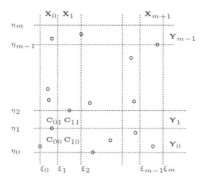

**Fig. 4.** Dividing the point set in horizontal and vertical strips

specific number $k_{ij}$ of selected points. The points from each cell $C_{ij}$ are selected in a way that minimizes the total distance to all other cells except for the $m-1$ cells in the same "horizontal" strip or the $m-1$ cells in the same "vertical" strip. As it turns out, this can be done in a way that the total neglected distance within the strips is bounded by a small fraction of the weight of an optimal solution, yielding the desired approximation property. See Figure 4 for the setup.

For ease of presentation, we assume that $k$ is a multiple of $m$ and $m > 2$. Approximation algorithms for other values of $k$ can be constructed in a similar fashion. Consider a division of the plane by a set of $m+1$ $x$-coordinates $\xi_0 \leq \xi_1 \leq \ldots \leq \xi_m$. Let $X_i := \{p = (x,y) \mid \xi_i \leq x \leq \xi_{i+1}\}$ be the vertical strip between coordinates $\xi_i$ and $\xi_{i+1}$. By enumeration of possible values of $\xi_0, \ldots, \xi_m$ we may assume that each of the $m$ strips $X_i$ contains precisely $k/m$ points of an optimal solution. (A small perturbation does not change optimality or approximation properties of solutions. Thus, without loss of generality, we assume that no pair of points share either $x$-coordinate or $y$-coordinate.)

In a similar manner, assume we know $m+1$ $y$-coordinates $\eta_0 \leq \eta_1 \leq \ldots \leq \eta_m$ so that an optimal solution has precisely $k/m$ points in each horizontal strip $Y_i := \{p = (x,y) \mid \eta_i \leq y \leq \eta_{i+1}\}$.

Let $C_{ij} := X_i \cap Y_j$, and let $k_{ij}$ be the number of points in $OPT$ that are chosen from $C_{ij}$. Since for all $i,j \in \{1, 2, \ldots, m\}$,

$$\sum_{0 \leq l < m} k_{lj} = \sum_{0 \leq l < m} k_{il} = k/m,$$

we may assume by enumeration over the $O(k^m)$ possible partitions of $k/m$ into $m$ pieces that we know all the numbers $k_{ij}$.

Finally, define the vector $\nabla_{ij} := ((2i + 1 - m)k/m, (2j + 1 - m)k/m)$. Our approximation algorithm is as follows: from each cell $C_{ij}$, choose $k_{ij}$ points that are minimum in direction $\nabla_{ij}$, i.e., select points $p = (x,y)$ for which $(x(2i+1-m)k/m, y(2j+1-m)k/m)$ is minimum. For an illustration, see Figure 5.

It can be shown that selecting points of $C_{ij}$ this way minimizes the sum of $x$-distances to points not in $X_i$ and the sum of $y$-distances to points not in $Y_j$.

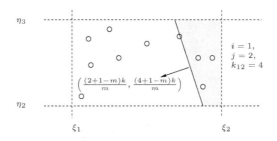

**Fig. 5.** Selecting points in cell $C_{12}$

The details are somewhat technical and are described in the full version of the paper [5]. We summarize:

**Theorem 2.** *The problem of selecting a subset of minimum total $L_1$ distance for a set of points in $\Re^2$ allows a PTAS.*

## 4    Higher-Dimensional Spaces

Using the same techniques, we also generalize our results to higher dimensions. We start by describing the performance of *MM*.

### 4.1    $\left(2 - \frac{1}{2d}\right)$ -Approximation

As in two-dimensional space, *MM* enumerates over the $O(n^d)$ possible medians. For each median, it constructs a candidate solution of the $k$ closest points.

**Lemma 4.** *MM is not better than a $2 - 1/(2d)$ approximation.*

*Proof.* We construct an example based on the cross-polytope in $d$ dimensions, i.e., the $d$-dimensional $L_1$ unit ball. Let $\varepsilon > 0$. Denote the origin with $O$ and the $i^{\text{th}}$ unit vector with $e_i$. The example has $k/2$ points at $O$ and $O + e_1$. In addition, there are $k/(4d)$ points at $O - (1 - \varepsilon)e_1$, $O + (2 - \varepsilon)e_1$, $O \pm (1 - \varepsilon)e_i$ for $i = 2, \ldots, d$, and $O + e_1 \pm (1 - \varepsilon)e_i$ for $i = 2, \ldots, d$. *MM* does best with $O$ or $O + e_1$ as median, giving a total distance of $(k^2/4)\,(2 - 1/(2d))\,(1 + \Theta(\varepsilon))$. Optimal is the points at $O$ and $O + e_1$, giving a total distance of $k^2/4$.    □

Establishing a matching upper bound can be done analogously to Section 2. Lemma 2 holds for general dimensions. The rest is based on the following lemma:

**Lemma 5.** *Worst-case arrangements for MM can be assumed to have all points at positions $(0, \ldots, 0)$ and $\pm e_i$, where $e_i$ is the ith unit vector.*

*Sketch of Proof.* Consider a worst-case arrangement within the cross-polytope centered at the origin with radius 1. Local moves consist of continuous changes in point coordinates, performed in such a way that the number of coordinate values is kept. This means that to move a point having a coordinate value different from

$0, 1, -1$, then all other points sharing that coordinate value are moved to keep the identical coordinates the same, analogous to the proof of Lemma 3.

Note that under these moves, the functions OPT and MM are locally linear, so the ratio of MM and OPT is locally constant, strictly increasing, or strictly decreasing. If a move decreases the ratio, the opposite move increases it, contradicting the assumption that the arrangement is worst-case.

If the ratio is locally constant during a move, it will continue to be extremal until an event occurs, i.e., when the number of coordinate identities between points increases, or the number of point coordinates at $0, 1, -1$ increase. While there are points with coordinates different from $0, 1, -1$, there is always a move that decreases the total degrees of freedom, until all $dn$ degrees of freedom have been eliminated. Thus, we can always reach an arrangement with point coordinates values from the set $\{0, 1, -1\}$. These leaves the origin and the $2d$ positions $\pm e_i$ as only positions within the cross-polytope.                                        □

The restricted set of arrangements can be evaluated with symmetry to yield

**Theorem 3.** *For points lying in $d$-dimensional space, MM is a $2-1/2d$-approximation algorithm, which is tight.*

### 4.2    PTAS for General Dimensions

**Theorem 4.** *For any fixed $d$, the problem of selecting a subset of minimum total $L_1$ distance for a set of points in $\Re^d$ allows a PTAS.*

*Sketch of Proof.* For $m = \Theta(1/\varepsilon)$, we subdivide the set of $n$ points with $d(m+1)$ axis-aligned hyperplanes, such that $(m + 1)$ are normal for each coordinate direction. Moreover, any set of $(m+1)$ hyperplanes normal to the same coordinate axis is assumed to subdivide the optimal solution into $k/m$ equal subsets, called *slices*. Enumeration of all possible structures of this type yields a total of $n^m$ choices of hyperplanes in each coordinate, for a total of $n^{md}$ possible choices. For each choice, we have a total of $m^d$ cells, each containing between 0 and $k$ points; thus, there are $O(m^{kd})$ different distributions of cardinalities to the different cells. As in the two-dimensional case, each cell picks the assigned number of points extremal in its gradient direction.

It is easily seen that for each coordinate $x_i$, the above choice minimizes the total sum of $x_i$-distances between points not in the same $x_i$-slice. The remaining technical part (showing that the sum of distances within slices are small compared to the distances between different slices) is analogous to the details described in the full version of the paper [5] and omitted.                                        □

## 5    Experiments

The work discussed so far is motivated by the allocation of a single job. In the following, we examine how well our algorithms allocate streams of jobs; now the set of free processors available for each job depends on previous allocations.

**Table 1.** Average sum of pairwise distances when the decision algorithm makes allocations with input provided by the situation algorithm

| Situation | Decision Algorithm | | | |
|-----------|------|------|--------|----------|
| Algorithm | MC1x1 | MM | MM+Inc | HilbertBF |
| MC1x1 | 5256 | 5218 | 5207 | 5432 |
| MM | 5323 | 5285 | 5276 | 5531 |
| MM+Inc | 5319 | 5281 | 5269 | 5495 |
| HilbertBF | 5090 | 5059 | 5046 | 5207 |

To understand the interaction between the quality of an individual allocation and the quality of future allocations, we ran a simulation involving pairs of algorithms. One algorithm, the *situation algorithm*, places each job. This determines the free processors available for the next job. Each allocation decision serves as an input to the other algorithm, the *decision algorithm*. Each entry in Table 1 represents the average sum of pairwise distances for the decision algorithm with processor availability determined by the situation algorithm.

Our simulation used the algorithms MC1x1, MM, MM+Inc, and HilbertBF. MM+Inc uses local improvement on the allocation of MM, replacing an allocated processor with an excluded processor that improves average pairwise distance until it reaches a local minimum. HilbertBF is the 1-dimensional strategy of Leung et al. [17] used on Cplant. The simulation used the LLNL Cray T3D trace[1] from the Parallel Workloads Archive [10]. This trace has 21323 jobs run on a machine with 256 processors, treated as a $16 \times 16$ mesh in the simulation.

In each row, the algorithms are ranked in the order MM+Inc, MM, MC1x1, and HilbertBF. This is consistent with the worst-case performance bounds; MM is a 7/4-approximation, MC1x1 is a 7/2-approximation, and HilbertBF has an unbounded ratio[2].

## 6    Conclusions

The algorithmic work described in this paper is one step toward developing algorithms for scheduling mesh-connected network-limited multiprocessors. We have given provably good algorithms to allocate a single job. The next step is to study the allocation of job sequences, a markedly different algorithmic challenge.

The difference between making a single allocation and a sequence of allocations is already illustrated by the diagonal entries in Table 1, where the free processors depend on the same algorithm's previous decisions. These give the ranking (from best to worst) HilbertBF, MC1x1, MM+Inc, and MM. The locally better decisions of MM+Inc seem to paint the algorithm into a corner over time. Figures 1 and 2 help explain why. When starting on an empty grid, MC

---

[1] We thank Moe Jette and Bill Nitzberg for providing the LLNL and NASA Ames iPSC/860 traces, respectively, to the Parallel Workloads Archive.

[2] On an $N \times N$ mesh, the approximation ratio can be $\Omega(N)$.

produces connected rectangular shapes. Locally, these shapes are slightly worse than the round shapes produced by MM, but rectangles have better packing properties because they avoid small patches of isolated grid nodes.

We confirmed this behavior over an entire trace using Procsimity [27, 28], which simulates messages moving through the network. We ran the NASA Ames iPSC/860 trace[1] from the Parallel Workloads Archive [10], scaling down the number of processors for each job by a factor of 4. This made the trace run on a machine with 32 processors, allowing us to find the greedy placement that minimizes average pairwise distance at that step. For average job flow time, MC1x1 was best, followed by MM, and then greedy. We did not run MM+Inc in this simulation. HilbertBF was much worse than all three of the algorithms mentioned in part due to difficulties using it on a nonsquare mesh.

Thus, the online problem in an iterated scenario is the most interesting open problem. We believe that a natural attack may be to consider online packing of rectangular shapes of given area. We plan to pursue this in future work.

## Acknowledgments

We thank Jens Mache for informative discussions on processor allocation. Michael Bender was partially supported by Sandia and NSF Grants EIA-0112849 and CCR-0208670. David Bunde was partially supported by Sandia and NSF grant CCR 0093348. Sándor Fekete was partially supported by DFG grants FE 407/7 and FE 407/8. Henk Meijer was partially supported by NSERC. Sandia is a multipurpose laboratory operated by Sandia Corporation, a Lockheed-Martin Company, for the United States Department of Energy under contract DE-AC04-94AL85000.

## References

1. A. Ahmadinia, C.Bobda, S. Fekete, J.Teich, and J. der Veen. Optimal routing-conscious dynamic placement for reconfigurable computing. In *International Conference on Field-Programmable Logic and its applications*, volume 3203 of *LNCS*, pages 847–851. Springer, 2004.
2. Y. Bartal, M. Charikar, and D. Raz. Approximating min-sum $k$-clustering in metric spaces. In *Proc. 33rd Symp. on Theory of Computation*, pages 11–20, 2001.
3. S. Baylor, C. Benveniste, and Y. Hsu. Performance evaluation of a massively parallel I/O subsystem. In R. Jain, J. Werth, and J. Browne, editors, *Input/Output in parallel and distributed computer systems*, volume 362 of *The Kluwer International Series in Engineering and Computer Science*, chapter 13, pages 293–311. Kluwer Academic Publishers, 1996.
4. C. M. Bender, M. A. Bender, E. D. Demaine, and S. P. Fekete. What is the optimal shape of a city? *J. Physics A: Mathematical and General*, 37:147–159, 2004.

---

[1] We thank Moe Jette and Bill Nitzberg for providing the LLNL and NASA Ames iPSC/860 traces, respectively, to the Parallel Workloads Archive.

5. M. A. Bender, D. P. Bunde, E. D. Demaine, S. P. Fekete, V. J. Leung, H. Meijer, and C. A. Phillips. Communication-aware processor allocation for supercomputers. Technical Report cs.DS/0407058, Computing Research Repository, http://arxiv.org/abs/cs.DS/0407058, 2004.
6. S. Bhattacharya and W.-T. Tsai. Lookahead processor allocation in mesh-connected massively parallel computers. In *Proc. 8th International Parallel Processing Symposium*, pages 868–875, 1994.
7. R. Brightwell, L. A. Fisk, D. S. Greenberg, T. Hudson, M. Levenhagen, A. B. Maccabe, and R. Riesen. Massively parallel computing using commodity components. *Parallel Computing*, 26(2-3):243–266, 2000.
8. C. Chang and P. Mohapatra. Improving performance of mesh connected multicomputers by reducing fragmentation. *Journal of Parallel and Distributed Computing*, 52(1):40–68, 1998.
9. P.-J. Chuang and N.-F. Tzeng. An efficient submesh allocation strategy for mesh computer systems. In *Proc. Int. Conf. Dist. Comp. Systems*, pages 256–263, 1991.
10. D. Feitelson. The parallel workloads archive. http://www.cs.huji.ac.il/labs/parallel/workload/index.html.
11. S. P. Fekete and H. Meijer. Maximum dispersion and geometric maximum weight cliques. *Algorithmica*, 38:501–511, 2004.
12. N. Guttmann-Beck and R. Hassin. Approximation algorithms for minimum sum $p$-clustering. *Disc. Appl. Math.*, 89:125–142, 1998.
13. P. Indyk. A sublinear time approximation scheme for clustering in metric spaces. In *Proc. 40th Ann. IEEE Symp. Found. Comp. Sci. (FOCS)*, pages 154–159, 1999.
14. R. M. Karp, A. C. McKellar, and C. K. Wong. Near-optimal solutions to a 2-dimensional placement problem. *SIAM Journal on Computing*, 4:271–286, 1975.
15. P. Krueger, T.-H. Lai, and V. Dixit-Radiya. Job scheduling is more important than processor allocation for hypercube computers. *IEEE Trans. on Parallel and Distributed Systems*, 5(5):488–497, 1994.
16. S. Krumke, M. Marathe, H. Noltemeier, V. Radhakrishnan, S. Ravi, and D. Rosenkrantz. Compact location problems. *Th. Comp. Sci.*, 181:379–404, 1997.
17. V. Leung, E. Arkin, M. Bender, D. Bunde, J. Johnston, A. Lal, J. Mitchell, C. Phillips, and S. Seiden. Processor allocation on Cplant: achieving general processor locality using one-dimensional allocation strategies. In *Proc. 4th IEEE International Conference on Cluster Computing*, pages 296–304, 2002.
18. K. Li and K.-H. Cheng. A two-dimensional buddy system for dynamic resource allocation in a partitionable mesh connected system. *Journal of Parallel and Distributed Computing*, 12:79–83, 1991.
19. V. Lo, K. Windisch, W. Liu, and B. Nitzberg. Non-contiguous processor allocation algorithms for mesh-connected multicomputers. *IEEE Transactions on Parallel and Distributed Computing*, 8(7), 1997.
20. J. Mache and V. Lo. Dispersal metrics for non-contiguous processor allocation. Technical Report CIS-TR-96-13, University of Oregon, 1996.
21. J. Mache and V. Lo. The effects of dispersal on message-passing contention in processor allocation strategies. In *Proc. Third Joint Conf. on Information Sciences, Sessions on Parallel and Distributed Processing*, volume 3, pages 223–226, 1997.
22. J. Mache, V. Lo, and K. Windisch. Minimizing message-passing contention in fragmentation-free processor allocation. In *Proc. 10th Intern. Conf. Parallel and Distributed Computing Systems*, pages 120–124, 1997.
23. S. Moore and L. Ni. The effects of network contention on processor allocation strategies. In *Proc. 10th Int. Par. Proc. Symp.*, pages 268–274, 1996.

24. S. Sahni and T. Gonzalez. *p*-complete approximation problems. *JACM*, 23(3):555–565, 1976.
25. Sandia National Laboratories. The Computational Plant Project. `http://www.cs .sandia.gov/cplant`.
26. V. Subramani, R. Kettimuthu, S. Srinivasan, J. Johnson, and P. Sadayappan. Selective buddy allocation for scheduling parallel jobs on clusters. In *Proc. 4th IEEE International Conference on Cluster Computing*, 2002.
27. University of Oregon Resource Allocation Group. Procsimity. `http://www.cs.uoregon.edu/research/DistributedComputing/ProcSimity.html%`.
28. K. Windisch, J. Miller, and V. Lo. Procsimity: An experimental tool for processor allocation and scheduling in highly parallel systems. In *Proc. Fifth Symp. on the Frontiers of Massively Parallel Computation*, pages 414–421, 1995.
29. Y. Zhu. Efficient processor allocation strategies for mesh-connected parallel computers. *J. Parallel and Distributed Computing*, 16:328–337, 1992.

# Dynamic Hotlinks

Karim Douïeb* and Stefan Langerman**

Département d'Informatique,
Université Libre de Bruxelles, CP212,
Boulevard du Triomphe, 1050 Bruxelles, Belgium

**Abstract.** Consider a directed rooted tree $T = (V, E)$ representing a collection $V$ of $n$ web pages connected via a set $E$ of links all reachable from a source home page, represented by the root of $T$. Each web page $i$ carries a weight $w_i$ representative of the frequency with which it is visited. By adding hotlinks, shortcuts from a node to one of its descendents, we are interested in minimizing the expected number of steps needed to visit pages from the home page. We give the first linear time algorithm for assigning hotlinks so that the number of steps to accede to a page $i$ from the root of the tree reaches the entropy bound, i.e. is at most $O(\log \frac{W}{w_i})$ where $W = \sum_{i \in T} w_i$. The best previously known algorithm for this task runs in time $O(n^2)$. We also give the first efficient data structure for maintaining hotlinks when nodes are added, deleted or their weights modified, in amortized time $O(\log \frac{W}{w_i})$ per update. The data structure can be made adaptive, i.e. reaches the entropy bound in the amortized sense without knowing the weights $w_i$ in advance.

## 1   Introduction

Since the discovery of the Internet by the general public, the growth of the *World Wide Web* reached an incredible speed and the quantity of information available for all became extraordinary large. By this fact, many inherent problems for consulting of this mass of data appeared, and methods were developed to facilitate and accelerate the search on the web, such as promoting and demoting pages, highlighting links, and clustering related pages in an adaptive fashion depending on user access patterns [15, 6]. In this article we consider the strategy of adding *hotlinks*, i.e. shortcuts from web pages to popular pages accessible from them.

A *web site* can be modeled as a directed graph $\mathcal{G} = (V, E)$ where the nodes $V$ correspond to the web pages and the edges $E$ represent the links. Each node carries a weight representative of its access frequency. We assume that all web pages are reached starting from the *homepage* $r$. Our goal in adding hotlinks (directed edges from a node to one accessible from it) is to minimize the expected number steps to reach a page from the homepage $r$.

The idea of hotlinks was suggested by Perkowitz and Etzioni [15] and studied later by Bose et al. [2] who proved that finding the optimal hotlink assignment

---

* Boursier FRIA, kdouieb@ulb.ac.be

** Chercheur qualifié du FNRS, stefan.langerman@ulb.ac.be

F. Dehne, A. López-Ortiz, and J.-R. Sack (Eds.): WADS 2005, LNCS 3608, pp. 182–194, 2005.

for a DAG is NP-hard, and analyzed several heuristics for assigning hotlinks. More recently, a 2-approximation algorithm for the archivable gain running in a polynomial time was presented by Matichin and Peleg [13].

The problem might become easier when the graph considered is a rooted tree. Kranakis, Krizanc and Shende [12] give a quadratic time algorithm for assigning one hotlink per node so that the expected number of steps to search a node from the root of the tree attain the entropy bound. Several results on adding hotlinks to nodes of $d$-regular complete trees are also reported by Fuhrmann et al. [8]. Recently, Gerstel et al.[10], and A.A. Pessoa et al. [16] independently discovered a polynomial time dynamic programming algorithm for finding the optimal placement of hotlinks on a tree whose depth is logarithmic in the number of nodes. Experimental results showing the validity of the hotlinks approach are given in [5], and a software tool to structure websites efficiently by automatic assignment of hotlinks has been developed [11].

The concept of hotlinks can be applied to other problems than that of web structuring. For instance, Bose et al.[3] use hotlink assignments to design efficient asymmetric communication protocols. Hotlinks can also be used to design data structures as was demonstrated by Brnnimann, Cazals and Durand [4] with their *jumplist* dynamic dictionary data structure. The jumplist structure can be seen as randomized hotlink assignment on a list, and is meant as a simplification of the skiplist structure [17]. A detreministic version of the randomized jumplist of Brnnimann was developed by Elmasry [7].

In this article, we consider rooted directed trees $T$ with $n$ nodes and maximum degree $d$. Every node $i$ in $T$ is associated with a weight $w_i$ representative of its access frequency, and $W = \sum_{i \in T} w_i$. Following the *greedy user* model assumption, we assume that the user always takes the hotlink from a node that leads him to a closer point on the path to the desired destination. Due to that, we consider that the assignment of one hotlink which points to a node $i$ can be see as the deletion of any other hyperlink that ends in $i$. Let $T^A$ be the tree resulting from an assignment $A$ of hotlinks. A measure of the average access time to the nodes is $E[T^A, p] = \sum_{i=1}^{n} d_A(i) p_i$, where $d_A(i)$ is the distance of the node $i$ from the root, and $p = < p_i = w_i/W : i = 1, \ldots, n >$ is the probability distribution on the nodes of the original tree $T$. We are interested in finding an assignment $A$ which minimizes $E[T^A, p]$.

A lower bound on the average access time $E[T^A, p]$ was given in [2] using information theory [14]. Let $H(p)$ be the entropy of the probability distribution $p$, defined by $H(p) = \sum_{i=1}^{n} p_i \log(1/p_i)$, then for any assignment of at most $\delta$ hotlinks per node the expected number of steps to reach a node from the root of the tree is at least $\frac{H(p)}{\log(d+\delta)}$. This bound is achieved up to a constant factor if $d_A(i) = O(\log(W/w_i))$. We show:

**Theorem 1.** *Given an arbitrary weighted rooted tree with $n$ nodes and of total weight $W$. There is an algorithm that runs in $O(n)$ time, which assigns one hotlink per node in such a way that the expected number of steps to reach a node $i$ of weight $w_i$ in the tree from the root is $O(\log \frac{W}{w_i})$.*

This algorithm constitutes a considerable improvement over the previous $O(n^2)$ time algorithm [12]. Furthermore, we present an efficient data structure for dynamically maintaining hotlinks on a tree:

**Theorem 2.** *There exists a data structure for maintaining hotlinks in a weighted tree $T$, allowing the insertion and deletion of leaves of weight 1 in $T$, and the incrementation or decrementation of the weight of any node of $T$. All updates on a node $i$ of weight $w_i$ run in amortized time $O(\log \frac{W}{w_i})$, and the shortest path to any node $i$ of weight $w_i$ is $O(\log \frac{W}{w_i})$ worst case, where $W = \sum_i w_i$.*

In particular, if the weight of a node is incremented every time that node is accessed, the running time of any sequence of accesses will be bounded by the entropy bound (amortized) without knowing the probability distribution in advance. The proof of the two preceding theorem will be given later in this paper. A weighted and amortized version of the Jumplist data structure is presented in the next section, it is developed for the application of hotlinks assignment to arbitrary trees. In section 3 we give a linear time algorithm to assign hotlinks to trees so that the number of steps to accede to a page from the root of the tree reaches the entropy bound. In Section 4, the dynamic hotlink assignment data structure is described.

## 2     Jumplists

The data structure named *Jumplist* [4] is a linked list whose nodes are endowed with an additional pointer, the *jump pointer*. Algorithms on the jumplist are based on the *jump-and-walk* strategy: whenever possible use the jump pointer to speed up the search, and walk along the list otherwise. This data structure provides the usual dictionary operations, i.e. SEARCH, INSERT and DELETE.

To each element $x$ of a jumplist is associated a $key[x]$, a $next[x]$ pointer like an ordinary list structure and also an additional $jump[x]$ pointer which points to a successor of $x$ in the list. Note that the jumplists we discuss here only allow insertions/deletions at the end of the list. This restriction greatly simplifies the presentation of the algorithm and is sufficient for its application to the hotlinks problem for trees.

The original version of the jumplists developed by Brnnimann et al. did not consider the access frequencies of the elements of a jumplist. In this paper we develop an enhanced jumplist structure by associating a weight $w_x$ with each element $x$, proportional to its access frequency, and where the access times reach the entropy bound. In such a situation we would like that the more frequently needed elements be accessed faster than the less frequently needed ones. A measure of the average access time for a jumplist $C$ is

$$\sum_{x \in C} \frac{w_x}{W} d_x,$$

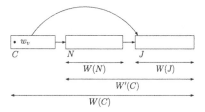

**Fig. 1.** Weight distribution of a weighted jumplist

where $W = \sum_{x \in C} w_x$ is the sum of the weights of the elements in the sublists of $C$ and $d_x$ is the depth in the jumplist of the element $x$, i.e. the minimum number of pointers needed to reach element $x$. We are interested in finding an assignment of the jump pointers for $C$ which minimizes the average access time.

We use the notation $C = (v, N, J)$ for the (sub)list $C$ with first element $v$ followed by the next sublist $N$, and the jump sublist $J$. We write $|C|$ for the number of elements in $C$, and $W(C)$ for the sum of the weights of the elements contained in $C$, i.e. $W(C) = w_v + W(N) + W(J)$. Also, $W'(C) = W(C) - w_v$. See figure 1.

## 2.1   Jump Pointer Assignment

**Lemma 1.** *Given an arbitrary weighted jumplist containing $n$ elements of weights $w_1, \ldots, w_n$, and $W = \sum_{i=1}^{n} w_i$. There is an algorithm which in $O(n)$ time assigns the jump pointers for the jumplist in such a way that $d_i \leq \lfloor \log_2 \frac{W}{w_i} \rfloor + 1$ for $i = 1, \ldots, n$.*

*Proof.* Constructing a jumplist from a list with weighted elements consists in choosing the jump pointer of the header, and recursively building the next and jump sublists. In order to reach the entropy bound, the jump pointer of the header will point to an element splitting the list into two sublists of roughly equal weight. In other words, the sum of the weights of the elements belonging to the next sublist and that of the jump sublist must be at most equal to half of the total weight of the sublists. Thus for the jumplist $C = (v, N, J)$, the condition of the weighted jumplists will be:

$$W(N) = \sum_{i=2}^{k-1} w_i \leq W/2 \qquad \text{and} \qquad W'(J) = \sum_{i=k+1}^{n} w_i \leq W/2. \qquad (1)$$

The problem is to efficiently determine a good element $k$ that satisfies the condition. For this, we first build a table from the jumplist, in time $O(n)$. The $i^{th}$ entry of the table will correspond to the $i^{th}$ element in the jumplist and will contain the value $s_i = \sum_{j=1}^{i} w_j$. Thus, the table is sorted in increasing order and has distinct elements.

Once the table built, we can use exponential search to find the $k^{th}$ element satisfying the condition (1). After this element is found, it will be necessary to reiterate the process recursively on both sublists. Note that it is not necessary to rebuild tables for that, it is sufficient to use segments of the table built during the first stage.

To improve the speed of the search in the table for the element satisfying the condition (1), we carry out a double exponential search in parallel from both sides of the table, in time $O(\log(\min\{k, n-k+1\}))$, where $k$ is the position of the sought element. We can consequently express the complexity of this algorithm, after construction of the table, by the recurrence $t(n) = O(\log(\min\{i, n-i+1\})) + t(i-1) + t(n-i) = O(n)$. $\qquad\square$

## 2.2  Dynamic Weighted Jumplists

One way to dynamize the weighted jumplist is to use the concept of tolerance. It is a method which consists in requiring that any jumplist $C = (v, N, J)$ satisfies the following relaxed version of condition (1):

$$W(N) \leq W'(C)(1+\tau)/2 \qquad \text{and} \qquad W'(J) \leq W'(C)(1+\tau)/2 \qquad (2)$$

where $0 < \tau < 1$ is a constant tolerance factor. The condition of the tolerant weighted jumplists, eq.(2) described above will be checked recursively by the sublists $N$ and $J$.

The methods used here are similar to the ones used by Elmasry [7], except for the fact that our structure doesn't need extra informations to maintain the jumplist (such as the number of elements in the next and the jump sublist of each element in the jumplist) and the elements are weighted. The only extra values the dynamic data structure needs to remember are $W$, the total weight of the entire jumplist, and $max\_W$, the maximal value of $W$ since the last time the jumplist was completely rebuilt.

**Searching.** The basic search algorithms on jumplist are based on the *jump-and-walk* strategy: Whenever possible use the jump pointer to speed up the search, and walk along the list otherwise (if the jumplist is ordered, it will be trivial to determine if the jump pointer improve or not the speed of the search. Else, in the case where all the elements are arbitrary ordered, we make the assumption that the user knows implicitly which pointer is good to use). For a tolerant weighted jumplist that observes the condition eq.(2), we can determine an upper bound to the number of steps to reach an element from the header of the jumplist.

**Theorem 3.** *Consider an arbitrary tolerant weighted jumplist $C$ of tolerance factor $\tau$ whose total sum of the weights of the elements is $W$, the number of steps to reach an element $i$ from the header of the jumplist is at most*

$$\lfloor \log(W/w_i)/\log(2/(1+\tau)) \rfloor + 1.$$

*Proof.* We know that the jumplist $C$ and all its sublists observe the condition of the tolerant weighted jumplist. Let us define $C_k = (v_k, N_k, J_k)$ as the sublist considered after the $k^{th}$ step of the search for element $i$. That is, $C_0 = C$, and if at step $k$ the element $i$ is in the next sublist then $C_k = N_{k-1}$, otherwise $C_k = J_{k-1}$. The value of $W'(C_k)$ can be bounded as a function of $W'(C_{k-1})$ using equation (2):

$$W'(N_{k-1}) \leq W(N_{k-1}) \leq W'(C_{k-1})\frac{1+\tau}{2} \text{ and } W'(J_{k-1}) \leq W'(C_{k-1})\frac{1+\tau}{2}$$

so

$$W'(C_k) \leq \max\{W'(J_{k-1}), W'(N_{k-1})\} \leq W'(C_{k-1})(1+\tau)/2.$$

The resolution of the recurrence gives $W'(C_k) \leq W\left(\frac{1+\tau}{2}\right)^k$. Step $k$ of the algorithm will not be performed unless, $w_i \leq W'(C_{k-1}) \leq W\left(\frac{1+\tau}{2}\right)^{k-1}$. This implies that the number of steps $k$ is bounded by $k \leq \log(W/w_i)/\log(2/(1+\tau))+1$.  $\square$

Thus, the maximum depth an element $x$ can have in a tolerant weighted jumplist of weight $W$ and tolerance factor $\tau$ is

$$d_\tau(x, W) = \lfloor \log(W/w_x)/\log(2/(1+\tau)))\rfloor + 1.$$

An element $x$ for which the depth exceeds the value $d_\tau(x, W)$ will be called a *deep* element. The presence of a deep element clearly implies that the jumplist does not satisfy the condition eq.(2).

**Insertion** We here describe how to insert an element of weight 1 at the end of the list. The insertion operation first uses the jump-and-walk algorithm to find the position of the last element in the jumplist. In the following, we will consider the search sequence $(C = C_0, C_1, \ldots, C_k)$, with $C_j = (x_j, N_j, J_j)$ and $C_{j+1} = J_j$ reaching the last element $x_k$ of the jumplist in $k$ steps (and so $N_k$ and $J_k$ are empty). During an insertion of a new element $z$, the element is placed in $N_k$, we increment $W$, and update $max\_W$ to the maximum of $W$ and $max\_W$. If the newly inserted element is deep, i.e. $k + 1 > d_\tau(z, W)$, then one of the lists $C_j$ containing it does not satisfy eq.(2). We reassign the jump pointers of the jumplist as follows. We climb the jumplist, examining $x_k, x_{k-1}, \ldots$ until we find an element $x_i$ whose sublist $C_i$ does not satisfy the condition (2). Since $x_k$ is at the end of the jumplist, $W(N_k) = 1$, and $W(C_k) = w_{x_k} + 1$. We compute $W(C_j)$ using the formula $W(C_j) = w_{x_j} + W(N_j) + W(C_{j+1})$, where $W(N_j)$ is computed in time $O(|N_j|)$ by walking the list from element $x_j$ to $x_{j+1}$. Thus the total cost for finding $x_i$ is $O(|C_i|)$.

We call $x_i$ the *scapegoat* element in reference to the lazy rebalancing schemes for binary search trees developped independently by Andersson and Lai [1], and by Galperin and Rivest [9]. Once the scapegoat element $x_i$ is found, we have to verify that the reconstruction of its sublist $C_i$ will not create deep nodes. If $i \leq \log(W/W(C_i))/\log(2/(1+\tau))$, then the reconstruction of the jump pointers in $C_i$ will not introduce new deep nodes since the number of links to follow from $x_i$ to any element $y$ will be at most $\log(W(C_i)/w_y)/\log(2/(1+\tau))$ in the reconstructed structure. The jump pointers of the sublist $C_i$ can then be reassigned in time $O(|C_i|)$ using Lemma 1. Otherwise, we continue the search for another scapegoat node $x_{i'}$ with $i' < i$.

Let us now consider a sequence of insert operations in a tolerant weighted jumplist whose total weight is $W$, we wish to show that the amortized complexity per insert is $O(\log(W/w_i))$. We begin by defining a nonnegative *potential function* for the jumplist. Consider the sublist $C = (v, N, J)$, and let $\Phi(v) = max(0, W(N) - W(C)/2, W(J) - W(C)/2)$ be the potential of the element $v$. We see that an element whose sublists are perfectly balanced have a potential of 0, and an element that does not satisfy the condition eq.(2) have a potential of $\Omega(W(C))$. The potential of the jumplist is the sum of the potentials of its elements.

It is easy to see that by increasing their cost by only a constant factor, the insertion operations pay for the increase in potential of the elements. That is, whenever we pass by an element $x$ to insert a new element as a descendant of $x$, we can pay for the increased potential in $x$ that may be required by the resulting increase in $\Phi(x)$.

The potential of the scapegoat element $x_i$, like all the elements that do not observe the condition eq.(2), is $\Omega(W(C_i))$. Therefore, this potential is sufficient to pay for finding the scapegoat element and reassigning the jump pointers of the sublist of which it is the header. These two operations have complexity $O(|C_i|) = O(W(C_i))$.

**Deletion.** The deletion operation consists of removing the last element of the jumplist, the weight of this element must be equal to 1. We will again use the *jump-and-walk* algorithm to reach this last element. Once the element removed, we update $W$. Then, if $W < max\_W(1+\tau)/2$, we reassign all the jump pointers of the entire jumplist[C], and we reset $max\_W$ to $W$. If we restate the analysis above ignoring the deletions, the search time is at most $d_\tau(x, max\_W) \le d_\tau(x, W) + 1$.

Since we perform $\Omega(n)$ operations between two successive rebuilds due to delete operations we can pay for them in the amortized sense (with $n$ equal to the number of elements in the jumplist). Thus for a sequence of delete operations, the amortized complexity per deletion of the last element $i$ of a tolerant weighted jumplist is equal to $O(\log(W/w_i))$.

**Reweighting.** The reweighting operation allows to increment or decrement the weight of an element by one unit. To find the element to be modified, we again use the jump-and-walk algorithm. Then, for incrementing, we use the same technique as during an insertion: We modify the weight of the element, we check that it does not become deep. If it does, we seek the scapegoat element and we reassign the jump pointers of its sublists. For decrementing, we act as during a deletion: We modify the weight of the element, we update $W$. Then, if $W < max\_W(1+\tau)/2$, we reassign all the jump pointers of the jumplist, and we reset $max\_W$ to $W$.

The reweight operation is based on the operation of insertion and deletion. We have by this fact same complexities as those, i.e. an amortized complexity per reweight of an element $i$ of a tolerant weighted jumplist equal to $O(\log(W/w_i))$.

# 3   Hotlinks

The hotlink assignment algorithm for a tree $T$ will proceed by first decomposing the tree into heavy paths (see fig. 2), and then finding hotlink (jump pointers) assignments on the paths viewed as weighted linked lists. In the following, we write $T_x$ for the subtree of $T$ rooted at $x$, $W(T_x)$ the sum of the weights of all elements in $T_x$, and $W'(T_x) = W(T_x) - w_x$.

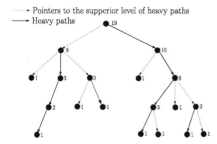

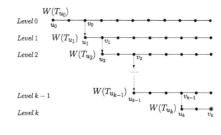

**Fig. 2.** Example of decomposition of a tree into heavy paths

**Fig. 3.** Example of search into heavy paths of a tree

## 3.1   Decomposition into Heavy Paths

Following the classical heavy path decomposition scheme [18], we connect each node to its heaviest child, i.e., we pick $next[x]$ among the children of $x$ if

$$W(T_{next[x]}) \geq W(T_y) \quad \forall y \text{ child of } x. \tag{3}$$

In particular, this implies $W(T_y) \leq W'(T_x)/2 \; \forall y \neq next[x]$ child of $x$. The chosen edges $(x, next[x])$ naturally decompose the tree into paths. This method of determination of lists is realized in time $O(n)$ with $n$ the number of elements in the tree $T$, because that decomposition can be done in a single bottom-up tranversal, at the same time as computing the weight of all subtrees.

## 3.2   Hotlink Assignment

Once the tree is decomposed into heavy paths, we must just apply Lemma 1 to assign the hotlinks (jump pointers) to the paths viewed as jumplists. The weight $z_x$ of an element $x$ in a heavy path will be equal to the weight $w_x$ of the node associated to it plus the sum of the nodes contained in all the subtrees indicated by the children of $x$ except $next[x]$ i.e. $z_x = W(T_x) - W(T_{next[x]})$. The weighted jumplist assignment algorithm is applied on each list using the weights $z_x$. This method is linear in the number of elements present in each list, so the sum of the assignment complexity for all heavy paths in the tree is $O(n)$. Theorem 4 in the next section shows that this hotlink assignment achieves the entropy bound.

## 4    Dynamic Hotlinks

In this section we present a data structure for maintaining hotlinks in a weighted tree when leaves are added or deleted and weights modified. Like in the previous section, the tree will be decomposed into paths, and each path will be managed like a jumplist. These jumplists will be managed dynamically using the structure described in section 2. The dynamic determination of the paths in the tree will require some extra work. Indeed, a sequence of update operations can lead the tree to stop satisfying condition (3). Similarily as for the tolerant jumplists, we will use a relaxed version of condition (3):

$$W(T_y) \leq W'(T_x)(1 + \tau)/2 \; \forall y \neq next[x] \text{ child of } x, \tag{4}$$

where $0 < \tau < 1$ is the tolerance factor.

**Lemma 2.** *The tolerant method of decomposition of the tree $T$ into heavy paths guarantee that maximum number of paths visited during a search for node $x$ is at most $\lfloor \log(W/w_x)/\log(2/(1 + \tau)) \rfloor$.*

*Proof.* To determine the maximum number of levels of paths in the decomposition of a tolerant hotlink tree, we must count the maximum number of times $k$ that we can pass from a list to another. Let $C_i$ be the $i^{th}$ list visited during a search. Every time we follow a link from a node $x$ from one list $C_i$ to the head $y$ of another list $C_{i+1}$, i.e. $y$ is a child of $x$ but $y \neq next[x]$, we know from condition (4) that $W(C_{i+1}) = W(T_y) \leq W'(T_x)(1 + \tau)/2 \leq W(C_i)(1 + \tau)/2$. The recurrence solves to: $W(C_k) \leq W\left(\frac{1+\tau}{2}\right)^k$ and $w_x \leq W(C_k)$.    □

**Searching.** An implicit assumption underlying the common hierarchical approach is that at any node along the search in the tree, the user is able to select the correct link leading towards the desired node. When hotlinks are added, there will exist multiple alternative paths for certain destinations. Again, an underlying assumption at the basis of hotlink idea is that faced with a hotlink in the current node, the user will be able to tell whether or not this hotlink may lead it to a closer point on the path to the desired destination. This has been referred to as the *greedy user* model. Otherwise, we can remark that with the *clairvoyant user* model, we make the assumption that the user somehow knows the topology of the enhanced structure. So, he will always choose the shortest path to reach the desired destination. But, with the method used in this paper, the two models will lead to the same choice of links because no two hotlinks will ever cross. We can now bound the maximum number of steps during a search operation:

**Theorem 4.** *Consider an arbitrary weighted rooted tree $T$ with $W$ the sum of weights of all its nodes. If one hotlink per node is assigned using a tolerant path decomposition and tolerant jumplists with tolerance $0 < \tau < 1$, then there is a constant $a_\tau$ so that the number of steps to reach an element $x$ is at most $d'_\tau(x, W) \leq a_\tau \log(W/w_x)$.*

*Proof.* The search of an node in a tree, can be seen as a succession of search operations in multiple heavy paths composing the tree. The complexity of a search operation in the heavy paths is given by theorem 3. Consider a search entering the $i^{th}$ path at node $u_i$, and leaving it at node $v_i$ to enter the $(i+1)^{th}$ path at node $u_{i+1}$, with $u_0$ being the root of $T$ and $v_k = x$ is the element we are looking for. See Fig.3. Then number of links followed on the $i^{th}$ path is at most $\lfloor \log(W(T_{u_i})/z_{v_i})/\log(2/(1+\tau)) \rfloor + 1$ and $z_{v_i} = W(T_{v_i}) - W(T_{next[v_i]}) \geq W(T_{u_{i+1}})$. So the total number of links followed along heavy paths and hotlinks is at most:

$$t \leq k + [\log \frac{W(T_{u_1})}{z_{v_1}} + \log \frac{W(T_{u_2})}{z_{v_2}} + \cdots + \log \frac{W(T_{u_k})}{z_{v_k}}]/\log(2/(1+\tau))$$

$$\leq k + [\log \frac{W}{W(T_{u_2})} + \log \frac{W(T_{u_2})}{W(T_{u_3})} + \cdots + \log \frac{W(T_{u_k})}{w_x}]/\log(2/(1+\tau))$$

$$= k + \log(W/w_x)/\log(2/(1+\tau)).$$

We must still add to that the number of links between the lists, which is also bounded by $k = \lfloor \log(W/w_x)/\log(2/(1+\tau)) \rfloor$ (see lemma 2). Thus $d'_\tau(x, W) \leq 2k + \log(W/w_x)/\log(2/(1+\tau)) = a_\tau \log(W/w_x)$. where $a_\tau = 3/\log(2/(1+\tau))$. □

To allow update operations, we must store in each node $x$ of the tree an integer between 1 and the outdegree of the node to identify $next[x]$. We furthermore maintain the global value $W$ which is the sum of the weight of the nodes present in the all tree and $max\_W$ which is the maximal value of $W$ since the last time that the hotlinks structure was completely rebuilt.

**Inserting.** We give in this section an algorithm to insert a leaf $x$ of weight $w_x = 1$. The shortest path $x_0, \ldots, x_k$ from the root to the leaf to be inserted is a succession of $k$ hotlinks, heavy tree links, and non-heavy tree links. After finding the shortest path to the parent of the leaf to be inserted, we create the leaf and we check if it is deep, that is, if $k > d'_\tau(x, W)$. If it is, then there must be some node on the path that does not satisfy one of the equations (2) or (4). We then walk up the path verifying those conditions. When walking up from node $x_{i+1}$ to node $x_i$ with $x_{i+1} = next[x_i]$ (heavy tree link) or when $x_{i+1} = jump[x_i]$ (hotlink) we verify eq.(2), and otherwise (non heavy tree link) we verify eq.(4).

To verify condition (2), we must first find the end of the sublist starting at $x_i$. Let $j$ be the largest integer $< i$ such that $x_{j+1} \neq jump[x_j]$. If $x_{j+1} = next[x_j]$, then the sublist starting at $x_i$ can be constructed by following the next pointers from $x_i$ until the element $jump[x_j]$ is found (see figure 4). Otherwise, $x_{j+1}$ is the head of the heavy path containing $x_i$, and the sublist starting at $x_i$ can be constructed by following the next pointers until a leaf is reached. Once the path constructed, the weights of the sublists can be computed by exploring exhaustively the subtrees of their elements. To verify condition (4), we explore the subtrees of the children of $x_i$. Once the scapegoat (node not satisfying one of the conditions) $x_i$ is found, we can consider reconstructing a sublist containing

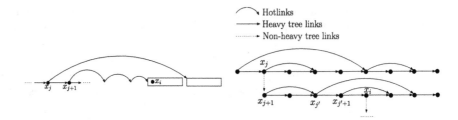

**Fig. 4.** How to find the end of a sublist    **Fig. 5.** Determination of the subtree to reconstruct

it. The sublist to reconstruct is the sublist starting at $x_i$ in the first two cases (jumplist violations). We still have to verify that for all next pointers in that sublist, eq.(4) is satisfied. If it is the case, only the jump pointers in that sublist will have to be reassigned. Otherwise, we are in the last case.

For the last case, eq.(4), we will have to reconstruct an entire subtree. Let $j$ be the largest integer $< i$ such that $(x_j, x_{j+1})$ is a non-heavy tree link, and let $j'$ be the smallest integer $j < j' < i$ such that $x_{j'+1} = next[x_{j'}]$ if it exists, and otherwise $j' = i$. That is, $x_{j'}$ is the first element in the heavy path of $x_i$ for which the next pointer is used, if it exists (See the figure 5). The subtree to reconstruct in this case is the subtree starting at $x_{j'}$. This is to ensure that no jump pointer will point to elements no longer in the same heavy path after the reconstruction. It is easy to see that in this case the weight of the subtree to reconstruct is no more than roughly twice the weight under the scapegoat element. Indeed, we know that the element $jump[x_{j'}]$ is a descendant of the node $x_i$, thus the jump sublist of $x_{j'}$ has a smaller weight than the weight under the scapegoat element. As the weighted jumplists guarantee a balance between the weight of the jump and the next sublists, we can conclude that the weight of the jumplist defined by $x_{j'}$ (equal to $W(T_{x_{j'}})$) is no more than roughly $2W(T_{x_i})$.

Let $C_i$ be the jump sublist or subtree starting at $x_i$ we want to reconstruct. Before reconstructing it, we have to verify that its reconstruction will not leave deep nodes in its subtree. If $i \leq a_\tau \log(W/W(C_i))$, then we know the reconstruction of the sublist will guarantee that no nodes be deep after the reconstruction, since the length of a search for $x$ in the reconstructed sublist/subtree will be at most $a_\tau \log(W(C_i)/w_x)$. Otherwise, we know there is another scapegoat element higher along the path to the root and we can afford to continue looking for it.

Let us now consider a sequence of insert operations beginning with a tree whose total weight is $W$, we wish to show that the amortized complexity per insert is $O(\log W)$. We begin by defining a nonnegative *potential function* for the hotlinks tree $T$. Let $\Phi(x) = \varphi_1 + \varphi_2$ be the potential of the element $x$. Where $\varphi_1$ is the potential function relative to the reassignation of hotlinks in heavy paths (see section 2.2) and $\varphi_2$ is the potential function relative to the reassignation of hotlinks in subtree. Thus let $y_1, \ldots, y_l$ be the children of $x$ and let its potential be equal to $\varphi_2(x) = \max(0, \max_i W(T_{y_i}) - W'(T_x)/2)$. Thus a node that indicates the heaviest tree as its next element has a potential of 0, and a node that does not satisfy the condition eq.(4) has a potential of $\Omega(W'(T_x))$. The potential of the tree is the sum of the potential of its nodes.

It is easy to see that by increasing their cost by only a constant factor, the insertion operations pay for the increase in potential of the nodes. That is, whenever we pass by an element $x$ to insert a new node as a descendant of $x$, we can pay for the increased potential in $x$ that may be required by the resulting increase in $\Phi(x)$. The potential of the scapegoat node $x_i$, like all the nodes that do not observe the condition eq.(4), is $\Theta(W'(T_{x_i}))$. Therefore, this potential is sufficient to pay for finding the scapegoat element and reassigning the hotlinks of the sublist or subtree that has to be reconstructed. These operations have complexity $\Theta(size(x_i)) < \Theta(W(x_i))$ (where $size$ is the number of elements in a sublist or a subtree).

**Deletion.** The deletion operation consists of removing a leaf node $x$ of the tree of weight 1. We first search the node $x$ in the tree then we removed it, and we update $W[C]$. Then, if $W < max\_W/2$, we reassign all the hotlinks of the tree, and we reset $max\_W$ to $W$. This method does not affect the search time $t$ by much: $a_\tau \log(max\_W/w_x) \leq a_\tau(\log(W/w_x) + 1)$.

Since we perform $\Omega(n)$ operations between two successive rebuilds due to delete operations we can pay for them in the amortized sense (with $n$ equal to the number of elements in the tree). Thus for a sequence of delete operations, the amortized complexity per deletion of a leaf node $i$ with the dynamic assignment method is equal to $O(\log(W/w_i))$.

**Reweighting.** This operation is exactly the same as the insertion and the deletion except that we do not actually insert or delete a node. See section 2.2.

# Acknowledgments

The authors thank Pat Morin for many stimulating discussions.

# References

1. A.Andersson and T.W.Lai. Fast updating of well-balanced trees. In *Proc. of the Second Scandinavian Worshop on Algorithm Theory*, volume 447 of *LNCS*, pages 111–121, 1990.
2. P. Bose, E. Kranakis, D. Krizanc, M. V. Martin, J. Czyzowicz, A. Pelc, and L. Gasieniec. Strategies for hotlink assignments. In *Proc. 11th Ann. Int. Symp. on Algoritms and Computation (ISAAC 2000)*, volume 1969 of *LNCS*, pages 23–34, 2000.
3. P. Bose, D. Krizanc, S. Langerman, and P. Morin. Asymmetric communication protocols via hotlink assignments. In *Proc. of the 9th Coll. on Structural Information and Communication Complexity (SIROCCO 2002)*, pages 33–40, 2002.
4. H. Brnnimann, F. Cazals, and M. Durand. Randomized jumplists : A jump-and-walk dictionary data structure. In *Proc. 20th Ann. Symp. on Theoretical Aspects of Computer Science (STACS 2003)*, volume 2607 of *LNCS*, pages 283–294, 2003.

5. J. Czyzowicz, E. Kranakis, D. Krizanc, A. Pelc, and M. V. Martin. Evaluation of hotlink assignment heuristics for improving web access. In *Proc. 2nd Int. Conf. on Internet Computing (IC'2001)*, pages 793–799, 2001.

6. M. Drott. Using web server logs to improving site design. In *Proc. ACM Conf. on Internet Computer Documentation*, pages 43–50, 1998.

7. A. Elmasry. Deterministic jumplists. Technical report, DIMACS, 2003.

8. S. Fuhrmann, S. O. Krumke, and H.-C. Wirth. Multiple hotlink assignment. In *Proc. 27th Int. Workshop on Graph-Theoric Concepts in Computer Science*, volume 2204 of *LNCS*, pages 189–200, 2001.

9. I. Galperin and L. Rivest. Scapegoat trees. In *Proc. 4th Ann. ACM-SIAM Symp. on Discrete Algorithms*, pages 165–174, 1993.

10. O. Gerstel, S. Kutten, R. Matichin, and D. Peleg. Hotlink enhancement algorithms for web directories. In *Proc. 14th Ann. Int. Symp. on Algorithms and Computation (ISAAC 2003)*, volume 2906 of *LNCS*, pages 68–77, 2003.

11. E. Kranakis, D. Krizanc, and M. V. Martin. The hotlink optimizer. In *Proc. 3rd Int. Conf. on Internet Computing (IC'2002)*, pages 33–40, 2002.

12. E. Kranakis, D. Krizanc, and S. Shende. Approximate hotlink assignment. In *Proc. 12th Ann. Int. Symp. on Algoritms and Computation (ISAAC 2001)*, volume 2223 of *LNCS*, pages 756–767, 2001.

13. R. Matichin and D. Peleg. Approximation algorithm for hotlink assignments in web directories. In *Proc. Workshop on Algorithms and Data Structures (WADS 2003)*, volume 2748 of *LNCS*, pages 271–280, 2003.

14. N.Abramson. Information theory and coding. *McGraw Hill*, 1963.

15. M. Perkowitz and O. Etzioni. Towards adaptive Web sites: conceptual framework and case study. *Computer Networks*, 31(11-16):1245–1258, 1999.

16. A. Pessoa, E. Laber, and C. de Souza. Efficient algorithms for the hotlink assignment problem: The worst case search. In *Proc. 15th Ann. Int. Symp. on Algorithms and Computation(ISAAC 2004)*, volume 3341 of *LNCS*, page 778, 2004.

17. W. Pugh. Skip lists: a probabilistic alternative to balanced trees. In F. Dehne, J.-R. Sack, and N. Santoro, editors, *Proc. Workshop on Algorithms and Data Structures (WADS 1989)*, volume 382 of *LNCS*, pages 437–449, 1989.

18. D. D. Sleator and R. E. Tarjan. A data structure for dynamic trees. *J. Comput. Syst. Sci.*, 26(3):362–381, 1983.

# The Minimum-Area Spanning Tree Problem

Paz Carmi[1,*], Matthew J. Katz[1,**], and Joseph S. B. Mitchell[2,***]

[1] Department of Computer Science, Ben-Gurion University of the Negev,
Beer-Sheva 84105, Israel
{carmip, matya}@cs.bgu.ac.il
[2] Department of Applied Mathematics and Statistics, Stony Brook University,
Stony Brook, NY 11794, USA
jsbm@ams.sunysb.edu

**Abstract.** Motivated by optimization problems in sensor coverage, we formulate and study the Minimum-Area Spanning Tree (MAST) problem: Given a set $\mathcal{P}$ of $n$ points in the plane, find a spanning tree of $\mathcal{P}$ of minimum "area," where the area of a spanning tree $\mathcal{T}$ is the area of the union of the $n-1$ disks whose diameters are the edges in $\mathcal{T}$. We prove that the Euclidean minimum spanning tree of $\mathcal{P}$ is a constant-factor approximation for MAST. We then apply this result to obtain constant-factor approximations for the Minimum-Area Range Assignment (MARA) problem, for the Minimum-Area Connected Disk Graph (MACDG) problem, and for the Minimum-Area Tour (MAT) problem. The first problem is a variant of the power assignment problem in radio networks, the second problem is a related natural problem, and the third problem is a variant of the traveling salesman problem.

## 1   Introduction

We introduce and study the Minimum-Area Spanning Tree (MAST) problem. Given a set $\mathcal{P}$ of $n$ points in the plane, find a spanning tree of $\mathcal{P}$ of minimum area, where the area of a spanning tree $\mathcal{T}$ of $\mathcal{P}$ is defined as follows. For each edge $e$ in $\mathcal{T}$ draw the disk whose diameter is $e$. The *area* of $\mathcal{T}$ is then the area of the union of these $n-1$ disks. Although this problem seems natural (see also applications below), we are not aware of any previous work on it.

One of the main results of this paper (presented in Section 2) is that the minimum spanning tree of $\mathcal{P}$ is a constant-factor approximation for MAST. This

---

[*] P. Carmi is partially supported by a Kreitman Foundation doctoral fellowship, and by the Lynn and William Frankel Center for Computer Sciences.

[**] M. Katz is partially supported by grant No. 2000160 from the U.S.-Israel Binational Science Foundation.

[***] J. Mitchell is partially supported by grant No. 2000160 from the U.S.-Israel Binational Science Foundation, NASA Ames Research (NAG2-1620), the National Science Foundation (CCR-0098172, ACI-0328930, CCF-0431030), and Metron Aviation.

F. Dehne, A. López-Ortiz, and J.-R. Sack (Eds.): WADS 2005, LNCS 3608, pp. 195–204, 2005.

is an important property of the minimum spanning tree as is shown below. (See, e.g., [7, 9] for background on the minimum spanning tree.)

We apply the result above to three problems from a class of problems that has received considerable attention. The first problem is a variant of the power assignment problem (also called the range assignment problem). Let $\mathcal{P}$ be a set of $n$ points in the plane, representing $n$ transmitters-receivers (or transmitters for short). In the standard version of the power assignment problem one needs to assign transmission ranges to the transmitters in $\mathcal{P}$, so that (i) the resulting communication graph is strongly connected (that is, the graph in which there exists a directed edge from $p_i \in \mathcal{P}$ to $p_j \in \mathcal{P}$ if and only if $p_j$ lies in the disk $D_{p_i}$ is strongly connected, where the radius of $D_{p_i}$ is the transmission range, $r_i$, assigned to $p_i$), and (ii) the total power consumption (i.e., the cost of the assignment of ranges) is minimal, where the total power consumption is $\sum_{p_i \in \mathcal{P}} \text{area}(D_{p_i})$.

The power assignment problem is known to be NP-hard (see Kirousis et al. [10] and Clementi et al. [6]). Kirousis et al. [10] also obtain a 2-approximation for this problem, based on the minimum spanning tree of $\mathcal{P}$, and this is the best approximation known.

Consider now the variant of the power assignment problem in which the second requirement above is replaced by (ii') the area of the union of the disks $D_{p_1}, \ldots, D_{p_n}$ is minimum. We refer to this problem as the Minimum-Area Range Assignment (MARA) problem. In general, the presence of a foreign receiver (whether friendly or hostile) in the region $D_{p_1} \cup \cdots \cup D_{p_n}$ is undesirable, and the smaller the area of this region, the lower the probability that such a foreign receiver is present. In Section 3 we prove that the range assignment of Kirousis et al. (that is based on the minimum spanning tree) is also a constant-factor approximation for MARA.

Another related and natural problem for which we obtain a constant-factor approximation (in Section 4) is the following. Let $\mathcal{P}$ be a set of $n$ points in the plane. For each point $p \in \mathcal{P}$, draw a disk $D_{p_i}$ of radius 0 or more, such that (i) the resulting disk graph is connected (that is, the graph in which there exists an edge between $p_i \in \mathcal{P}$ and $p_j \in \mathcal{P}$ if and only if $D_{p_i} \cap D_{p_j} \neq \emptyset$ is connected), and (ii) the area of the union of the disks $D_{p_1}, \ldots, D_{p_n}$ is minimized. We refer to this problem as the Minimum-Area Connected Disk Graph (MACDG) problem. (See, e.g., [8, 11] for background on intersection graphs and on disk graphs in particular.)

The last problem for which we obtain a constant-factor approximation (in Section 5) is a variant of the well-known traveling salesman problem. Given a set $\mathcal{P}$ of $n$ points in the plane, find a tour of $\mathcal{P}$ of minimum area, where the area of a tour $T$ is the area of the $n$ disks whose diameters are the edges of the tour. We refer to this problem as the Minimum-Area Tour (MAT) problem. The constant-factor approximation that we obtain for this problem is also based on results concerning the traveling salesman problem with a parameterized triangle inequality.

A potentially interesting property concerning the area of the minimum spanning tree that is obtained as an intermediate result in Section 2 is that the depth of the arrangement of the disks corresponding to the edges of the minimum spanning tree is bounded by some constant. Notice that this property does not follow immediately from the fact that the degree of the minimum spanning tree is at most 6, as is shown in Figure 2.

Finally, all the above results hold in any fixed dimension $d$ (with small modifications).

## 2    MST is a Constant-Factor Approximation for MAST

Let $\mathcal{T}$ be any spanning tree of $\mathcal{P}$. For an edge $e$ in $\mathcal{T}$, let $D(e)$ denote the disk whose diameter is $e$. Put $D(\mathcal{T}) = \{D(e) \,|\, e \text{ is an edge in } \mathcal{T}\}$, $\bigcup_{\mathcal{T}} = \bigcup_{e \in \mathcal{T}} D(e)$, and $\sigma_{\mathcal{T}} = \sum_{e \in \mathcal{T}} \text{area}(D(e))$. Let MST be a minimum spanning tree of $\mathcal{P}$. MST is not necessarily a solution for the Minimum-Area Spanning Tree (MAST) problem; see Figure 1. In this section we prove that MST is a constant-factor approximation for MAST, that is, $\text{area}(\bigcup_{\text{MST}}) = O(\text{area}(\bigcup_{\text{OPT}}))$, where OPT is an optimal spanning tree, i.e., a solution to MAST.

We begin by showing another interesting property of MST, namely, that the depth of any point $p$ in the interior of a cell of the arrangement of the disks in $D(\text{MST})$ is bounded by a small constant. This property does not follow directly from the fact that the degree of MST is bounded by 6; see Figure 2. Let $\text{MST}_p$ be a minimum spanning tree for $\mathcal{P} \cup \{p\}$. We need the following known and easy claim.

**Claim 1.** *We may assume that there is no edge $(a, b)$ in $\text{MST}_p$, such that $(a, b)$ is not in MST and both $a$ and $b$ are points of $\mathcal{P}$.*

*Proof.* Assume there is such an edge $(a, b)$ in $\text{MST}_p$. Consider the path in MST between $a$ and $b$. At least one of the edges along this path is not in $\text{MST}_p$. Let

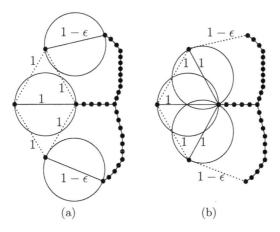

(a)                              (b)

**Fig. 1.** A minimum spanning tree is not necessarily a minimum-area spanning tree. (a) The minimum spanning tree. (b) A minimum-area spanning tree

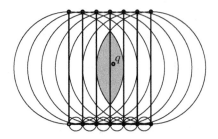

**Fig. 2.** A spanning tree $\mathcal{T}$ of degree 3, and a point $q$ (in the interior of a cell of the arrangement of the disks in $D(\mathcal{T})$) of depth $O(n)$

$e$ be such an edge. $|e| \leq |(a,b)|$, since otherwise $(a,b)$ would have been chosen by the algorithm that computed MST (e.g., Kruskal's minimum spanning tree algorithm [5]). Therefore, we may replace the edge $(a,b)$ in $\text{MST}_p$ by $e$, without increasing the total weight of the tree.

An immediate corollary of this claim is that we may assume that if $e$ is an edge in $\text{MST}_p$ but not in MST, then one of $e$'s endpoints is $p$.

**Lemma 1.** $\sigma_{\text{MST}} \leq 5\,\text{area}(\bigcup_{\text{MST}})$.

*Proof.* We prove that $p$ belongs to at most 5 of the disks in $D(\text{MST})$. Let $D(q_1, q_2)$ be a disk in $D(\text{MST})$, such that $p \in D(q_1, q_2)$. (Notice that $p$ is not on the boundary of $D(q_1, q_2)$, since $p$ is in the interior of a cell of the arrangement of the disks in $D(\text{MST})$.) We show that the edge $(q_1, q_2)$ is not in $\text{MST}_p$. If it is, then either the path from $q_1$ to $p$ or the path from $q_2$ to $p$ includes the edge $(q_1, q_2)$ (but not both). Assume, e.g., that the path from $q_1$ to $p$ includes the edge $(q_1, q_2)$. Then, since $(q_1, p)$ is shorter than $(q_1, q_2)$, we can decrease the total weight of $\text{MST}_p$ by replacing $(q_1, q_2)$ in $\text{MST}_p$ by $(q_1, p)$. We conclude that $(q_1, q_2)$ is not in $\text{MST}_p$.

Thus, by the corollary immediately preceding the lemma, each disk $D \in D(\text{MST})$ such that $p \in D$, induces a distinct edge in $\text{MST}_p$ that is connected to $p$. But the degree of $p$ is at most 6 (this is true for any vertex of any Euclidean minimum spanning tree), so there can be at most 5 disks covering $p$, since one of the edges connected to $p$ is present due to the increase in the number of points (i.e., $p$ was added to $\mathcal{P}$).

**Remark.** Ábrego et al. [1] have shown that the constant 5 can be improved to a constant 3, with a significantly more delicate argument. Their result appeared in an earlier (unpublished) draft of their manuscript.

Let OPT be an optimal spanning tree of $\mathcal{P}$, i.e., a solution to MAST. We use OPT to construct another spanning tree, ST, of $\mathcal{P}$. Initially ST is empty. Let $e_1$ be the longest edge in OPT. Draw two concentric disks $C_1$ and $C_1^3$ around the mid point of $e_1$ of diameters $|e_1|$ and $3|e_1|$, respectively. Compute a minimum spanning tree of the points of $\mathcal{P}$ lying in $C_1^3$, using Kruskal's algorithm [5]. Whenever an edge is chosen by Kruskal's algorithm, it is immediately added to

ST. See Figure 3. Let $S_1$ denote the set of edges that have been added to ST in this (first) iteration.

Next, let $e_2$ be the longest edge in OPT, such that at least one of its endpoints lies outside $C_1^3$. As for $e_1$, draw two concentric disks $C_2$ and $C_2^3$ around the mid point of $e_2$ of diameters $|e_2|$ and $3|e_2|$, respectively. Apply Kruskal's minimum spanning algorithm to the points of $\mathcal{P}$ lying in $C_2$ with the following modification. The next edge in the sorted list of potential edges is chosen by the algorithm if and only if it is not already in ST and its addition to ST does not create a cycle in ST. Moreover, when an edge is chosen by the algorithm it is immediately added

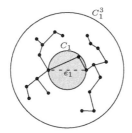

**Fig. 3.** ST after choosing $e_1$

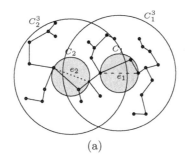

(a)

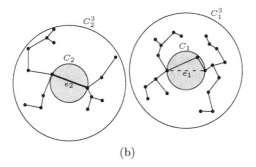

(b)

**Fig. 4.** ST after choosing $e_1$ and $e_2$. (a) One of the end points of $e_2$ is in $C_1^3$. (b) Both endpoints of $e_2$ are not in $C_1^3$

to ST; see Figure 4 (a) and (b). Let $S_2$ denote the set of edges that have been added to ST in this iteration.

In the $i$'th iteration, let $e_i$ be the longest edge in OPT, such that there is no path yet in ST between its endpoints. Draw two concentric circles $C_i$ and $C_i^3$ around the mid point of $e_i$, and apply Kruskal's minimum spanning tree algorithm with the modification above to the points of $\mathcal{P}$ lying in $C_i^3$. Let $S_i$ denote the set of edges that have been added to ST in this iteration. The process ends when for each edge $e$ in OPT there already exists a path in ST between the endpoints of $e$.

**Claim 2.** *For each $i$, $S_i$ is a subset of the edge set of the minimum spanning tree $\mathrm{MST}_i$ that is obtained by applying Kruskal's algorithm, without the modification above, to the points in $C_i^3$.*

*Proof.* Let $e$ be an edge that was added to ST during the $i$'th iteration. If $e$ is not chosen by Kruskal's algorithm (without the modification above), it is only because, when considering $e$, a path between its two endpoints already existed in $\mathrm{MST}_i$. But this implies that $e$ could not have been added to ST, since, any edge already in $\mathrm{MST}_i$ was either also added to ST or was not added since there already existed a path in ST between its two endpoints. Thus, when $e$ was considered by the modified algorithm it should have been rejected. We conclude that $e$ must be in $\mathrm{MST}_i$.

**Claim 3.** *ST is a spanning tree of $\mathcal{P}$.*

*Proof.* Since only edges that do not create a cycle in ST were added to ST, there are no cycles in ST. Also, ST is connected, since otherwise there still exists an edge in OPT that forces another iteration of the construction algorithm.

Let $\mathcal{C}$ denote the set of the disks $C_1, \ldots, C_k$, and let $\mathcal{C}^3$ denote the set of the disks $C_1^3, \ldots, C_k^3$, where $k$ is the number of iterations in the construction of ST.

**Claim 4.** *For any pair of disks $C_i, C_j$ in $\mathcal{C}$, $i \neq j$, it holds that $C_i \cap C_j = \emptyset$.*

*Proof.* Let $C_i$ be any disk in $\mathcal{C}$. We show that for any disk $C_j \in \mathcal{C}$ such that $j > i$, $C_i \cap C_j = \emptyset$. From the construction of ST it follows that $|e_j|$, the diameter of $C_j$, is smaller or equal to $|e_i|$, the diameter of $C_i$. Moreover, at least one of the endpoints of $e_j$ lies outside $C_i^3$ (since if both endpoints of $e_j$ lie in $C_i^3$, then, by the end of the $i$'th iteration, a path connecting between these endpoints must already exist in ST). Therefore, $C_j$ whose center coincides with the mid point of $e_j$, cannot intersect $C_i$.

**Claim 5.** $\sigma_{\mathrm{ST}} = O(\mathrm{area}(\bigcup_{\mathrm{OPT}}))$.

*Proof.* Recall that $\sigma_{\mathrm{ST}} = \Sigma_i \sigma_{S_i}$, where $\sigma_{S_i} = \Sigma_{e \in S_i} \mathrm{area}(D(e))$. We first show by the sequence of inequalities below that $\sigma_{S_i} = O(\mathrm{area}(C_i))$.

$$\sigma_{S_i} \leq^1 \sigma_{\mathrm{MST}_i} \leq^2 5\,\mathrm{area}(\bigcup_{\mathrm{MST}_i}) =^3 O(\mathrm{area}(C_i^3)) =^4 O(\mathrm{area}(C_i)).$$

The first inequality follows immediately from Claim 2. The second inequality is true by Lemma 1. Consider Equality 3. Since all edges in $\text{MST}_i$ are contained in $C_i^3$, it holds that $\bigcup_{\text{MST}_i}$ is contained in a disk that is obtained by expanding $C_i^3$ by some constant factor. It follows that $\text{area}(\bigcup_{\text{MST}_i}) = O(\text{area}(C_i^3)) = O(\text{area}(C_i))$. Therefore,

$$\sigma_{\text{ST}} = \Sigma_i \sigma_{S_i} = \Sigma_i O(\text{area}(C_i)).$$

But according to Claim 4, the latter expression is equal to $O(\text{area}(\bigcup_{\mathcal{C}}))$, and, since $\mathcal{C}$ is a subset of $D(\text{OPT})$, we conclude that $\sigma_{\text{ST}} = O(\text{area}(\bigcup_{\text{OPT}}))$.

We are now ready to prove the main result of this section.

**Theorem 1.** MST *is a constant-factor approximation for* MAST, *i.e., area* $(\bigcup_{\text{MST}}) \leq c \cdot \text{area}(\bigcup_{\text{OPT}})$, *for some constant $c$.*

*Proof.*

$$\text{area}(\bigcup_{\text{MST}}) \leq^1 \sigma_{\text{MST}} \leq^2 \sigma_{\text{ST}} \leq^3 c \cdot \text{area}(\bigcup_{\text{OPT}}).$$

The first inequality is trivial. The second inequality holds for any spanning tree of $\mathcal{P}$; that is, for any spanning tree $\mathcal{T}$, $\sigma_{\text{MST}} \leq \sigma_{\mathcal{T}}$. (Since if the lengths $|e|$ of the edges are replaced with weights $\pi|e|^2/2$, we remain with the same minimum spanning tree.) The third inequality is proven in Claim 5.

# 3   A Constant-Factor Approximation for MARA

MST induces an assignment of ranges to the points of $\mathcal{P}$. Let $p_i \in \mathcal{P}$ and let $r_i$ be the length of the longest edge in MST that is connected to $p_i$, then the range that is assigned to $p_i$ is $r_i$. Put $\text{RA} = \{D_{p_1}, \ldots, D_{p_n}\}$, where $D_{p_i}$ is the disk of radius $r_i$ centered at $p_i$. In this section we apply the main result of the previous section (i.e., MST is a constant-factor approximation for MAST), in order to prove that the range assignment that is induced by MST is a constant-factor approximation for the Minimum-Area Range Assignment (MARA) problem. That is, (i) the corresponding (directed) communication graph is strongly connected, and (ii) the area of the union of the disks in RA is bounded by some constant times the area of the union of the transmission disks in an optimal range assignment, i.e., a solution to MARA.

The first requirement above was already proven by Kirousis et al. [10], who showed that the range assignment induced by MST is a 2-approximation for the standard range assignment problem. Let $\text{OPT}^R$ denote an optimal range assignment, i.e., a solution to MARA. It remains to prove the second requirement above.

**Claim 6.** $\text{area}(\bigcup_{RA}) \leq 9\,\text{area}(\bigcup_{\text{MST}})$.

*Proof.* We define an auxiliary set of disks. For each edge $e$ in MST, draw a disk of diameter $|3e|$ centered at the mid point of $e$. Let $D^3(\text{MST})$ denote the set of these $n-1$ disks; see Figure 5. We now observe that $\text{area}(\bigcup_{RA}) \leq \text{area}(\bigcup_{D^3(\text{MST})})$.

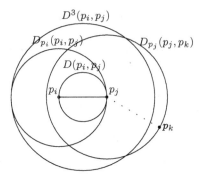

**Fig. 5.** $(p_i, p_j) \in \text{MST}$; $D(p_i, p_j) \in D(\text{MST})$; $D_{p_i}(p_i, p_j), D_{p_j}(p_j, p_k) \in \text{RA}$; $D^3(p_i, p_j) \in D^3(\text{MST})$

This is true since for each $p_i \in \mathcal{P}$, $D_{p_i} = D_{p_i}(p_i, p_j)$ for some point $p_j \in \mathcal{P}$ that is connected to $p_i$ (in MST) by an edge, and $D_{p_i}(p_i, p_j)$ is contained in the disk of $D^3(\text{MST})$ corresponding to the edge $(p_i, p_j)$. Finally, clearly $\text{area}(\bigcup_{D^3(\text{MST})}) \leq 9\,\text{area}(\bigcup_{\text{MST}})$.

**Theorem 2.** RA *is a constant-factor approximation for* MARA, *i.e.*, $\text{area}(\bigcup_{\text{RA}}) \leq c' \cdot \text{area}(\bigcup_{\text{OPT}^R})$, *for some constant $c'$.*

*Proof.* The proof is based on the observation that the (directed) communication graph corresponding to $\text{OPT}^R$ contains a spanning tree, and on the main result of Section 2. Let $p$ be any point in $\mathcal{P}$. We construct a spanning tree $\mathcal{T}$ of $\mathcal{P}$ as follows. For each point $q \in \mathcal{P}$, $q \neq p$, compute a shortest (in terms of number of hops) directed path from $q$ to $p$, and add the edges in this path to $\mathcal{T}$. Now make all edges in $\mathcal{T}$ undirected. $\mathcal{T}$ is a spanning tree of $\mathcal{P}$. For each edge $(p_i, p_j)$ in $\mathcal{T}$, the disk $D(p_i, p_j)$ is contained either in the transmission disk of $p_i$ (in $\text{OPT}^R$), or in the transmission disk of $p_j$ (in $\text{OPT}^R$). Hence, $\bigcup_{\mathcal{T}} \subseteq \bigcup_{\text{OPT}^R}$.

The following sequence of inequalities completes the proof. (OPT denotes a solution to MAST.)

$$\text{area}(\bigcup_{\text{RA}}) \leq^1 9\,\text{area}(\bigcup_{\text{MST}}) \leq^2 9c \cdot \text{area}(\bigcup_{\text{OPT}}) \leq^3 9c \cdot \text{area}(\bigcup_{\mathcal{T}}) \leq^4 9c \cdot \text{area}(\bigcup_{\text{OPT}^R}).$$

The first inequality follows from Claim 6; the second inequality follows from Theorem 1; the third inequality follows from the definition of OPT; the fourth inequality was shown above.

## 4    A Constant-Factor Approximation for MACDG

MST induces an assignment of radii to the points of $\mathcal{P}$. Let $p_i \in \mathcal{P}$ and let $r_i$ be the length of the longest edge in MST connected to $p_i$, then the radius that is assigned to $p_i$ is $r_i/2$. Put $\text{DG} = \{D_{p_1}, \ldots, D_{p_n}\}$, where $D_{p_i}$ is the disk of radius $r_i/2$ centered at $p_i$. In this section we apply the main result of Section 2, in order to prove that DG is a constant-factor approximation for the Minimum-Area Connected Disk Graph (MACDG) problem. That is, (i) viewing DG as an intersection graph, DG is connected, and (ii) the area of the union of the disks in DG is bounded by some constant times the area of the union of the disks in an optimal assignment of radii, i.e., a solution to MACDG.

The first requirement above clearly holds, since each edge in MST is also an edge in DG. Let $\text{OPT}^D$ denote an optimal assignment of radii, i.e., a solution to MACDG. It remains to prove the second requirement above.

**Theorem 3.** DG *is a constant-factor approximation for* MACDG, *i.e., area* $(\bigcup_{\text{DG}}) \leq c'' \cdot \text{area}(\bigcup_{\text{OPT}^D})$, *for some constant* $c''$.

*Proof.* We only outline the proof, since it is very similar to the proof of the previous section. We first claim that $\text{area}(\bigcup_{\text{DG}}) \leq 9\,\text{area}(\bigcup_{\text{MST}})$. This follows immediately from Claim 6, since $\bigcup_{\text{DG}} \subseteq \bigcup_{\text{RA}}$. Next, we observe that if one doubles the radius of each of the disks in $\text{OPT}^D$, then the resulting set of disks contains the set of disks of some spanning tree $\mathcal{T}$ of $\mathcal{P}$. Thus, by Theorem 1, $\text{area}(\bigcup_{\text{MST}}) \leq c \cdot \text{area}(\bigcup_{\text{OPT}^D})$. We complete the proof by putting the two inequalities together.

## 5   A Constant-Factor Approximation for MAT

Consider the complete graph induced by $\mathcal{P}$. We assign weights to the edges of the graph, such that the weight $w(e)$ of an edge $e$ is $|e|^2$. Let $G^2$ denote this graph. Define the weight $w(F)$ of a subset $F$ of the edge-set of $G^2$ to be the sum of the weights of the edges in $F$.

Notice that the triangle inequality does not hold in $G^2$. However, the triangle inequality "almost" holds, in that $|uv|^2 \leq 2 \cdot (|uw|^2 + |wv|^2)$. For distance functions such that $d(u,v) \leq \tau \cdot (d(u,w) + d(w,v))$, constant-factor approximation algorithms for the TSP are known: Andreae and Bandelt [3] give a $(3\tau^2/2 + \tau/2)$-approximation, which was refined by Andrea [2] to a $(\tau^2 + \tau)$-approximation, and Bender and Chekuri [4] give a $4\tau$-approximation. For our case ($\tau = 2$), this implies that there is a 6-approximation.

Andreae and Bandelt actually compute a tour $T$ in $G^2$, such that $w(T) \leq c \cdot w(\text{MST}_{G^2})$, where $\text{MST}_{G^2}$ is the minimum spanning tree of $G^2$ and $c$ is some constant. We show that $T$ is a constant-factor approximation for the Minimum-Area Tour (MAT) problem.

For an edge $e$ in $T$, let $D(e)$ denote the disk whose diameter is $e$. Put $D(T) = \{D(e) \mid e \text{ is an edge in } T\}$, $\bigcup_T = \bigcup_{e \in T} D(e)$, and $\sigma_T = \sum_{e \in T} \text{area}(D(e))$. Let $\text{OPT}^T$ be an optimal tour, i.e., a solution to MAT. Clearly $\text{area}(\bigcup_{\text{OPT}^T}) \geq \text{area}(\bigcup_{\text{OPT}^S})$, where $\text{OPT}^S$ is a solution to the Minimum Area Spanning Tree (MAST) problem. We need to show that $\text{area}(\bigcup_T) = O(\text{area}(\bigcup_{\text{OPT}^T}))$. Indeed

$$\text{area}(\bigcup_T) \leq \sigma_T \leq w(T) \leq c \cdot w(\text{MST}_{G^2}).$$

But $w(\text{MST}_{G^2}) = \sum_{e\in\text{MST}} |e|^2$, where MST is the minimum spanning tree of $\mathcal{P}$ (since both trees are identical in terms of edges). So

$$\text{area}(\bigcup_T) = O(\sum_{e\in\text{MST}} |e|^2) = O(\sigma_{\text{MST}}) = O(\text{area}(\bigcup_{\text{MST}})),$$

where the latter equality follows from Lemma 1. And, by the main result of Section 2,

$$O(\text{area}(\bigcup_{\text{MST}})) = O(\text{area}(\bigcup_{\text{OPT}^S})) = O(\text{area}(\bigcup_{\text{OPT}^T})).$$

The following theorem summarizes the result of this section.

**Theorem 4.** $T$ is a constant-factor approximation for MAT, i.e., $\text{area}(\bigcup_T) \leq \hat{c} \cdot \text{area}(\bigcup_{\text{OPT}^T})$, for some constant $\hat{c}$.

# References

1. B. Ábrego, G. Araujo, E. Arkin, S. Fernández, F. Hurtado, M. Kano, J.S.B. Mitchell, E. O. na Pulido, E. Rivera-Campo, J. Urrutia, and P. Valencia. Matching points with geometric objects. Manuscript, Universitat Politècnica de Catalunya, Nov. 2003.
2. T. Andreae. On the traveling salesman problem restricted to inputs satisfying a relaxed triangle inequality. Tech. Report 74, Mathematisches Seminar, University of Hamburg, 1998.
3. T. Andreae and H.-J. Bandelt. Performance guarantees for approximation algorithms depending on parameterized triangle inequalities. *SIAM Journal of Discrete Mathematics*, 8(1):1–16, 1995.
4. M.A. Bender and C. Chekuri. Performance guarantees for the tsp with a parameterized triangle inequality. *Inf. Process. Lett.*, 73(1-2):17–21, 2000.
5. T.H. Cormen, C.E. Leiserson, R.L. Rivest, and C. Stein. *Introduction to Algorithms*, Second Edition. *MIT Press*, 2001.
6. A.E.F. Clementi, P. Penna, and R. Silvestri. On the power assignment problem in radio networks. *Electronic Colloquium on Computational Complexity*, 2000.
7. D. Eppstein. Spanning trees and spanners. *Handbook of Computational Geometry*, J.-R. Sack and J. Urrutia, eds., Elsevier, 1999, pp. 425–461.
8. M.C. Golumbic. *Algorithmic Graph Theory and Perfect Graphs. Academic Press*, New York, 1980.
9. R.L. Graham and P. Hell. On the history of the minimum spanning tree problem. *Annals of the History of Computing*, 7:43–57, 1985.
10. L.M. Kirousis, E. Kranakis, D. Krizanc, and A. Pelc. Power consumption in packet radio networks. *14th Annual Sympos. Theoretical Aspects of Computer Science (STACS)*, LNCS 1200, 363–374, 1997.
11. T.A. McKee and F. R. McMorris. Topics in intersection graph theory. Volume 2 of *Monographs on Discrete Mathematics and Applications*, SIAM, 1999.

# Hinged Dissection of Polypolyhedra

Erik D. Demaine[1,*], Martin L. Demaine[1],
Jeffrey F. Lindy[2,**], and Diane L. Souvaine[3,***]

[1] Computer Science and Artificial Intelligence Laboratory,
Massachusetts Institute of Technology, Cambridge, MA, USA
{edemaine, mdemaine}@mit.edu
[2] Courant Institute of Mathematical Sciences,
New York University, New York, NY, USA
lindy@cs.nyu.edu
[3] Department of Computer Science,
Tufts University, Medford, MA, USA
dls@cs.tufts.edu

**Abstract.** This paper presents a general family of 3D hinged dissections for *polypolyhedra*, i.e., connected 3D solids formed by joining several rigid copies of the same polyhedron along identical faces. (Such joinings are possible only for reflectionally symmetric faces.) Each hinged dissection consists of a linear number of solid polyhedral pieces hinged along their edges to form a flexible closed chain (cycle). For each base polyhedron $P$ and each positive integer $n$, a single hinged dissection has folded configurations corresponding to all possible polypolyhedra formed by joining $n$ copies of the polyhedron $P$. In particular, these results settle the open problem posed in [7] about the special case of polycubes (where $P$ is a cube) and extend analogous results from 2D [7]. Along the way, we present hinged dissections for polyplatonics (where $P$ is a platonic solid) that are particularly efficient: among a type of hinged dissection, they use the fewest possible pieces.

## 1   Introduction

A *dissection* of a set of figures (solid 2D or 3D shapes, e.g., polygons or polyhedra) is a way to cut one of the figures into finitely many (compact) pieces such that it can be transformed into any other of the figures by moving the pieces rigidly. Dissections have been studied extensively, particularly in 2D [12, 15]. It is well-known that any two polygons of the same area have a dissection [5, 12, 16]. By transitivity, it is easy to extend this result to a dissection

---

* Supported in part by NSF CAREER award CCF-0347776 and DOE grant DE-FG02-04ER25647.
** Work begun when author was at Tufts University. Supported in part by NSF grant EIA-99-96237.
*** Supported in part by NSF grants EIA-99-96237 and CCF-0431027.

F. Dehne, A. López-Ortiz, and J.-R. Sack (Eds.): WADS 2005, LNCS 3608, pp. 205–217, 2005.
© Springer-Verlag Berlin Heidelberg 2005

of any finite set of polygons. Thus, in this context, the main interest is in finding the dissection of the polygons that uses the fewest possible pieces. On the other hand, not every two polyhedra of the same volume have a dissection: for example, there is no dissection of a regular tetrahedron and an equal-volume cube [5]. This result was a solution to Hilbert's Third Problem [5].

A *hinged dissection* of a set of figures is a dissection in which the pieces are hinged together at points (in 2D or 3D) or along edges (in 3D), and there is a motion between any two of the figures that adheres to the hinging, keeping the hinge connections between pieces intact. While a few hinged dissections such as the one in Figure 1 are quite old [8], hinged dissections have received most of their study in the last few years [3,7,9,13]. It remains open whether every two polygons of the same area have a hinged dissection, or whether every two polyhedra that have a dissection also have a hinged dissection. It also remains open whether hinge-dissectability is transitive.

**Fig. 1.** Hinged dissection of square and equilateral triangle [8]. Different shades show different folded states

In this paper we develop a broad family of 3D hinged dissections for a class of polyhedra called poly-polyhedra. For a polyhedron $P$ with labeled faces, a *polypolyhedron of type $P$* is an interior-connected non-self-intersecting solid formed by joining several rigid copies of $P$ wholly along identically labeled faces. See Figure 2. These joinings must perfectly match two opposite orientations of the same face of $P$, so joinings can occur only along faces with reflectional symmetry. We call $P$ the *base polyhedron*. If a polypolyhedron consists of $n$ rigid copies of $P$, we call it an *$n$-polyhedron of type $P$*. Examples of polypolyhedra include *polycubes* (where $P$ is a cube) or more generally *polyplatonics* (where $P$ is any fixed platonic solid); in any of these cases, any pair of faces can be joined because of the regular symmetry of the platonic solids. See Figure 3 for some examples of polycubes.

For every polyhedron $P$ and positive integer $n$, we develop one hinged dissection that folds into all (exponentially many) $n$-polyhedra of type $P$. This result is superior to having one hinged dissection between every pair of $n$-polyhedra of type $P$. The number of pieces in the hinged dissection is linear in $n$ and the combinatorial complexity of $P$. For polyplatonics, we give particularly efficient hinged dissections, tuning the number of pieces to the minimum possible among a natural class of "regular" hinged dissections

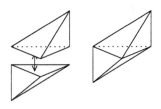

**Fig. 2.** Joining two rigid copies of a tetrahedron. The face of joining is reflectionally symmetric

of polypolyhedra. For polyparallelepipeds (where $P$ is any fixed parallelepiped), we give hinged dissections in which every piece is a scaled copy of $P$. All of our hinged dissections are hinged along edges and form a cyclic chain of pieces, which can be broken into a linear chain of pieces.

Our solution combines several techniques to obtain increasingly more general families of hinged dissections. We reduce the problem of finding a hinged dissection of polypolyhedra of type $P$ to finding a hinged dissection of $P$ that has "exposed hinges" at certain locations on its surface. We find the first such hinged dissection for every platonic solid, exploiting that such a solid is star-shaped and has a Hamiltonian cycle on its faces. Then we relax the star-shaped constraint, generalizing $P$ to be any solid with a Hamiltonian cycle on its faces, using a more general re-

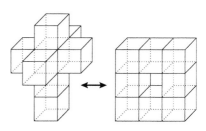

**Fig. 3.** Two polycubes of order 8, which have a 24-piece edge-hinged dissection by our results

finement scheme based on the straight skeleton. Then we relax the Hamiltonicity constraint by using a Hamiltonian refinement scheme. Finally, we show how faces with more than a single reflectional symmetry can be joined even when their labeled rotations are not equal. This step uses a general "twister" gadget, a hinged dissection that can rotate by any angle that is a multiple of $360°/k$ for fixed $k$.

Our results generalize analogous results about hinged dissections of "polyforms" in 2D [7]. For a polygon $P$ with labeled edges, a *polyform of type $P$* is an interior-connected non-self-intersecting planar region formed by joining several rigid copies of $P$ wholly along identically labeled edges. In particular, polyforms include polyominoes (where $P$ is a square) and polyiamonds (where $P$ is an equilateral triangle). In 2D, edges are always reflectionally symmetric (about their midpoint), so a polyform can join any pair of identically labeled edges. For any polygon $P$ and positive integer $n$, [7] develops a single vertex-hinged dissection that folds into all $n$-forms of type $P$. The same paper asks whether analogous dissections exist in 3D, in particular for polycubes; we solve this open problem, building on the general inductive approach of [7].

We do not know whether our hinged dissections can be folded from one configuration to another without self-intersection. (The same is true of most previous theoretical work in hinged dissections [3, 7, 9].) However, we demonstrate such motions for the most complicated gadget, the twister.

Our results have applications in self-assembly and nanomanufacturing, and may find applications in self-reconfigurable robotics. Existing reconfigurable robots (see, e.g., [19]) consist of units that can attach and detach from each other, and this mechanism is complicated; 3D hinged dissection may offer a way to avoid this complication and still achieve arbitrary reconfiguration.[1] In self-assembly, recent progress has enabled chemists to build millimeter-scale "self-

---

[1] This idea was suggested by Joseph O'Rourke in personal communication, Nov. 2004.

working" 2D hinged dissections [17]. An analog for 3D hinged dissections may enable building a complex 3D structure out of a chain of units. If the process is programmable, we could even envision an object that can re-assemble itself into different 3D structures on demand. These directions have recently been explored (so far at a more macroscale) using ideas from this paper [14].

## 2    Polyplatonics

In this section we demonstrate our approach for constructing a hinged dissection of polypolyhedra of type $P$ in the special case that $P$ is a platonic solid. Although several of the details change in more general settings in later sections, the overall approach remains the same.

First, we find a suitable hinged dissection of the base polyhedron $P$. The exact constraints on this dissection vary, but two necessary properties are that the hinged dissection must be (1) *cyclic*, forming a closed chain (cycle) of pieces in which there is a single hinge connecting every consecutive pair of pieces and there are no other hinges, and (2) *exposed* in the sense that, for every face of $P$, there is a hinge in $H$ that lies on the face (either interior to the face or on its boundary). For platonic solids, these hinges will be edges of the polyhedron. Second, we repeat $n$ copies of this hinged dissection of $P$, spliced together into one long closed chain. Finally, we prove that this new hinged dissection can fold into all $n$-polyhedra of type $P$, by induction on $n$.

### 2.1    Exposed Cyclic Hinged Dissections of Platonic Solids

We construct an exposed cyclic hinged dissection of any platonic solid as follows. First we carve the platonic solid into a cycle of face-based pyramids with the platonic solid's centroid as the apex. Thus, a refined tetrahedron consists of four triangle-based pyramids (irregular tetrahedra); a refined cube consists of six square-based pyramids; a refined octahedron consists of eight triangle-based pyramids; a refined dodecahedron consists of twelve pentagon-based pyramids; and a refined icosahedron consists of twenty triangle-based pyramids. Every platonic solid has a Hamiltonian cycle on its faces. Consequently, the pieces in the refinement can be hinged together in a cycle, following the Hamiltonian path on the faces. Figure 5 shows unfoldings of these hinged dissections, in particular illustrating the Hamiltonian cycle.

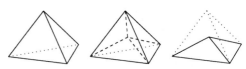

**Fig. 4.** Carving a regular tetrahedron into four face-based pyramids

Because there is a hinge dual to every edge in the Hamiltonian path on the faces, every face of the platonic solid has exactly two hinges. Therefore, the hinged dissection is exposed. Even more, we can merge adjacent pairs of pyramids along a face, halving the number of pieces, and leave exactly one hinge

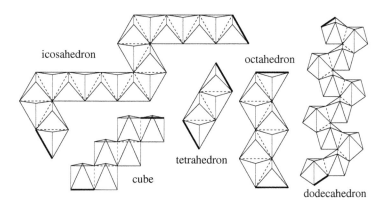

**Fig. 5.** Unfolded exposed cyclic hinged dissections of the platonic solids. The bold lines indicate a pair of edges that are joined by a hinge but have been separated in this figure to permit unfolding. The dashed lines denote all other hinges between pieces. In the unfolding, the bases of all of the pyramid pieces lie on a plane, and the apexes lie above that plane (closer to the viewer)

per face of the platonic solid. Now two faces share every hinge, but still the hinged dissection is exposed because every face has a hinge along its boundary. Thus we have proved

**Theorem 1.** *The platonic solid with $f$ faces has an exposed cyclic hinged dissection of $f/2$ pieces in which every hinge is an edge of the platonic solid.*

These exposed hinged dissections have the fewest possible pieces, subject to the exposure constraint, because a hinge can simultaneously satisfy at most two faces of the original polyhedron.

## 2.2   Inductive Hinged Dissection

Next we show how to build a hinged dissection of all $n$-platonics of type $P$ based on a repeatable hinged dissection of a platonic solid $P$. The hinged dissection is essentially $n$ repetitions of the exposed hinged dissection from the previous section. Specifically, the *$n$th repetition* of a cyclic hinged dissection is the result of cutting the cyclic hinged dissection at an arbitrary hinge to form an open chain, repeating this open chain $n$ times, and then reconnecting the ends to restore a closed chain. Thus, if there are $k$ pieces $H_1, H_2, \ldots, H_k$ connected in that order (and cyclically) in a cyclic hinged dissection, then the $n$th repetition consists of $nk$ pieces $H_1, \ldots, H_k, H_1, \ldots, H_k, \ldots \ldots H_1, \ldots, H_k$ connected in that order (and cyclically). (Although the order $H_1, \ldots, H_k$ depends on where we cut the cyclic order, the resulting $n$th repetition is independent of this cut.)

We prove that this hinged dissection has the desired foldings by an inductive/incremental construction based on the following tool, similar to [7, Prop. 1]:

**Lemma 1.** *Every $n$-polyhedron of type $P$ has a copy of $P$ whose removal results in a (connected) $(n-1)$-polyhedron, provided $n > 1$.*

*Proof.* The graph of adjacencies between copies of $P$ in an $n$-polyhedron is a connected graph on $n$ vertices. Any spanning tree of this graph has at least two leaves, and the removal of either leaf leaves the original graph connected. The resulting pruned graph is the adjacency graph of a $(n-1)$-polyhedron. □

Reversing the inductive process of this lemma implies that any $n$-polyhedron of $P$ can be built up by adding one copy of $P$ at a time, yielding a connected 1-, 2-, ..., and $(n-1)$-polyhedron along the way.

**Theorem 2.** *Given an exposed cyclic hinged dissection of the platonic solid $P$ in which exactly one piece is incident to each face of $P$, the nth repetition of this hinged dissection can fold into any n-platonic of type $P$.*

*Proof.* The proof is by induction. The base case of $n = 1$ is trivial: there is only one 1-platonic of type $P$, namely $P$ itself. The exposed hinged dissection satisfies all the desired properties.

Consider an $n$-platonic $Q$ of type $P$. By Lemma 1, one copy $P_1$ of $P$ can be removed from $Q$ to produce an $(n-1)$-platonic $Q'$. By induction, the $(n-1)$st repetition of the exposed hinged dissection can fold into $Q'$. Also, $P_1$ itself can be decomposed into an instance of the exposed hinged dissection. Our goal is to merge these two hinged dissections.

Let $P_2$ denote a copy of $P$ in $Q'$ that shares a face $f$ with $P_1$. Suppose the exposed cyclic hinged dissection of $P$ consists of pieces $H_1, H_2, \ldots, H_k$ in that order. Let $H_i$ denote the piece in the hinged dissection of $P_2$ incident to face $f$. Let $h$ be a hinge incident to $f$ (which must be an edge of $f$) and thus incident to $H_i$. Suppose by symmetry that the other piece in $Q'$ incident to hinge $h$ is $H_{i+1}$.

Then we rotate $P_1$ so that its piece $H_{i+1}$ is flush against the $H_i$ piece in $P_2$, along the shared face $f$ between $P_1$ and $P_2$. We further rotate $P_1$ so that the hinge $h'$ between pieces $H_i$ and $H_{i+1}$ in $P_1$ aligns with the hinge $h$ between pieces $H_i$ and $H_{i+1}$ in $P_2$. We then replace hinges $h$ and $h'$ with two hinges, one from $H_i$ in $P_2$ to $H_{i+1}$ in $P_1$, and the other from $H_i$ in $P_1$ to $H_{i+1}$ in $P_2$. The resulting hinged dissection is a single cycle, and every instance of piece $H_i$ hinges to pieces $H_{i-1}$ and $H_{i+1}$, so the resulting hinged dissection is a folding of the $n$th repetition of $H_1, H_2, \ldots, H_k$ as desired. □

**Corollary 1.** *If $P$ is the platonic solid with $f$ faces, then there is an $(nf/2)$-piece cyclic hinged dissection that can fold into all n-platonics of type $P$.*

## 3   Generalized Interior Dissection

The proof of hinged dissections for polyplatonics consists of two main parts: (1) the construction of an exposed cyclic hinged dissection of a single platonic solid, with the property that at most one piece is incident to each face, and (2) an inductive argument about the $n$th repetition. In this section we generalize the first part to any polyhedron with a Hamiltonian cycle on its faces. The second part will remain restrictive until future sections.

## 3.1    Exposed Cyclic Hinged Dissections of Hamiltonian Polyhedra

The exposed cyclic hinged dissection for platonic solids from Section 2.1 essentially exploited that platonic solids, like all convex polyhedra, are "star-shaped". A polyhedron is *star-shaped* if it has at least one point $c$ in its interior from which the line segment to any point on the polyhedron's surface remains interior to the polyhedron. Any star-shaped polyhedron can be carved into face-based pyramids with apexes at $c$. These pyramids can be hinged together cyclically at the edges of the polyhedron crossed by the Hamiltonian cycle on the faces.

Dissection of a polyhedron into face-based pyramids with a common apex is possible precisely when the polyhedron is star-shaped. However, it is not hard to obtain a dissection of an arbitrary polyhedron into one piece per face, though the pieces are no longer pyramids. One approach is to use the *straight skeleton* [2, 1, 10, 6]. The straight skeleton is normally defined as a particular one-dimensional tree structure contained in a given two-dimensional polygon. For our purposes, the relevant property is that the tree structure subdivides the polygon into exactly one region per polygon edge, and only that region is incident to that polygon edge [2].

The straight skeleton can be generalized to 3D as a decomposition of a given polyhedron into exactly one cell per facet, and only that cell is incident to that facet. We imagine sweeping every facet perpendicularly inwards at the same speed in parallel. Faces change geometry as they are inset by clipping or extending to where they meet adjacent faces. Faces may become disconnected, in which case the sweep continues with each piece, or disappear, in which case the sweep continues without that face. In the end, the entire polyhedron is swept, and the regions swept by individual faces form a partition with the desired property that exactly one region is incident to each facet. Erickson [11] points out that the straight skeleton is no longer well-defined in 3D: there are choices during the offset process that can be resolved multiple ways. However, for our purposes, we just need a single straight skeleton, with an arbitrary decision for each choice, for a suitable decomposition.

As before, the pieces can be hinged together cyclically at the edges of the polyhedron crossed by the Hamiltonian cycle. Thus, for any polyhedron with a Hamiltonian cycle on its $n$ faces, we obtain an $n$-piece exposed cyclic hinged dissection with the property that each face of the polyhedron is incident to exactly one piece.

## 3.2    Inductive Hinged Dissection

The second part of the argument is the inductive construction. The key steps here are the two rotations of an added piece $P_1$. The first rotation ensures that the next piece in the hinging of $P_1$ ($H_{i+1}$) is against the piece to which we want to join $P_1$ ($H_i$ of $P_2$). The second rotation ensures that the exposed hinges of these two pieces coincide.

These rotations enforce restrictions on what types of polypolyhedra we can build. The first rotation essentially requires that all faces of $P$ "look the same" (in addition to having the same shape): the rotation that brings any face to any other face should result in an identical copy of $P$ (but with faces relabeled). The second rotation requires that all orientations of a face look the same. Unfortunately, these two restrictions force $P$ to be a platonic solid. The goal of the remaining sections is to remove these restrictions, in addition to the restriction that $P$ has a Hamiltonian cycle on its faces.

## 4    Surface Refinement

In this section we remove two constraints on the base polyhedron $P$: the requirement that $P$ has a Hamiltonian cycle on its faces, and the requirement that all faces of $P$ look the same. We achieve both of these generalizations by subdividing each face of $P$ by a collection of linear cuts.

First, we divide each reflectionally symmetric face of $P$ along one of its lines of symmetry. Recall that joinings between copies of $P$ are possible only along reflectionally symmetric faces. Now if we can arrange for these symmetry lines to be hinges in an exposed cyclic hinged dissection of the new polyhedron $P'$, then whenever we attempt to attach a new piece $P_1'$, we are guaranteed that the two consecutive pieces $H_i$ and $H_{i+1}$ of the hinging that we need to place against

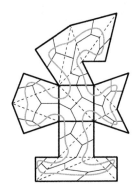

**Fig. 6.** Hamiltonian refinement of five faces in a hypothetical polyhedron, shown here unfolded. Bold lines outline faces. Dashed lines show triangulations and are not cuts. Thin solid lines are cuts. The curved line shows a Hamiltonian cycle induced by the spanning tree of this unfolding

each other are in fact the two reflectional halves of the original face. Thus the first rotation in the induction construction does exactly what we want: it brings together the two identically labeled faces of $P$.

Second, we divide each face of $P'$ so that any spanning tree of the faces in $P'$ translates into a Hamiltonian cycle in the resulting polyhedron $P''$. This reduction is similar to the Hamiltonian triangulation result of [4] as well as a refinement for hinged dissection of 2D polyforms [7, Section 6]. We conceptually triangulate each face $f$ of $P'$ using chords (though we do not cut along the edges of that triangulation). Then, for each triangle, we cut from an arbitrarily chosen interior point to the midpoints of the three edges. Figure 6 shows an example of this process. For any spanning tree of the faces of $P'$, we can walk around the tree (i.e., follow an Eulerian tour) and produce a Hamiltonian cycle on the faces of $P''$.

In particular, we can start from the matching on the faces of $P'$ from the reflectionally symmetric pairing, and choose a spanning tree on the faces of $P'$ that contains this matching. Then the resulting Hamiltonian cycle in $P''$ crosses a subdivided edge of every line of symmetry. (In fact, the Hamiltonian cycle crosses every subdivided edge of every line of symmetry.) Thus, in the exposed cyclic hinged dissection of the Hamiltonian polyhedron $P''$, there is an exposed hinge along every line of symmetry. Therefore all joinings between copies of $P''$ can use these hinges, which means that the first rotation in the induction construction happens automatically from joining along corresponding faces.

## 5    Mutually Rotated Base Polyhedra: Twisters

The last generalization concerns the second rotation in the inductive construction. If every reflectionally symmetric face has only one line of symmetry, this second rotation is automatic just from making the faces meet geometrically. However, if a face has more than one line of symmetry, the polypolyhedron may require different rotations of the two base polyhedra around their common face.

To enable these kinds of joinings, we introduce the *twister gadget* shown in Figure 7. This gadget allows the top face to rotate by any integer multiple of $360°/k$ with respect to the bottom face. The volume occupied by the twister gadget is a prism with a regular $k$-gon as a base.

To construct the pieces, we slice this prism in half parallel to the base, leaving two identical prisms, one stacked atop the other. Then we divide each prism by making several planar cuts perpendicular to the base: in projection of a regular $k$-gon, we cut from the center to every vertex, to the midpoint of every edge, and to each quarter point between a vertex and an edge midpoint. The resulting $8k$ pieces are all triangular prisms.

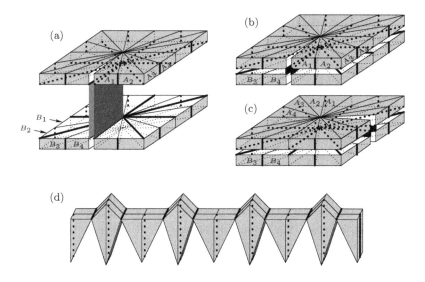

**Fig. 7.** The twister gadget with $k = 4$: 32 pieces allowing any between none and three quarter turns. For visual clarity, the two layers are drawn substantially separated in (a) and slightly separated in (b) and (c); in fact they are flush. (d) shows the result of unfolding along the perimeter hinges. (c) shows a refolding that achieves a half turn

We hinge these prisms together cyclicly as follows. Two hinges connect the top and bottom levels, lying (in projection) along a cut from the center to an edge midpoint. For each remaining cut from the center to an edge midpoint (in projection), and for each cut from the center to a vertex (in projection), there is a hinge connecting the two incident pieces on the "inside" (on the bottom of the top prism and on the top of the bottom prism). For each cut from the center to a quarter point (in projection), there is a hinge connecting the two incident pieces on the perimeter of the regular $k$-gon.

The perimeter hinges enable the twister to unfold as shown in Figure 7(d) to make all the inside hinges parallel. The inside hinges allow the twister to be further unfolded from this state into a convex three-dimensional "ring". Then we can reverse the process, collapsing the 3D ring back down along the inside hinges to a nearly flat unfolding like Figure 7(d), and folding it back along the perimeter hinges into the regular $k$-gon configuration. In between the unfolding and folding, by rotating the ring state, we can change which pieces are ultimately on which layer as shown in Figure 7(c).

Specifically, by this continuous folding process, we can move any multiple of 4 pieces from the top layer to the bottom layer on one side of the gap where the layers connect, and the same number of pieces from the bottom layer to the top layer on the other side of the gap. If we move $4j$ pieces on either side, we rotate the top regular $k$-gon by $j \cdot 360°/k$ relative to the bottom regular $k$-gon. If we restrict $j$ to satisfy $0 \leq j < k$ (which suffices for the desired set of $k$ possible rotations), then there are four pieces $A_1, A_2, A_3, A_4$ that always remain on the top layer and four pieces $B_1, B_2, B_3, B_4$ that always remain on the bottom layer.

To allow the twister gadget to attach to other pieces on its top and bottom, we need to add exposed hinges. We remove the inner hinge connecting $A_2$ and $A_3$, which in projection connects the center to a vertex of the regular $k$-gon, and replace it with a corresponding outer hinge on the top side of the twister gadget. Similarly, we remove the inner hinge connecting $B_2$ and $B_3$, whose projection connects the center to the same vertex of the regular $k$-gon, and replace it with a corresponding outer hinge on the bottom side of the twister gadget. The modified twister gadget can be folded continuously as before, except that now we keep $A_2$ rigidly attached to $A_3$ and $B_2$ rigidly attached to $B_3$ when opening up into a three-dimensional ring, not folding the two outer hinges at all.

We embed the modified twister gadget in each face of the base polyhedron $P$ that has $k$-fold symmetry for $k \geq 3$. More precisely, we carve out of $P$ a thin prism with a small regular $k$-gon base, centered at the symmetry center of the face, and infuse this carved space with a twister gadget. Then we construct the refinement $P''$ of $P$ as before, choosing an arbitrary line of symmetry of a $k$-fold symmetric face for the subdivision and resulting matching. The line of symmetry actually now "bends" slightly to dip underneath the thin twister gadget at the center. Normally the hinged dissection of $P''$ would have a hinge along this line of symmetry, connecting the two incident pieces $C$ and $D$. Instead, we rotate the embedded twister gadget so that its outer hinges (those between $A_2$ and $A_3$ and between $B_2$ and $B_3$) align with this chosen line of symmetry, and so that

$B_2$ is atop $C$ and $B_3$ is atop $D$. Then we replace the outer hinge between $B_2$ and $B_3$ with a hinge between $B_2$ and $C$ and a hinge between $B_3$ and $D$. (All three of these hinges lie geometrically along the same line segment in the folded configuration.)

In the inductive construction of an $n$-polyhedron of type $P$, we use the outer hinge between pieces $A_2$ and $A_3$ to combine two copies of $P'''$ along a $k$-fold symmetric face, $k \geq 3$. This hinge lies along the chosen line of symmetry, in the middle of the face, and therefore can be aligned between the two copies. Note that the resulting construction has two copies of the twister gadget joined along their top sides, which is redundant because it allows up to two full turns of the faces, but we cannot easily remove this redundancy while having two identical copies of a single hinged dissection.

Two copies of $P'''$ joined along a face of $k$-fold symmetry can now rotate with respect to each other by $j \cdot 360°/k$, for any desired $0 \leq j < k$. This property is exactly what we need to perform the second rotation in the inductive argument of hinged dissectibility.

This completes our construction of a hinged dissection that folds into all $n$-polyhedra of type $P$, for any positive integer $n$ and for any polyhedron $P$.

## 6    Self-similar Hinged Dissections

This section considers a related side problem from the main line of the paper, called "self-similar hinged dissections". A hinged dissection is *self-similar* if every piece is similar to (a scaled copy of) the base polyhedron $P$. Self-similar dissections (without hingings) are well-studied in recreational mathematics, usually in 2D, so it is natural to consider their hinged, 3D counterparts.

Figure 8 gives a self-similar exposed hinged dissection of a cube, which by our techniques leads to a self-similar hinged dissection of all $n$-cubes, for any $n$. The dissection is simple, dividing the cube into a $2 \times 2 \times 2$ array of identical subcubes. The hinging is less trivial because of the requirement that every face of the original cube has an exposed hinge. The hinges are always between the midpoint of an original edge to the center of an original face, so two hinges between adjacent cubes can always be brought into alignment, after possible rotation around the shared face, during the merging process in the inductive construction.

The resulting dissection of $n$-cubes uses $8n$ pieces (compared to $3n$ pieces from Corollary 1):

**Theorem 3.** *The $n$th repetition of the cyclic hinged dissection in Figure 8 consists of $8n$ identical cubes and folds into all $n$-cubes.*

This hinged dissection of a cube is clearly the smallest exposed self-similar hinged dissection of the cube, and hence is optimal among such dissections. The hinged dissection also applies more generally to any parallelepiped (e.g., an $x \times y \times z$ box) as the base shape $P$.

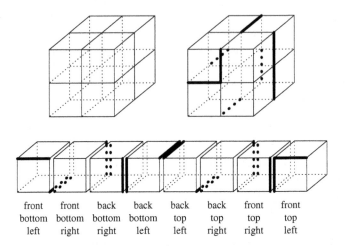

front   front   back    back    back    back    front   front
bottom  bottom  bottom  bottom  top     top     top     top
left    right   right   left    left    right   right   left

**Fig. 8.** A hinged dissection of a cube into a $2 \times 2 \times 2$ array of 8 subcubes. This hinged dissection can be used in place of that in Figure 5; every face has (at least) one exposed hinge. Top-left: The dissection. Top-right: The cyclic hinging. Bottom: Unfolded after cutting one hinge. Hinges are drawn bold

This extension has been used in an interactive sculpture [18] consisting of roughly a thousand identical wooden blocks (boxes) hinged together according to Figure 8. (For manipulation purposes, the chain was broken into small segments.)

# References

1. O. Aichholzer and F. Aurenhammer. Straight skeletons for general polygonal figures in the plane. In *Proc. 2nd Annual Internat. Computing and Combinatorics Conf.*, LNCS 1090, pp. 117–126, 1996.
2. O. Aichholzer, F. Aurenhammer, D. Alberts, and B. Gärtner. A novel type of skeleton for polygons. *Journal of Universal Computer Science*, 1(12):752–761, 1995.
3. J. Akiyama and G. Nakamura. Dudeney dissection of polygons. In *Rev. Papers, Japan Conf. Discrete and Computational Geometry*, LNCS 1763, pp. 14–29, 1998.
4. E.M. Arkin, M. Held, J.S.B. Mitchell, and S.S. Skiena. Hamiltonian triangulations for fast rendering. *The Visual Computer*, 12(9):429–444, 1996.
5. V.G. Boltianskii. *Hilbert's Third Problem*. V. H. Winston & Sons, 1978.
6. S.-W. Cheng and A. Vigneron. Motorcycle graphs and straight skeletons. In *Proc. 13th Ann. ACM-SIAM Sympos. Discrete Algorithms*, pp. 156–165, 2002.
7. E.D. Demaine, M.L. Demaine, D. Eppstein, G.N. Frederickson, and E. Friedman. Hinged dissection of polyominoes and polyforms. *Computational Geometry: Theory and Applications*. To appear. http://arXiv.org/abs/cs.CG/9907018.
8. H.E. Dudeney. Puzzles and prizes. *Weekly Dispatch*, April 6, 1902.
9. D. Eppstein. Hinged kite mirror dissection. June 2001. http://arXiv.org/abs/cs.CG/0106032.
10. D. Eppstein and J. Erickson. Raising roofs, crashing cycles, and playing pool: Applications of a data structure for finding pairwise interactions. *Discrete & Computational Geometry*, 22(4):569–592, 1999.

11. J. Erickson. Personal communication, February 2000.
12. G.N. Frederickson. *Dissections: Plane and Fancy.* Cambridge Univ. Press, 1997.
13. G.N. Frederickson. *Hinged Dissections: Swinging & Twisting.* Cambridge Univ. Press, 2002.
14. S. Griffith. *Growing Machines.* PhD thesis, MIT Media Laboratory, Sept. 2004.
15. E. Kranakis, D. Krizanc, and J. Urrutia. Efficient regular polygon dissections. *Geometriae Dedicata*, 80:247–262, 2000.
16. M. Lowry. Solution to question 269, [proposed] by Mr. W. Wallace. In T. Leybourn, ed., *Mathematical Repository*, vol. 3, part 1, pp. 44–46. W. Glendinning, 1814.
17. C. Mao, V.R. Thallidi, D.B. Wolfe, S. Whitesides, and G.M. Whitesides. Dissections: Self-assembled aggregates that spontaneously reconfigure their structures when their environment changes. *J. Amer. Chemical Soc.*, 124:14508–14509, 2002.
18. L. Palmer. The helium stockpile: Under shifting conditions of heat and pressure. Installation, Radcliffe College, Cambridge, Massachusetts, April 2004.
19. D. Rus, Z. Butler, K. Kotay, and M. Vona. Self-reconfiguring robots. *Communications of the ACM*, 45(3):39–45, 2002.

# Convex Recolorings of Strings and Trees: Definitions, Hardness Results and Algorithms*

Shlomo Moran[1] and Sagi Snir[2]

[1] Computer Science dept., Technion, Haifa 32000, Israel
`moran@cs.technion.ac.il`
[2] Mathematics dept. University of California, Berkeley, CA 94720, USA
`ssagi@math.berkeley.edu`

**Abstract.** A coloring of a tree is convex if the vertices that pertain to any color induce a connected subtree. Convex colorings of trees arise in areas such as phylogenetics, linguistics, etc. e.g., a perfect phylogenetic tree is one in which the states of each character induce a convex coloring of the tree.

When a coloring of a tree is not convex, it is desirable to know "how far" it is from a convex one, and what are the convex colorings which are "closest" to it. In this paper we study a natural definition of this distance - the *recoloring distance*, which is the minimal number of color changes at the vertices needed to make the coloring convex. We show that finding this distance is NP-hard even for a path, and for some other interesting variants of the problem. In the positive side, we present algorithms for computing the recoloring distance under some natural generalizations of this concept: the *uniform weighted* model and the *non-uniform* model. Our first algorithms find optimal convex recolorings of strings and bounded degree trees under the non-uniform model in linear time for any fixed number of colors. Next we improve these algorithms for the uniform model to run in linear time for any fixed number of *bad* colors. Finally, we generalize the above result to hold for trees of unbounded degree.

## 1 Introduction

A phylogenetic tree is a tree which represents the course of evolution for a given set of species. The leaves of the tree are labelled with the given species. Internal vertices correspond to hypothesized, extinct species. A *character* is a biological attribute shared among all the species under consideration, although every species may exhibit a different *character state*. Mathematically, if $X$ is the set of species under consideration, a character on $X$ is a function $C$ from $X$ into a set $C$ of character states. A character on a set of species can be viewed as a *coloring* of the species, where each color represents one of the character's states. A natural

---

* This research was supported by the Technion VPR-fund and by the Bernard Elkin Chair in Computer Science.

biological constraint is that the reconstructed phylogeny have the property that each of the characters could have evolved without reverse or convergent transitions: In a reverse transition some species regains a character state of some old ancestor whilst its direct ancestor has lost this state. A convergent transition occurs if two species posses the same character state, while their least common ancestor possesses a different state.

In graph theoretic terms, the lack of reverse and convergent transitions means that the character is *convex* on the tree: for each state of this character, all species (extant and extinct) possessing that state induce a single *block*, which is a maximal monochromatic subtree. Thus, the above discussion implies that in a phylogenetic tree, each character is likely to be convex or "almost convex". This make convexity a fundamental property in the context of phylogenetic trees to which a lot of research has been dedicated throughout the years. The *Perfect Phylogeny* (PP) problem, whose complexity was extensively studied (e.g. [10, 12, 1, 13, 4, 17]), receives a set of characters on a set of species and seeks for a phylogenetic tree on these species, that is simultaneously convex on each of the characters. *Maximum parsimony* (MP) [8, 15] is a very popular tree reconstruction method that seeks for a tree which minimizes the parsimony score defined as the number of mutated edges summed over all characters (therefore, PP is a special case of MP). [9] introduce another criterion to estimate the distance of a phylogeny from convexity. They define the *phylogenetic number* as the maximum number of connected components a single state induces on the given phylogeny (obviously, phylogenetic number one corresponds to a perfect phylogeny). However, both the parsimony score and the phylogenetic number of a tree do not specify a distance to some *concrete* convex coloring of the given tree: there are colored trees with large phylogenetic numbers (and large parsimony scores) that can be transformed to convex coloring by changing the color of a single vertex, while other trees with smaller phylogenetic numbers can be transformed to convex colorings only by changing the colors of many vertices.

Convexity is a desired property in other areas of classification, beside phylogenetics. For instance, in [3, 2] a method called *TNoM* is used to classify genes, based on data from gene expression extracted from two types of tumor tissues. The method finds a separator on a binary vector, which minimizes the number of "1" in one side and "0" in the other, and thus defines a convex vector of minimum Hamming distance to the given binary vector. Algorithms which finds this distance for vectors with any number of letters, in order to handle more types of tumor tissues, are given by the optimal string recoloring algorithms in this paper. In [11], distance from convexity is used (although not explicitly) to show strong connection between strains of Tuberculosis and their human carriers.

In this work we define and study a natural distance from a colored tree to a convex one: the *recoloring distance*. In the simplest, unweighted model, this distance is the minimum number of color changes at the vertices needed to make the given coloring convex (for strings this reduces to Hamming distance from a closest convex coloring). This measure generalizes to a weighted model, where changing the color of vertex $v$ costs a nonnegative weight $w(v)$. These weighted and unweighted models are *uniform*, in the sense that the cost of changing the color of a vertex is independent of the colors involved. The most general model we study is the *non-uniform* model, where the cost of coloring vertex $v$ by a color $d$ is an arbitrary nonnegative number $cost(v, d)$.

We show that finding the recoloring distance in the unweighted model is NP-hard even for a string (a tree with two leaves), and also for the case where character states are given only at the leaves (so that changes on extinct species are not counted); we also address a variant of the problem, in which a block-recoloring is considered as an atomic operation. This operation changes the color of all the vertices in a given input block. We show that finding the minimum number of block-recolorings needed to obtain convexity is NP-Hard as well.

On the positive side, we present few algorithms for minimal convex recoloring of strings and trees. The first algorithms solve the problem in the non-uniform model. The running time of these algorithms for bounded degree trees is exponential in the number of colors, but for each fixed number of colors is linear in the input size. Then we improve these algorithms for the uniform model, so that the running time is exponential only in the number of *bad* colors, which are colors that violate convexity (to be defined precisely). These algorithms are noted to be fixed parameter tractable algorithms ([5]) for bounded degree trees, where the parameter is taken to be the recoloring distance. Finally, we eliminate the dependence on the degree of the tree in both the non-uniform and the uniform versions of the algorithms.

Due to space limitation, figures, proofs of theorems and some of the results were removed. However, all of these can be found at the full paper at     *http://www.cs.technion.ac.il/users/wwwb/cgi-bin/tr-info.cgi?2004/CS/CS-2004-14*

The rest of the paper is organized as follows. In the next section we present the notations used and define the unweighted, weighted and non-uniform versions of the problem. In Section 3 we show our NP-Hardness results and in Section 4 we present the algorithms. We conclude and point out future research directions in Section 5.

## 2    Preliminaries

A colored tree is a pair $(T, C)$ where $T = (V, E)$ is a tree with vertex set $V = \{v_1, \ldots, v_n\}$, and $C$ is a *coloring* of $T$, i.e. - a function from $V$ onto a set of colors $\mathcal{C}$. For a set $U \subseteq V$, $C|_U$ denotes the restriction of $C$ to the vertices of $U$, and $C(U)$ denotes the set $\{C(u) : u \in U\}$. For a subtree $T' = (V(T'), E(T'))$ of $T$, $C(T')$ denotes the set $C(V(T'))$. A *block* in a colored tree is a maximal set of vertices which induces a monochromatic subtree. A *$d$-block* is a block of color $d$. The number of $d$-blocks is denoted by $n_b(C, d)$, or $n_b(d)$ when $C$ is clear from the context. A coloring $C$ is said to be *convex* if $n_b(C, d) = 1$ for every color $d \in \mathcal{C}$. The number of *$d$-violations* in the coloring $C$ is $n_b(C, d) - 1$, and the total number of *violations* of $C$ is $\sum_{c \in \mathcal{C}} (n_b(C, d) - 1)$. Thus a coloring $C$ is convex iff the total number of violations of $C$ is zero (in [7] the above sum, taken over all characters, is used as a measure of the distance of a given phylogenetic tree from perfect phylogeny).

The definition of convex coloring is extended to *partially colored* trees, in which the coloring $C$ assigns colors to some subset of vertices $U \subseteq V$, which is denoted by $Domain(C)$. A partial coloring is said to be convex if it can be extended to a total convex coloring (see [16]). Convexity of partial and total coloring have simple characterization by the concept of *carriers*: For a subset $U$ of $V$, $carrier(U)$ is the minimal subtree that contains $U$. for a colored tree $(T, C)$ and a color $d \in C$, $carrier_T(C, d)$ (or $carrier(C, d)$ when $T$ is clear) is the carrier of $C^{-1}(d)$. We say that $C$ has the *disjointness property* if for each pair of colors $\{d, d'\}$ it holds that $carrier(C, d) \cap carrier(C, d') = \emptyset$. It is easy to see that a total or partial coloring $C$ is convex iff it satisfies the disjointness property (in [6] convexity is actually defined by the disjointness property).

When some (total or partial) input coloring $(C, T)$ is given, any other coloring $C'$ of $T$ is viewed as a *recoloring* of the input coloring $C$. We say that a recoloring $C'$ of $C$ *retains* (the color of) a vertex $v$ if $C(v) = C'(v)$, otherwise $C'$ *overwrites* $v$. Specifically, a recoloring $C'$ of $C$ overwrites a vertex $v$ either by changing the color of $v$, or just by *uncoloring* $v$. We say that $C'$ retains (overwrites) a set of vertices $U$ if it retains (overwrites resp.) every vertex in $U$. For a recoloring $C'$ of an input coloring $C$, $\mathcal{X}_C(C')$ (or just $\mathcal{X}(C')$) is the set of the vertices overwritten by $C'$, i.e.

$$\mathcal{X}_C(C') = \{v \in V : [v \in Domain(C)] \bigwedge [(v \notin Domain(C')) \vee (C(v) \neq C'(v))]\}.$$

With each recoloring $C'$ of $C$ we associate a *cost*, denoted as $cost_C(C')$ (or $cost(C')$ when $C$ is understood), which is the number of vertices overwritten by $C'$, i.e. $cost_C(C') = |\mathcal{X}_C(C')|$. A coloring $C^*$ is an *optimal convex recoloring of $C$*, or in short an *optimal recoloring of $C$*, and $cost_C(C^*)$ is denoted by $OPT(T, C)$, if $C^*$ is a convex coloring of $T$, and $cost_C(C^*) \leq cost_C(C')$ for any other convex coloring $C'$ of $C$.

The above cost function naturally generalizes to the *weighted* version: the input is a triplet $(T, C, w)$, where $w : V \to \mathbb{R}^+ \cup \{0\}$ is a weight function which assigns to each vertex $v$ a nonnegative weight $w(v)$. For a set of vertices $X$, $w(X) = \sum_{v \in X} w(v)$. The cost of a convex recoloring $C'$ of $C$ is $cost_C(C') = w(\mathcal{X}(C'))$, and $C'$ is an optimal convex recoloring if it minimizes this cost.

The above unweighted and weighted cost models are *uniform*, in the sense that the cost of a recoloring is determined by the set of overwritten vertices, regardless the specific colors involved. A yet further generalization allows *non-uniform* cost functions. This version, motivated by weighted maximum parsimony [15], assumes that the cost of assigning color $d$ to vertex $v$ is given by an arbitrary nonnegative number $cost(v, d)$ (note that, formally, no initial coloring $C$ is assumed in this cost model). In this model $cost(C')$ is defined only for a total recoloring $C'$, and is given by the sum $\sum_{v \in V} cost(v, C'(v))$. The non-uniform cost model appears to be more subtle than the uniform ones. Unless otherwise stated, our results assume the uniform, weighted and unweighted, models.

We complete this section with a definition and a simple observation which will be useful in the sequel. Let $(T, C)$ be a colored tree. A coloring $C^*$ is an *expanding* recoloring of $C$ if in each block of $C^*$ at least one vertex $v$ is retained (i.e., $C(v) = C^*(v)$).

**Observation 1.** *let $(T, C)$ be a colored tree. Then there exists an expanding optimal convex recoloring of $C$.*

*Proof.* Let $C'$ be an optimal recoloring of $C$ which uses a minimum number of colors (i.e. $|C'(V)|$ is minimized). We shall prove that $C'$ is an expanding recoloring of $C$.

If $C'$ uses just one color $d$, then by the optimality of $C'$, there must be a vertex $v$ such that $C(v) = d$ and the claim is proved. Assume for contradiction that $C'$ uses at least two colors, and that for some color $d$ used by $C'$, there is no vertex $v$ s.t. $C(v) = C'(v) = d$. Then there must be an edge $(u, v)$ such that $C'(u) = d$ but $C'(v) = d' \neq d$. Therefore, in the uniform cost model, the coloring $C''$ which is identical to $C'$ except that all vertices colored $d$ are now colored by $d'$ is an optimal recoloring of $C$ which uses a smaller number of colors - a contradiction.

# 3    NP-Hardness Results

The main result of this section is that unweighted minimum convex recoloring of strings is NP-Hard. Then we use reductions from this problem to prove that the unweighted versions of minimal convex recoloring of leaves, and a natural variant of the problem called minimal convex *block recoloring*, in which an atomic operation changes the color of a complete block, are NP-Hard as well.

## 3.1    Minimal Convex Recoloring of Strings is NP-Hard

A string $S = (v_1, \ldots, v_n)$ is a simple tree with $V = \{v_1, \ldots, v_n\}$ and $E = \{(v_i, v_{i+1}) | i = 1, \ldots, n - 1\}$. In a colored string $(S, C)$, a $d$-block is simply a maximal sequence of consecutive vertices colored by $d$.

A nice property of optimal convex recoloring of strings is given below:

*Claim.* Let $(S, C)$ be a colored string, and let $C^*$ be an optimal recoloring of $C$. Then each block of $C$ is either completely retained or completely overwritten by $C^*$.

*Proof.* Suppose, for contradiction, that $B'$ is a $d$-block in $C$ that is partially overwritten by $C^*$. Let $C'$ be a recoloring identical to $C^*$ except that $C'$ retains the block $B'$. Then $C'$ is convex and $cost(C') < cost(C^*)$ - a contradiction.

We prove that the problem is NP-Hard by reducing the 3 satisfiability problem to the following decision version of minimal convex recoloring:

## Minimal Convex Recoloring of Strings:

*Input:* A colored string $(S, C)$ and an integer $k$.

*Question:* Is there a convex recoloring $C^*$ of $C$ such that $cost_C(C^*) \leq k$.

Let formula $F$ be an input to the 3 satisfiability problem, $F = D_1 \wedge \ldots \wedge D_m$, where $D_i = (l_{i1} \vee l_{i2} \vee l_{i3})$ is a clause of three literals, each of which is either a variable $x_j$ or its negation $\neg x_j$, $1 \leq j \leq n$. We describe below a polynomial time reduction of $F$ to a colored string $(S, C)$ and an integer $k$, such that there is a convex coloring $C^*$ of $C$ with $cost_C(C^*) \leq k$ iff $F$ is satisfiable.

In the reduction we define block sizes using parameters $A$ and $B$, where $A$ and $B$ are integers satisfying $A > m - 2$ and $B > 2mA$. $k$ is set to $n(2m+1)B + 2mA$ (e.g., possible values are $A = 3m$, $B = 9m^2$, and $k = 3m^2(6mn + 3n + 2)$).

We describe the coloring $C$ of $S$ as a sequence of *segments*, where each segment consists of one or more consecutive blocks. There will be $2n + m$ *informative* segments: one for each clause and one for each literal, and $2n + m - 1$ *junk* segments separating the informative segments. Each junk segment consists of a unique block of $k + 1$ vertices colored by a distinct color, thus $2n + m - 1$ colors are used for the junk segments. It is easy to see that in any convex recoloring that is at most at distance $k$ from $C$, non of the junk segments is recolored, what implies that the order of the segments, informative and non informative, does not matter. The informative segments will use additional $n$ *variable colors* $d_1, \ldots, d_n$ and $2nm$ *literal colors* $\{c_{i,x_j}, c_{i,\neg x_j} | i = 1, \ldots, m; j = 1, \ldots n\}$.

For each clause $D_i = (l_1 \vee l_2 \vee l_3)$ there is a *clause segment* $S_{D_i}$ of size $3A$, obtained by $A$ repetitions of the pattern $c_{i,l_1}, c_{i,l_2}, c_{i,l_3}$

for each non-negated literal $x_j$ there is a *literal segment* $S_{x_j}$, which consists of $2m + 1$ consecutive blocks of the same size $B$. All the $m + 1$ odd numbered blocks are $d_j$-blocks, called *variable blocks*. The $m$ even numbered blocks are *literal blocks*, colored by $c_{i,x_j}$, $i = 1, \ldots, m$. Similarly, for each negated literal $\neg x_j$ we have a literal segment $S_{\neg x_j}$, which is similar to $S_{x_j}$ except that the colors of the literal blocks are $c_{i,\neg x_j}$, $i = 1, \ldots, m$ (note that each of the literal segments $S_{x_j}$ and $S_{\neg x_j}$ contain $m + 1$ $d_j$-blocks).

**Theorem 2.** *Let $(S, C)$ be the colored string defined by the above reduction. Then $OPT(S, C) \leq k$ iff $F$ is satisfiable.*

The following two results pertain to restricted versions of the original problem which model specific problem in phylogenetics:

**Theorem 3.** *Minimal unweighted convex recoloring of a tree is NP-Complete even when the coloring is restricted to the leaves of the tree only.*

**Theorem 4.** *Minimal unweighted convex recoloring of a totally colored tree is NP-Hard even when recoloring a complete block is considered as a single operation.*

**Note:** In a Zebra string, overwriting a single vertex is also a block recoloring. Thus Theorem 4 also implies that the problem of minimizing the total number of vertex recoloring *and* block recoloring needed to transform a colored string to convex one is NP-Hard.

# 4     Optimal Convex Recoloring Algorithms

In this section we present dynamic programming algorithms for optimal convex recoloring of totally colored strings or trees. The input is either a totally colored string $(S, C)$ or a totally colored tree $(T, C)$, which will be clear from the context. The optimal convex recolorings returned by the algorithms will be either total or partial, as will be detailed.

The basic ingredient in all the algorithms is coloring with forbidden colors: A convex recoloring of the whole tree is constructed by extending convex recolorings of smaller subtrees, and in order to maintain convexity of the coloring, in each subtree certain colors cannot be used.

The computational costs of the algorithms depend either on $n_c$, the total number of colors used, or on $n_c^*$, the number of colors which violate convexity in the input tree, defined as follows: A color $d$ is a *good color* for a totally colored tree $(T, C)$ if $(T, C)$ contains a unique $d$-block. Else $d$ is a *bad color*. The set of bad colors for the input $(T, C)$ is denoted by $\mathcal{C}^*$, and $|\mathcal{C}^*|$ is denoted by $n_c^*$.

We start with basic algorithms which are valid for the general non-uniform cost model, and their time complexity in bounded degree trees is $Poly(n)Exp(n_c)$. We then modify these algorithms to run in time $Poly(n)Exp(n_c^*)$ in the uniform weighted model. Finally, we remove the degree bound and modify the algorithms to run in $Poly(n)Exp(n_c^*)$ time for arbitrary trees.

## 4.1     Basic Algorithms for the Non-uniform Cost Model

Our first algorithms find optimal convex recoloring of strings and trees in the non-uniform model, where for each vertex $v$ and each color $d \in \mathcal{C}$, the cost of coloring $v$ by $d$ is an arbitrary nonnegative number $cost(v, d)$. The running times of both algorithms are governed by $2^{n_c}$, the number of subsets of the set of colors $\mathcal{C}$. First we present an algorithm for colored strings, and then extend it to colored trees.

**Non-uniform Optimal Convex Recoloring of Strings.** Throughout this section $(S, C)$ is a fixed, $n$-long input colored string, where $S = (v_1, \ldots, v_n)$. The algorithm scans the string from left to right. After processing vertex $v_i$, it keeps for each subset of colors $\mathcal{D} \subseteq \mathcal{C}$, and for each color $d \notin \mathcal{D}$, the cost of the optimal coloring of the $i$ leftmost vertices $v_1, \ldots, v_i$ which *does not* use colors from $\mathcal{D}$, and the rightmost vertex $v_i$ is colored by $d$. We define this more formally now:

**Definition 1.** *Let $\mathcal{D} \subseteq \mathcal{C}$ be a set of colors and $i \in \{1, \ldots, n\}$. A coloring $C'$ is a $(\mathcal{D}, i)$-coloring (of the string $S = (v_1, \ldots, v_n)$) if it is a convex coloring of $(v_1, \ldots, v_i)$, the $i$ leftmost vertices of $S$, such that $C'(\{v_1, \ldots, v_i\}) \cap \mathcal{D} = \emptyset$. $opt(\mathcal{D}, i)$ is the cost of an optimal $(\mathcal{D}, i)$-recoloring of $(S, C)$.*

It is easy to see that by the above definition, $opt(\emptyset, n)$ is the cost of an optimal convex recoloring of $(S, C)$.

**Definition 2.** *For a set of colors $\mathcal{D}$, a color $d$, and $i \in \{1, \ldots, n\}$, a coloring $C'$ is a $(\mathcal{D}, d, i)$-coloring if it is a $(\mathcal{D}, i)$-coloring and $C'(v_i) = d$. $opt(\mathcal{D}, d, i)$ is the cost of an optimal $(\mathcal{D}, d, i)$-coloring. $opt(\mathcal{D}, d, i) = \infty$ when no $(\mathcal{D}, d, i)$-coloring exists (eg when $d \in \mathcal{D}$).*

**Observation 5.** $opt(\mathcal{D}, i) = \min_{d \in \mathcal{C}} opt(\mathcal{D}, d, i)$.

For the recursive calculation of $opt(\mathcal{D}, d, i)$ we use the following function $R$, defined for a color set $\mathcal{D} \subseteq \mathcal{C}$, a color $d \in \mathcal{C}$ and $i \in \{1, \ldots, n\}$:

$$R(\mathcal{D}, d, i) = \min\{opt(\mathcal{D} \cup \{d\}, i), opt(\mathcal{D} \setminus \{d\}, d, i)\}$$

That is, $R(\mathcal{D}, d, i)$ is the minimal cost of a convex recoloring of the leftmost $i$ vertices, which does not use colors from $\mathcal{D} \setminus \{d\}$, and may use the color $d$ only as the color of the last (rightmost) block in $(v_1, \ldots, v_i)$. By convention, $opt(\mathcal{D}, d, 0) = 0$ for all $\mathcal{D} \subseteq \mathcal{C}$ and $d \notin \mathcal{D}$. Note that $R(\mathcal{D}, d, i) = R(\mathcal{D} \cup \{d\}, d, i) = R(\mathcal{D} \setminus \{d\}, d, i)$; we will usually use this function when $d \notin \mathcal{D}$.

**Theorem 6.** *For a color set $\mathcal{D}$, a color $d \notin \mathcal{D}$ and $i \in \{1, \ldots, n\}$:*

$$opt(\mathcal{D}, d, i) = cost(v_i, d) + R(\mathcal{D}, d, i - 1)$$

Theorem 6 yields the following dynamic programming algorithm for the minimal convex string recoloring:

---

Non-Uniform Optimal Convex String Recoloring

1. for every $\mathcal{D} \subseteq \mathcal{C}$ and for every $d \notin \mathcal{D}$, $opt(\mathcal{D}, d, 0) \leftarrow 0$
2. for $i = 1$ to $n$
   for every $\mathcal{D} \subseteq \mathcal{C}$
   (a) for every $d \notin \mathcal{D}$, $opt(\mathcal{D}, d, i) \leftarrow cost(v_i, d) + R(\mathcal{D}, d, i - 1)$
   (b) $opt(\mathcal{D}, i) \leftarrow \min_d opt(\mathcal{D}, d, i)$.
3. return $opt(\emptyset, n)$

---

Each of the $n$ iterations of the algorithms requires $O(n_c \cdot 2^{n_c})$ time. So the running time of the above algorithm is $O(n \cdot n_c 2^{n_c})$.

**Non-uniform Optimal Convex Recoloring of Trees.** We extend the algorithm of the previous section for optimal convex recoloring of trees. First, we root the tree at some vertex $r$. For each vertex $v \in V$, $T_v$ is the subtree rooted at $v$. A convex recoloring of $T_v$ denotes a convex recoloring of the colored subtree $(T_v, C|_{V(T_v)})$. We extend the definitions of the previous section to handle trees:

**Definition 3.** *Let $\mathcal{D} \subseteq \mathcal{C}$ be a set of colors and $v \in V$. Then a coloring $C'$ is a $(\mathcal{D}, T_v)$-coloring if it is a recoloring of $T_v$ s.t. $C'(V(T_v)) \cap \mathcal{D} = \emptyset$. $opt(\mathcal{D}, T_v)$ is the cost of an optimal $(\mathcal{D}, T_v)$-coloring.*

Again, a $(\mathcal{D}, T_v)$-coloring is a (convex) coloring on $T_v$ that does *not* use any color of $\mathcal{D}$. Thus $opt(\emptyset, T_r)$ is the cost of an optimal coloring of $T = T_r$.

**Definition 4.** *For a set of colors $\mathcal{D} \subseteq \mathcal{C}$, a color $d \in V$ and $v \in V$, a coloring $C'$ is a $(\mathcal{D}, d, T_v)$-coloring if it is a $(\mathcal{D}, T_v)$-coloring such that $C'(v) = d$. $opt(\mathcal{D}, d, T_v)$ is the cost of an optimal $(\mathcal{D}, d, T_v)$-coloring; in particular, if $d \in \mathcal{D}$ then $opt(\mathcal{D}, d, T_v) = \infty$.*

If $v$ is a leaf and $d \notin \mathcal{D}$, then $opt(\mathcal{D}, d, T_v) = cost(v, d)$. For the recursive calculation of $opt(\mathcal{D}, d, T_v)$ at internal vertices we need the following generalization of the function $R$ used for the string algorithm:

$$R(\mathcal{D}, d, T_v) = \min\{opt(\mathcal{D} \cup \{d\}, T_v), opt(\mathcal{D} \setminus \{d\}, d, T_v)\}$$

That is, $R(\mathcal{D}, d, T_v)$ is the minimal cost of a convex recoloring of $T_v$, which uses no colors from $\mathcal{D} \setminus \{d\}$ and does not include a $d$-block which is disjoint from the root $v$.

The calculation of $opt(\mathcal{D}, d, T_v)$ at an internal vertex with $k$ children $v_1, \dots, v_k$ uses the notion of *$k$-ordered partition* of a set $S$, which is a $k$-tuple $(S_1, \dots, S_k)$, where each $S_i$ is a (possibly empty) subset of $S$, s.t. $S_i \cap S_j = \emptyset$ for $i \neq j$ and $\cup_{i=1}^{k} S_i = S$. The set of $k^{|S|}$ $k$-ordered partitions of a set $S$ is denoted by $\mathcal{PART}_k(S)$.

**Theorem 7.** *Let $v$ be an internal vertex with children $v_1, \dots, v_k$. Then, for a color set $\mathcal{D}$ and a color $d \notin \mathcal{D}$:*

$$opt(\mathcal{D}, d, T_v) = cost(v, d) + \min_{(\mathcal{E}_1, \dots, \mathcal{E}_k) \in \mathcal{PART}_k(\mathcal{C} \setminus (\mathcal{D} \cup \{d\})} \sum_{i=1}^{k} R(\mathcal{C} \setminus \mathcal{E}_i, d, T_{v_i})$$

Theorem 7 above leads to a straightforward dynamic programming algorithm. In order to compute $opt(\mathcal{D}, d, T_v)$ for each $\mathcal{D} \subseteq \mathcal{C}$ and $d \notin \mathcal{D}$, we only need the corresponding values at $v$'s children. This can be achieved by a post order visit of the vertices, starting at $r$. To evaluate the complexity of the algorithm, we first note that each subset of colors $\mathcal{D}$ and a $k$-ordered partition $(\mathcal{E}_1, \dots, \mathcal{E}_k)$ of $\mathcal{C} \setminus (\mathcal{D} \cup \{d\})$ corresponds to the $(k + 1)$-ordered partition $(\mathcal{D}, \mathcal{E}_1, \dots, \mathcal{E}_k)$ of $\mathcal{C} \setminus \{d\}$. For each such ordered partition, $O(k)$ computation step are needed. As there are $n_c$ colors, the total time for the computation at vertex $v$ with $k$ children is $O(kn_c(k+1)^{n_c-1})$. Since $k \leq \Delta - 1$, the time complexity of the algorithm for trees with bounded degree $\Delta$ is $O(n \cdot n_c \cdot \Delta^{n_c})$.

We conclude this section by presenting a simpler linear time algorithm for optimal recoloring of a tree by two colors $d_1, d_2$. For this, we compute for $i = 1, 2$ the minimal cost convex recoloring $C_i$ which sets the color of the root to $d_i$ (i.e. $C_i(r) = d_i$). The required optimal convex recoloring is either $C_1$ or $C_2$. The computation of $C_1$ can be done as follows:
Compute for each vertex $v \neq r$ a cost defined by

$$cost(v) = \sum_{v' \in T_v} cost(v', d_2) + \sum_{v' \notin T_v} cost(v', d_1))$$

This can be done by one post order traversal of the tree. Then, select the vertex $v_0$ which minimizes this cost, and set $C_1(w) = d_2$ for each $w \in T_{v_0}$, and $C_1(w) = d_1$ otherwise.

## 4.2   Enhanced Algorithms for the Uniform Cost Model

The running times of the algorithms in Section 4.1 do not improve even when the input coloring is convex. However, for the uniform cost model, we can modify these algorithms so that their running time on convex or nearly convex input (string or tree) is substantially smaller. The new algorithms, instead of returning a total coloring, return a convex partial coloring, in which some of the new colors assigned to the vertices are unspecified. For the presentation of the algorithms we need the notion of convex cover which we define next.

A set of vertices $X$ is a *convex cover* (or just a cover) for a colored tree $(T, C)$ if the (partial) coloring $C_X = C|_{[V \setminus X]}$ is convex (i.e., $C$ can be transformed to a convex coloring by overwriting the vertices in $X$). Thus, if $C'$ is a convex recoloring of $(T, C)$, then $\mathcal{X}_C(C')$, the set of vertices overwritten by $C'$, is a cover for $(T, C)$. Moreover, deciding whether a subset $X \subseteq V$ is a cover for $(T, C)$, and constructing a total convex recoloring $C'$ of $C$ such that $\mathcal{X}(C') \subseteq X$ in case it is, can be done in $O(n \cdot n_c)$ time. Also, in the uniform cost model, the cost of a recoloring $C'$ is $w(\mathcal{X}(C'))$. Therefore, in this model, finding an optimal convex total recoloring of $C$ is polynomially equivalent to finding an optimal cover $X$, or equivalently a partial convex recoloring $C'$ of $C$ so that $w(\mathcal{X}(C')) = w(X)$ is minimized.

### Optimal String Recoloring via Relaxed Convex Recoloring

The enhanced algorithm for the string, makes use of the fact that partially colored strings can be characterized by the following property of "local convexity":

**Definition 5.** *A color $d$ is* locally convex *for a partially colored tree $(T, C)$ iff $C(carrier(C, d)) = \{d\}$, that is $carrier(C, d)$ does not contain a vertex of color different from $d$.*

**Observation 8.** *A partially colored string $(S, C)$ is convex iff it is locally convex for each color $d \in \mathcal{C}$.*

Note that Observation 8 does not hold for partially colored trees, since every leaf-colored tree is locally convex for each of its colors.

Given a colored string $(S, C)$ and a color $d$, $(S, C)$ is a *d-relaxed convex coloring* if it can be completed to total coloring such that for every color $d' \neq d$ there is a unique $d'$-block.

**Observation 9.** *$C$ is a $d$-relaxed convex coloring of a string $S$ if and only if each color $d' \neq d$ is locally convex for $(S, C)$.*

Given a colored string $(S, C)$, we transform $C$ to a coloring $\hat{C}$ as follows: For every vertex $v \in V(S)$:

$$\hat{C}(v) = \begin{cases} \hat{d} & \text{if } C(v) \text{ is a good color} \\ C(v) & \text{otherwise.} \end{cases}$$

where $\hat{d}$ is a new color.

A set of vertices $X \subseteq V$ is a *d-relaxed cover* of $(S, C)$ if the partial coloring $C|_{V \setminus X}$, denoted $C_X$, is a $d$-relaxed convex coloring of $(S, C)$.

**Theorem 10.** *Let $(S, C)$ and $\hat{C}$ be as above. Then $X \subseteq V$ is a cover for $(S, C)$ if and only if $X$ is a $\hat{d}$-relaxed cover for $(S, \hat{C})$.*

Theorem 10 implies that an optimal convex cover (and hence an optimal convex recoloring) of $(S, C)$ can be obtained as follows: transform $C$ to $\hat{C}$, and then compute an optimal $\hat{d}$-relaxed convex recoloring, $C'$, for $(S, \hat{C})$. The $\hat{d}$-relaxed cover defined by $C'$ is an optimal cover of $(S, C)$. An optimal convex recoloring of $(S, \hat{C})$ can be obtained by replacing step 2(a)of the non-uniform string recoloring algorithm of Section 4.1 by:

$$opt(\mathcal{D}, d, i) \leftarrow w(v)\delta_{C(v_i),d} + \begin{cases} opt(\mathcal{D}, i-1) & \text{if } d = \hat{d} \\ R(\mathcal{D}, d, i-1) & \text{otherwise.} \end{cases}$$

where $R$ is defined in Section 4.1, and where $\delta_{d,d'}$ is the complement of Kronecker delta:

$$\delta_{d,d'} = \begin{cases} 1 & \text{if } d \neq d' \\ 0 & \text{otherwise} \end{cases}$$

The improved algorithm has running time of $O\left(n_c^* n 2^{n_c^*}\right)$. In particular, for each fixed value of $n_c^*$ the running time is polynomial in the input size.

**Extension for Trees**

The technique of getting convex recoloring by treating all good colors as a special color $\hat{d}$ and then finding a $\hat{d}$-relaxed cover does not apply to trees.

Let $(T = (V, E), C)$ be a colored tree. For a vertex $v \in V$, let $\mathcal{C}_v^* = \mathcal{C}^* \cup \{C(v)\}$ (note that if $C(v) \in \mathcal{C}^*$ then $\mathcal{C}_v^* = \mathcal{C}^*$). Assume that the children of $v$ are $v_1, \ldots, v_k$. The crucial observation for our improved algorithm for convex recoloring of trees is that only colors from $\mathcal{C}_v^*$ may appear in more than one subtree $T_{v_i}$ of $T_v$. This observation enables us to modify the recursive calculation of the algorithm of Section 4.1 so that instead of computing $opt(\mathcal{D}, d, T_v)$ for all subsets $\mathcal{D}$ of $\mathcal{C}$ and each $d \notin \mathcal{D}$, it computes similar values only for subsets $\mathcal{D} \subseteq \mathcal{C}_v^*$ and $d \in \mathcal{C}_v^* \setminus \mathcal{D}$, and thus to reduce the exponential factor in the complexity bound from $2^{n_c}$ to $2^{n_c^*}$.

To enable the bookkeeping needed for the algorithm, it considers only optimal *partial* recolorings of $(T, C)$, which use good colors in a very restricted way: no vertex is overwritten by a good color (ie vertices are either retained, or uncolored, or overwritten by bad colors), and good colors are either retained or overwritten (by bad colors), but are never uncolored. The formal definition is given below.

**Definition 6.** *A partial convex recoloring $C'$ of the input coloring $C$ is conservative if it satisfies the following:*

1. *If $C'(v) \neq C(v)$ then $C'(v) \in \mathcal{C}^*$ (a color can be overwritten only by a bad color).*
2. *If $C(v) \notin \mathcal{C}^*$ then $v \in Domain(C')$ and $C'(v) \in \{C(v)\} \cup \mathcal{C}^*$ (a good color is either retained or overwritten by a bad color, but not uncolored).*
3. *For every $d \in \mathcal{C}$, $C'^{-1}(d)$ is connected (if a vertex is left uncolored then it does not belong to any carrier of $C'$).*

The fact that a conservative recoloring of minimum possible cost is an optimal convex recoloring follows from the following lemma, which seems to be of independent interest:

**Lemma 1.** *Let $X$ be a convex cover of a colored tree $(T, C)$. Then there is a convex total recoloring $\hat{C}$ of $(T, C)$ so that $X(\hat{C}) \subseteq X$ and for each vertex $v$ for which $C(v) \notin C^*$, $\hat{C}(v) = C(v)$ or $\hat{C}(v) \in C^*$ (that is, $\hat{C}$ does not overwrite a good color by another good color). In particular, there is an optimal total recoloring $\hat{C}$ of $(T, C)$ which never overwrites a good color by another good color.*

Let $\hat{C}$ be a convex total recoloring satisfying Lemma 1. Then it can be easily verified that the partial coloring obtained from $\hat{C}$ by *uncoloring* all the vertices $v$ for which $\hat{C}(v) \neq C(v)$ and $\hat{C}(v) \notin C^*$, is a conservative recoloring. Hence a conservative recoloring of minimum possible cost is an optimal convex recoloring.

For our algorithm we need variants of the functions $opt$ and $R$, adapted for conservative recolorings, which we define next. A coloring $C'$ is a $(\mathcal{D}, T_v)$-conservative recoloring if it is a conservative recoloring of $T_v$ which does not use colors from $\mathcal{D}$. If in addition $C'(v) = d$, then $C'$ is a $(\mathcal{D}, d, T_v)$-conservative recoloring; a $(\mathcal{D}, T_v)$-conservative recoloring in which $v$ is uncolored is a $(\mathcal{D}, *, T_v)$-conservative recoloring. Note that for certain combinations of $\mathcal{D} \subseteq C, f \in (C \setminus \mathcal{D}) \cup \{*\}$, and $v \in V$, no $(\mathcal{D}, f, T_v)$-conservative recoloring exists (eg, when $C(v)$ and $f$ are two distinct good colors).

For $f \in C \cup \{*\}$, a set of colors $\mathcal{D} \subseteq C$ and $v \in V$, $\widehat{opt}(\mathcal{D}, f, T_v)$ is the cost of an optimal $(\mathcal{D}, f, T_v)$-conservative recoloring ($\widehat{opt}(\mathcal{D}, f, T_v) = \infty$ if no $(\mathcal{D}, f, T_v)$-conservative recoloring exists). $opt(\mathcal{D}, T_v)$, the optimal cost of a conservative recoloring of $T_v$ which does not use colors from $\mathcal{D}$, is given by $\min_f \widehat{opt}(\mathcal{D}, f, T_v)$. By Lemma 1, the cost of an optimal recoloring of a colored tree $(T, C)$ is given by $opt(T_r, \emptyset)$, where $r$ is the root of $T$. The recursive computation of this value uses the function $\hat{R}$, given by

$$\hat{R}(\mathcal{D}, d, T_v) = \min\{\widehat{opt}(\mathcal{D} \cup \{d\}, T_v), \widehat{opt}(\mathcal{D} \setminus \{d\}, d, T_v)\}$$

Recall that $C_v^* = C^* \cup \{C(v)\}$. Rather than computing the functions $\widehat{opt}$ (and $\hat{R}$) at each vertex $v$ for all subsets $\mathcal{D}$ of $C$, our algorithm computes $\widehat{opt}(\mathcal{D}, f, T_v)$ at a vertex $v$ only for subsets of $C_v^*$. The correctness and complexity of the algorithm follows from following two lemmas.

**Lemma 2.** *For a vertex $v$ with children $v_1, \ldots, v_k$, a set of colors $\mathcal{D} \subseteq C_v^*$, and a color $d \in C_v^*$:*

*1. If $d \in \mathcal{D}$ then $\widehat{opt}(\mathcal{D}, d, T_v) = \infty$. If $d \in C_v^* \setminus \mathcal{D}$ then:*

$$\widehat{opt}(\mathcal{D}, d, T_v) = w(v)\delta_{C(v),d} + \min_{(\mathcal{E}_1,\ldots,\mathcal{E}_k)\in\mathcal{PART}_k\left(C_v^*\setminus(\mathcal{D}\cup\{d\})\right)} \sum_{i=1}^{k} \hat{R}(C_v^* \setminus \mathcal{E}_i, d, T_{v_i})$$

2. If $C(v) \notin \mathcal{C}^*$ then $\widehat{opt}(\mathcal{D}, *, T_v) = \infty$. Else (ie $C(v) \in \mathcal{C}^*$ and $\mathcal{C}_v^* = \mathcal{C}^*$):

$$\widehat{opt}(\mathcal{D}, *, T_v) = w(v) + \min_{(\mathcal{E}_1, \ldots, \mathcal{E}_k) \in \mathcal{PART}_k(\mathcal{C}_v^* \setminus \mathcal{D})} \sum_{i=1}^{k} \widehat{opt}(\mathcal{C}_v^* \setminus \mathcal{E}_i, T_{v_i})$$

Lemma 2 implies a dynamic programming algorithm similar to the one presented in Section 4.1. The algorithm computes for each vertex $v$, for each subset of colors $\mathcal{D} \subseteq \mathcal{C}_v^*$ and for each $f \in (\mathcal{C}_v^* \setminus \mathcal{D}) \cup \{*\}$, the values of $\widehat{opt}(\mathcal{D}, d, T_v)$. when $v$ is a leaf, this value for each $\mathcal{D} \subseteq \mathcal{C}_v^*$ and each $d \in \mathcal{D}$ is given by $\widehat{opt}(\mathcal{D}, d, T_v) = w(v)\delta_{C(v),d}$, and the value of $\widehat{opt}(\mathcal{D}, *, T_v)$ when $C(v) \in \mathcal{C}^*$ is $w(v)$. So it remains to show that these values can be computed at internal vertices, assuming they were previously computed at their children.

For an internal vertex $v$ with children $v_1, \ldots, v_k$, the algorithm uses Lemma 2(1) to compute the values $\widehat{opt}(\mathcal{D}, d, T_v)$ for each $\mathcal{D} \subseteq \mathcal{C}_v^*$ and for each $d \in \mathcal{C}_v^* \setminus \mathcal{D}$. If $C(v) \in \mathcal{C}^*$, then Lemma 2(2) is used to compute the value of $\widehat{opt}(\mathcal{D}, *, T_v)$. There is however a subtle point in the realization of this algorithm, which stems from the fact that the sets $\mathcal{C}_v^*$ which define the values computed at each vertex $v$ may vary from vertex to vertex. The following claim guarantees that all the values needed for the calculations at an internal vertex $v$ are calculated by its children $v_1, \ldots, v_k$.

**Lemma 3.** Let $v$ be an internal vertex with children $v_1, \ldots, v_k$, and assume that $v$ is visited by the algorithm after its children. Then for each subset of colors $\mathcal{D} \subseteq \mathcal{C}_v^*$ and each $f \in \mathcal{C}_v^* \cup \{*\}$, all the values required for computing $\widehat{opt}(\mathcal{D}, f, T_v)$ by Lemma 2 (1) and (2) are computed by $v_1, \ldots, v_k$.

Combining the results so far, we have

**Theorem 11.** Optimal convex recoloring of totally colored trees with $n$ vertices can be computed in $O(n \cdot n_c^* \Delta^{n_c^*+2})$ time, where $n_c^*$ is the number of bad colors and $\Delta$ is the maximum degree of vertices in $T$.

## 5    Discussion and Future Work

In this work we studied the complexity of computing the distance from a given coloring of a tree or string to a convex coloring, motivated by the scenario of introducing a new character to an existing phylogenetic tree. We considered few natural definitions for that distance, along with few model variants of the problem, and proved that the problem is NP-Hard in each of them. We then presented exact algorithms to solve the problem under the non-uniform and the uniform cost models.

Few interesting research directions which suggest themselves are:

- Similarly to the generalization of the small parsimony question to the general one: Given a set of characters (colorings) such that the number of colors of each character is bounded by a fixed small constant, is there an efficient algorithm which computes a phylogenetic tree of minimum distance from a perfect phylogeny, where the distance is taken as the number of color changes needed to achieve perfect phylogeny? Note that, as in maximum parsimony, this problem is trivial for one character.

– Similarly to the above, but rather than bounding the number of colors, the bound now is on the number of color changes, which is the recoloring distance from convexity. The goal is to decide whether there is a tree within this distance from a perfect phylogeny over the given set of characters. This corresponds to a fixed parameter tractable algorithm for constructing an optimal tree.

– Can our results for the uniform cost model from Section 4.2 be extended for the non-uniform cost model.

– Phylogenetic network are accumulating popularity as a model for describing evolutionary history. This trend, motivates the extension of our problem to more generic cases such are directed acyclic graphs or general graphs. It would be interesting to explore the properties of convexity on these types of graphs.

## Acknowledgments

We would like to thank Mike Fellows, Shiri Moran, Rami Reshef, Mike Steel, Zohar Yakhni and Irad Yavneh For enlightening discussions and helpful comments.

## References

1. R. Agrawala and D. Fernandez-Baca. Simple algorithms for perfect phylogeny and triangulating colored graphs. *International Journal of Foundations of Computer Science*, 7(1):11–21, 1996.
2. A. Ben-Dor, N. Friedman, and Z. Yakhini. Class discovery in gene expression data. In *RECOMB*, pages 31–38, 2001.
3. M. Bittner and et.al. Molecular classification of cutaneous malignant melanoma by gene expression profiling. *Nature*, 406(6795):536–40, 2000.
4. H.L. Bodlaender, M.R. Fellows, and T. Warnow. Two strikes against perfect phylogeny. In *ICALP*, pages 273–283, 1992.
5. R. G. Downey and M. R. Fellows. newblock *Parameterized Complexity*. Springer, 1999.
6. A. Dress and M.A. Steel. Convex tree realizations of partitions. *Applied Mathematics Letters,*, 5(3):3–6, 1992.
7. D. Fernndez-Baca and J. Lagergren. A polynomial-time algorithm for near-perfect phylogeny. *SIAM Journal on Computing*, 32(5):1115–1127, 2003.
8. W. M. Fitch. A non-sequential method for constructing trees and hierarchical classifications. *Journal of Molecular Evolution*, 18(1):30–37, 1981.
9. L.A. Goldberg, P.W. Goldberg, C.A. Phillips, Z Sweedyk, and T. Warnow. Minimizing phylogenetic number to find good evolutionary trees. *Discrete Applied Mathematics*, 71:111–136, 1996.
10. D. Gusfield. Efficient algorithms for inferring evolutionary history. *Networks*, 21:19–28, 1991.

11. A. Hirsh, A. Tsolaki, K. DeRiemer, M. Feldman, and P. Small. From the cover: Stable association between strains of mycobacterium tuberculosis and their human host populations. *PNAS*, 101:4871–4876, 2004.

12. S. Kannan and T. Warnow. Inferring evolutionary history from DNA sequences. *SIAM J. Computing*, 23(3):713–737, 1994.

13. S. Kannan and T. Warnow. A fast algorithm for the computation and enumeration of perfect phylogenies when the number of character states is fixed. *SIAM J. Computing*, 26(6):1749–1763, 1997.

14. S. Moran and S. Snir. Convex recoloring of strings and trees. Technical Report CS-2003-13, Technion, November 2003.

15. D. Sankoff. Minimal mutation trees of sequences. *SIAM Journal on Applied Mathematics*, 28:35–42, 1975.

16. C. Semple and M.A. Steel. *Phylogenetics*. Oxford University Press, 2003.

17. M. Steel. The complexity of reconstructing trees from qualitative characters and subtrees. *Journal of Classification*, 9(1):91–116, 1992.

# Linear Time Algorithms for Generalized Edge Dominating Set Problems

André Berger and Ojas Parekh

Department of Mathematics and Computer Science,
Emory University, Atlanta, GA, 30322, USA
aberge2@emory.edu, ojas@mathcs.emory.edu

**Abstract.** In this paper we consider a generalization of the edge dominating set (EDS) problem, in which each edge $e$ needs to be covered $b_e$
times and refer to this as the $b$-EDS problem. We present an exact linear
time primal dual algorithm for the weighted $b$-EDS problem on trees with
$b_e \in \{0, 1\}$, and our algorithm generates an optimal dual solution as well.
We also present an exact linear time algorithm for the unweighted $b$-EDS
problem on trees. For general graphs we exhibit a relationship between
this problem and the maximum weight matching problem. We exploit
this relationship to show that a known linear time $\frac{1}{2}$-approximation algorithm for the weighted matching problem is also a 2-approximation
algorithm for the unweighted $b$-EDS problem on general graphs.

## 1 Introduction

Domination problems in graphs have been subject of many studies in graph theory, and have many applications in operations research, e.g. in resource allocation
and network routing as well as in coding theory.

In this paper we consider a generalization of the *edge dominating set (EDS)
problem*. Given a graph $G = (V, E)$, a function $b : E \to \mathbb{N}$ and a weight function
$c : E \to \mathbb{Q}^+$, a $b$-*EDS* is a subset $F \subseteq E$ together with a multiplicity $m_e \in \mathbb{N}^+$
for each $e \in F$, so that each edge in $E$ is adjacent to at least $b_e = b(e)$ edges
in $F$, counting multiplicities. The $b$-*EDS problem* is then to find a $b$-EDS which
minimizes $\sum_{e \in F} m_e$ in the unweighted and $\sum_{e \in F} c(e) \cdot m_e$ in the weighted case.
The $b$-EDS problem generalizes the EDS problem in much the same way that
the set multicover problem generalizes the set cover problem [17].

When $b_e = 1$ for all $e \in E$ this is the edge dominating set problem (EDS),
which is one of the four natural covering problems in graphs: edge cover (cover $V$
with elements from $E$), vertex cover ($E$ with $V$), dominating set ($V$ with $V$), and
EDS ($E$ with $E$). In fact, weighted EDS is a common generalization of weighted
edge cover and weighted vertex cover [1] and is equivalent to a restricted total
covering problem in which $E \cup V$ must be covered by a minimum weight set of
elements from $E \cup V$ [13].

The unweighted version of EDS is NP-complete even for planar and bipartite
graphs of maximum degree 3 [18] as well as for several other families of graphs [8].

F. Dehne, A. López-Ortiz, and J.-R. Sack (Eds.): WADS 2005, LNCS 3608, pp. 233–243, 2005.
© Springer-Verlag Berlin Heidelberg 2005

However, there are also families of graphs for which the unweighted EDS problem is polynomial-time solvable [8, 16]. In particular, linear time algorithms for the unweighted version are known for trees [10] and block graphs [9].

Much less is known about the weighted version of the problem. Recently Fujito and Nagamochi [4] and Parekh [12] independently discovered a 2-approximation for the weighted EDS problem; the latter also showed that weighted EDS restricted to bipartite graphs is no easier to approximate than weighted vertex cover, which is MAX-SNP-hard [11] and is suspected to have no polynomial time approximation algorithm with approximation ratio asymptotically less than 2. Thus a 2-approximation for $b$-EDS may be the best we can hope for.

When $b_e \in \{0, 1\}$ for all $e \in E$ we call the resulting problem $\{0, 1\}$-EDS. The weighted version of $\{0, 1\}$-EDS is particularly interesting since it is equivalent to the generalization of weighted vertex cover in which in addition to single vertices, weights may also be assigned to pairs, $\{u, v\}$, of vertices (by adding an edge $uv$ with $b_{uv} = 0$ if one does not already exist). This generalization may be used to model a limited economy of scale in existing applications of vertex cover: for a pair of vertices $\{u, v\}$ one may stipulate that selecting both $u$ and $v$ costs less than the sum of the individual costs of $u$ and $v$.

Our main contributions are linear-time algorithms for three special cases of the $b$-EDS problem. To the best of our knowledge an exact linear time algorithm was not known for even the special case of weighted EDS on trees. Table 1 gives an overview of known results and new results from this paper.

**Table 1.** Approximation ratios for variants of the EDS problem ($^*$ denotes a linear time algorithm)

| | unweighted EDS | weighted EDS | unweighted $b$-EDS | weighted $b$-EDS |
|---|---|---|---|---|
| general graphs | $2^*$ | 2 [4, 12] | $2^*$ (Cor. 2) | 8/3 [13] |
| bipartite graphs | ,, | ,, | ,, | 2 [13] |
| trees | $1^*$ [10] | $1^*$ (Thm. 3) | $1^*$ (Thm. 2) | 1 (Thm. 1) |

In Section 2 we expose a relationship between the maximum weighted matching problem and the unweighted $b$-EDS problem; we use this relationship to analyze an algorithm of Preis [14] and show that it is also a linear time 2-approximation for the unweighted $b$-EDS problem. This generalizes a known relationship between maximal matchings and (unweighted) edge dominating sets.

In Section 3 we show that the weighted $b$-EDS is solvable in polynomial time on trees. We also present exact linear-time algorithms which solve the unweighted $b$-EDS problem, and the weighted $\{0, 1\}$-EDS problem on trees. The latter is a primal dual algorithm which also generates an optimal dual solution. If the weights $c_e$ are integral for all $e \in E$, then the dual solution is integral as well and is a maximum size set of edges such that each edge $e$ has at most $c_e$ edges adjacent to it. This problem is a common generalization of the maximum independent set problem and the maximum strong matching problem. An exact linear time algorithm for the latter on trees is known [2].

*Notation.* We will use the following notation for a simple undirected graph $G = (V, E)$. The neighbors of a vertex $v \in V$ are denoted by $\delta(v) = \{u \in V : \exists uv \in E\}$. The edges incident upon the vertex $v \in V$ are denoted by $N(v)$ and the set of edges adjacent with an edge $e \in E$ plus the edge $e$ itself is denoted by $N(e)$, i.e. $N(uv) = N(u) \cup N(v)$ for any $uv \in E$.

## 2    The Unweighted $b$-EDS Problem for General Graphs

For general graphs, the weighted EDS problem admits a $2\frac{1}{10}$-approximation based on a natural linear program relaxation of the problem, whose integrality gap is also $2\frac{1}{10}$ [1]. A corresponding linear relaxation for the weighted $b$-EDS problem yields an $\frac{8}{3}$-approximation for general graphs and a 2-approximation for bipartite graphs [13]. This relaxation can be strengthened to yield a 2-approximation for weighted EDS [4, 12]; however, the corresponding strengthening fails to deliver a 2-approximation for weighted $b$-EDS.

The unweighted EDS and $b$-EDS problems, however, can be approximated more easily due to their relation to matching problems. Harary's book [6] demonstrates that there always exists a minimum cardinality EDS which is also a maximal matching. Since any maximal matching is also an EDS, the minimum cardinality maximal matching and the unweighted EDS problems are equivalent. In contrast Fujito [3] showed that the minimum weighted maximal matching problem is much more difficult than weighted EDS. Any maximal matching in a graph has size at least one half times the size of a maximum matching. Therefore, finding any maximal matching, which can be easily done in linear time, yields a 2-approximation for the unweighted EDS problem.

An issue with extending the relationship described above to the minimum unweighted $b$-EDS problem is that a maximal matching is not necessarily a feasible $b$-EDS. Using the resemblance of the maximum weight matching problem to the dual of a natural linear formulation for $b$-EDS, we exhibit a connection between weighted matchings and $b$-edge dominating sets that generalizes the relationship between maximal matchings and edge dominating sets. Given a matching $M$ and a vector $b \in \mathbb{N}^{|E|}$, let $b|_M \in \mathbb{N}^{|E|}$ denote the vector which for each component $e \in M$ has value $b_e$, and has value 0 for all other components.

**Lemma 1.** *For any matching $M$ and any vector $b \in \mathbb{N}^{|E|}$, $\sum_{e \in M} b_e$ is at most twice the weight of a minimum size (counting multiplicities) $b$-EDS.*

*Proof.* Consider the following pair of dual LP's, LP 1 being the linear programming relaxation for the unweighted $b$-EDS problem and LP 2 being the relaxation for the weighted strong matching problem.

We call $P$ and $R$ the sets of feasible fractional solutions of LP 1 and LP 2, respectively. By setting the dual variables to $y_e = 1/2$ for each $e \in M$, and to

LP 1: Min $\mathbb{1} \cdot x$, subject to          LP 2: Max $b \cdot y$, subject to

$$x(N(e)) \geq b_e \text{ for all } e \in E \qquad\qquad y(N(e)) \leq 1 \text{ for all } e \in E$$
$$x_e \geq 0 \;\; \text{for all } e \in E \qquad\qquad\qquad y_e \geq 0 \text{ for all } e \in E$$

$y_e = 0$ for each $e \in E \setminus M$, we obtain such a feasible solution $y \in R$ to LP 2, since each edge $e \in E$ can be adjacent with at most two edges from $M$. Using duality we have

$$b(M)/2 = b \cdot y \leq Max_{y \in R} b \cdot y = Min_{x \in P} \mathbb{1} \cdot x \leq OPT.$$

Hence the weight $b(M)$ of the solution we return is at most $2 \cdot OPT$.     □

**Corollary 1.** *For any matching $M$ and any vector $b \in \mathbb{N}^{|E|}$, if $b|_M$ is a feasible b-EDS then it is a 2-approximate unweighted b-EDS.*

Corollary 1 motivates the following definition: we say a matching $M$ is $b$-feasible if $b|_M$ is a feasible $b$-EDS. Thus any matching algorithm that returns a $b$-feasible matching $M$ is a 2-approximation for the unweighted $b$-EDS problem. Before presenting a linear time algorithm, we present a very simple $O(|E| \cdot \log |V|)$ 2-approximation for unweighted $b$-EDS.

**Proposition 1.** *The greedy algorithm for the maximum weight matching problem (with weights $b \in \mathbb{N}^{|E|}$) that repeatedly selects the edge of greatest weight that maintains a matching, always returns a matching that is also b-feasible.*

*Proof.* The greedy algorithm also satisfies Lemma 2 below and thus the proof is the same as the proof of Lemma 3.     □

It is not difficult to see that any matching of maximum weight with respect to $b$ is $b$-feasible. Lemma 1 also implies that any feasible $b$-EDS has size at least $\frac{1}{2}$ the cost of a maximum weight matching with respect to $b$, thus if a $b$-EDS algorithm always returns a $b$-EDS that is a matching when copies of an edge are removed, then the algorithm is a $\frac{1}{2}$-approximation for the maximum weight matching problem.

Preis [14] gave a linear time algorithm, which, given a weighted graph, computes a maximal matching with weight at least one half times the weight of any matching. The algorithm incrementally adds edges to a matching $M$. A vertex $u$ is called free (w.r.t. to $M$), if $u$ is not incident with any edge in $M$. We will use the following lemma, which gives a necessary condition for an edge to be added to the matching, to show that Preis's algorithm always generates a $b$-feasible matching.

**Lemma 2** ([14–Lemma 3]). *If an edge $uv$ is added to $M$ during the algorithm, then $u$ and $v$ are free and neither $u$ nor $v$ are adjacent to a free vertex with an edge of higher weight than the weight of the edge $uv$.*

**Lemma 3.** *Preis's algorithm always generates a b-feasible matching.*

*Proof.* First, note that each edge $e \in M$ is covered $b_e$ times by itself. Since $M$ is a maximal matching, any edge $e = uv$ in $E \setminus M$ must be adjacent to some edge $f = vw$ in $M$. If $b_f \geq b_e$, then $e$ is covered. Otherwise, Lemma 2 says that $u$ was not a free vertex at the time when $f$ was added to $M$. But then there must be an edge $tu \in M$, which was added to $M$ before $f$, i.e. $t$, $u$ and $v$ were free vertices at the time $tu$ was considered. Using Lemma 2 again, we must have $b_{tu} \geq b_e$.     □

**Corollary 2.** *The unweighted b-EDS problem on general graphs can be 2-approximated in linear time.*

# 3 The $b$-EDS Problem for Trees

## 3.1 The General Case

Many problems which are hard to solve optimally or even approximate for general graphs become a lot easier when restricted to a small family of graphs. The same is true for the weighted $b$-EDS problem when we restrict the possible inputs to trees.

In this section we will show that the weighted $b$-EDS problem on trees can be solved optimally as a linear program. A square matrix $A$ is called *totally unimodular* if every square sub-matrix of $A$ has determinant +1, -1 or 0. Totally unimodular matrices play an important role in linear programming due to the following lemma.

**Lemma 4 ([7, 15]).** *If $A \in \mathbb{Z}^{m \times n}$ is a totally unimodular matrix and $b \in \mathbb{Z}^m$, then every extreme point of the polyhedron $\{x \in \mathbb{R}^n : Ax \leq b, x \geq 0\} \neq \emptyset$ has integer coordinates.*

Ghouila-Houri [5] gave the following sufficient condition for a matrix to be totally unimodular.

**Lemma 5.** *A $m \times n$ matrix $A$ with entries in $\{1, 0, -1\}$ is totally unimodular if for every $J \subseteq \{1, \ldots, n\}$ there exist a partition $J = J_1 \cup J_2$ such that for any $1 \leq j \leq m$ it holds that $|\sum_{i \in J_1} a_{ij} - \sum_{i \in J_2} a_{ij}| \leq 1$.*

For completeness we include a proof for the fact that the constraint matrix for the EDS problem for any tree is totally unimodular.

**Lemma 6.** *Let $T = (V, E)$ be a tree on $n$ vertices. Let $A = (a_{ij})$ be the edge-edge adjacency matrix of $T$ with 1's on the diagonal, i.e. $a_{ij} = 1$ for all $ij \in E$, $a_{ij} = 0$ for all $ij \notin E$ and $a_{ii} = 1$ for all $1 \leq i \leq n-1$, $1 \leq j \leq n-1$ and $i \neq j$. Then $A$ is totally unimodular.*

*Proof.* According to Lemma 5 it is enough to show that we can partition any $E' \subseteq E$ into two sets of edges $E_1$ and $E_2$, so that for any $e \in E$ we have $|N(e) \cap E_1| - |N(e) \cap E_2| \in \{-1, 0, 1\}$. This is equivalent to say that for any $E' \subseteq E$ there is a labeling $\pi : E \to \{1, 0, -1\}$ such that $\pi(e) = 0$ if and only if $e \in E \setminus E'$ and such that $\Pi(e) := |\sum_{f \in N(e)} \pi(f)| \leq 1$ for any $e \in E$.

We will prove by induction on $|V|$ that such a labeling exists for any $E' \subseteq E$ and that it also satisfies $|\Pi(v)| \leq 1$ where $\Pi(v) := \sum_{f \in N(v)} \pi(f)$ for any $v \in V$. Note that if such a labeling $\pi$ exists, than $-\pi$ clearly also satisfies the condition.

The base case $|V| = 1$ is trivial. Let now $|V| \geq 2$ and let $v_0 \in V$ be an arbitrary vertex of $T$. Let $T_i = (V_i, E_i)$ $(1 \leq i \leq k)$ be the connected components of $T - \{v\}$, which are trees as well. Let $v_i$ be the neighbor of $v$ in $T_i$ and $e_i = v_0 v_i$. For every $1 \leq i \leq k$ let $\pi_i$ be a labeling of $E_i$ with $\pi_i(e) = 0$ if and only if $e \in E_i \setminus E'$, which exist by the inductive hypothesis. They also satisfy $|\Pi_i(v_i)| \leq 1$ for every $1 \leq i \leq k$. If we set $\pi(e_i) = 0$ for $1 \leq i \leq k$ and $\pi(e) = \pi_i(e)$ whenever $e \in E_i$, we obtain a labeling $\pi : E \to \{1, 0, -1\}$ such that $\Pi(e_i) = |\Pi_i(v_i)| \leq 1$ and $\Pi(v_0) = 0$, i.e. $\pi$ satisfies the claim. However, the edges incident with $v_0$ which are in $E'$ are falsely labeled 0. We will show how to label these edges with 1 or $-1$ and maintain the desired properties.

We can assume w.l.o.g. that for some $0 \leq s \leq k$ we have that $\{e_1, \ldots, e_s\} \subseteq E'$ and that $\{e_{s+1}, \ldots, e_k\} \subseteq E \setminus E'$. Further assume that $|\Pi(v_i)| = 1$ for $1 \leq i \leq s_0$ and that $|\Pi(v_i)| = 0$ for $s_0 < i \leq s$ for some $0 \leq s_0 \leq s$.

We will switch the labelings $\pi_i$ for $1 \leq i \leq s_0$ if necessary to obtain $\Pi(v_i) = 1$ for $1 \leq i \leq \lfloor s_0/2 \rfloor$ and $\Pi(v_i) = -1$ for $\lfloor s_0/2 \rfloor < i \leq s_0$. Then we define $\pi(e_i) = -1$ for $1 \leq i \leq \lfloor s_0/2 \rfloor$ and $\pi(e_i) = 1$ for $\lfloor s_0/2 \rfloor < i \leq s_0$. The edges $e_{s_0+1}, \ldots, e_s$ will be labeled with 1 and $-1$ so that we have $\Pi(v_0) \in \{0, 1\}$. The numbers of edges labeled 1 and $-1$, respectively, will depend on the parity of $s$ and $s_0$ and differ by at most 1. If $\Pi(v_0) = 1$, then we also switch the labelings $\pi_i$ of those trees $T_i$ with $i > s$ for which $\Pi(v_i) = 1$ to $-\pi_i$ to ensure that $\Pi(e_i) \leq 1$ for those indices $i$.

It is easy to check that the conditions on $\pi$ remain true for all edges and vertices of $T$.    □

Using Lemma 4 and Lemma 6 we immediately have

**Theorem 1.** *The b-EDS problem on weighted trees can be solved optimally in strongly polynomial time.*

The algorithm to solve the $b$-EDS problem on trees relies on solving a linear program. However, we would prefer a combinatorial algorithm, ideally running in linear time. This is indeed possible if we restrict the trees to have either uniform weights (Section 3.2) or if we restrict ourselves to the $\{0, 1\}$-EDS problem on weighted trees.

## 3.2    The Unweighted b-EDS Problem for Trees

A linear time algorithm for the unweighted EDS problem on trees was first given by Mitchell and Hedetniemi [10] and later simplified by Yannakakis and Gavril [18]. The unweighted $b$-EDS problem on trees can also be solved by an easy greedy algorithm in linear time. Call an edge $e$ of a tree a *leaf edge* if it is incident with a leaf.

For any tree $T$ there will always be an optimal solution to the $b$-EDS problem which does not use any leaf edges (unless $T$ is a star), since any edge adjacent with a leaf edge covers at least those edges covered by the leaf edge.

Therefore we can recursively solve the problem by first finding a vertex $v$ which is incident with exactly one non-leaf edge $e$, then setting the multiplicity of $e$ to the maximum $b$-value of the leaf edges incident with $v$ and finally removing those leaf edges and updating the $b$-values of $e$ and those edges adjacent with $e$ in the remaining tree.

Any optimal solution to the $b$-EDS problem on that updated tree plus the multiplicity of the edge $e$ as determined before will give an optimal solution to the original instance. A formal proof of this fact is straightforward and we omit it in this abstract.

**Theorem 2.** *The unweighted b-EDS problem on trees can be solved optimally in linear time.*

## 3.3   The Weighted $\{0, 1\}$-EDS Problem for Trees

To the best of our knowledge no linear time algorithm for the weighted EDS problem on trees has appeared in the literature. Algorithm 1 is a linear time primal-dual algorithm which solves the weighted $\{0,1\}$-EDS problem on trees (indeed a generalization of the weighted EDS problem) optimally in linear time. This problem generalizes the weighted $b$-vertex cover problem on trees as follows:

**Lemma 7.** *The weighted vertex cover problem for trees can be solved optimally in linear time by solving a weighted $\{0,1\}$-EDS instance on a tree in linear time.*

*Proof.* Let $T = (V, E)$, $c_v \in \mathbb{R}^+$ for all $v \in V$ be an instance of the weighted vertex cover problem for trees. We build a tree $T' = (V \cup V', E \cup E')$ with $V' = \{v' : v \in V\}$ and $E' = \{vv' : v \in V\}$, i.e. we add an extra edge incident with each vertex of $T$. We set $b_e = 1$ for every $e \in E$ and $b_e = 0$ for every $e \in E'$. Furthermore, we set $c'_{vv'} = c_v$ for all $v \in V$ and $c'_e = \infty$ for all $e \in E$. Then any $b$-EDS of $T'$ of finite weight corresponds to a vertex cover of $T$ having the same weight, and vice versa (an edge $vv' \in E'$ is in the $b$-EDS if and only if $v$ is in the vertex cover). □

We now present our primal-dual algorithm for the weighted $\{0, 1\}$-EDS problem for trees. In a nutshell the algorithm works as follows. We first pick some arbitrary vertex of the tree as the root. Then we determine an optimal dual solution by raising dual variables from the leaves up to the root, making at least one constraint of the dual problem tight whenever we raise a dual variable. Finally, we recover a primal solution from the root down to the leaves, which satisfies the complementary slackness conditions with the dual solution.

We denote by $d_T(v, u)$ the (combinatorial) distance between $v$ and $u$ in $T$, i.e. the number of edges on the path between $v$ and $u$ in $T$. If $T$ is rooted at $v_0$, then by denoting an edge by $e = vu$ we implicitly mean that $v$ is closer to the root, i.e. $d_T(v, v_0) = d_T(u, v_0) - 1$. For a vertex $v \neq v_0$ $p(v)$ denotes the *parent* of $v$, i.e. the unique vertex on the path from $v$ to $v_0$ which is adjacent to $v$. The set of children of $v$ is denoted $\bar{\delta}(v) = \delta(v) \setminus \{p(v)\}$.

**Theorem 3.** *Algorithm 1 solves the weighted $\{0,1\}$-EDS problem on trees optimally in linear time.*

*Proof.* We will argue that both $x$ and $y$ are feasible solutions to the following LP's and that they satisfy complementary slackness and hence are optimal solutions. Here we let $D = \{e \in E : b_e = 1\}$.

---

**Algorithm 1: $\{0,1\}$-EDS on weighted trees**

Input: A tree $T = (V, E)$, $c : E \to \mathbb{R}^+ \cup \{0\}$,
.        $b : E \to \{0, 1\}$ and a root $v_0 \in V$.

1. Set $K := \max_{v \in V} d_T(v, v_0)$.
% *Construct the dual solution from the leaves to the root.*
2. FROM $i = K$ DOWNTO 0 DO
      FOR ALL $v \in V$ with $d_T(v, v_0) = i$ DO
         IF $v$ is a leaf THEN $y_v := c_{p(v)v}$
         ELSE
            $\underline{c} := \min_{u \in \delta(v)} c_{vu}$
            FOR EVERY $u \in \bar{\delta}(v)$ with $b_{vu} = 1$ DO
            $y_{vu} := \min\{y_u, \underline{c}\}$
            $\underline{c} := \underline{c} - y_{vu}$
            $\bar{y} := \sum_{u \in \bar{\delta}(v)} y_{vu}$
            $y_v := \min_{u \in \bar{\delta}(v)} (c_{vu} - \bar{y})$
            IF $v \neq v_0$ THEN $c_{p(v)v} := c_{p(v)v} - \bar{y}$
% *Construct the primal solution from the root to the leaves.*
% $e \in E$ *is 'tight', if* $y(N(e)) = c_e$
3. $F := \emptyset$
4. Whenever an edge $vu$ is added to $F$, set $x_v = 1$ and $x_u = 1$.
5. IF $y_{v_0 v} = 0$ for all $v \in \delta(v_0)$ THEN add all tight edges incident with $v_0$ to $F$ ELSE add one tight edge incident with $v_0$ to $F$.
6. FROM $i = 1$ to $K - 1$ DO
      FOR ALL $v \in V$ with $d_T(v, v_0) = i$ DO
         $e := p(v)v$
  Case 1 IF $y_e > 0$ and $x_e = 0$ and $x_{p(v)} = 0$ THEN add ONE arbitrary tight edge incident with $v$ to $F$
  Case 2 IF $y_{vu} = 0$ for all $u \in \bar{\delta}(v)$ and ($b_e = 0$ or $y_e = 0$) THEN add ALL tight edges incident with $v$ to $F$
  Case 3 IF $y_{vu} > 0$ for some $u \in \bar{\delta}(v)$ and $\Big( (b_e = 0$ and $x_e = 0)$ or $y_e = 0 \Big)$ add ONE arbitrary tight edge incident with $v$ to $F$
  Case 4 In the remaining cases no edges are added to the primal solution $F$.
      % *Remark: If in any of the cases there is no tight edge incident upon $v$ then $F$ remains unchanged.*
7. RETURN $F$.

---

First note that $y$ is feasible for LP 2, i.e. $y \in R$. The variable $y_v$ always contains the maximum value that any $y$-value of an edge incident with $v$ can be increased by to maintain a feasible solution to LP 2. Using this and the fact that any positively set $y$-value for an edge $vu$ is at most $c_{vu'}$, where $u' \in \delta(v)$, we see that $y$ is a feasible solution to LP 2.

The primary solution $x$ constructed in steps 5 and 6 is chosen so that it satisfies the complementary slackness conditions with $y$. First, only *tight* edges are chosen to be in the solution $F$, i.e. whenever $x_e > 0$ then $y(N(e)) = c_e$. Second, if $y_e > 0$ for some $e = p(v)v$ with $b_e = 1$, and $v$ is considered in step 6, then we only add edges to $F$ in Case 2 with the additional conditions $x_e = 0$ and $x_{p(v)} = 0$. But this means no edge incident with $p(v)$ is already in $F$ and we add at most one edge incident with $v$ to $F$, i.e. $x(N(e)) \leq 1$. As we will show below $x$ is a feasible solution to LP 1 and therefore $x(N(e)) = 1$. Thus $x$ and $y$ satisfy the complementary slackness conditions.

LP 1: Min $c \cdot x$, subject to                     LP 2: Max $\mathbb{1} \cdot y$, subject to

$x(N(e)) \geq 1$ for all $e \in D$                     $y(N(e)) \leq c_e$ for all $e \in E$
$x_e \geq 0$ for all $e \in E$                              $y_e \geq 0$  for all $e \in D$

To show that $x$ is a feasible solution for the primal LP let $e = vu \in D$ where $v = p(u)$ and assume to the contrary that for all $x(N(e)) = 0$. For now assume $e$ is not incident with a leaf or with the root.

Let $f = p(v)v$ denote the *parent edge* of $e$. The *sibling edges* of $e$ are all edges $vu' \in E$ with $u' \neq u$ and $u' \neq p(v)$; the *children edges* of $e$ are all edges $uw \in E$ with $w \neq v$. We claim that neither any of the children edges and sibling edges of $e$ nor $e$ itself have a tight dual inequality. If one of them did, then it was not added to $F$ during step 6 because $f$ imposed a constraint on the complementary slackness condition, i.e. $f \in D$ and $y_f > 0$. However, we can only have $y_f > 0$ if none of the sibling edges of $e$ and $e$ itself were tight when $y_f$ was considered to be increased in step 2. This means at least one of the children edges of $e$ must be tight and if neither $f$ nor $e$ nor any of the sibling edges of $e$ are in $F$, then this child edge of $e$ must be in $F$, a contradiction to our assumption.

Hence none of the sibling edges and children edges of $e$ and $e$ itself are tight. Therefore, the only reason $y_e$ was not increased any further must be that $f$ was tight already after $e$ was considered. Consequently, none of the sister edges of $f$ which are in $D$ nor the parent of $f$ (if it has one) can have a positive $y$-value. This finally contradicts our assumption, since then we should have picked $f$ for the primal solution in step 6 (Case 2).

We now consider the cases that $e$ is incident with a leaf or with the root. If $e$ is incident with a leaf, then certainly one of its sister edges or $e$ itself must be tight. Hence for the parent edge $f$ of $e$ either $f \notin D$ or $y_f = 0$, hence at least one of the tight children edges of $f$ must be in $F$ and hence we again have a contradiction.

Finally, when $e$ is incident with the root and neither $e$ nor any of the other edges incident with the root are in $F$, then it must be that all edges incident with the root are not tight. But then, if $e$ is not incident with a leaf at the same time, at least one of $e$'s children must be tight and should be added to $F$ during the algorithm.

Noting that each edge of the tree is considered at most three times in step 2 and at most twice in step 6, we conclude that the algorithm runs in $O(|E|) = O(|V|)$ time.                                                                                    □

The ideas from Section 3.2 for the $b$-EDS problem on unweighted trees also lead to a recursive linear time algorithm for the weighted $\{0,1\}$-EDS problem on trees. It is conceptually a bit easier than Algorithm 1, but the proof for the optimality of the solution involves some case analysis as well. Moreover, it does not explicitly provide an optimal dual solution and as with Algorithm 1, it cannot be generalized to solve the general weighted $b$-EDS problem on trees. However, we do conjecture the following.

*Conjecture 1.* The weighted $b$-EDS problem on trees can be solved optimally in linear time.

# References

1. Robert Carr, Toshihiro Fujito, Goran Konjevod, and Ojas Parekh. A 2 1/10-approximation algorithm for a generalization of the weighted edge-dominating set problem. *J. Comb. Optim.*, 5:317–326, 2001.
2. Gerd Fricke and Renu Laskar. Strong matchings on trees. In *Proceedings of the Twenty-third Southeastern International Conference on Combinatorics, Graph Theory, and Computing (Boca Raton, FL, 1992)*, volume 89, pages 239–243, 1992.
3. Toshihiro Fujito. On approximability of the independent/connected edge dominating set problems. *Inform. Process. Lett.*, 79(6):261–266, 2001.
4. Toshihiro Fujito and Hiroshi Nagamochi. A 2-approximation algorithm for the minimum weight edge dominating set problem. *Discrete Appl. Math.*, 118(3): 199–207, 2002.
5. Alain Ghouila-Houri. Caractérisation des matrices totalement unimodulaires. *C. R. Acad. Sci. Paris*, 254:1192–1194, 1962.
6. Frank Harary. *Graph Theory*. Addison-Wesley, Reading, MA, 1969.
7. Alan Jerome Hoffman and Joseph Bernard Kruskal. Integral boundary points of convex polyhedra. In *Linear inequalities and related systems*, Annals of Mathematics Studies, no. 38, pages 223–246. Princeton University Press, Princeton, N. J., 1956.
8. Joseph D. Horton and Kyriakos Kilakos. Minimum edge dominating sets. *SIAM J. Discrete Math.*, 6(3):375–387, 1993.
9. Shiow Fen Hwang and Gerard J. Chang. The edge domination problem. *Discuss. Math. Graph Theory*, 15(1):51–57, 1995.
10. Sandra L. Mitchell and Stephen T. Hedetniemi. Edge domination in trees. In *Proc. of the 8th Southeastern Conference on Combinatorics, Graph Theory and Computing (Louisiana State Univ., Baton Rouge, La., 1977)*, pages 489–509, 1977.
11. Christos H. Papadimitriou and Mihalis Yannakakis. Optimization, approximation, and complexity classes. *J. Comput. System Sci.*, 43(3):425–440, 1991.
12. Ojas Parekh. Edge dominating and hypomatchable sets. In *Proceedings of the 13th Annual ACM-SIAM Symposium On Discrete Mathematics (SODA-02)*, pages 287–291, New York, 2002. ACM Press.
13. Ojas Parekh. *Polyhedral techniques for graphic covering problems*. PhD thesis, Carnegie Mellon University, 2002.
14. Robert Preis. Linear time $\frac{1}{2}$-approximation algorithm for maximum weighted matching in general graphs. In *STACS 99 (Trier)*, volume 1563 of *Lect. Notes in Comp. Sci.*, pages 259–269. Springer, Berlin, 1999.

15. Alexander Schrijver. *Combinatorial optimization. Polyhedra and efficiency.*, volume 24 of *Algorithms and Combinatorics*. Springer-Verlag, Berlin, 2003.
16. Anand Srinivasan, K. Madhukar, P. Nagavamsi, C. Pandu Rangan, and Maw-Shang Chang. Edge domination on bipartite permutation graphs and cotriangulated graphs. *Inform. Process. Lett.*, 56(3):165–171, 1995.
17. Vijay Vazirani. *Approximation Algorithms*. Springer-Verlag, 2001.
18. Mihalis Yannakakis and Fanica Gavril. Edge dominating sets in graphs. *SIAM J. Appl. Math.*, 38(3):364–372, 1980.

# On Geometric Dilation and Halving Chords

Adrian Dumitrescu[1], Annette Ebbers-Baumann[2], Ansgar Grüne[2,***],
Rolf Klein[2,†], and Günter Rote[3]

[1] Computer Science, University of Wisconsin-Milwaukee, 3200 N. Cramer Street,
Milwaukee, WI 53211, USA
ad@cs.uwm.edu
[2] Universität Bonn, Institut für Informatik I, D-53117 Bonn, Germany
{ebbers, gruene, rolf.klein}@cs.uni-bonn.de
[3] Freie Universität Berlin, Institut für Informatik,
Takustraße 9, D-14195 Berlin, Germany
rote@inf.fu-berlin.de

**Abstract.** Let $G$ be an embedded planar graph whose edges may be
curves. The *detour* between two points, $p$ and $q$ (on edges or vertices) of
$G$, is the ratio between the shortest path in $G$ between $p$ and $q$ and their
Euclidean distance. The supremum over all pairs of points of all these
ratios is called the *geometric dilation* of $G$. Our research is motivated by
the problem of designing graphs of low dilation. We provide a characterization of closed curves of constant halving distance (i.e., curves for which
all chords dividing the curve length in half are of constant length) which
are useful in this context. We then relate the halving distance of curves
to other geometric quantities such as area and width. Among others,
this enables us to derive a new upper bound on the geometric dilation
of closed curves, as a function of $D/w$, where $D$ and $w$ are the diameter
and width, respectively. We further give lower bounds on the geometric
dilation of polygons with $n$ sides as a function of $n$. Our bounds are tight
for centrally symmetric convex polygons.

## 1   Introduction

Consider a planar graph $G$ embedded in $\mathbb{R}^2$, whose edges are curves that do not
intersect. Such graphs arise naturally in the study of transportation networks,
like waterways, railroads or streets. For two points, $p$ and $q$ (on edges or vertices)
of $G$, the *detour* between $p$ and $q$ in $G$ is defined as $\delta_G(p,q) = \frac{d_G(p,q)}{|pq|}$ where
$d_G(p,q)$ is the shortest path length in $G$ between $p$ and $q$ and $|pq|$ denotes the
Euclidean distance. Good transportation networks should have small detour values. To measure the quality of e.g. a network of streets in a city we have to take
into account not only the vertices of the graph but all the points on its edges,

*** Ansgar Grüne was partially supported by a DAAD PhD grant.
† Rolf Klein was partially supported by DFG grant number KL 655 / 14-1.

F. Dehne, A. López-Ortiz, and J.-R. Sack (Eds.): WADS 2005, LNCS 3608, pp. 244–255, 2005.
© Springer-Verlag Berlin Heidelberg 2005

because access to the streets is possible from everywhere. The resulting supremum value is the *geometric dilation* of $G$. For example the geometric dilation of a square, or that of a square divided into 100 congruent squares is 2.

Ebbers-Baumann et al. [5] recently considered the problem of constructing a graph of lowest possible geometric dilation containing a given finite point set on its edges. They pointed out that even for three points this is a difficult task and proved that there exist point sets which require graphs with dilation at least $\pi/2$. They conjectured that this lower bound is not best possible, which was recently confirmed by Dumitrescu et al. in [4].

Ebbers-Baumann et al. have also shown that for any finite point set there exists a grid-like planar graph that contains the given points and whose geometric dilation is at most 1.678, thereby improving on $\sqrt{3}$, the geometric dilation obtained by embedding the points in a hexagonal grid. Their design uses a certain closed curve of constant halving distance, see Figure 4. Understanding such curves and their properties is a key point in designing networks with a small geometric dilation and is our current focus in this paper. Due to space limitations, some of our proofs are omitted.

## 2     Basic Definitions and Properties

Throughout this paper we consider finite, simple[1], closed curves in the Euclidean plane. We call them *closed curves* or *cycles* for short. For simplicity, we assume that they are piecewise continuously differentiable, but most of the proofs work for less restrictive differentiability conditions.

By $|C|$ we denote the length of a closed curve. Shortest path distance $d_C(p, q)$, detour $\delta_C(p, q)$ and geometric dilation $\delta(C)$ are defined like in the case of arbitrary graphs.

Ebbers-Baumann, Grüne and Klein [6] introduced halving pairs to facilitate the dilation analysis of closed curves. For a given point $p \in C$, the unique *halving partner* $\hat{p}$ of $p$ is given by $d_C(p, \hat{p}) = |C|/2$. This means that both paths connecting $p$ and $\hat{p}$ on $C$ have equal length. The pair $(p, \hat{p})$ is called *halving pair* and the connecting line segment $p\hat{p}$ is a *halving chord*. The length of a halving chord is the corresponding *halving distance*. By $h$ and $H$ we will denote the *minimum* and *maximum halving distance* of a given closed curve.

Furthermore, we will consider the *diameter* $D := \max\{|pq|, p, q \in C\}$ of a closed curve $C$ and the *width* $w$ of a convex cycle $C$ which is the minimum distance of two parallel lines enclosing $C$.

The following lemma is the main reason why halving pairs play a crucial role in the dilation analysis of closed curves.

**Lemma 1.** [6–Lemma 11] *If $C$ is a closed convex curve, its dilation $\delta(C)$ is attained by a halving pair, i.e. $\delta(C) = |C|/2h$.*

---
[1] A curve is called *simple* if it has no self-intersections.

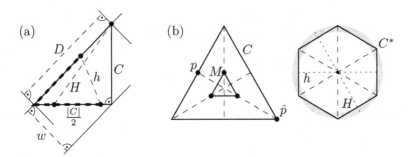

**Fig. 1.** (a) Diameter $D$, width $w$, minimum and maximum halving distance $h$ and $H$ of an isosceles, right-angled triangle (b) An equilateral triangle $C$ and the derived curves $M$ and $C^*$

In [4] Dumitrescu, Grüne and Rote have introduced a decomposition of a cycle $C$ into two curves $C^*$ and $M$, see Figure 1(b) for an illustration. Let $c : [0, |C|) \rightarrow C$ be an arc-length parameterization of $C$. Then, the two curves are defined by the parameterizations

$$m(t) := \frac{1}{2}\left(c(t) + c\left(t + \frac{|C|}{2}\right)\right), \quad c^*(t) := \frac{1}{2}\left(c(t) - c\left(t + \frac{|C|}{2}\right)\right) \quad (1)$$

The *midpoint curve* $M$ is formed by the midpoints of the halving chords. It will turn out to be useful in the analysis of curves of constant halving distance. The curve $C^*$ is the result of applying the *halving pair transformation* (defined in [6]) to $C$. It is obtained by moving the midpoint of every halving chord to the origin.

## 3 Closed Curves of Constant Halving Distance

Closed curves of constant halving distance turn up naturally if one wants to construct graphs of low dilation (compare to [5]). Lemma 1 shows that the dilation of any convex curve of constant halving distance is attained by all its halving pairs. Hence, it is difficult to improve (decrease) the dilation of such cycles, because local changes decrease $h$ or they increase $|C|$.

Theorem 21 in [6] or the proof of Lemma 1 in [4] show that only curves with constant $m(t)$ and constant halving distance can attain the global dilation minimum of $\pi/2$. It is easy to see that only circles satisfy both conditions (compare to [6–Corollary 23], [1–Corollary 3.3], [9], [7]).

What happens if only one of the conditions is satisfied? Clearly, $m(t)$ is constant if and only if $C$ is centrally symmetric. The class of closed curves of constant halving distance is not as easy to describe. One could guess — incorrectly — that it consists only of circles. The "Rounded Triangle" $C_\triangle$ shown in Figure 2 is a counterexample, and could be seen as an analogy to the Reuleaux triangle [2], the most popular representative of curves of constant

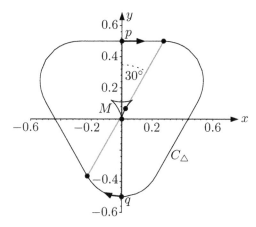

**Fig. 2.** The "Rounded Triangle", a curve of constant halving distance

width. It seems to be a somehow prominent example, because two groups of the authors of this paper discovered it independently.

We construct $C_\triangle$ by starting with a pair of points $p := (0, 0.5)$ and $q := (0, -0.5)$. Next, we move $p$ to the right along a horizontal line. Simultaneously, $q$ moves to the left such that the distance $|pq| = 1$ is preserved and both points move with equal speed. It can be shown that these conditions lead to a differential equation whose solution defines the path of $q$ uniquely. We move $p$ and $q$ like this until the connecting line segment $pq$ forms an angle of $30°$ with the $y$-axis. Next, we swap the roles of $p$ and $q$. Now, $q$ moves along a line with the direction of its last movement, and $p$ moves with equal speed on the unique curve which guarantees $|pq| = 1$, until $pq$ has rotated with another $30°$. In this way we concatenate six straight line and six curved pieces to build the Rounded Triangle $C_\triangle$ depicted in Figure 2.

We have to omit the details of the differential equation and its solution. Here, we mention only that the perimeter of $C_\triangle$ equals $3\ln 3$. By Lemma 1 this results in

$$\delta(C_\triangle) = |C_\triangle|/(2h(C_\triangle)) = \frac{3}{2}\ln 3 \approx 1.6479 .$$

The midpoint curve of $C_\triangle$ is built from six congruent pieces that are arcs of a tractrix, which we will discuss in the end of this section. First, we give a necessary and sufficient condition for curves of constant halving distance.

**Theorem 1.** *Let $C$ be a planar closed curve, and let $c : [0, |C|) \to C$ be an arc-length parameterization. Then, the following two statements are equivalent:*

1. *If $c$ is differentiable in $t$ and in $t + |C|/2$, $\dot{m}(t) \neq 0$, and $\ddot{c}^*(t) \neq 0$, then the halving chord $c(t)c(t + |C|/2)$ is tangent to the midpoint curve at $m(t)$. And if the midpoint stays at $m \in \mathbb{R}^2$ on a whole interval $(t_1, t_2)$, the halving pairs are located on the circle with radius $h(C)/2$ and center point $m$.*
2. *The closed curve $C$ is a cycle of constant halving distance.*

*Proof.* "*2.* ⇒ *1.*"

Let $C$ have constant halving distance. If $c$ is differentiable in $t$ and $t + |C|/2$, $c^*$ and $m$ are differentiable in $t$. And due to $|c^*| \equiv h(C)/2$ it follows that $\dot{c}^*(t)$ must be orthogonal to $c^*(t)$ which can be shown by

$$0 = \frac{\mathrm{d}}{\mathrm{d}t}|c^*(t)|^2 = \frac{\mathrm{d}}{\mathrm{d}t}\langle c^*(t), c^*(t)\rangle = 2\langle c^*(t), \dot{c}^*(t)\rangle . \tag{2}$$

On the other hand, by using the linearity of the scalar product and $|\dot{c}(t)| = 1$, we obtain

$$\langle \dot{m}(t), \dot{c}^*(t)\rangle \stackrel{(1)}{=} \frac{1}{4}\left\langle \dot{c}(t) + \dot{c}\left(t + \frac{|C|}{2}\right), \dot{c}(t) - \dot{c}\left(t + \frac{|C|}{2}\right)\right\rangle \tag{3}$$

$$= \frac{1}{4}\left(|\dot{c}(t)|^2 - \left|\dot{c}\left(t + \frac{|C|}{2}\right)\right|^2\right) = \frac{1}{4}(1 - 1) = 0.$$

The derivative vectors $\dot{m}(t)$ and $\dot{c}^*(t)$ are orthogonal. Hence, $\dot{m}(t) \neq 0 \neq \dot{c}^*(t)$ implies $\dot{m}(t) \parallel c^*(t)$ and the first condition of 1. is proven. The second condition follows trivially from $c(t) = m(t) + c^*(t)$.

"*1.* ⇒ *2.*"

Let us assume that both conditions of 1. hold. We have to show that $|c^*(t)|$ is constant.

First, we consider an interval $(t_1, t_2) \subseteq [0, |C|)$, where $m(t)$ is constant $(= m)$ and the halving pairs are located on a circle with radius $h(C)/2$ and center $m$. This immediately implies that $|c^*|$ is constant on $(t_1, t_2)$.

If $(t_1, t_2) \subseteq [0, |C|)$ denotes an interval where $|c^*(t)| = 0$, then obviously $|c^*|$ is constant.

Now, let $(t_1, t_2) \subseteq [0, |C|)$ be an open interval where $c(t)$ and $c\left(t + \frac{|C|}{2}\right)$ are differentiable and $\dot{m}(t) \neq 0$ and $\dot{c}^*(t) \neq 0$ for every $t \in (t_1, t_2)$. We follow the proof of "*2.* ⇒ *1.*" in the opposite direction. Equation (3) shows that $\dot{c}^*(t) \perp \dot{m}(t)$ and the first condition of 1. gives $c^*(t) \parallel \dot{m}(t)$. Combining both statements results in $\dot{c}^*(t) \perp c^*(t)$ which by (2) yields that $|c^*(t)|$ is constant.

The range $[0, |C|/2)$ can be divided into countably many disjoint intervals $[t_i, t_{i+1})$ where $m$ and $c^*$ are differentiable on the open interval $(t_i, t_{i+1})$, and one of the three conditions $\dot{m}(t) = 0$, $\dot{c}^*(t) = 0$ or $\dot{m}(t) \neq 0 \neq \dot{c}^*(t)$ holds for the whole interval $(t_i, t_{i+1})$. We have shown that $|c^*|$ must be constant on all these open intervals. Thus, due to $c^*$ being continuous on $[0, |C|/2)$, $|c^*|$ must be globally constant.  □

The theorem shows that curves of constant halving distance can consist of three types of parts; parts where the halving chords lie tangentially to the midpoint curve, circular arcs of radius $h(C)/2$, and parts where $\dot{c}^*(t) = 0$ and the halving pairs are only moved by the translation due to $m$, i.e., for every $\tau_1$ and $\tau_2$ within such a part we have $c(\tau_2) - c(\tau_1) = c(\tau_2 + |C|/2) - c(\tau_1 + |C|/2) = m(\tau_2) - m(\tau_1)$. For convex cycles of constant halving distance, the translation parts cannot occur:

**Lemma 2.** *Let $C$ be a closed convex curve of constant halving distance. Then there exists no non-empty interval $(t_1, t_2) \subset [0, |C|)$ such that $c^*$ is constant on $(t_1, t_2)$.*

*Proof.* Assume that $c^*$ is constant on $(t_1, t_2)$ and choose $s_1, s_2$ with $t_1 < s_1 < s_2 < t_2$ and $s_2 < s_1 + |C|/2$. If the four points $p_1 = C(s_1)$, $p_2 = C(s_2)$, $p_3 = C(s_2 + |C|/2)$, $p_4 = C(s_1 + |C|/2)$ don't lie on a line, they form a parallelogram in which $p_1 p_4$ and $p_2 p_3$ are parallel sides. However, these points appear on $C$ in the cyclic order $p_1 p_2 p_4 p_3$, which is different from their convex hull order $p_1 p_2 p_3 p_4$ (or its reverse), a contradiction. The case when the four points lie on a line $\ell$ can be dismissed easily (convexity of $C$ implies that the whole curve $C$ would have to lie on $\ell$, but then $C$ could not be a curve of constant halving distance). □

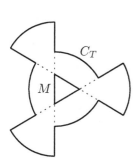

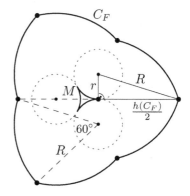

**Fig. 3.** $C_T$ consists of translation parts and circular arcs

**Fig. 4.** The "Flower" from [5] is a non-convex cycle of constant halving distance

Figures 3 and 4 show examples of non-convex cycles of constant halving distance. The first one, $C_T$, demonstrates that such closed curves can indeed include translation parts. The second one, $C_F$, was used in [5] to build a grid of low dilation. It turned out that the non-convex parts were useful for this purpose although the dilation of this "Flower" is[2] $\delta(C_F) = 1.6787\ldots$, which is somewhat larger than the dilation of the Rounded Triangle.

Now we show that the midpoint curve of the Rounded Triangle $C_\triangle$ is built from six tractrix pieces. The tractrix is illustrated in Figure 5. A watch is placed on a table, say at the origin $(0, 0)$ and the end of its watchchain of length 1 is pulled along the horizontal edge of the table starting at $(0, 1)$, either to the left or to the right. As the watch is towed in the direction of the chain, the chain is always tangential to the path of the watch, the tractrix. There are several known

---

[2] Of course, the dilation of the "Flower" depends on the values of $r$ and $R$. The radii chosen in [5] result in the cited dilation value.

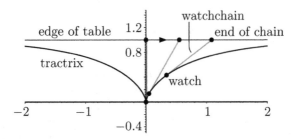

**Fig. 5.** The tractrix, the curve of a watch on a table town with its watchchain (the curve is symmetric about the $y$-axis)

parameterizations. We will use one of them in the end of section 4.1 to calculate the area of $C_\triangle$.

From the definition it is clear that the midpoint curve of the cycle $C_\triangle$ consists of such tractrix pieces, scaled by $1/2$, because by definition and Theorem 1 its halving chords are always tangential to the midpoint curve, always one of the points of these pairs is moving on a straight line, and its distance to the midpoint curve stays $1/2$.

# 4   Relating Halving Distance to Other Geometric Quantities

One of the most important topics in convex geometry is the relation between different geometric quantities of convex bodies like area $A$ and diameter $D$. Scott and Awyong [10] give a short survey of basic inequalities in $\mathbb{R}^2$. For example, it is known that $4A \leq \pi D^2$, and equality is attained only by circles, the so-called *extremal set* of this inequality.

In this context the minimum and maximum halving distance $h$ and $H$ give rise to some new interesting questions, namely the relation to other basic quantities. As the inequality $h \leq w$ is immediate from definition, the known upper bounds on $w$ hold for $h$ as well. However, not all of them are tight for $h$. One counter-example ($A \geq w^2/\sqrt{3} \geq h^2/\sqrt{3}$) will be discussed in the following subsection.

## 4.1   Minimum Halving Distance and Area

Here, we consider the relation between the minimum halving distance $h$ and the area $A$ (for convex cycles). Clearly, the area can get arbitrarily big while $h$ stays constant. For instance this is the case for a rectangle of smaller side length $h$ where the bigger side length tends to infinity.

How small the area $A$ can get for a given minimum halving distance $h$? A first answer $A \geq h^2/\sqrt{3}$ is easy to prove, because it is known [11–ex. 6.4, p.221] that $A \geq w^2/\sqrt{3}$, and we combine this with $w \geq h$. This bound is not tight since the equilateral triangle is the only closed curve attaining $A = w^2/\sqrt{3}$ and its

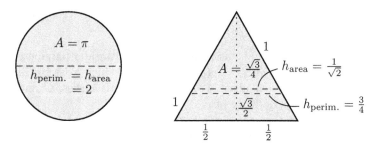

**Fig. 6.** The equilateral triangle has a smaller ratio $A/h^2_{\text{perim.}} = 4/(3\sqrt{3}) \approx 0.770$ and a bigger ratio $A/h^2_{\text{area}} = \sqrt{3}/2 \approx 0.866$ than the circle ($\pi/4 \approx 0.785$)

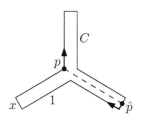

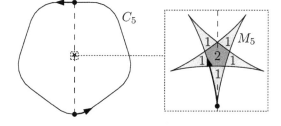

**Fig. 7.** By decreasing $x$ we can make the area of this closed curve arbitrarily small while $h$ stays bounded

**Fig. 8.** The midpoint curve of a rounded pentagon, constructed analogously to $C_\triangle$ of Figure 2, contains regions with winding number 2 and regions with winding number 1

width $w = \sqrt{3}/2 \approx 0.866$ (for side length 1) is strictly bigger than its minimum halving distance $h = 3/4 = 0.75$.

For the analogous problem considering chords bisecting the *area* instead of chords halving the *perimeter*, Santaló conjectured[3] that $A \geq (\pi/4)h^2_{\text{area}}$ (see [3-A26, p.37]). Note that equality is attained by a circular disk. As pointed out earlier, in the case of perimeter halving distance this inequality does not hold: the equilateral triangle gives a counterexample, $A/h^2 = \frac{\sqrt{3}}{4}/\frac{9}{16} \approx 0.770 < 0.785 \approx \frac{\pi}{4}$, see Figure 6. But we do not know if the equilateral triangle is the convex cycle minimizing $A/h^2$. On the other hand $A/h^2$ can become arbitrarily small if we drop the convexity condition, see Figure 7.

Not only the equilateral triangle attains a smaller ratio $A/h^2$ than the circle, so does every curve of constant halving distance.

**Lemma 3.** *If $C$ is a convex cycle of constant halving distance $h$, its area satisfies $A = (\pi/4)h^2 - 2A(M)$ where $A(M)$ denotes the area bounded by the midpoint curve $M$. In $A(M)$, the area of any region encircled several times by $M$ is counted*

---

[3] We would like to thank Salvador Segura Gomis for pointing this out.

with the multiplicity of the corresponding winding number, see Figure 8 for an example. In particular, $A \leq (\pi/4)h^2$.

The idea (proof omitted here) is: assuming $h = 2$, we consider parameterizations $c^*(\alpha) = (\cos\alpha, \sin\alpha)$ and $\dot{m}(\alpha) = v(\alpha)(\cos\alpha, \sin\alpha)$ which exist by Theorem 1 and Lemma 2. Then, we calculate $A = \int x\,dy$ for both curves, $C$ and $M$, and take advantage of the periodicity of $v$.

The theorem shows that the circle is the cycle of constant halving distance attaining maximum area. But which cycle of constant halving distance attains minimum area? We conjecture that the answer is the Rounded Triangle $C_\triangle$. Lemma 3 helps us to calculate its area $A(C_\triangle)$. The tractrix-construction of the midpoint curve $M$ makes it possible to get a closed form for $A(M)$. It results in

$$A(C_\triangle) = \pi\frac{h^2}{4} - 2A(M) = (\pi - 2 \cdot 0.01976\ldots)\frac{h^2}{4} = 0.7755\ldots \cdot h^2.$$

## 4.2   Minimum Halving Distance and Width

In order to achieve a lower bound to $h$ in terms of $w$, we examine the relation of both quantities to the area $A$ and the diameter $D$. The following inequality was first proved by Kubota [8] in 1923 and is listed in [10].

**Lemma 4 (Kubota [8]).** *If $C$ is a convex curve, then $A \geq Dw/2$.*

We will combine this known inequality with the following new result.

**Lemma 5.** *If $C$ is a convex curve, then $A \leq hD$.*

*Proof.* Without loss of generality we assume that a halving chord $pq$ of minimum length $h$ lies on the $y$-axis, $p$ on top and $q$ at the bottom (see Figure 9). Let $C_-$ be the part of $C$ with negative $x$-coordinate and let $C_+ := C \setminus C_-$ be the remainder. We have $|C_-| = |C_+| = |C|/2$ because $pq$ is a halving chord.

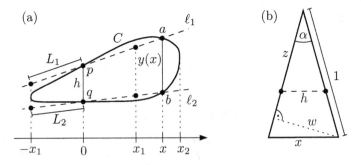

**Fig. 9.** (a) Proving by contradiction that $y(x) \leq h$ for every $x$ in $[x_1, x_2]$. (b) In a thin isosceles triangle $h/w \searrow 1/2$ if $\alpha \to 0$

Let $-x_1$ and $x_2$ denote the minimum and maximum $x$-coordinate of $C$. Note that $x_1$ has a positive value. We assume that $x_2 > x_1$. Otherwise we could mirror the situation at the $y$-axis. Let $y(x)$ be the length of the vertical line segment of $x$-coordinate $x$ inside $C$, for every $x \in [-x_1, x_2]$. These definitions result in $x_1 + x_2 \leq D$ and $A = \int_{-x_1}^{x_2} y(x)\mathrm{d}x$. Furthermore, the convexity of $C$ implies

$$\forall x \in [0, x_1] : \quad y(-x) + y(x) \leq 2h . \tag{4}$$

As a next step, we want to show that

$$\forall x \in [x_1, x_2] : \quad y(x) \leq h . \tag{5}$$

We assume that $y(x) > h$. Let $ab$ be the vertical segment of $x$-coordinate $x$ inside $C$, $a$ on top and $b$ at the bottom. Then, we consider the lines $\ell_1$ through $p$ and $a$ and $\ell_2$ through $q$ and $b$. Let $L_1$ ($L_2$) be the length of the piece of $\ell_1$ ($\ell_2$) in the $x$-interval $[0, x_1]$. By construction the corresponding lengths in the $x$-interval $[-x_1, 0]$ are equal. Then, by the convexity of $C$, we have $|C_-| \leq L_1 + L_2 + h < L_1 + L_2 + y(x) \leq |C_+|$. This contradicts to $pq$ being a halving chord, and the proof of (5) is completed.

Now we can plug everything together and get

$$A = \int_{-x_1}^{x_2} y(x)\mathrm{d}x = \int_{0}^{x_1} y(-x) + y(x)\mathrm{d}x + \int_{x_1}^{x_2} y(x)\mathrm{d}x$$

$$\overset{(4),(5)}{\leq} x_1 \cdot 2h + (x_2 - x_1)h = (x_1 + x_2)h \leq Dh . \qquad \square$$

Finally, we obtain the desired inequality relating $h$ and $w$.

**Lemma 6.** *If $C$ is a convex curve, then $h \geq w/2$. This bound cannot be improved.*

*Proof.* The inequality follows directly from Lemma 4 and Lemma 5. To see that the bound is tight, consider a thin isosceles triangle like that depicted in Figure 9(b) and let $\alpha$ tend to 0. $\qquad \square$

## 5 Dilation Bounds

### 5.1 Upper Bound on Geometric Dilation

Our Lemma 6 leads to a new upper bound depending only on the ratio $D/w$. This complements the lower bound

$$\delta(C) \geq \arcsin\left(\frac{w}{D}\right) + \sqrt{\left(\frac{D}{w}\right)^2 - 1} \tag{6}$$

of Ebbers-Baumann et al.[6–Theorem 22]. The new upper bound is stated in the following theorem.

**Theorem 2.** *If C is a convex curve, then*

$$\delta(C) \le 2\left(\frac{D}{w}\arcsin\left(\frac{w}{D}\right) + \sqrt{\left(\frac{D}{w}\right)^2 - 1}\right).$$

*Proof.* Kubota [8] (see also [10]) showed that

$$|C| \le 2D\arcsin\left(\frac{w}{D}\right) + 2\sqrt{D^2 - w^2}.$$  (7)

Combining this with Lemma 6 and Lemma 1 yields

$$\delta(C) \overset{\text{Lem.1}}{=} \frac{|C|}{2h} \overset{\text{Lem.6}}{\le} \frac{|C|}{w} \overset{(7)}{\le} 2\left(\frac{D}{w}\arcsin\left(\frac{w}{D}\right) + \sqrt{\left(\frac{D}{w}\right)^2 - 1}\right). \qquad \square$$

## 5.2     Lower Bounds on the Geometric Dilation of Polygons

In this subsection we apply the lower bound (6) of Ebbers-Baumann et al.[6] to deduce lower bounds on the dilation of polygons with $n$ sides (in special cases we proceed directly). We start with the case of a triangle (and skip the easy proof):

**Lemma 7.** *For any triangle C, $\delta(C) \ge 2$. This bound cannot be improved.*

Note that plugging the well-known inequality $D/w \ge 2/\sqrt{3}$ into (6) would only give $\delta(C) \ge \pi/3 + 1/\sqrt{3} \approx 1.624$. We continue with the case of centrally symmetric convex polygons, for which we obtain a tight bound.

**Theorem 3.** *If C is a centrally symmetric convex n-gon (n even), then*

$$\delta(C) \ge \frac{n}{2}\tan\frac{\pi}{n}.$$

*This bound cannot be improved.*

*Proof.* We adapt the proof of Theorem 22 in [6], which proves inequality (6) for closed curves. Since $C$ is centrally symmetric, it must contain a circle of radius $r = h/2$. It can easily be shown (using the convexity of the tangent function) that the shortest $n$-gon containing such a circle is a regular $n$-gon. Its length equals $2rn\tan\pi/n$ which further implies that

$$\delta(C) \overset{\text{Lemma 1}}{=} \frac{|C|}{2h} \ge \frac{2rn\tan\frac{\pi}{n}}{2r} = \frac{n}{2}\tan\frac{\pi}{n}.$$  (8)

The bound is tight for a regular $n$-gon.  $\square$

In the last part of this section we address the case of arbitrary (not necessarily convex) polygons. Let $C$ be a polygon with $n$ vertices, and let $C' = conv(C)$. Clearly $C'$ has at most $n$ vertices. By Lemma 9 in [6], $\delta(C) \ge \delta(C')$. Further on, consider $C'' = \frac{C'+(-C')}{2}$, the convex curve obtained by *central symmetrization* from $C'$ (see [11,6]). It is easy to check that $C''$ is a convex polygon, whose number of vertices is at most twice that of $C'$, therefore at most $2n$. One can now replace $n$ by $2n$ in (8) to obtain a lower bound on the geometric dilation for any polygon with $n$ sides.

**Corollary 1.** *The geometric dilation of any polygon $C$ with $n$ sides satisfies*

$$\delta(C) \geq n \tan \frac{\pi}{2n}.$$

# References

1. A. Abrams, J. Cantarella, J. Fu, M. Ghomi, and R. Howard. Circles minimize most knot energies. *Topology*, 42(2):381–394, 2002.
2. G.D. Chakerian and H. Groemer. Convex bodies of constant width. *Convexity and its Applications*, pages 49–96, 1983.
3. H.P. Croft, K. J. Falconer, and R.K. Guy. *Unsolved Problems in Geometry*. Springer-Verlag, 1991.
4. A. Dumitrescu, A. Grüne, and G. Rote. On the geometric dilation of curves and point sets. preprint, July 2004.
5. A. Ebbers-Baumann, A. Grüne, and R. Klein. On the geometric dilation of finite point sets. In *14th Annual International Symposium on Algorithms and Computation*, volume 2906 of *LNCS*, pages 250–259. Springer, 2003.
6. A. Ebbers-Baumann, A. Grüne, and R. Klein. Geometric dilation of closed planar curves: New lower bounds. *to appear in special issue of CGTA dedicated to Euro-CG '04*, 2004.
7. M. Gromov, J. Lafontaine, and P. Pansu. Structures métriques pour les variétés riemanniennes. volume 1 of *Textes math.* CEDIC / Fernand Nathan, Paris, 1981.
8. T. Kubota. Einige Ungleichheitsbeziehungen über Eilinien und Eiflächen. *Sci. Rep. Tôhoku Univ.*, 12:45–65, 1923.
9. R.B. Kusner and J.M. Sullivan. On distortion and thickness of knots. In S. G. Whittington, D. W. Sumners, and T. Lodge, editors, *Topology and Geometry in Polymer Science*, volume 103 of *IMA Volumes in Math. and its Applications*, pages 67–78. Springer, 1998.
10. P.R. Scott and P.W. Awyong. Inequalities for convex sets. *Journal of Inequalities in Pure and Applied Mathematics*, 1(1), 2000. http://jipam.vu.edu.au/article.php?sid=99.
11. I.M. Yaglom and V.G. Boltyanski. *Convex Figures*. English Translation, Holt, Rinehart and Winston, New York, NY, 1961.

# Orthogonal Subdivisions with Low Stabbing Numbers

Csaba D. Tóth

MIT, Cambridge, MA 02139, USA
toth@math.mit.edu

**Abstract.** It is shown that for any orthogonal subdivision of size $n$ in a $d$-dimensional Euclidean space, $d \in \mathbb{N}$, $d \geq 2$, there is an axis-parallel line that stabs at least $\Omega(\log^{1/(d-1)} n)$ boxes. For any integer $k$, $1 \leq k < d$, there is also an axis-aligned $k$-flat that stabs at least $\Omega(\log^{1/\lfloor (d-1)/k \rfloor} n)$ boxes of the subdivision. These bounds cannot be improved.

## 1 Introduction

We say that a line (or a $k$-flat[1]) *stabs* an object $b$ in the $d$-dimensional Euclidean space $\mathbb{R}^d$ if it intersects the relative interior of $b$ but does not contain $b$. The *stabbing number* ($k$-stabbing number) of a family of objects in $\mathbb{R}^d$ is the maximal number of objects stabbed by any line ($k$-flat). Structures such as spanning trees, triangulations, $R$-trees, among others, of low stabbing number have versatile applications in computational geometry. They were used for devising efficient algorithms for simplex range searching [11], ray shooting [1], motion planning and collision detection [1, 17], dynamic point location [7]. They are also related to geometric tomography [9] and the shadow problem [4].

Geometric data structures are often represented as subdivisions of the space (e.g., triangulations, Kd-trees, BSPs, etc.). Instead of studying specific subdivisions with low stabbing numbers obtained for some geometric input, we focus on a more fundamental, combinatorial problem: What is the minimum stabbing number of a subdivision of size $n$? Chazelle, Edelsbrunner, and Guibas [10] considered convex subdivisions. They showed that for any subdivision of the plane into $n$ convex cells, there is a line that stabs $\Omega(\log n / \log \log n)$ cells; and for any $n \in \mathbb{N}$, there is a subdivision of $n$ convex cells where every line stabs $O(\log n / \log \log n)$ cells. No nontrivial bound is known for higher dimensional convex subdivisions.

In this paper, we obtain tight bounds on the axis-aligned stabbing numbers of orthogonal subdivisions in $d$-dimensional Euclidean spaces, for any fixed $d \in \mathbb{N}$. For a subdivision $B$ of $\mathbb{R}^d$ into axis-aligned boxes, we denote by $s_{d,k}(B)$ the *axis-aligned $k$-stabbing number* of $B$, which is the maximum number of boxes stabbed by an *axis-aligned $k$-flat*. For an integer $n \in \mathbb{N}$, we let

---

[1] A $k$-flat is a $k$-dimensional affine subspace in $\mathbb{R}^d$.

F. Dehne, A. López-Ortiz, and J.-R. Sack (Eds.): WADS 2005, LNCS 3608, pp. 256–268, 2005.

$s_{d,k}(n) = \min\{s_{d,k}(B) : |B| = n, B$ is an axis-aligned subdivision of $\mathbb{R}^d\}$.

We summarize our results as follows. We start out Section 2 with a tight bound $s_{2,1}(n) = \Theta(\log n)$ that follows from the Erdős-Szekeres Theorem and a result by de Berg and van Kreveld [8]. We then generalize the proof technique to arbitrary dimensions:

**Theorem 1.** *For any fixed $d, k \in \mathbb{N}$, $d/2 \le k < d$, we have $s_{d,k}(n) = \Theta(\log n)$.*

In Section 3, we pursue the axis-parallel 1-stabbing number of higher dimensional orthogonal subdivisions. We develop nontrivial techniques to prove our upper and lower bounds. Our result here is the following.

**Theorem 2.** *For any fixed $d \in \mathbb{N}$, $d \ge 2$, we have $s_{d,1}(n) = \Theta(\log^{\frac{1}{d-1}} n)$.*

In other words, if any axis-parallel line stabs at most $i$ boxes of a $d$-dimensional orthogonal subdivision $B$, then $|B| = O(2^{i^{d-1}})$, and this bound is tight in the worst case. Theorem 1 indicates that we cannot prove Theorem 2 by simple induction on the dimension: By Theorem 1, there is an orthogonal subdivision of size $n$ in $\mathbb{R}^3$ such that any axis-aligned plane intersects only $O(\log n)$ boxes, but Theorem 2 says that there is an axis-parallel line stabbing $\Omega(\sqrt{\log n})$ boxes. Simple orthogonal projection is not a promising proof technique, either: The projection of a $d$-dimensional axis-aligned subdivision along a coordinate axis is not a subdivision in $\mathbb{R}^{d-1}$.

We generalize our techniques for the axis-aligned $k$-stabbing number for any $k \in \mathbb{N}$, $1 \le k < d/2$. Combined with Theorem 1, our bounds can be formulated for any $k$ and $d$ in the following theorem.

**Theorem 3.** *For any fixed $d, k \in \mathbb{N}$, $1 \le k < d$, we have*

$$s_{d,k}(n) = \Theta\left(\log^{\lfloor \frac{d-1}{k} \rfloor^{-1}} n\right).$$

## 1.1    Related Work

In a pivotal result, Chazelle and Welzl [11, 23] proved that for any $n$ points in $\mathbb{R}^d$, for any fixed $d \in \mathbb{N}$, there is a spanning tree such that any hyperplane crosses $O(n^{1-\frac{1}{d}})$ edges, and this bound is best possible. Matoušek [20] proved that for any $r$, $1 \le r^d \le n$, a set of $n$ points in $\mathbb{R}^d$ can be partitioned into $r^d$ subsets of size $\Theta(\frac{n}{r^d})$ such that every hyperplane stabs the convex hull of $O(r^{d-1})$ subsets. If the points are uniformly distributed in a cube, then a subdivision into $r^d$ congruent cubes gives such a partition. The partition in [20], however, does not typically correspond to a subdivision of the space.

Researchers have also considered subdivisions with low stabbing number. Agarwal, Aronov, and Suri [2] proved that $n$ points in $\mathbb{R}^2$ and in $\mathbb{R}^3$ can be triangulated (with Steiner points) such that any line stabs $O(\sqrt{n} \cdot \log n)$ simplices. They also presented point sets in general position where the stabbing number of any triangulation is $\Omega(\sqrt{n})$. The stabbing number of a *Delaunay* triangulation of $n$ points in $d$-space, however, may be $\Theta(n^{\lceil d/2 \rceil})$ in the worst case [22].

Hershberger and Suri [18] studied a stabbing number restricted to polygons: They showed that every simple polygon with $n$ vertices has a triangulation (with Steiner points) of size $O(n)$ such that any segment lying entirely in the polygon stabs $O(\log n)$ triangles. De Berg and van Kreveld [8] proved that every simple rectilinear polygon with $n$ vertices can be subdivided into $O(n)$ rectangles such that any axis-parallel line segment lying in the polygon stabs $O(\log n)$ rectangles. The restrictions on the stabbing segment are necessary: A line may cross $\Theta(n)$ triangles in any triangulation of a comb polygon with $n$ vertices, and a line segment with slope 1 may intersect $\Theta(n)$ tiles in any rectangular subdivision of a staircase polygon with $n$ vertices.

Instead of the stabbing number of triangulations in $\mathbb{R}^d$, one can consider the stabbing number of lower dimensional but non-crossing simplices on a fixed vertex set. Pach [21] asked what is the $k$-stabbing number of $e$ distinct $r$-simplices with disjoint relative interiors and with $n$ distinct vertices in $\mathbb{R}^d$, for $1 \leq r < k \leq d$. Some initial bounds were obtained by Dey and Pach [12].

Orthogonal box configurations with low axis-aligned stabbing number were studied in various contexts. Agarwal et al. [3] constructed an axis-aligned rectangle hierarchy with stabbing number $O(\log^2 n)$ for an input of $n$ disjoint rectangles in the plane. Gaur et al. [16] gave a $d$-approximation algorithm for computing the minimum number of axis-aligned hyperplanes that stab a set of input rectangles in $\mathbb{R}^d$. De Berg et al. [7] studied axis-aligned $R$-trees with low stabbing number for moving objects in $\mathbb{R}^d$.

Our bounds for the stabbing number depend on the size of the subdivisions only. We do not attempt to compute or approximate the stabbing number of specific orthogonal subdivisions. Fekete, Lübbecke, and Meijer [15] have recently proved that it is NP-complete to compute the (axis-parallel) stabbing number of matchings or triangulations for an input set of points in the plane. Aronov and Fortune [6] and Aronov et al. [5] consider the average stabbing number of triangulations of polygonal scenes instead of the worst case stabbing number.

## 1.2    Definitions

Every logarithm in this paper has base 2, and we define the exponential function as $\exp(x) = 2^x$. For two integers, $n_1, n_2 \in \mathbb{N}$, $n_1 \leq n_2$, we denote by $[n_1, n_2] = \{i \in \mathbb{N} : n_1 \leq i \leq n_2\}$ the set of integers from $n_1$ to $n_2$. For two reals $a, b \in \mathbb{R}$, $I = (a, c)$ denotes the open interval between $a$ and $c$, and $\overline{I}$ is the closure of an interval $I$. An *axis-aligned box* (for short, *box*) in $\mathbb{R}^d$ is the cross product of $d$ (possibly infinite) closed intervals: $b = \prod_{i=1}^{d} \overline{(a_i, c_i)}$, where $a_i \in \{-\infty\} \cup \mathbb{R}$ and $c_i \in \mathbb{R} \cup \{\infty\}$ for $i \in [1, d]$. The interval $\overline{(a_i, c_i)}$, $i \in [1, d]$, is the $i$-th *extent* of $b$.

A subdivision $B$ in $\mathbb{R}^d$ is a set of interior disjoint boxes whose union is $\mathbb{R}^d$. We denote by $|B|$ the size of $B$. For a subdivision $B$ and a box $b$, let $B|b := \{b \cap b' : b' \in B, \operatorname{int}(b \cap b') \neq \emptyset\}$ denote the restriction of $B$ to the box $b$. For any $k \in [1, d-1]$, we say that a $k$-flat $f$ *stabs* a $d$-dimensional box $b$ if $f$ intersects the interior of $b$. In our upper bound constructions, we define subdivisions restricted to a cube—of course, all such subdivisions can be extended to fill the space with the same number of boxes and the same axis-aligned stabbing numbers.

# 2    $k$-Stabbing Number for $d/2 \leq k < d$

In this section we show that $s_{2,1}(n) = \Theta(\log n)$ by an easy combination of known results, then we generalize the proof to obtain Theorem 1.

## 2.1    Stabbing Number of Planar Subdivisions

De Berg and van Kreveld [8] proved that a staircase polygon with $n$ vertices can be subdivided into at least $n/2$ rectangles so that the stabbing number of the subdivision is $O(\log n)$, and the stabbing number of every orthogonal subdivision of the staircase polygon with $n$ vertices is $\Omega(\log n)$.

For $s_{2,1}(n) = \Omega(\log n)$, we consider an orthogonal subdivision of the plane into $n$ boxes. Choose a reference point on the boundary of each box. If there are $\log n$ collinear reference points, then $s_{2,1}(B) \geq \frac{1}{2} \log n$, so we may assume that there are $n/\log n$ reference points with distinct $x$- and $y$-coordinates. We apply the Erdős-Szekeres Theorem [14], which says that every set of $N$ points in the plane contains a sequence of $\lceil \sqrt{N} \rceil$ points with monotone increasing $x$-coordinates and monotone (increasing or decreasing) $y$-coordinates. It follows that there is a sequence $R$ of $\lceil \sqrt{n/\log n} \rceil$ reference points with *strictly* monotone (increasing or decreasing) $x$- and $y$-coordinates. Let $\gamma$ be a rectilinear curve with $2\lceil \sqrt{n/\log n} \rceil - 1$ vertices whose upper-left vertices are the points of $R$ and that partition the plane into two region. Let $F$ be the region below $\gamma$. Observe that every tile of $B|F$ is a rectangle. By [8], there is an axis-parallel line stabbing $\Omega(\log(2\lceil \sqrt{n} \rceil / \log n - 1)) = \Omega(\log n)$ rectangles of $B|F$.

For $s_{2,1}(n) = O(\log n)$, partition the plane into two halfplanes by a rectilinear staircase curve with $n$ vertices. Subdivide either halfplane into at least $n/2$ rectangles with stabbing number $O(\log n)$ by [8]. The union $B$ of the two subdivisions has size at least $n$ and stabbing number $O(\log n)$. For a suitable rectangle $b$, the subset $B|b$ has size exactly $n$ and stabbing number $O(\log n)$.

## 2.2    Proof of Theorem 1

Both the Erdős-Szekeres Theorem and the argument of De Berg and van Kreveld generalize to higher dimensions if $d/2 \leq k$. De Bruijn [19] generalized the Erdős-Szekeres Theorem (his result follows from Dilworth's Theorem [13], as well):

**Theorem 4 (De Bruijn).** *Every $n$ element sequence of $m$-dimensional vectors has a subsequence of size $n^{2^{-m}}$ which is monotone in each coordinate. This bound is tight in the worst case.*

**Corollary 1.** *For any orthogonal subdivision $B$ of size $n$ in $\mathbb{R}^d$, $d \geq 2$, there is a coordinate-wise strictly monotone sequence $R = (r_i : i = 1, 2, \ldots, \lceil n^{2^{1-d}} \rceil)$ of points lying in the interior of distinct boxes of $B$.*

*Proof.* Choose a *reference point* in the interior of every box such that all coordinates are pairwise distinct. Order the reference points according to their first coordinate. By Theorem 4, there is a subsequence of size $\lceil n^{2^{1-d}} \rceil$ which is strictly monotone in each coordinate.                                                                    □

**Lemma 1.** *For any fixed $d, k \in \mathbb{N}$, $d/2 \leq k < d$, we have $s_{d,k} = \Omega(\log n)$.*

*Proof.* Consider a coordinate-wise strictly monotone sequence $R = (r_i : i = 1, 2, \ldots, \lceil n^{2^{1-d}} \rceil)$ of reference points provided by Corollary 1. We choose $m$ to be the largest power of 2 such that $m \leq \lceil n^{2^{1-d}} \rceil < 2m$. Let $b(i, j)$ denote the box spanned by $r_j$ and $r_{j+2^i-1}$ for every $i \in [0, \log m]$ and $j \in [1, m - 2^i + 1]$. (See Fig. 1, left.) We choose two axis-aligned $k$-flats $X$ and $Y$ such that they span the entire space $\mathbb{R}^d$. Finally, let $x(i, j)$ (resp., $y(i, j)$) denote the maximal number of boxes of $B|b(i, j)$ stabbed by a $k$-flat which is parallel to $X$ (resp., $Y$) and passes through a reference point of $R \cap b(i, j)$.

We show by induction that $i \leq x(i, j) + y(i, j)$ for every $i$ and $j$. This immediately implies that $s_{d,k}(B) \geq \max(x(\log m, 1), y(\log m, 1)) \geq \frac{1}{2}(x(\log m, 1) + y(\log m, 1)) \geq \frac{1}{2} \log m = \Omega(\log n)$.

We proceed by induction on $i$. Since $b(0, j)$ is a single point in the interior of one box of $B$, we have $x(0, j) = y(0, j) = 1$ for every $j \in [1, m]$. Suppose that $i' \leq x(i', j) + y(i', j)$ holds for every $i' \leq i$. Consider a box $b(i+1, j)$ for some $j \in [1, m-2^{i+1}+1]$. It contains the disjoint boxes $b(i, j)$ and $b(i, j+2^i)$. Let $h_1$ and $v_1$ be $k$-flats parallel to $X$ and $Y$ through reference points of $R \cap b(i, j)$ attaining $x(i, j)$ and $y(i, j)$. Similarly, let $h_2$ and $v_2$ be $k$-flats through reference points of $R \cap b(i, j+2^i)$, that attain $x(i, j+2^i)$ and $y(i, j+2^i)$. If $h_1$ and $v_1$ jointly stab at least $i+1$ boxes in $B|b(i+1, j)$, then $i+1 \leq x(i+1, j)+y(i+1, j)$ and the induction step is complete. Otherwise, $h_1$ and $v_1$ jointly stab exactly $x(i, j) + y(i, j)$ boxes in $B|b(i + 1, j)$. In this case, the portion $h_1 \cap (b(i + 1, j) \setminus b(i, j))$ of the $k$-flat $h_1$ lies in a box $b \in B$ which intersects $b(i, j)$. Observe that $b$ is disjoint from $b(i, j + 2^i)$ because any box intersecting both $b(i, j)$ and $b(i, j + 2^i)$ must contain the reference points $r_{j+2^i-1}$ and $r_{j+2^i}$, which are in distinct boxes of $B$. Note also that the intersection of $h_1$ and $v_2$ is a point in $b(i + 1, j)$ but not in $b(i, j)$. Therefore $v_2$ must intersect $b \in B$ in $b(i + 1, j) \setminus b(i, j)$. We conclude that $v_2$ intersects at least $y(i, j + 2^i) + 1$ boxes of $B|b(i + 1, j)$. So $h_2$ and $v_2$ jointly intersects at least $i + 1$ boxes in $B|b(i + 1, j)$.     $\square$

The upper bound $s_{d,k}(n) = O(\log n)$ follows from a trivial construction obtained by repeatedly subdividing a box along one of its diagonals.

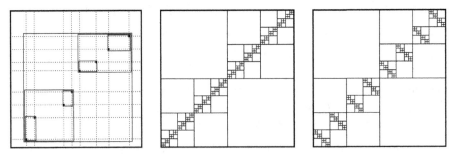

**Fig. 1.** Boxes $b(i, j)$ spanned by reference points (left). Upper bound construction (middle). The size of a monotone sequence of reference points is at most $\sqrt{n}$ (right)

**Lemma 2.** $s_{d,k}(n) = O(\log n)$ *for every* $1 \le k \le d$.

*Proof.* We iteratively construct orthogonal subdivisions $B_i$, $i \in \mathbb{N}$, of a $d$-dimensional cube $C$. Let $B_1 = \{C\}$, and let $e$ be a diagonal of $C$. Once $B_i$ is defined, we can construct $B_{i+1}$: Consider every box $b \in B_i$ whose interior intersects $e$. Choose a point $p_b \in \text{int}(b) \cap e$ and split $b$ into $2^d$ boxes along the $d$ axis-aligned hyperplanes through $p_b$. (Fig. 1, middle.)

By induction, the line $e$ intersects the interior of $2^{i-1}$ boxes of $B_i$. In one iteration, each of these boxes is split into $2^d$ pieces, so $|B_{i+1}| - |B_i| = (2^d - 1)2^{i-1}$. We have, by induction, $|B_i| = (2^d - 1)(2^{i-1} - 1) + 1$. On the other hand, every axis-aligned $k$-flat, $0 \le k \le d - 1$, stabs only $i(2^k - 1) + 1$ boxes of $B_i$.    □

# 3    Stabbing Number in Higher Dimensions

In this section we prove Theorem 2 about the axis-parallel stabbing number of orthogonal subdivisions.

## 3.1    Lower Bound

**Lemma 3.** *For every fixed* $d \in \mathbb{N}$, $2 \le d$, *we have* $s_{d,1} = \Omega(\log^{\frac{1}{d-1}} n)$.

*Proof.* We apply induction on $d$. The base case $d = 2$ is established in Lemma 1. Assume that $s_{d',1} = \Omega(\log^{\frac{1}{d'-1}} n)$ for every $d' \in \mathbb{N}$, $2 \le d' < d$, and consider an orthogonal subdivision $B$ of size $n$ in $\mathbb{R}^d$. Let $R = (r_i : i \in [1, \lceil n^{2^{1-d}} \rceil])$ be a coordinate-wise strictly monotone sequence of reference points chosen from the interior of distinct boxes as in Corollary 1.

We choose $m$ to be the largest power of 2 such that $m \le \lceil n^{2^{1-d}} \rceil < 2m$. For $i \in [0, \log m]$ and $j \in [1, m - 2^i + 1]$, we denote by $b(i,j)$ the box spanned by the reference points $r_j$ and $r_{j+2^i-1}$. For every box $b(i,j)$, we define a function $\beta(i,j)$ as follows:

Fix a box $b(i,j)$. Let $H(i,j) = H$ be a set of lines $h_\ell$ parallel to the $x_1$-axis through the reference points $r_\ell \in R \cap b(i,j)$ for $\ell \in [j, j + 2^i - 1]\}$. Let $V_l$ and $V_r$ be hyperplanes orthogonal to the $x_1$-axis through the reference points $r_j$ and $r_{j+2^i-1}$. We define an equivalence relation on the lines of $H(i,j)$: We say that $h_s, h_t \in H$ are *equivalent* in $b(i,j)$ if and only if the both intersection points $h_s \cap V_1$ and $h_t \cap V_1$ lie in one box of $B$ *and* also both $h_s \cap V_2$ and $h_t \cap V_2$ lie in one box of $B$. This equivalence relation partitions $H$ into $u = u(i,j) \in \mathbb{N}$ equivalence classes $H = H_1 \cup^* H_2 \cup^* \ldots \cup^* H_u$. Clearly, every equivalence class contains elements with consecutive indices: $H_k = \{h_\ell : \ell \in I_k\}$ for every $k \in [1, u(i,j)]$. For every equivalence class $H_k$, we define $y(k)$ to be the average number of boxes of $B|b(i,j)$ stabbed by a line in $H_k$. Now we let

$$\beta(i,j) := \sum_{k=1}^{u(i,j)} \exp\left(y(k) \cdot \log^{\frac{d-2}{d-1}} n\right) = \sum_{k=1}^{u(i,j)} \left(2^{\log^{\frac{d-2}{d-1}} n}\right)^{y(k)}. \tag{1}$$

We show by induction the following claim.

**Claim 1.** At least one of the following two statements holds true:
(a) a hyperplane orthogonal to the $x_1$-axis stabs at least $\exp(\log^{\frac{d-2}{d-1}} n)$ boxes of $B$;
(b) For every $i \in [0, \log m]$ and $j \in [1, m - 2^i + 1]$, we have

$$2^i \leq \beta(i, j). \tag{2}$$

We first show that Claim 1 implies $s_{d,1}(B) = \Omega(\log^{\frac{1}{d-1}} n)$. If a hyperplane stabs at least $\exp(\log^{\frac{d-2}{d-1}} n)$ boxes of $B$, then by induction on the dimension, there is a line in that hyperplane that stabs $\Omega(\log^{\frac{1}{d-2}} \exp(\log^{\frac{d-2}{d-1}} n)) = \Omega(\log^{\frac{1}{d-1}} n)$ boxes of $B$. Otherwise, there are at most $\exp(\log^{\frac{d-2}{d-1}} n)$ equivalence classes in $H(\log m, 1)$ and Eq. (2) holds for $\beta(\log m, 1)$. We have

$$\Theta(n) = 2^{\log m} \leq \sum_{k=1}^{u(i,j)} \exp\left(y(k) \cdot \log^{\frac{d-2}{d-1}} n\right) \leq \exp\left((1 + \max_{k=1}^{u(i,j)} y(k)) \cdot \log^{\frac{d-2}{d-1}} n\right).$$

$$\Theta(\log n) \leq \left(1 + \max_{k=1}^{u(i,j)} y(k)\right) \log^{\frac{d-2}{d-1}} n. \tag{3}$$

Therefore, $\Theta(\log^{\frac{1}{d-1}} n) \leq \max_{k=1}^{u(i,j)} y(k)$, and so there is an equivalence class $H_k$ where $\Theta(\log^{\frac{1}{d-1}} n) \leq y(k)$. By definition of $y(k)$, there is a line parallel to the $x_1$-axis stabbing $\Theta(\log^{\frac{1}{d-1}} n)$ boxes of $B$.

If Claim 1 (a) does not hold, we prove Claim 1 (b) by induction on $i$. In the base case that $i = 0$, we have $u(0, j) = 1$ equivalence class and the line $h_1$ stabs a unique box of $b(0, j)$. So $y(1) = 1$ and Eq. 2 trivially holds. Suppose that it holds for every $i'$, $0 \leq i' \leq i$. Consider a box $b(i+1, j)$ for some $j \in [1, m - 2^{i+1} + 1]$. It contains the disjoint boxes $b(i, j)$ and $b(i, j + 2^i)$. If the set of equivalence classes of $H(i+1, j)$ is the union the equivalence classes of $H(i, j)$ and $H(i, j + 2^i)$ then Eq. (2) holds because the values $y(k)$ cannot decrease in any class.

We now consider the case where some of the classes of $H(i+1, j)$ are different from the classes of $H(i, j)$ and $H(i, j + 2^i)$ (e.g., several classes of $H(i, j)$ and $H(i, j + 2^i)$ are merged into one class of $H(i + 1, j)$). Since we assumed that Claim 1 (a) does not hold, there are at most $\exp(\log^{\frac{d-2}{d-1}} n)$ classes in each of $H(i, j)$ and $H(i, j + 2^i)$. Let $F \subset H(i + 1, j)$ be a minimal set of lines such that $F$ is the union of equivalence classes of $H(i + 1, j)$, and it is also the union of equivalence classes of $H(i, j)$ and $H(i, j + 2^i)$.

**Claim 2.** If $F$ is not a single equivalence class of both $H(i + 1, j)$ and $H(i, j) \cup H(i, j+2^i)$, then every line of $F$ stabs more boxes in $B|b(i+1, j)$ than in $B|b(i, j) \cup B|b(i, j + 2^i)$.

Claim 2 immediately establishes the induction step and proves Claim 1 because, by our assumption, at most $\exp(\log^{\frac{d-2}{d-1}} n)$ classes can merge into one class.

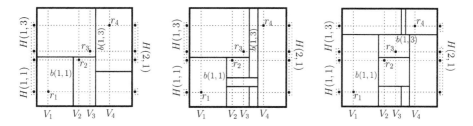

**Fig. 2.** The equivalence classes of $H(1,1)$ and $H(1,3)$ on the left side of the bounding box, the equivalence classes of $H(2,1)$ on the right for the cases *(i)*, *(ii)*, and *(iii)*.

We prove Claim 2 by elementary geometry. *Case (i)*: $F$ is a single equivalence class in $H(i,j)$ or $H(i,j+2^i)$, which is split into several classes in $H(i+1,j)$. Suppose w.l.o.g. that $F$ is a class of $H(i,j)$. None of the boxes $b \in B$ containing the intersection points $\{V_{j+2^i+1-1} \cap h_\ell : h_\ell \in F\}$, intersect $b(i,j)$, because all points $\{V_{j+2^i-1} \cap h_\ell : h_\ell \in F\}$ are in one box of $B$. So every such $h_\ell$ stabs one more box in $b(i+1,j)$ then in $b(i,j)$.

*Case (ii)*: $F$ contains two consecutive classes $H_a$ and $H_c$ of $H(i,j)$. (We can argue analogously if both $H_a$ and $H_c$ are in $H(i,j+2^i)$.) By the minimality of $F$, a class of $H(i+1,j)$ contains lines from both $H_a$ and $H_c$. That is, a box $b \in B$ contains $V_{j+2^{i+1}-1} \cap h_a$ and $V_{j+2^{i+1}-1} \cap h_c$ for some $h_a \in H_a$ and $h_c \in H_c$. Therefore none of the boxes of $B$ containing the intersection points $\{V_j \cap h_\ell : h_\ell \in H_a \cup H_c\}$ intersect $b(i,j)$. So every line in $H_a \cup H_c$ stabs one more box in $b(i+1,j)$ than in $b(i,j)$.

*Case (iii)*: $F$ contains two consecutive classes $H_a \subseteq H(i,j)$ and $H_c \subseteq H(i,j+2^i)$. By the minimality of $F$, a class of $H(i+1,j)$ contains lines $h_a \in H_a$ and $h_c \in H_c$. There are boxes $b_1, b_2 \in B$ such that $\{V_j \cap h_a, V_j \cap h_c\} \subset b_1$ and $\{V_{j+2^{i+1}-1} \cap h_a, V_{j+2^{i+1}-1} \cap h_c\} \subset b_2$. Notice that $V_j \cap h_a \in b(i,j)$ and $V_{j+2^{i+1}-1} \cap h_c \in b(i,j+2^i)$, so $b_1 \cap b(i,j) \neq \emptyset$ and $b_2 \cap b(i,j+2^i) \neq \emptyset$. Neither $b_1$ nor $b_2$ can intersect both $b(i,j)$ and $b(i,j+2^i)$, because any box intersecting both would contain the reference points $r_{j+2^i-1}$ and $r_{j+2^i}$, which lie in distinct boxes of $B$. Therefore none of the boxes of $B$ containing the intersection points $\{V_j \cap h_\ell : h_\ell \in H_c\}$ can intersect $b(i,j)$; and none of the boxes of $B$ containing the intersection points $\{V_{j+2^{i+1}-1} \cap h_\ell : h_\ell \in H_a\}$ can intersect $b(i,j+2^i)$. Every line in $H_a \cup H_c$ stabs one more box in $b(i+1,j)$ than in $b(i,j)$ and in $b(i,j+2^i)$. □

## 3.2   Upper Bound Construction

Our Lemma 2 provided a planar construction for $s_{2,1}(n) = O(\log n)$. In higher dimensions, we construct orthogonal subdivisions with low stabbing numbers by induction on the dimension. In every dimension $d$, we define a set of auxiliary subdivisions $B_{d,i}$, $i \in \mathbb{N}$, which are then used to construct subdivision $D_{d,n}$ with $n$ boxes and $s_{d,1}(D_{d,n}) = O(\log^{\frac{1}{d-1}} n)$. A crucial property of the auxiliary subdivisions relies on the concept of *shadows* defined here.

**Definition 1.** *For two axis-aligned boxes* $C$ *and* $U$, $U \subset C$, *we call the* shadow *of* $U$ *the volume between* $U$ *and its orthogonal projection to a side of* $C$. $U$ *may have* $2d$ *shadows (one for each side of* $U$). *A shadow of* $U$ *to a side of* $C$ *is* empty *if this side contains a side of* $U$.

We collect the properties of the two families of subdivisions in two lemmas.

**Lemma 4.** *For every* $d, i \in \mathbb{N}$, $3 \le d$, *a* $d$-*dimensional cube has an orthogonal subdivision* $B_{d,i}$ *of size* $\Omega(2^i)$ *with the following properties:* (i) *a diagonal* $e$ *stabs* $2^i - 1$ *congruent cubes of* $B_{d,i}$, *which we call* diagonal boxes; (ii) *every shadow of every diagonal box is an element of* $B_{d,i}$; (iii) $s_{d,1}(B_{d,i}) = O(i^{\frac{1}{d-2}})$.

**Lemma 5.** *For every* $d, n \in \mathbb{N}$, $3 \le d$, *a* $d$-*dimensional cube has an orthogonal subdivision* $D_{d,n}$ *with the following properties:* (a) *a diagonal* $e$ *stabs* $n$ *congruent cubes of* $D_{d,n}$; (b) $s_{d,1}(D_{d,n}) = O(\exp(\log^{\frac{1}{d-1}} n))$.

We construct both subdivisions by double induction. The induction on the dimension $d$ is interdependent in the two proofs: For every $d \in \mathbb{N}$, $4 \le d$, the construction $B_{d,i}$ assumes that Lemma 5 holds for $d' = d - 1$ and for every $n$. The construction of $D_{d,n}$ assumes that Lemma 4 holds for the same $d \in \mathbb{N}$ and for every $i \in \mathbb{N}$.

*Proof (of Lemma 4).* We apply double induction in $d$ and $i$. For $i = 0$, let $B_{d,0}$ be the $d$-dimensional unit cube. When constructing $B_{d,i}$ for $3 \le d$ and $1 \le i$, we assume that Lemma 4 holds for $d$ and $i - 1$; and Lemma 5 hold for $d - 1$ if $4 \le d$ and for $n = 2^{i-1} - 1$.

We construct $B_{d,i}$ as an orthogonal subdivision of the cube $C_{d,i} = (0, 2^i - 1)^d$. Let $e$ be a diagonal of $C_{d,i}$ and let $U$ be the central unit cube of $C_{d,i}$. The $2d$ hyperplanes along the sides of $U$ partition $C_{d,i}$ into $3^d$ regions. (See Fig. 3, left.) Let $R_0$, and $R_1$ denote the regions adjacent to the two endpoints of the diagonal $e$. Let $S_0$ denote the union of $R_0$ and its $2d$ nonempty shadows to sides of $C_{d,i}$ (we define $S_1$ analogously for $R_1$). Notice that $S_0$ (resp., $S_1$) is the union of $1 + 2d$ regions out of $3^d$. We further subdivide $S_0$ and $S_1$; while the remaining $3^d - 2(2d + 1)$ regions of $C_{d,i}$ (including $U$) are boxes of $B_{d,k}$.

We describe the subdivision of $S_0$ (we can analogously subdivide $S_1$). We place a copy of $B_{d,i-1}$ into $R_0$. $S_0 \setminus R_0$ consists of $d$ nonempty shadows of $R_0$, each having one common face with $R_0$. We describe the subdivision in one of these boxes, $R_{0,1}$, whose common face with $R_0$ is orthogonal to the $x_1$-axis. The subdivisions in the other $d - 1$ components of $S_0 \setminus R_0$ are analogous.

Consider the copy of $B_{d,i-1}$ in $R_0$. Along the diagonal $e$, it contains a set of $2^{i-1} - 1$ unit cubes, which we denote by $\{U_\ell : \ell \in [1, 2^{k-1} - 1]\}$. For each unit cube $U_\ell$, let us denote by $U_{\ell,1}$ the shadow of $U_\ell$ between $U_\ell$ and its $x_1$-projection to the common side of $R_0$ and $R_{0,1}$. (Note that the shadow may have empty interior if $U_\ell$ has a common side with $R_0$.) We extend the $x_1$ extent of $U_{\ell,1} \in B_{d,i-1}$ by the full $x_1$-extent of $R_{0,1}$. (Fig. 3, right.) Let the extended boxes $\{U'_{\ell,1} : \ell \in [1, 2^{k-1} - 1]\}$ be elements of $B_{d,i}$. It remains to describe the subdivision of $R_{0,1} \setminus \left( \bigcup_{\ell=1}^{2^{k-1}-1} U'_\ell \right)$. We subdivide this volume into boxes that

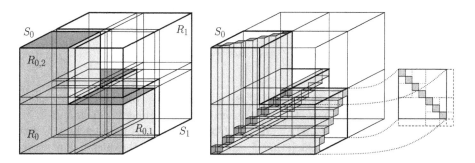

**Fig. 3.** The partition of $C_{3,4}$ into 27 regions (left). The diagonal boxes in $S_0$ and their shadows extended to $R_{0,1}$, $R_{0,2}$, and $R_{0,3}$ for the construction of $B_{3,4}$ (right).

are fully contained in $R_{0,1}$ and whose $x_1$-extent is the same as that of $R_{0,1}$. Since their $x_1$ extent is the same, it is enough to describe their $x_1$-projections.

Consider the $(d-1)$-dimensional orthogonal $x_1$-projection of $R_{0,1}$. The projection of $\{U'_{\ell,1} : \ell \in [1, 2^{i-1}-1]\}$ is a set of $2^{i-1}-1$ unit cubes along the diagonal of $(0, 2^{i-1} - 1)^{d-1}$. If $d = 3$, then the complement of these unit squares can be subdivided into $O(2^i)$ boxes such that any axis-parallel line stabs at most $O(i)$ of them. If $4 \leq d$, then we can apply Lemma 5 in dimension $d-1$ with $n = 2^{i-1}-1$ by induction. The complement of the diagonal unit cubes in $(d-1)$-space can be subdivided into boxes such that any axis-parallel line crosses at most $O(i^{\frac{1}{d-2}})$ of them. This completes the description of $B_{d,i}$.

Finally we check that that $B_{d,i}$ satisfies conditions *(i)–(iii)*. Conditions *(i)* and *(ii)* are clearly satisfied. For *(iii)*, consider an axis-parallel line $f$ stabbing $C_{d,i}$. If $f$ stabs the unit cube $U$, then it stabs exactly 3 boxes of $B_{d,i}$. If $f$ stabs $R_0$, then it stabs $O\left((i-1)^{\frac{1}{d-2}}\right)$ boxes of $B_{d,i-1}$ in $R_0$ by induction on $i$, and at most one more box of $B_{d,i}$ in $S_0 \setminus R_0$. We argue similarly if $f$ stabs $R_1$. Otherwise, $f$ may stab both $S_0 \setminus R_0$ and $S_1 \setminus R_1$. In either region, $f$ stabs $O\left(\log^{\frac{1}{d-2}}(2^{i-1}-1)\right) = O\left((i-1)^{\frac{1}{d-1}}\right)$ boxes by Lemma 5; and $f$ can also stab at most 3 boxes in $C_{d,i} \setminus (S_0 \cup S_1)$. □

*Proof (of Lemma 5).* For a given $n \in \mathbb{N}$ and $d \in \mathbb{N}$, $3 \leq d$, we assume that Lemma 4 holds for $d$ and for every $i$. If $n < 2^{2^{d-1}}$, then we let $D_{d,n}$ be a subconstruction of the subdivision described in Lemma 2 spanned by $n$ congruent cubes along the diagonal. The stabbing number in this case is at most $2\lceil \log n \rceil 2 \cdot 2^{d-1} = O(1)$. For the following iterative construction, we assume $n \geq 2^{2^{d-1}}$.

Let $n'$ be the smallest integer such that $n \leq n'$ and $\log^{\frac{1}{d-1}} n' \in \mathbb{N}$. Let $m = \log^{\frac{1}{d-1}} n'$ and $M = m^{d-2}$. For every $i \in \mathbb{N}$, we define iteratively a subdivision $D(i)$. Let $D(0) = B_{d,M}$. For every $i$, $1 \leq i$, we obtain $D(i)$ from $D(i-1)$ as follows. Consider a construction $B_{d,M}$ and place a scaled copy of $D(i-1)$ in each of the $2^M - 1$ unit boxes along the diagonal of $B_{d,M}$.

Now, we are in the position to define $D_{d,n}$: Consider the subdivision $D(2m)$. The number of congruent cubes along the diagonal of $D(2m)$ is

$$\left( 2^{(\log^{\frac{d-2}{d-1}} n)} - 1 \right)^{2 \log^{\frac{1}{d-1}} n} \geq 2^{\log n} \geq n.$$

Let $D_{d,n}$ be a subdivision spanned by $n$ consecutive congruent cubes along the diagonal of $D(2m)$. Now $D_{d,n}$ has exactly $n$ cubes along its diagonal.

We show that $s_{d,1}(D(2m)) = O(m)$. Consider an axis-parallel line $f$. Let $j \in \mathbb{N}$ be the smallest index such that $f$ stabs a box $U$ along the diagonal in a copy of $B_{d,M}$ (which was placed into a unit cube in the $j$-th iteration). By Lemma 4, $f$ stabs $O(M^{\frac{1}{d-2}}) = O(m)$ boxes inside $U$. Outside $U$, it can stab $2(m-j) = O(m)$ boxes because it may stab two new shadow boxes constructed in each of the last $m - j$ iterations. Finally, $s_{d,1}(D_{d,n}) \leq s_{d,1}(D_{d,n'}) = s_{d,1}(D(2m)) = O(m) = O(\exp(\log^{\frac{1}{d-1}} n')) = O(\exp(\log^{\frac{1}{d-1}} n))$, as required. □

## 4   $k$-Stabbing Number in Higher Dimensions

The proof techniques of the previous section generalize to the case where we count the number of boxes stabbed by axis-aligned $k$-flats. Combined with Theorem 1, we obtain a formula for the axis-aligned $k$-stabbing number of a $d$-dimensional orthogonal subdivision. For any fixed $k, d \in \mathbb{N}$, $1 \leq k < d$, we have
$s_{d,k} = \Theta\left( \log^{\lfloor \frac{d-1}{k} \rfloor^{-1}} n \right).$

For the lower bound, we follow the proof of Lemma 3. We choose an axis-aligned $k$-flat $X$, and let $h_\ell$ be the $k$-flat parallel to $X$ through the reference point $r_\ell$. Similarly, we let $V_\ell$ be the $(d-k)$-flat orthogonal to $h_\ell$ though $r_\ell$.

For the upper bound, we extend the proof of Lemmas 4 and 5. Along with Definition 1 (the one-dimensional shadow of a cube $U$ in a box $C$), we define inductively $j$-shadows for $j \in [1, k]$. For any $j > 1$, the $j$-shadow of $U$ is the 1-shadow of the $(j-1)$-shadow of $U$. We can define the subdivisions $B_{d,i}$ similarly to Lemma 4. We split a $d$-dimensional cube $C_{d,i}$ into $3^d$ regions and place copies of $B_{d,i-1}$ into two regions at opposite corners. Either region then determines the subdivision in their respective $k$-shadows, which are disjoint as long as $k < d/2$.

## 5   Possible Extensions and Open Problems

In this paper, we studied the *axis-aligned* $k$-stabbing number of *orthogonal* subdivisions. One could, as well, consider the $k$-stabbing number with arbitrary $k$-flats, and convex subdivisions instead of orthogonal subdivisions. It is easy to determine the axis-aligned stabbing number of arbitrary convex subdivisions. For any $d \leq \ell$ and $n \in \mathbb{N}$, there are convex subdivisions with $n$ cells in $\mathbb{R}^d$ whose axis-parallel stabbing number is $k$. (E.g., consider a thin cylinder along a diagonal of a cube: partition the cylinder into $d-1$ tiny cells, and fill the complement of the cylinder by $d$ convex cells.)

For orthogonal subdivisions and arbitrary stabbing $k$-flats, all our lower bounds remain valid. It is possible that they are tight for this variant of the problem. For convex subdivisions and arbitrary stabbing $k$-flats, there are better bounds than ours already in the plane: Chazelle, Edelsbrunner, and Guibas [10] proved that the stabbing number of any convex subdivision in the plane is $\Theta(\log n / \log \log n)$. No nontrivial bound is known for the stabbing numbers of convex subdivision in 3- or higher dimensions.

# References

1. P.K. Agarwal, Ray shooting and other applications of spanning trees with low stabbing number, *SIAM J. Comput.* **21** (1992), 540–570.
2. P.K. Agarwal, B. Aronov, and S. Suri, Stabbing triangulations by lines in 3D, in *Proc. 11th Symp. Comput. Geom. (Vancouver, BC)*, ACM, 1995, pp. 267–276.
3. P.K. Agarwal, M. de Berg, J. Gudmundsson, M. Hammar, and H.J. Haverkort, Box-trees and R-trees with near-optimal query time, *Discrete Comput. Geom* **28** (2002), 291–312.
4. N. Amenta and G.M. Ziegler, Shadows and slices of polytopes, in *Advances in Discrete & Computational Geometry*, vol 223 of Contemp. Math., AMS, 1999, pp. 57-90.
5. B. Aronov, H. Brönnimann, A.Y. Chang, and Y-J. Chiang, Cost-driven octree construction schemes: an experimental study, *Comput. Geom. Theory Appl.* **31** (2005), 127–148.
6. B. Aronov and S. Fortune, Approximating minimum-weight triangulations in three dimensions, *Discrete Comput. Geom.* **21** (4) (1999), 527–549.
7. M. de Berg, J. Gudmundsson, M. Hammar, and M. H. Overmars, On R-trees with low stabbing number, *Comput. Geom. Theory Appl.* **24** (2003), 179–195.
8. M. de Berg and M. van Kreveld, Rectilinear decompositions with low stabbing number, *Inform. Process. Lett.* **52** (4) (1994), 215–221.
9. P. Bose, F. Hurtado, H. Meijer, S. Ramaswami, D. Rappaport, V. Sacristan, T.C. Shermer, and G.T. Toussaint, Finding specified sections of arrangements: 2D results. *J. Math. Model. Algorithms* **1**(2002), 3–16.
10. B. Chazelle, H. Edelsbrunner, L.J. Guibas, The complexity of cutting complexes, *Discrete Comput. Geom.* **4** (1989), 139–181.
11. B. Chazelle and E. Welzl, Quasi-optimal range searching in space of finite VC-dimension, *Discrete Comput. Geom.* **4**, (1989), 467–489.
12. T. K. Dey and J. Pach, Extremal problems for geometric hypergraphs, *Discrete Comput. Geom.* **19** (4) (1998), 473–484.
13. R. Dilworth, A decomposition theorem for partially ordered sets, *Ann. of Maths.* **51** (1950), 161–166.
14. P. Erdős and G. Szekeres, A combinatorial theorem in geometry, *Compositio Math.* **2** (1935), 463–470.
15. S.P. Fekete, M.E. Lübbecke, and H. Meijer, Minimizing the stabbing number of matchings, trees, and triangulations, in *Proc. 15th Sympos. on Discrete Algorithms (New Orleans, LA)*, ACM, 2004, pp. 437–446.
16. D. Gaur, T. Ibaraki, and R. Krishnamurti, Constant ratio approximation algorithms for the rectangle stabbing problem and the rectilinear partitioning problem, *J. Algorithms* **43** (2002), 138–152.

17. M. Held, J. T. Klosowski, and J.S.B. Mitchell, Evaluation of collision detection methods for virtual reality fly-throughs, in *Proc. 7th Canadian Conf. Comput. Geom.*, Quebec City, QC, 1995, pp. 205–210.

18. J. Hershberger and S. Suri, A pedestrian approach to ray shooting: shoot a ray, take a walk, *J. Algorithms* **18** (3) (1995), 403–431.

19. J.B. Kruskal, Monotonic subsequences, *Proc. Amer. Math. Soc.* **4** (1953), 264–274.

20. J. Matoušek, Efficient partition trees. *Discrete Comput. Geom.* **8** (1992), 315–334.

21. J. Pach, Notes on geometric graph theory, in *Discrete and computational geometry*, vol. 6 of DIMACS Ser. Discrete Math. Theor. Comp. Sci., AMS, 1991, pp 273–285.

22. J. R. Shewchuk, Stabbing Delaunay tetrahedralizations, manuscript, 2004.

23. E. Welzl, On spanning trees with low crossing numbers, in *Data Structures and Efficient Algorithms*, vol. 594 of LNCS, Springer, Berlin, 1992, pp. 233–249.

# Kinetic and Dynamic Data Structures for Convex Hulls and Upper Envelopes

Giora Alexandron[1], Haim Kaplan[1], and Micha Sharir[1,2]

[1] School of Computer Science, Tel Aviv University, Tel Aviv 69978, Israel
{gioraa, haimk, michas}@post.tau.ac.il
[2] Courant Institute, New York University, New York, NY 10012, USA

**Abstract.** Let $S$ be a set of $n$ moving points in the plane. We present a kinetic and dynamic (randomized) data structure for maintaining the convex hull of $S$. The structure uses $O(n)$ space, and processes an expected number of $O(n^2 \beta_{s+2}(n) \log n)$ critical events, each in $O(\log^2 n)$ expected time, including $O(n)$ insertions, deletions, and changes in the flight plans of the points. Here $s$ is the maximum number of times where any specific triple of points can become collinear, $\beta_s(q) = \lambda_s(q)/q$, and $\lambda_s(q)$ is the maximum length of Davenport-Schinzel sequences of order $s$ on $n$ symbols. Compared with the previous solution of Basch et al. [2], our structure uses simpler certificates, uses roughly the same resources, and is also dynamic.

## 1    Introduction

The *Kinetic Data Structure* (KDS) framework, introduced by Basch et al. [2], proposes an algorithmic approach, together with several quality criteria, for maintaining certain geometric configurations determined by a set of objects, each moving along a semi-algebraic trajectory of constant description complexity (see below for a precise definition); see [5].

A kinetic data structure maintains some configuration (e.g. convex hull, closest pair) of moving objects. It does so by finding a set of *certificates* that, on one hand, ensure the correctness of the configuration currently being maintained, and, on the other hand, are inexpensive to maintain. When the motion starts, we can compute the closest failure time of any of the certificates, and insert these times into a global event queue. When the time of the next event in the queue matches the current time, we invoke the KDS repair mechanism, which fixes the configuration and the failing certificate(s). In doing so, the mechanism will typically delete from the queue failure times that are no longer relevant, and insert new failure times into it.

To analyze the efficiency of a KDS, we distinguish between two types of events: *internal* and *external*. *External events* are events associated with real (combinatorial) changes in the configuration, thus forcing a change in the output. *Internal events*, on the other hand, are events where some certificate fails, but the overall desired configuration still remains valid. These events arise because

F. Dehne, A. López-Ortiz, and J.-R. Sack (Eds.): WADS 2005, LNCS 3608, pp. 269–281, 2005.

of our specific choice of the certificates, and are essentially an overhead incurred by the data structure. If the ratio between the number of internal events to (an upper bound on) the number of external events is no more than polylogarithmic in the number of input objects, the KDS is said to be *efficient*.

Other parameters of the KDS that one would like to minimize are the following. We say that the KDS is *responsive* if the processing time of a critical event by the repair mechanism is polylogarithmic in the number of input objects. The KDS is *local* if each object is associated with a polylogarithmic number of events in the event queue. Locality allows efficient flight plan changes. The KDS is *compact* if it occupies space which is larger than the number of input objects by at most a polylogarithmic factor.

In addition, which is one of the central issues considered in this paper, one might wish to design a KDS that is also *dynamic*, meaning that it can also efficiently support insertions and deletions of objects.

In their paper, Basch et al. [2] developed a KDS that maintains the convex hull of a set of moving points in the plane, which is compact, efficient, local, and responsive. Specifically, their structure processes $O(n^{2+\varepsilon})$ events, for any $\varepsilon > 0$, each in $O(\log^2 n)$ time. (The number of events has been slightly improved in a later work [1], to $O(n\lambda_s(n))$, where $s$ is the number of times any fixed triple of points can become collinear.) To achieve locality, their algorithm uses a fairly complicated set of certificates. Furthermore, Basch et al. have focussed only on kinetization, and did not consider insertions and deletions of points. The motivation for our work has been twofold: (i) to simplify the certificates used by [2], and (ii) to obtain a dynamic algorithm that still meets the four quality criteria mentioned above.

**Our results.** In this paper we present an efficient *dynamic* KDS for maintaining the convex hull of a set of $n$ moving points in the plane. Our certificates are simpler than those of [2], and the performance of our algorithm is comparable with that of [2]. Specifically, write $\beta_q(n) = \lambda_q(n)/n$, where $q$ is any constant, and where $\lambda_q(n)$ is the maximum length of a Devenport-Schinzel sequence of order $q$ on $n$ symbols (see [8]). We show that, for $m \geq n$ insertions and deletions, our structure processes an expected number of $O(mn\beta_{s+2}(n)\log n)$ events, each in $O(\log^2 n)$ expected time, that it has size $O(n)$, and that each line participates in only $O(\log n)$ "certificates" maintained by the structure. In the terminology defined above, our structure is compact, efficient, local, and responsive.

We assume that each moving point $i$ is given as a pair $(a_i(t), b_i(t))$ of semi-algebraic functions of time of *constant description complexity*. That is, each function is defined as a Boolean combination of a constant number of predicates involving polynomials of constant maximum degree. We present our result in the dual plane, where each point is mapped to the moving non-vertical line $y = a_i(t)x + b_i(t)$, and the goal is to maintain the upper and lower envelopes of this set of moving lines. For simplicity, and without loss of generality, we will only consider the maintenance of the upper envelope.

The main idea in our solution is to maintain the lines sorted by slope in a data structure similar to the stationary data structure of Overmars and van

Leeuwen [6]. This is in contrast with the data structure of Basch et al. [2] that keeps the lines in a tree in some arbitrary order.

Because of some technical difficulties in the analysis, which are discussed at the end of Section 3 (these difficulties arise due to lack of tight bounds on the complexity of a single level in planar arrangements), we have to use a *treap* [7] as the underlying tree. Our data structure is therefore randomized, and its performance bounds hold only in expectation.

We present the algorithm in three stages. First, we describe the classical dynamic algorithm of Overmars and van Leeuwen for stationary lines [6], upon which our solution is built. Second, we make this algorithm kinetic, by designing a set of simple certificates and an efficient algorithm for maintaining them as the lines move. Third, we make the algorithm dynamic, by showing how to perform insertions and deletions efficiently, adapting and enhancing the basic technique of [6]. Due to space limitations some proofs are omitted.

## 2    Preliminaries

In this section we introduce our framework and notation, by briefly reviewing the data structure of Overmars and van Leeuwen [6] for dynamically maintaining the upper envelope of a set of lines. We describe this structure here in its original stationary context.

We denote by $S = \{\ell_1, \ldots \ell_n\}$ the set of lines in the data structure, sorted in order of increasing slopes, so that $\ell_k$ is the line with the $k$-th smallest slope. We store the lines at the leaves of a balanced binary search tree $T$ in this order. Slightly abusing the notation, we also use $\ell_k$ to denote the node of $T$ containing $\ell_k$. Later, we take $T$ to be a *treap* (see [7] and below), but for now any kind of balanced search tree will do. Denote the root of $T$ by $r$. For a node $v \in T$, denote the left and right children of $v$ by $\ell(v)$ and $r(v)$, respectively, and denote the parent of $v$ by $p(v)$. Denote the set of lines in the leaves of the subtree of $v$ by $S(v)$.

Each node $v \in T$ stores a sorted list of the lines that appear in the upper envelope $E(v)$ of $S(v)$, in their left-to-right order along the envelope, which is the same as the increasing order of their slopes. To facilitate fast implementation of searching, splitting, and concatenation of upper envelopes, we represent each such sorted list as a balanced search tree. Abusing the notation slightly, we denote by $E(v)$ both the upper envelope of the lines in $S(v)$ and the tree representing it. Overmars and van Leeuwen [6] exploit the simple property that, for two sets of lines $L$ and $R$, such that any line in $L$ has a smaller slope than that of any line in $R$, the upper envelopes of $L$ and $R$ have exactly one common intersection point $q$. The envelope is attained by lines of $L$ to the left of $q$, and by lines of $R$ to the right of $q$.

After sorting the lines of $S$ in the increasing order of their slopes, we build $T$ and the secondary structures $E(v)$, for each $v \in T$, in the following bottom-up recursive manner. For a node $v$, we build $E(v)$ from $E(\ell(v))$ and $E(r(v))$: First we compute the intersection $q(v)$ of $E(\ell(v))$ and $E(r(v))$, by simultaneous binary search over $E(\ell(v))$ and $E(r(v))$, in the manner described in [6]. Then we split $E(\ell(v))$ and $E(r(v))$ at $q(v)$, and concatenate the part of $E(\ell(v))$ that lies to the left of $q(v)$ with the part of $E(r(v))$ that lies to the right of $q(v)$, to obtain $E(v)$.

To save space, Overmars and van Leeuwen [6] store at each node $v$ *only the part of $E(v)$ that does not appear on $E(p(v))$*. One can then reconstruct $E(v)$ on the fly from $E(p(v))$, and from the piece stored at $v$.

The operations of finding $q(v)$, splitting and concatenating $E(\ell(v))$ and $E(r(v))$, take $O(\log n)$ time each. Therefore, we can build the entire structure in $O(n \log n)$ time. The size of the primary tree $T$, including the portions of the envelopes $E(v)$ stored at each node $v$, is $O(n)$ [6].

To support insertions and deletions of lines, each time we traverse an edge of the tree from a node $v$ to one of its children, we construct the envelopes $E(\ell(v))$ and $E(r(v))$ from $E(v)$. Later on when we traverse the same edge going from the child back to $v$ we reconstruct $E(v)$ from the potentially new values of $E(\ell(v))$ and $E(r(v))$. The overall cost of an insertion or deletion is $O(\log^2 n)$.

To simplify the presentation in the subsequent sections, we will consider upper envelopes stored at various nodes of the structure as if they are stored there in full, and will ignore the issues related to this more space-efficient representation. Nevertheless, the bounds that we will state will take this improved representation into account.

## 3    Making the Data Structure Kinetic

We now show how to maintain the upper envelope $E$ of $S$, using the structure of Section 2, when the lines are moving along trajectories known to the algorithm. Note that now the increasing slope order of the lines $\ell_1, \ldots, \ell_n$ may change over time. So when we refer to $\ell_k$ we mean the line with the $k^{th}$ smallest slope at some particular time, which will always be clear from the context.

Fix an internal node $v \in T$. Denote the two lines from $E(\ell(v))$ and $E(r(v))$ that intersect at $q(v)$ as $\mu_\ell(v)$ and $\mu_r(v)$, respectively. Denote the line in $E(\ell(v))$ immediately preceding (resp., succeeding) $\mu_\ell(v)$ as $\mu_\ell^-(v)$ (resp., $\mu_\ell^+(v)$). Similarly, we denote the lines immediately preceding and succeeding $\mu_r(v)$ in $E(r(v))$ by $\mu_r^-(v)$ and $\mu_r^+(v)$, respectively; see Figure 1(a). We denote the intersection point of two lines $a$ and $b$ by $ab$. We write $ab <_x cd$ if the $x$-coordinate of $ab$ is smaller than the $x$-coordinate of $cd$.

To ensure the validity of the structure as the lines are moving, we use two types of certificates, denoted by CT and CE. For each pair of consecutive lines $\ell_k, \ell_{k+1}$ in $T$, we have a *CT-certificate* that asserts that the slope of $\ell_k$ is smaller than or equal to the slope of $\ell_{k+1}$. For each node $v$, we maintain the following (at most) four *CE-certificates* that assert the following inequalities (1) $\mu_\ell(v)\mu_r(v) <_x \mu_r(v)\mu_r^+(v)$, (2) $\mu_\ell(v)\mu_r(v) <_x \mu_\ell(v)\mu_\ell^+(v)$, (3) $\mu_\ell(v)\mu_r(v) >_x \mu_r(v)\mu_r^-(v)$, (4) $\mu_\ell(v)\mu_r(v) >_x \mu_\ell(v)\mu_\ell^-(v)$. (see Figure 1(a)); recall that $\mu_\ell(v)\mu_r(v) = q(v)$. The proof of the following lemma is straightforward.

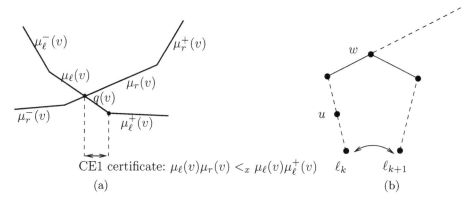

CE1 certificate: $\mu_\ell(v)\mu_r(v) <_x \mu_\ell(v)\mu_\ell^+(v)$

(a)                                                                 (b)

**Fig. 1.** (a) One of the four CE-certificates for guaranteeing the validity of $E(v)$. (b) Handling a CT event, at which the slope order at two consecutive leaves changes

**Lemma 1.** *As long as all CT and CE certificates are valid, the lines are stored at the leaves of $T$ from left to right in increasing order of their slopes, and, for each node $v \in T$, $E(v)$ stores the correct upper envelope of $S(v)$.* □

Each certificate contributes a *critical event* to a global event queue $Q$, which is the first future time when the certificate becomes invalid (if there is such a time). In the next section we describe how to handle each such event.

**Handling critical events:** By a CT or CE *critical event* we mean a failure event of one of the current CT or CE certificates. A CT certificate fails when the slopes of two consecutive lines $\ell_k$ and $\ell_{k+1}$ in $T$ become equal. If, right after the failure, the slope of $\ell_{k+1}$ becomes smaller than the slope of $\ell_k$, we have to update $T$ as follows. Let $w = LCA(\ell_k, \ell_{k+1})$ be the lowest common ancestor of the two leaves containing $\ell_k$ and $\ell_{k+1}$ (see Figure 1(b)).

We swap $\ell_k$ and $\ell_{k+1}$, and then delete from $Q$ the two CT events associated with $\ell_k$ and $\ell_{k-1}$, and with $\ell_{k+1}$ and $\ell_{k+2}$, and add to $Q$ up to three new CT events: between $\ell_{k-1}$ and the new $\ell_k$, between the new $\ell_{k+1}$ and $\ell_{k+2}$, and between $\ell_k$ and $\ell_{k+1}$, if their slopes become equal again at some future time. In addition, this swap might affect upper envelopes at the nodes on the two paths from $\ell_k$ and from $\ell_{k+1}$ to the root. Hence, for each node $u$ on either of the paths, we recompute $E(u)$ *from scratch* in a bottom-up fashion. In particular, it means that, for each such node $u$ at which $E(u)$ has changed, we may have to delete from $Q$ the at most four CE events associated with $u$, and replace them by at most four new CE events.

When a CE certificate fails at some node $v$, $E(v)$ is no longer valid. The following changes can take place: If the certificate $\mu_\ell(v)\mu_r(v) <_x \mu_r(v)\mu_r^+(v)$ fails, the line $\mu_r(v)$ is removed from $E(v)$. If $\mu_\ell(v)\mu_r(v) <_x \mu_\ell(v)\mu_\ell^+(v)$ fails, the line $\mu_\ell^+(v)$ is added to $E(v)$ between $\mu_\ell(v)$ and $\mu_r(v)$. Similarly, if $\mu_\ell(v)\mu_r(v) >_x \mu_r(v)\mu_r^-(v)$ fails, the line $\mu_r^-(v)$ is added to $E(v)$ between $\mu_\ell(v)$ and $\mu_r(v)$, and if $\mu_\ell(v)\mu_r(v) >_x \mu_\ell(v)\mu_\ell^-(v)$ fails, the line $\mu_\ell(v)$ is removed from $E(v)$. Because of the continuity of the motion of the lines, only these local changes can occur at a failure of a CE certificate.

We restore $E(v)$ by inserting or deleting the appropriate line at the appropriate location. We replace the four old CE certificates associated with $v$ by four new certificates, to reflect the fact that either $\mu_r(v)$ or $\mu_\ell(v)$ has changed, as did its predecessor and successor in the respective sub-envelope. We also delete from $Q$ the failure times of the old certificates, and insert into $Q$ the failure times of the new certificates.

The change in $E(v)$ may also cause $E(w)$ to change at ancestors $w$ of $v$. We propagate the change from $v$ up towards the root, until we reach an ancestor $w$ of $v$ for which $E(w)$ is not affected by the change at $v$. Let $w$ be an ancestor of $v$ at which $E(w)$ changes. Let $p(w)$ be the parent of $w$, and let $s(w)$ be the sibling of $w$. If the line that joins or leaves $E(w)$ also joins or leaves $E(p(w))$, we change $E(p(w))$ accordingly. In addition, if the change replaces the line in $E(w)$ on which the intersection of $E(w)$ and $E(s(w))$ occurs, or one of lines adjacent to it on $E(w)$, we also replace the CE certificates associated with $p(w)$, and replace the corresponding failure times in $Q$.

**Performance analysis:** Using the terminology of [2], it is quite easy to show that the data structure is *compact, local, and responsive*. Clearly, we have a linear number of CT certificates, and a linear number of CE certificates, so our event queue $Q$ is of linear size. The size of the primary tree $T$ and of all the trees $E(v)$ is $O(n)$, if we store partial envelopes in the manner outlined in Section 2. Therefore our KDS can be implemented in linear space, and is thus *compact*. *Locality* follows since each line $\ell$ participates in only two CT certificates, and $O(\log n)$ CE-certificates in its ancestors in $T$. *Responsiveness* follows since the time needed to process a critical event is $O(\log^2 n)$. This includes the time to update $O(\log n)$ envelopes that are affect by the change, and the time for replacing $O(\log n)$ events in $Q$.

The most interesting and involved part of the analysis is to show that the data structure is *efficient*, in the sense of obtaining an upper bound on the total number of critical events that is comparable with the bound on the total number of real combinatorial changes in the overall upper envelope.

**Bounding the number of critical events:** To analyze the total number of critical events that our data structure processes, we refine a technique of Basch et al. [2], in which time is considered as an additional (static) dimension, which allows us to represent each critical event as a vertex of an appropriate upper envelope of *bivariate* functions, where these envelopes are the graphs of the sub-envelopes $E(v)$, as they evolve over time.

In more detail, we parameterize the moving lines as surfaces in the 3-dimensional $xty$-space. For each line $\ell \in S$, its surface $\sigma_\ell$ is the locus of all points $(x, t, y)$, such that $(x, y)$ lies on $\ell$ at time $t$. Note that $\sigma_\ell$ is a ruled surface, and that it is *xt-monotone*, so that we can regard it as the graph of a function of $x$ and $t$, which, with a slight abuse of notation, we denote as $y = \sigma_\ell(x, t)$. For any node $v$ of $T$, we denote by $E_3(v)$ the upper envelope of the bivariate functions $\sigma_\ell$, for $\ell \in S(v)$.

If we assume that the motions of the lines are semi-algebraic of constant description complexity, then the surfaces $\sigma_\ell$ are also semi-algebraic of constant description complexity. The intersection curve of a pair of surfaces is the trace of the moving intersection point between the two respective lines, and an intersection point of three surfaces represents an event where the three respective lines become concurrent. It follows that the number of changes in the time-evolving upper envelope of the lines is upper bounded by the combinatorial complexity of the upper envelope of their surfaces. Note that the above assumptions on the motion of the lines, including the assumption of general position, imply that any triple of surfaces intersect in at most $s$ points, where $s$ is some constant.

Using standard techniques, we prove that the complexity of the upper envelope of $n$ such surfaces is $O(n^2\beta_{s+2}(n))$. (In case the lines correspond to moving points in the dual plane this can be reduced to $O(n^2\beta_s(n))$ as in [1].)

We next derive an upper bound on the number of events that our data structure handles (the so-called internal events), which is not much larger than $O(n^2\beta_{s+2}(n))$. By our assumption on the motion, the slopes of two lines can coincide at most $O(1)$ times. Therefore, the total number of CT events is $O(n^2)$.

The main part of the analysis is to bound the number of CE events. Consider such an event that occurs when a CE certificate at some node $v$ of $T$ fails. Note that, at this event, three lines of $S(v)$ become concurrent, and the point of concurrency lies on the upper envelope $E(v)$. Hence, we can charge the event to a vertex of the corresponding bivariate upper envelope $E_3(v)$.

Not every vertex of $E_3(v)$ corresponds to a CE event at $v$. Each charged vertex is an intersection of three surfaces, such that *at least one* of them corresponds to a line in $S(\ell(v))$, and *at least one* of them corresponds to a line in $S(r(v))$. Recall that the intersection corresponds to the event where the three lines defining these surfaces become concurrent, at a point on the upper envelope $E(v)$. To bound the number of CE events at $v$, we need to bound the number of such "bichromatic" vertices of $E_3(v)$.

Let $P(v)$ denote the multiset of pairs of lines $(\ell, \ell')$, for which there exists some time at which $\ell \in S(\ell(v))$ and $\ell' \in S(r(v))$ simultaneously. The multiplicity of a pair $(\ell, \ell')$ in $P(v)$ is taken to be the number of maximal connected time intervals during which $(\ell, \ell') \in S(\ell(v)) \times S(r(v))$. The following main technical lemma, whose proof is omitted, bounds the total number of events encountered in $v$, in terms of $|P(v)|$.

**Lemma 2.** *Let $P(v)$ be the multiset of pairs of lines $(\ell, \ell') \in S(\ell(v)) \times S(r(v))$, as defined above. Let $m$ be the maximum number of lines under $v$ at any fixed time. Let $s$ be the maximum number of times a triple of lines become concurrent. Then the total number of CE events that are encountered at $v$ is $O(|P(v)|\beta_{s+2}(m))$.*

**A technical difficulty:** The next goal is to bound the quantities $|P(v)|$. Since the lines keep swapping between the nodes of $T$, the sets $P(v)$ keep acquiring new pairs. The difficulty in the analysis stems from the fact that if a line $\ell$ enters, say, the left subtree $\ell(v)$ of a node $v$, it creates $|S(r(v))|$ new pairs with the lines stored at $r(v)$, all of which are to be added to $P(v)$. That is the sum $\sum_v |P(v)|$, over all nodes $v$ of $T$, is $O\left(\sum_v M(v)|S(v)|\right)$, where $M(v)$ is the number of swaps performed between the left and right subtrees of $v$.

To appreciate the difficulty in bounding this sum, consider the slopes of the lines as functions of time. These $n$ functions define an arrangement $\mathcal{A}$ in the slope-versus-time plane. Each swap of lines between the $k$-th and the $(k+1)$-st leaves of $T$ corresponds to a vertex of $\mathcal{A}$ where the $k$-th and the $(k+1)$-st levels of $\mathcal{A}$ meet. Now even in the simplest case, where the slopes are linear functions of time, the best upper bound known for the complexity of the $k$-th level is $O(nk^{1/3})$ [4], and the situation becomes much worse for classes of more general curves (see, e.g., [3]). Thus, at the root $v$, this leads, by Lemma 2, to a too weak upper bound of $O(n^{7/3})$ or worse, which is much larger than the near-quadratic bound on the number of external events.

### 3.1   Treaps

The preceding discussion means that, with a lack of good bounds on the complexity of any single level in an arrangement $\mathcal{A}$ of functions of low complexity in the plane (namely, our slope-versus-time functions), our approach falls short of proving a good bound on the number of internal events, if the underlying tree $T$ causes levels of $\mathcal{A}$ with large complexity to appear near the root. To overcome this difficulty, and exploit the fact that, on average, levels have linear size, we make $T$ a *treap* [7]. Intuitively, using a treap allows us to make the height of a "bad level" in $T$ a random variable, so that, on average (over the choice of the priorities that define the treap), swaps at that level would occur rather low in the tree, and consequently would not be too expensive.

In more details, a *treap* is a randomized search tree with optimal *expected* behavior. Each node $v$ in the treap has two fields $rank(v)$ and $priority(v)$. The treap is a search tree with respect to the *ranks*, and a heap with respect to the *priorities*. We use integer ranks from 1 to $n$, that index the given lines in the increasing order of their slopes. We assume that the priorities are drawn independently and uniformly at random from an appropriate continuous distribution, so that with probability 1, the set of priorities defines a random permutation of the nodes, and therefore the resulting treap $T$ is uniquely determined.

We turn our underlying tree $T$ into a treap as follows. A node $v$ of rank $k$ stores the line $\mu(v) = \ell_k$, which is the line with the $k$-th smallest slope. We now denote by $S(v)$ the set of lines stored at all nodes in the subtree rooted at $v$, including $\mu(v)$ itself, and we define $E(v)$ to be the upper envelope of the new set of lines $S(v)$.

Since now every node of $T$ stores a line, rather than just the leaves, we need to slightly modify the algorithm, so that each node $v$ maintains certificates that encode the interaction between $E(\ell(v))$, $E(r(v))$, and $\mu(v)$; we omit the easy details in this version.

**Handling critical events:** The main modification of the preceding analysis for the case of treaps is in handling CT events. Consider a CT event, involving a swap between two lines $\ell_k$ and $\ell_{k+1}$ whose slopes are equal at the critical time $t$. Let $v$ be the node containing $\ell_k$, and $v'$ the node containing $\ell_{k+1}$; then $rank(v) = k$ and $rank(v') = k + 1$. It follows that either $v'$ is the leftmost descendant of $r(v)$, or $v$ is the rightmost descendant of $\ell(v')$. When processing the swap, we place $\ell_{k+1}$ in $v$ and $\ell_k$ in $v'$, without changing the structure of the treap. Then we recompute the envelopes $E(w)$, for all nodes $w$ on the path between $v$ and $v'$, and update the CE events associated with each such node $w$. Finally, we delete from $Q$ the CT events previously associated with $\ell_k$ and $\ell_{k+1}$, and insert into $Q$ new CT events between $\ell_k$ and $\ell_{k+2}$, between $\ell_{k+1}$ and $\ell_{k-1}$, and between $\ell_k$ and $\ell_{k+1}$ (if their slopes become equal again at some future time). Handling CE events is done in essentially the same way as in Section 3, and we omit the easy details.

**Performance analysis for treaps:** The same argument as in Section 3 shows that our data structure can be implemented in linear space. The analysis of [7] shows that the depth of any node in a treap is on average (over the draw of the priorities) $O(\log n)$. This fact immediately implies that any line participates in an expected number of $O(\log n)$ certificates at any given time, and that, in any CT or CE critical event, the expected number of nodes $v$ that need to be updated is $O(\log n)$, and thus the expected time it takes to process a critical event, or a change in the flight plan of a line, is $O(\log^2 n)$. Hence, the new data structure is compact, local, and responsive, in an expected sense.

**Number of critical events in the case of treaps:** We bound the expected number of critical events using the approach suggested in Section 3. The following version of Lemma 2 holds when a line is stored at every node of $T$. The proof follows that of Lemma 2.

**Lemma 3.** *Let $P(v)$ be the multiset of pairs of lines $(\ell, \ell')$, such that (i) $\ell \neq \ell'$, (ii) $\ell \in S(\ell(v))$ or $\ell = \mu(v)$, and (iii) $\ell' \in S(r(v))$ or $\ell' = \mu(v)$, where the multiplicity of a pair is the number of maximal connected time intervals during which $(\ell, \ell')$ satisfy (i)–(iii). Let $m$ be the maximum number of lines under $v$ at any fixed time (including also $\mu(v)$). Let $s$ be the maximum number of times where any fixed triple of lines becomes concurrent. Then the total number of CE events that are encountered at $v$ is $O(|P(v)|\beta_{s+2}(m))$.*

This lemma reduces the problem of bounding the expected number of events to the problem of bounding the expected value of the sum $\mathcal{P} = \sum_{v \in T} |P(v)|$ of the sizes of the multisets $P(v)$, over all nodes $v$. Recall that the sets $P(v)$ are affected only by swaps that take place at CT critical events.

We perform the analysis in two steps. First, we bound the expected initial value of $\mathcal{P}$. Then we bound the expected contribution of each swap to $\mathcal{P}$. We denote by $\pi$ the permutation of the nodes when we order them by increasing priority. Specifically, $\pi(v)$ is the number of nodes with priorities smaller than the priority of $v$. In the following we refer to a line $\ell_k$ simply by its index $k$, that is, by its rank in the list of lines sorted by slope. We also denote by $v(k)$ the node containing line $k$, which is the node of rank $k$ of the treap. Note that $v(k)$ is always the same node but the line that it contains may change over time through swaps.

Bounding the initial value of $\mathcal{P}$ is trivial. Indeed, a pair $(i, j)$, with $i < j$, appears in exactly one set $P(v)$: If $i$ is a descendant of $j$ then $(i, j)$ belongs (only) to $P(v(j))$. Symmetrically, if $j$ is a descendant of $i$ then $(i, j)$ belongs (only) to $P(v(i))$. Finally, if neither of them is a descendant of the other, then $(i, j)$ belongs only to $P(v)$, where $v$ is the lowest common ancestor of $i$ and $j$. Hence, initially, we have $\sum_v |P(v)| = \binom{n}{2}$.

We now estimate the expected contribution of a swap between two consecutive lines, say, $m - 1$ and $m$, to $\mathcal{P}$ (recall that $m - 1$ and $m$ denote the *ranks* of these lines, and not the lines themselves). Clearly, either $m - 1$ is the rightmost leaf descendant of $\ell(m)$, or $m$ is the leftmost leaf descendant of $r(m - 1)$. The two cases are symmetric, so we only handle the first case. See Figure 2(a).

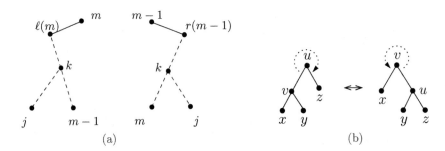

**Fig. 2.** (a) $m$ creates new pairs with lines in the subtree of $\ell(m)$. (b) A rotation around the edge $(u, v)$. The new nodes $u$ and $v$ retain the identities of the old nodes, as shown

The swap of $m - 1$ and $m$ creates a new pair of $m$ with every line $j$ in the subtree of $\ell(m)$, other than $m - 1$, which are added to $P(v(k))$, where $k$ is the lowest common ancestor of $j$ and $m$ after the swap. Similarly, for every $j$ in the subtree of $\ell(m)$, other than $m - 1$, we get a new pair of $m - 1$ and $j$ that contributes to $P(v(m))$ after the swap. Therefore we will estimate the expected number of new pairs created by $m - 1$ (in its new location); the expected number of new pairs created by $m$ (in its new location) will be the same. Moreover, as is easily verified, no new pairs are formed with elements $j > m$, nor with elements $j$ to the left of the subtree of $\ell(m)$.

For $j < m - 1 < m$, define $A_{j,m}$ to be an indicator random variable, that is 1 if and only if $v(j)$ is a descendant of $v(m)$. As just argued, if $j$ and $m - 1$ form a new pair in $P(v(m))$ (again, recall that line $m - 1$ resides in $v(m)$ after the swap), then $j$ must be a descendant of that node, and vice versa. Hence, the expected number of new pairs created by $m$ is $\sum_{j | j < m - 1} \mathbf{E}(A_{j,m})$. To compute $\mathbf{E}(A_{j,m})$, we have to calculate the probability that $v(j)$ is a descendant of $v(m)$. This happens if and only if $\pi(y) < \pi(m)$, for all nodes $y$ such that $j \le rank(y) \le m - 1$. This probability is equal to the probability that the nodes of ranks between $j$ and $m$ (inclusive) are arranged in $\pi$ such that $m$ is last. That is, $\mathbf{E}(A_{j,m}) = (m - j)!/(m - j + 1)! = 1/(m - j + 1)$. The expected number of new pairs is then

$$\sum_{j|j<m-1} \mathbf{E}(A_{j,m}) = \sum_{j|j<m-1} \frac{1}{m-j+1} = \sum_{3\leq j<m} \frac{1}{j} = O(\log m) = O(\log n).$$

Since the number of new pairs created by $m-1$ is the same as the number of new pairs created by $m$, we conclude that the expected contribution of each swap to $\mathcal{P}$ is $O(\log n)$.

Our assumptions on the motion implies that the total number swaps is $O(n^2)$. Therefore, we get that all swaps generate $O(n^2 \log n)$ additional pairs, so in total $\mathcal{P} = O(n^2 \log n)$. Using Lemma 3, our structure thus processes $O(n^2 \beta_{s+2}(n) \log n)$ events, and is therefore *efficient*.

So far, this matches (but, as we argue, simplifies) the KDS structure of [2]. In the main contribution of the paper, presented in the next section, we make this KDS dynamic.

## 4   Making the Data Structure Kinetic and Dynamic

First, we review the algorithms in [7] for inserting and deleting elements into/from a treap. To insert a new line $\ell$, we create a new leaf, in a position determined by its rank. Then we draw a random priority for $\ell$ from the given distribution, and rotate the node storing $\ell$ up the tree, as long as its priority is larger than the priority of its parent. While rotating the node of $\ell$ up, we also re-compute the envelope of every node involved in a rotation from the envelopes of its children and from the line that it stores. After $\ell$ is located in its right place, we recompute the envelopes on the path from $\ell$ to the root in a bottom-up manner. The expected logarithmic depth of the treap implies that insertion takes $O(\log^2 n)$ expected time.

The implementation of a delete operation is similar. Let $m$ be the line to be deleted. We keep rotating the edge connecting $m$ to its child of largest priority, until $m$ becomes a leaf. We then discard $m$ and recompute the envelopes of all nodes involved in the rotations, in a bottom-up manner, until we reach the first node that did not contain $m$ on it envelope or we reach the root. As in the case of insertion, deletion also takes $O(\log^2 n)$ expected time.

Consider for example a right rotation around an edge $(v, u = p(v))$ as shown in Figure 2(b). Node $v$ before the rotation changes its right child to be $u$, and node $u$ changes its left child to be $y$, previously the right child of $v$. The rotation introduces new pairs associated with $v$ (and removes pairs associated with $u$). Therefore, we need to re-analyze the *efficiency* of the data structure, when insertions and deletions are allowed, to take these changes into account. Using Lemma 3, we need to estimate the number of new pairs that are generated during an insertion or a deletion. We show below that the expected number of such pairs is $O(n)$, for each update operation. Hence, if $m$ such operations are performed, starting with the empty set, they generate an expected number of $O(mn)$ pairs, and thus create only $O(mn\beta_{s+2}(n))$ new CE events. This bound applies to both internal and external events.

We analyze deletion in detail; the analysis of insertion is analogous and hence omitted. Assume that $m$ is the line to be deleted. We examine the rotations that bring $m$ down, and bound the expected number of new pairs created by these rotations. Let $v$ be the node containing the line $m$. Let $\sigma^\ell$ denote the rightmost path from $\ell(v)$ to a leaf, and let $\sigma^r$ denote the leftmost path from $r(v)$ to a leaf. Each edge on $\sigma^\ell$ and $\sigma^r$ corresponds to a rotation. That is, a right rotation around the edge between $v$ and $\ell(v)$ changes the left child of $v$ to the next node on $\sigma^\ell$. In this case, for each line $x$ in $S(\ell(\ell(v)))$, including $\mu(\ell(v))$, and for each line $y$ in $S(r(v))$, the rotation introduces a new pair $(x,y)$ in $P(\ell(v))$. These are the only new pairs that are generated. A left rotation around $(v, r(v))$ has a symmetric effect, and it changes the right child of $v$ to be the next node along $\sigma^r$. Therefore, regardless of the order of the rotations, the total number of new pairs is dominated by $|S(\ell(v)) \times S(r(v))|$.

For $i < m < j$, define $B_{i,m,j}$ to be an indicator random variable, which is 1 if and only node $v(m)$ is the lowest common ancestor of $v(i)$ and $v(j)$. The expected size of $S(\ell(v(m))) \times S(r(v(m)))$, for a fixed node $v(m)$, is $\sum_{i,j|i<m<j} \mathbf{E}(B_{i,m,j})$. For $B_{i,m,j}$ to be 1, $v(m)$ must have the largest priority among all nodes $x$, such that $i \leq rank(x) \leq j$. The probability of this event is equal to the probability that $v(m)$ ends up last in a random permutation of the nodes $\{x \mid i \leq rank(x) \leq j\}$. That is, $\mathbf{E}(B_{i,m,j}) = (j-i)!/(j-i+1)! = 1(j-i+1)$, for any $i < m < j$. Summing up over all such pairs $i$ and $j$, we get that

$$\sum_{i,j|i<m<j} \mathbf{E}(B_{i,m,j}) = \sum_{i,j|i<m<j} \frac{1}{j-i+1} \leq \sum_{2\leq k\leq n} (k-1)\frac{1}{k+1} = O(n).$$

That is, we have shown that the expected increase in the sum $\sum_v |P(v)|$, caused by inserting or deleting an element (at place $m$) is $O(n)$. Following the preceding discussion, and the duality between points and lines, we thus obtain:

**Theorem 1.** *Let $S$ be a fully dynamic set of $n$ lines (points) moving in the plane. Assuming that $S$ undergoes $m \geq n$ insertions and deletions, and that the motion of each line (point) is semi-algebraic of constant description complexity, we can maintain the upper envelope (convex hull) of $S$ in a randomized structure of linear size, that processes an expected number of $O(mn\beta_{s+2}(n)\log n)$ events, each in $O(\log^2 n)$ expected time, where $s$ is the number of times where any fixed triple of lines (points) can become concurrent (collinear). Each line (point) participates at any given time in $O(\log n)$ certificates that the structure maintains.*

# References

1. P. K. Agarwal, L. Guibas, J. Hershberger, and E. Veach, Maintaining the extent of a moving point set, *Discrete Comput. Geom.* 26 (2001), 353–374.
2. J. Basch, L. J. Guibas, and J. Hershberger, Data structures for mobile data, *J. Algorithms* 31 (1999), 1–28.
3. T.M. Chan, On levels in arrangements of curves, II: A simple inequality and its consequences, *Proc. 44th IEEE Sympos. Foundat. Comput. Sci.*, 2003, 544–550.
4. T. Dey, Improved bounds for planar $k$-sets and related problems, *Discrete Comput. Geom.* 19 (1998), 373–382.

5. L. Guibas, Kinetic data structures: a state of the art report. *Robotics: the Algorithmic Perspective* (WAFR 1998), 191–209, A.K. Peters, Natick, MA, 1998.

6. M. H. Overmars and J. van Leeuwen, Maintenance of configurations in the plane, *J. Computer Syst. Sci.* 23 (1981), 166–204.

7. R. Seidel and C. R. Aragon, Randomized search trees, *Algorithmica* 16 (1996), 464-497.

8. M. Sharir and P.K. Agarwal, *Davenport-Schinzel Sequences and Their Geometric Applications*, Cambridge University Press, New York, 1995.

# Approximation Algorithms for Forests Augmentation Ensuring Two Disjoint Paths of Bounded Length

Victor Chepoi, Bertrand Estellon, and Yann Vaxès

Laboratoire d'Informatique Fondamentale de Marseille,
Faculté des Sciences de Luminy, Universitée de la
Méditerranée, F-13288 Marseille Cedex 9, France
{chepoi, estellon, vaxes}@lif.univ-mrs.fr

**Abstract.** Given a forest $F = (V, E)$ and a positive integer $D$, we consider the problem of finding a minimum number of new edges $E'$ such that in the augmented graph $H = (V, E \cup E')$ any pair of vertices can be connected by two vertex-disjoint paths of length $\leq D$. We show that this problem and some of its variants are NP-hard, and we present approximation algorithms with worst-case bounds 6 and 4.

## 1  Introduction and Preliminaries

Biconnectivity is a fundamental requirement to the topology of communication networks: a biconnected network survives any single link or node failure (the probability of two or several simultaneous failures in most networks is reasonably small). On the other hand, the communication performances of a network depend of the communication delay between any two nodes of the network. Since the delay of sending a message from one node to another is roughly proportional to the number of nodes (or links) the message has to traverse, it is desirable to route the messages along short paths or paths of bounded length. Therefore a network in which any pair of nodes can be connected by two disjoint paths of bounded length ensure a low communication delay even in case of a single link or node failure. In this paper, we consider the problem of optimal augmentation of networks (more precisely, of their underlying graphs) so that the resulting networks satisfy this connectivity requirement. We show that this augmentation problem is NP-hard even in the case of forests. On the other hand, we provide efficient approximation algorithms for this problem and its variants if the input graph is a forest. Our work continues the research started in [4, 10].

Several other models have been proposed in the literature to study fault-tolerant networks whose reliability and communication performances survive node or edge failures. For instance, Farley and Proskurowski [8] study the class of self-repairing graphs which consists of 2-connected graphs such that the removal of any single vertex results in no increasing in distance between any pair of vertices in the graph. Another interesting model has been proposed by

F. Dehne, A. López-Ortiz, and J.-R. Sack (Eds.): WADS 2005, LNCS 3608, pp. 282–293, 2005.

Dolev et al. [6]. Given a graph $G$, a fixed routing and a set of faults $F$, they define a surviving route graph consisting of all non-faulty nodes in the network with two nodes being connected by an edge if and only if the route between them avoid $F$. Then, the problem is to obtain a routing such that for any set of faults of a given cardinality, the surviving route graph has a small diameter. Note that, since this diameter represents the number of routes along which a message must travel between any two non-faulty nodes, it can be viewed as an estimate of the fault-tolerance of the routing.

The problem of augmenting a graph to reach biconnectivity by adding a minimum number of new edges is an important graph-algorithmic problem with applications to network reliability and fault-tolerant computing. Eswaran and Tarjan [7] introduced this problem and established that its basic version can be solved efficiently. Subject to additional constraints, the biconnectivity augmentation problem becomes difficult: for example, both the weighted augmentation problem and the optimal augmentation of a planar graph to a biconnected planar graph are NP-hard [7, 11]; for both problems there exist constant factor approximation polynomial algorithms.

The problem of augmenting a graph $G = (V, E)$ to a graph $H = (V, E \cup E')$ of a given diameter $D$ by adding a minimum number of edges is NP-hard for any $D \geq 2$ [4, 12, 13] (and is at least as difficult to approximate as SET COVER). The complexity status of this problem is unknown if the input graph is a forest (or a tree). In this case, [4] presents a factor 2 algorithm for even $D$ and [10] presents a factor 8 algorithm for odd $D$ (recently, a factor $2 + \frac{1}{\delta}$ for any $\delta > 0$ approximation algorithm in the case of odd $D$ was proposed in [3]). Chung and Garey [5] established that if $G$ is a path with $n$ vertices, then the minimum number of added edges is at least $(n - D - 1)/(D + 1)$ and at most $(n - D + 2)/(D - 2)$; for some other related bounds see [1]. If, additionally to be of diameter $D$, the resulting graph $H$ must be biconnected, then the resulting augmentation problem is NP-hard even if the input graph is a tree [4]. For forests, [4] presents a factor 3 approximation algorithm in case of even $D$ and a factor 6 approximation algorithm for odd $D$. The last result has been improved by Ishii, Yamamoto, and Nagamochi [10] to a factor 4 (plus 2 edges) approximation algorithm. Notice that for trees the performance guarantees of all mentioned algorithms should be much better, however the bottleneck in analyzing them is the difficulty of establishing better lower bounds for the minimum number of added edges; for example, the proof of the above mentioned lower bound for paths [5] is already quite involved.

In this note, we consider three variants of the augmentation problem with additional distance constraints:

**Problem A2VDBP (Augmentation with 2 Vertex-Disjoint Bounded length Paths):** *given a graph $G = (V, E)$ and a positive integer $D$, add a minimum number of new edges $E'$ such that any pair of vertices can be connected in the augmented graph $H = (V, E \cup E')$ by two vertex-disjoint paths of length $\leq D$.*

**Problem A2EDBP (Augmentation with 2 Edge-Disjoint Bounded length Paths):** *given a graph $G = (V, E)$ and a positive integer $D$, add a minimum* number of new edges $E'$ such that any pair of vertices can be connected in the augmented graph $H = (V, E \cup E')$ by two edge-disjoint paths of length $\leq D$.

**Problem ADCE (Augmentation with Diameter Constraints in the augmented graph minus an Edge):** *given a graph $G = (V, E)$ and a positive integer $D$, add a minimum number of new edges $E'$ such that for any edge $e \in E \cup E'$ the diameter of the augmented graph $H$ minus $e$ is at most $D$.*

The problems A2VDBP, A2EDBP, and ADCE are not equivalent: any feasible augmentation for A2VDBP is a feasible solution to A2EDBP and any feasible solution to A2EDBP is a feasible solution of ADCE but not vice versa, as the following example shows. Let $H$ be a graph consisting of a cycle of length $2R + 2$ plus a diagonal $cc'$ connecting two opposite vertices $c$ and $c'$ of this cycle. Removing any edge from $H$ results into a graph of diameter at most $2R - 1$, however the neighbors $a$ and $b$ of $c$ different from $c'$ cannot be connected in $H$ by two vertex- or edge-disjoint paths of length $\leq 2R - 1$.

In Section 2 we will prove that the problems A2VDBP, A2EDBP, and ADCE are NP-hard already when the input graph is a forest. Based on lower bounds established in Section 3, in Section 4 we present a factor 6 approximation algorithm for all three problems. In Section 5, this algorithm is improved to a factor 4 approximation algorithm for ADCE. Due to space limitations, we will present full proofs and analysis of both algorithms only in the case $D = 2R - 1$.

We conclude this introductory section with a few necessary definitions. A polynomial algorithm is called an $\alpha$-*factor approximation* algorithm for a minimization problem $\Pi$ if for each instance $I$ of $\Pi$, it returns a solution whose value is at most $\alpha$ times the optimal value $\mathrm{OPT}_\Pi(I)$ of $\Pi$ on $I$ plus a constant not depending of $I$; see [14]. For a graph $G = (V, E)$, the *length* of a path between two vertices is the number of edges in this path. The *distance* $d(u, v) := d_G(u, v)$ between two vertices $u, v$ of $G$ is the length of the shortest path between these vertices (if $u$ and $v$ are in distinct connected components of $G$ we will set $d(u, v) = \infty$). The *diameter* $\mathrm{diam}(G)$ of $G$ is the largest distance between two vertices of $G$. For a positive integer $k$ and a vertex $u \in V$ let $B(u, k) = \{v \in V : d(u, v) \leq k\}$ denote the *ball* of radius $k$ centered at $u$. (For other graph-theoretical notions and notations used but not defined in this text, see [15].) For two vertices $u, v$ of a tree $T = (V, E)$ denote by $P(u, v)$ the unique path of $T$ between $u$ and $v$. For a vertex $x$ in a rooted tree $T$ with root $r$, any vertex $y \neq x$ on the path $P(r, x)$ is called an *ancestor* of $x$. If $y$ is an ancestor of $x$, then $x$ is a *descendant* of $y$. For a subset $S$ of vertices of $T$, the set of *direct descendants* of a vertex $v \in V$ consists of all descendants $u \in S$ of $v$ such that the path $P(u, v)$ does not contain any other vertex from $S$.

## 2    NP-Completeness

The decision variants of the problems A2VDBP, A2EDBP, and ADCE belong to the class NP. To establish that these problems are NP-complete on forests, we present pseudo-polynomial transformations from the strongly NP-complete problem 3-PARTITION. Then Lemma 4.1 of [9] implies that the augmentation problems are NP-complete as well. Our construction is similar to that used in [4] with one difference: we use an additional path in order to force the structure of the augmented graph.

**Theorem 1.** *The problems A2VDBP, A2EDBP, and ADCE are NP-complete on forests.*

*Proof.* We will describe a pseudo-polynomial transformation from 3-PARTITION [9] to A2VDBP. Let an instance of 3-PARTITION be given, i.e., a set $A$ of $3m$ elements $a_1, \ldots, a_{3m}$, a bound $(8 \leq)B \in \mathbb{Z}^+$, and a size $s(a_i) \in \mathbb{Z}^+$ for each $a_i \in A$ such that $B/4 < s(a_i) < B/2$ and $\sum_{a_i \in A} s(a) = mB$. We construct a forest $F = (V, E)$ as follows: for each $a_i \in A$ introduce a path $P_i$ of length $s(a_i)(> 2)$, additionally consider a path $P_0$ of length $B$, and a "bistar" formed by a path $P$ of length $B + 6$ plus $m + 1$ leaves at each end of this path; see Fig. 1. We assert that the set $A$ can be partitioned into $m$ disjoint sets $A_1, \ldots, A_m$ such that $\sum_{a_i \in A_j} s(a_i) = B$ for every $1 \leq j \leq m$ if and only if there exists a feasible solution to the problem A2VDBP with $D := 2B + 10$ using at most $4m + 2$ edges.

The forest $F$ has $8m + 4$ leaves. Since a feasible augmentation of $F$ results into a biconnected graph, any leaf of $F$ must be incident to a new edge, therefore this augmentation must contain at least $4m + 2$ edges. If the optimal solution $E'$ of A2VDBP uses exactly $4m + 2$ edges, then both ends of added edges are leaves of $F$. It can be easily seen that the graph $H = (V, E \cup E')$ obtained from $F$ by adding $4m + 2$ edges $E'$ consists of the path $P$ and $m + 1$ ears. An *ear* Ear$_i$ is a path of $H$ between $x$ and $y$ consisting of the edges $xx_i, yy_i$ and 0,1, or several paths $P_j$ ($j \in \{0, 1, \ldots, 3m\}$) and some new edges connecting either the end-vertices of two paths or an end-vertex of a path with $x_i$ or $y_i$.

Let Ear$_0$ be the ear containing the path $P_0$. We assert that Ear$_0$ does not contain other paths of $F$. By using the feasibility of $E'$, one can prove that Ear$_0$ does not contain other paths of $F$, and thus, Ear$_0$ has length $B + 4$. We will show now that each of the ears Ear$_1, \ldots,$Ear$_m$ has length $B + 6$. Suppose by way of contradiction that Ear$_i$ has length $\geq B + 7$. Due to the structure of $H$, there exists a unique path between $x_i$ and $x'$ not passing via $x$ : it consists of the subpath of $P$ between $x'$ and $y$, and the subpath of Ear$_i$ between $y$ and $x_i$. The length of this path is $\geq B + 5 + B + 6 = 2B + 11 > D$, therefore all paths of length $\leq D$ between $x_i$ and $x'$ pass via $x$, contrary to the admissibility of $H$. Thus every Ear$_i$ has length $B + 6$, i.e., Ear$_i$ is composed of exactly three paths of $F$ of total length $B$ (recall that the length of every path of the forest $F$ is comprised between $B/4$ and $B/2$). The corresponding triplets of $A$ yield a feasible 3-partition.

Conversely, given a feasible solution to 3-PARTITION, we can biconnect $F$ by adding $4m + 2$ edges in such a way that in the augmented graph $H$, Ear$_0$ has length $B + 4$ and every other ear Ear$_i$ ($i = 1, \ldots, m$) has length $B + 6$. We claim that $H$ is a feasible solution of the problem A2VDBP. This establishes the NP-completeness of A2VDBP.

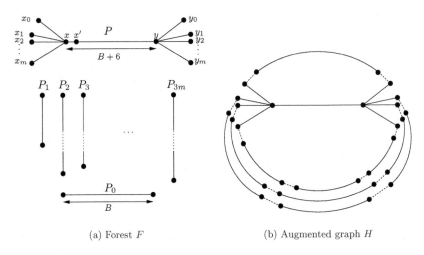

(a) Forest $F$                    (b) Augmented graph $H$

**Fig. 1.** NP-completeness of the problem A2VDBP for forests

The NP-completeness of ADCE and A2EDBP can be also established by using a pseudo-polynomial transformation from 3-PARTITION.    □

## 3    Lower Bounds

For a graph $G = (V, E)$ and a positive integer $k$, a *vertex $k$-dominating set* is a set of vertices $C \subseteq V$ such that $\cup_{c \in C} B(c, k) = V$. Analogously, $C \subseteq V$ is an *edge $k$-dominating set* if the two end-vertices of any edge of $G$ belong to a common ball of radius $k$ centered at a vertex of $C$. Finding a minimum vertex or edge $k$-dominating set in a graph is NP-hard, however for trees (and forests) these problems can be solved in linear time [2]. The algorithm can be easily modified to find in linear time a minimum vertex $k$-dominating set $VC_k$ or a minimum edge $k$-dominating set $EC_k$ of a forest $F$ with the additional constraint that $C$ contains the set $L$ of all leaves of $F$. Let $vc_k(F) := |VC_k|$ and $ec_k(F) := |EC_k|$.

**Proposition 1.** *For a forest $F$,*

*(i)* $vc_{R-1}(F)/2 \leq OPT_{ADCE}(F) \leq OPT_{A2EDBP}(F) \leq OPT_{A2VDBP}(F)$ *if* $D = 2R - 1$.
*(ii)* $ec_R(F)/2 \leq OPT_{ADCE}(F) \leq OPT_{A2EDBP}(F) \leq OPT_{A2VDBP}(F)$ *if* $D = 2R$.

*Proof.* In both cases it suffices to establish only the leftmost inequality, because the inequalities $\text{OPT}_{ADCE}(F) \leq \text{OPT}_{A2EDBP}(F) \leq \text{OPT}_{A2VDBP}(F)$ trivially hold for all graphs. Let $E'$ be an optimal augmentation for the problem ADCE, and let $C$ denote the set of end-vertices of the edges from $E'$. Notice that any leaf $x$ of the forest $F$ belongs to $C$, otherwise the paths in the graph $H = (V, E \cup E')$

issued from $x$ will use the unique edge $e$ of $H$ incident to $x$, contrary to the fact that $H$ is a solution of ADCE. Thus $L \subseteq C$. We assert that if $D = 2R - 1$, then every vertex of $F$ is covered by a ball of radius $R - 1$ centered at $C$, and if $D = 2R$, then the end-vertices of every edge of $F$ are at distance $\leq R$ from a vertex of $C$. Suppose by way of contradiction that in the case $D = 2R - 1$ there exists a vertex $u \notin \cup\{B_{R-1}(c) : c \in C\}$. Obviously, $u$ is not a leaf of $F$. Pick a neighbor $v$ of $u$ in $F$ and let $e = uv$. From the choice of $u$ one conclude that the closest to $v$ vertex of $C$ is at distance $\geq R - 1$ in $F$. Since the closest to $u$ vertex of $C$ is at distance $\geq R$ and any path between $u$ and $v$ of the graph $H - e$ uses at least one added edge, we deduce that the distance between $u$ and $v$ in the graph $H - e$ is at least $R + 1 + (R - 1) = 2R$, contrary to the assumption that $E'$ is a feasible augmentation for ADCE. Hence $\cup\{B_{R-1}(c) : c \in C\} = V$, yielding $vc_{R-1}(F) \leq |C|$ (analogously one can show that $ec_R(F) \leq |C|$ if $D = 2R$). Since $|C| \leq 2|E'|$ (the worst case occurs when $E'$ is a matching on $C$), we obtain the required inequality. $\qquad\square$

An immediate consequence of Proposition 3.1 is that any feasible augmentation for A2VDBP using at most $3vc_{R-1}(F)$ edges for $D = 2R - 1$ and at most $3ec_R(F)$ edges for $D = 2R$ would provide a factor 6 approximation algorithm for each of the problems A2VDBP, A2EDBP, and ADCE. Next section is devoted to the description and analysis of such an algorithm.

## 4    A Factor 6 Approximation Algorithm

We describe and justify the augmentation algorithm for $D = 2R - 1$ and only outline the changes for the case $D = 2R$. Assume without loss of generality that the input forest $F = (V, E)$ contains at least one edge, otherwise we simply run the algorithm on the forest obtained from $F$ by adding an arbitrary edge. Let $L$ be the set of leaves of $F$ and let $L_0 \subset L$ be the set of leaves constituting one-vertex trees of $F$. For the rest of this paper, we denote by $d(u, v)$ *the distance in the forest $F$* between two vertices $u, v \in V$. Suppose that every tree of the forest $F$ containing at least two vertices is rooted at some leaf. Let $S$ be the set of all such roots. The algorithm picks an arbitrary root $r \in S$ and the neighbor $r'$ of $r$. At the next stage, a minimum vertex $(R - 1)$-dominating set $C$ of the forest $F$ containing $r'$ and the set $L$ of leaves is computed. The algorithm proceeds each rooted tree level-by-level starting from its root, and for current vertex $c \in (C - L) \cup S$ it computes the list $D_c$ of its direct descendants in $C$ sorted in increasing order with respect to the distances to $c$. For each vertex $c' \in D_c$, the algorithm selects 0,1, or 2 vertices (this number depends of the distance $d(c', c)$) on the path $P(c, c')$ between $c$ and $c'$. The set consisting of $C$ and all selected vertices is grouped into two classes $A$ and $A'$ such that every vertex of the forest $F$ can be connected by two vertex-disjoint paths of length $\leq R - 2$, one going to a vertex of $A$ and another to a vertex of $A'$ (the one-vertex components of $F$ are included in both $A$ and $A'$, while $S$ is included only in $A$). The algorithm returns the augmentation consisting of all edges of the form $ra$ for $a \in A$ and

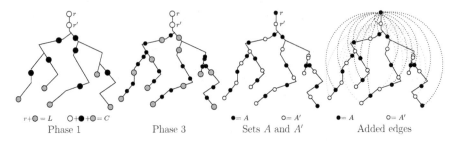

**Fig. 2.** Illustration of Algorithm A2VDBP(F,D)

$r'a'$ for $a' \in A'$. In the case $D = 2R$, the algorithm is completely analogous, except that $C$ is the minimum edge $R$-dominating set of $F$ containing $L \cup \{r'\}$.

---

**Algorithm A2VDBP(F,D)**

**Input:** A forest $F = (V, E)$ with $E \neq \emptyset$ and an odd integer $D = 2R - 1 \geq 9$.

**Output:** A set $E'$ of new edges, such that any $u, v \in V$ can be connected in $H = (V, E \cup E')$ by two vertex-disjoint paths of length $\leq D$ and $|E'| \leq 6 \cdot OPT_{A2VDBP}(F)$.

**Phase 0:** Root every tree with at least two vertices at some leaf and denote by $S$ the set of such roots. Set $A := L_0 \cup S$ and $A' := L_0$. Pick $r \in S$ and the neighbor $r'$ of $r$.

**Phase 1:** Compute a minimum vertex $(R-1)$-dominating set $C$ with $L \cup \{r'\} \subseteq C$.

**Phase 2:** For each $c \in (C - L) \cup S$ compute the list $D_c$ of direct descendants of $c$ in $C$ and sort $D_c$ in increasing order of the distances to $c$.

**Phase 3:** Proceed every rooted tree level-by-level. For current vertex $c \in (C-L) \cup S$ traverse the list $D_c$ and to each vertex $c' \in D_c$ apply one of the following rules:
   **case $\mathbf{d(c, c') \leq R - 2}$:** if $c \in A$, then insert $c'$ in $A'$, else insert $c'$ in $A$.
   **case $\mathbf{R - 1 \leq d(c, c') \leq 2R - 4}$:** pick the vertex $c_1 \in P(c, c')$ at distance $R-2$ to $c'$. If $c \in A$, then insert $c_1$ in $A'$ and $c'$ in $A$, else insert $c_1$ in $A$ and $c'$ in $A'$.
   **case $\mathbf{d(c, c') \geq 2R - 3}$:** pick the vertices $c_1, c_2 \in P(c, c')$, $c_1$ at distance $R-2$ and $c_2$ at distance $2R-4$ to $c'$. Find a closest to $c_2$ vertex $c''$ of $C$ which is not a descendant of $c_2$. If $c'' \in A$, then insert $c_2, c'$ in $A'$, and $c_1$ in $A$, else insert $c_2, c'$ in $A$, and $c_1$ in $A'$.

**Phase 4:** Return $E' = \{ra : a \in A - \{r\}\} \cup \{r'a' : a' \in A' - \{r'\}\} \cup \{rr'\}$.

---

Next we establish that any pair $u, v$ of vertices can be connected in the augmented graph $H = (V, E \cup E')$ by two vertex-disjoint paths of length $\leq D$.

**Lemma 1.** *For any vertex $x \in V$ there exists two vertices $a \in A$ and $a' \in A'$ such that $d(x, a) \leq R - 2$, $d(x, a') \leq R - 2$, $P(a, a') \cap C \subseteq \{a, a'\}$, and $x$ belongs to the (possibly degenerated) path $P(a, a')$.*

*Proof.* The result trivially holds if $x \in L_0$, because $L_0 \subseteq A \cap A'$ and we can set $a := x =: a'$. If $x \in S$, then $x \in A$ and we can set $a := x$. Let $c'$ be the closest to $x$ vertex of $C$. According to the algorithm, we can set $a' := c'$ if $d(x, c') \leq R-2$ and

$a' := c_1$ if $R - 1 \leq d(x, c') \leq 2R - 4$. In the remaining case $d(c, c') \geq 2R - 3$, the role of $a'$ is played by the vertex $c_2$ described in the algorithm. If $x \in C - S$, then $x$ is a direct descendant of some vertex $c \in C$, and, according to the algorithm, we can define $\{a, a'\} := \{x, c\}$ if $d(x, c) \leq R - 2$ and $\{a, a'\} := \{x, c_1\}$ otherwise.

So, assume that $x \notin C \cup L_0$. Let $c$ be the closest to $x$ ancestor from $C$ and let $c'$ be the closest to $x$ descendant from $C$. Clearly $c'$ is a direct descendant of $c$. If $d(c, c') \leq R - 2$, then we can set $\{a, a'\} := \{c, c'\}$ because $c$ and $c'$ belong to distinct sets $A, A'$. On the other hand, if $R - 1 \leq d(c, c') \leq 2R - 4$, then set $\{a, a'\} := \{c, c_1\}$ if $x \in P(c, c_1)$ and set $\{a, a'\} := \{c_1, c'\}$ if $x \in P(c_1, c')$. From the algorithm we infer that the vertices $c, c'$ belong to one set $A, A'$ and $c_1$ belongs to another set, establishing the assertion.

Finally suppose that $d(c, c') \geq 2R - 3$, i.e., we are in the third case of Phase 3. Recall that the algorithm picks two vertices $c_1, c_2 \in P(c, c')$, $c_1$ at distance $R - 2$ and $c_2$ at distance $2R - 4$ to $c'$ and considers a closest in $C$ non-descendant $c''$ of $c_2$. The vertices $c', c_2$ belong to one set $A, A'$, while $c_1$ and $c_2$ (as well as $c''$ and $c_2$) belong to distinct sets. Therefore we can assume that $x \in P(c_2, c)$, otherwise the proof is immediate by setting $\{a, a'\} = \{c', c_1\}$ if $x \in P(c', c_1)$ and $\{a, a'\} = \{c_1, c_2\}$ if $x \in P(c_1, c_2)$. Let $z$ be the vertex of $P(x, c')$ at distance $R$ to $c'$ ($z \in P(c_1, c_2)$ because $R \geq 5$, thus $z \neq x$). Denote by $\hat{c} \in C$ any closest to $z$ center of an $(R - 1)$-ball covering $z$. Since $d(z, \hat{c}) \leq R - 1 < R = d(z, c')$, the choice of $c'$ implies that $\hat{c}$ cannot be a descendant of $z$. For the same reason, $\hat{c}$ cannot be a descendant of $x$ either. Hence $d(z, \hat{c}) = d(z, c_2) + d(c_2, \hat{c})$, $d(z, c'') = d(z, c_2) + d(c_2, c'')$, and $d(c_2, c'') \leq d(c_2, \hat{c})$ by the choice of $c''$, yielding $d(z, c'') \leq d(z, \hat{c}) \leq R - 1$. The choice of $\hat{c}$ implies $d(z, c'') = d(z, \hat{c})$, therefore $c''$ can play the role of $\hat{c}$. In particular, this implies that $c''$ is not a descendant of $x$. Thus $R - 1 \geq d(z, c'') = d(z, c_2) + d(c_2, x) + d(x, c'') \geq R - 4 + d(c_2, x)$ because $d(z, c_2) = R - 4$, yielding $d(c_2, x) \leq 3 \leq R - 2$. Notice also that $d(x, c'') < d(z, c'') \leq R - 1$ because $x \in P(z, c'') - \{z\}$, i.e., $d(x, c'') \leq R - 2$. Finally, we assert that $d(c, c'') < d(c, c')$. This is obviously true if $c'' \in P(c, x) \subset P(c, c')$ (in fact, $c'' = c$.) Otherwise, let $y$ be the nearest common ancestor of $c''$ and $x$. Then $d(y, c'') \leq R - 3$ while $d(y, c') = d(y, x) + d(x, c_2) + d(c_2, c') \geq 1 + 2R - 4 = 2R - 3 > R - 3$. Thus $d(c, c'') < d(c, c')$, and at the moment when the algorithm analyzes the pair $c, c'$, the vertex $c''$ is already affected to $A$ or $A'$. Since $c_2$ and $c''$ belong to distinct sets $A, A'$, we can set $\{a, a'\} := \{c'', c_2\}$. Indeed $d(x, c_2) \leq R - 2$, $d(x, c'') \leq R - 2$, and $x \in P(c_2, c'')$, thus completing the proof.    □

**Lemma 2.** *Any pair $u, v$ of distinct vertices can be connected in the augmented graph $H = (V, E \cup E')$ by two vertex-disjoint paths $P_1$ and $P_2$ of length at most $D = 2R - 1$.*

*Proof.* By Lemma 4.1, there exist four vertices $a, b \in A$ and $a', b' \in A'$ such that $\max\{d(a, u), d(a', u), d(b, v), d(b', v)\} \leq R - 2$ and $u \in P(a, a'), v \in P(b, b')$. Moreover, $P(a, a') \cap C \subseteq \{a, a'\}$ and $P(b, b') \cap C \subseteq \{b, b'\}$. Then the vertices $r$ and $r'$ may occur in the paths $P(a, a')$ and $P(b, b')$ only as their end-vertices, namely, $r' \in P(a, a')$ implies that $r' = a'$, while $r \in P(a, a')$ implies that $r = a$ and $u \in \{r, r'\}$.

If $\{u,v\} = \{r,r'\}$, then $P_1$ and $P_2$ are the two parallel edges between $r$ and $r'$. On the other hand, if $u \in \{r,r'\}$, $v \notin \{r,r'\}$, then as $P_1$ and $P_2$ we take the two paths between $u$ and $v$ in the simple cycle of length at most $2R - 4 + 3 = 2R - 1$ formed by the path $P(b,b')$ and the new edges $br, rr'$, and $r'b'$ (if $r' \neq b'$). So, assume that $r, r' \notin \{u,v\}$. If the paths $P(a,a')$ and $P(b,b')$ are disjoint, then as $P_1$ we take the path formed by $P(u,a), P(b,v)$, and the new edges $ar, rb$, while as $P_2$ we take the path formed by $P(u,a'), P(b',v)$, and the new edges $a'r'$ (if $r' \neq a'$), $r'b'$ (if $r' \neq b'$). These paths have length at most $2R - 4 + 2 = 2R - 2 < D$. They are disjoint because they may intersect only in $r$ and $r'$ and it is easy to see that $r \notin P_2$ and $r' \notin P_1$. Now, consider the case when $P(a,a') \cap P(b,b') \neq \emptyset$. Then the length of the path $P(u,v)$ is at most $D$ because $d(u,t) + d(t,v) \leq R - 2 + R - 2 < D$ for any vertex $t \in P(a,a') \cap P(b,b')$. Set $P_1 := P(u,v)$. It remains to specify the second path $P_2$. First suppose that $r' \in P(u,v)$. Then $a' = r' = b'$ because $r'$ may appear only as an end-vertex on the paths $P(a,a')$ and $P(b,b')$. Therefore $(P(u,a) \cup P(v,b)) \cap P(u,v) = \{u,v\}$. Since $r \notin P_1$, we take as $P_2$ the path of $H$ consisting of $P(u,a)$, the new edges $ar$ and $rb$, and the path $P(b,v)$. Its length is at most $R - 2 + 2 + R - 2 = 2R - 2 < D$. Hence, let $r' \notin P(u,v)$. If $P(u,v) \cap P(u,a) = \{u\}$, then set $Q_1 := P(u,a), \alpha_1 := a, \beta_1 := r$. Otherwise, we have $P(u,v) \cap P(u,a') = \{u\}$, and set $Q_1 := P(u,a'), \alpha_1 := a', \beta_1 := r'$. In both cases, $Q_1 \cap P(u,v) = \{u\}$ and the length of $Q_1$ is at most $R - 2$. Analogously, if $P(u,v) \cap P(v,b) = \{v\}$, then set $Q_2 := P(v,b), \alpha_2 := b, \beta_2 := r$ otherwise set $Q_2 := P(v,b'), \alpha_2 := b', \beta_2 := r'$. Again, in both cases $Q_2 \cap P(u,v) = \{v\}$ and the length of $Q_2$ is at most $R - 2$. The paths $Q_1$ and $Q_2$ are disjoint, because all vertices of $Q_1 \cap Q_2$ would lie on the unique path of $F$ connecting $u$ and $v$ and we know that $P(u,v)$ intersects $Q_1 \cup Q_2$ only in the vertices $u, v$. We take as $P_2$ the path of $H$ consisting of $Q_1$, the edges $\alpha_1\beta_1, \beta_1\beta_2$ (if $\beta_1 \neq \beta_2$), $\alpha_2\beta_2$, and the path $Q_2$. Its length is at most $2(R - 2) + 3 = 2R - 1 = D$ and $P_1 \cap P_2 = \{u,v\}$. This establishes that indeed $H$ is a feasible solution to the problem A2VDBP. □

**Lemma 3.** $|A| + |A'| \leq 3|C| - 4$.

*Proof.* In Phase 3, for vertex $c \in (C - L) \cup S$ and each its direct descendant $c'$ in $C$, the algorithm insert in $A \cup A'$ at most two new vertices (if $c = r$ and $c' = r'$, then no new vertex is added). Any vertex $c' \in C - S$ either belongs to $L_0$ or is a direct descendant of a unique vertex of $(C - L) \cup S$. Hence the number of new vertices is at most $2(|C| - |S| - |L_0|) - 2$. Since the vertices of $L_0$ are included in both sets $A$ and $A'$ and the remaining vertices of $C$ in only one such set, we conclude that $|A| + |A'| \leq 2(|C| - |S| - |L_0|) - 2 + 2|L_0| + (|C| - |L_0|) = 3|C| - 2|S| - |L_0| - 2 \leq 3|C| - 4$. □

Hence the number of edges added by the algorithm is $|A_1| + |A_2| + 1 \leq 3|C| - 3$. From Proposition 3.1 and Lemma 4.2 we obtain the following result:

**Theorem 2.** *Algorithm A2VDBP is a factor 6 approximation algorithm for the problems A2VDBP, A2EDBP, and ADCE on forests $F = (V, E)$ for any $D \geq 9$.*

# 5    A Factor 4 Approximation Algorithm for ADCE

In this section, we modify (and simplify) the algorithm A2VDBP in order to return feasible solutions of smaller size for the problem ADCE (we assume $D = 2R-1$). We use the same notations and conventions as in the algorithm A2VDBP, except that we do not need the vertex $r'$ and instead of sets $A, A'$ we will use one multiset $A$ ($A$ contains two copies of each vertex from $L_0 \cup S - \{r\}$). Namely, given a minimum vertex $(R - 1)$-dominating set $C$ of $F$ containing the set of leaves $L$, we proceed each rooted tree of $F$ and complete $C$ to a set $A$ with the property that for every vertex $x$ there exist two vertices $a, a' \in A$ such that $x \in P(a, a')$ and $d(x, a) \le R - 2, d(x, a') \le R - 1$. The algorithm returns the augmentation $E' = \{ar : a \in A\}$. Every element of $A$ gives rise to an added edge, therefore $r$ will be connected with every vertex of $L_0 \cup (S - \{r\})$ by two parallel edges.

---

**Algorithm ADCE(F,D)**

**Input:** A forest $F = (V, E)$ and an odd integer $D = 2R - 1 \ge 5$.

**Output:** A set $E'$ of new edges, such that $\mathrm{diam}(H - e) \le D$ for any $e \in E \cup E'$ where $H = (V, E \cup E')$ and $|E'| \le 4 \cdot OPT_{ADCE}(F)$.

**Phase 0:** Root every tree containing at least two vertices at some leaf. Denote by $S$ the list of such roots and pick $r \in S$.

**Phase 1:** Compute a minimum vertex $(R - 1)$-dominating set $C$ with $L \subseteq C$. Set $A := (C - \{r\}) \cup (L_0 \cup (S - \{r\}))$.

**Phase 2:** For each vertex $c \in C - (L_0 \cup S)$, find the closest in $C$ ancestor $c'$ of $c$. If $d(c, c') \ge R$, then find $c_1 \in P(c, c')$ at distance $R - 1$ to $c$ and insert $c_1$ in $A$.

**Phase 3:** Return $E' = \{ra : a \in A\}$.

---

We present several auxiliary results establishing the feasibility of the augmentation $E'$.

**Lemma 4.** *For any vertex $x \in V - \{r\}$, there exists two different elements $a, a'$ of the multiset $A$ such that $x \in P(a, a')$ and $d(x, a) \le d(x, a') \le R - 1$.*

*Proof.* First, let $x \in V - S$. The result straightforwardly follows from the algorithm if $x$ is a vertex of $C$ or if $x$ has been inserted in $A$. So, let $x \notin A$, and let $c$ be a closest to $x$ vertex of $C$ (clearly, $d(c, x) \le R - 1$). If $x$ is an ancestor of $c$, then $c$ and one of the vertices $c'$ or $c_1$ described in Phase 2 form the required pair $\{a, a'\}$. Now, assume that $c$ is not a descendant of $x$. Suppose by way of contradiction that all descendants of $x$ in $A$ are located at distance $\ge R$ from $x$. Consider an arbitrary direct descendant $a$ of $x$ in $A$. Since $d(x, a) \ge R$, from the algorithm we infer that $a \in A - C$. Hence $a$ is at distance $R - 1$ from some descendant $b \in C$, i.e. $d(x, b) \ge R + (R - 1) = 2R - 1$. From this we can deduce that $x$ is at distance $\ge 2R-1$ from all its descendants from $C$. Pick a descendant $x'$ of $x$ satisfying $d(x, x') = R - 1$. Since any descendant of $x'$ in $C$ is at distance $\ge (2R - 1) - (R - 1) = R$ from $x'$ and $d(x, x') = R - 1$, the center $c'' \in C$ of any $(R - 1)$-ball covering $x'$ is a descendant of $x$ but not a descendant of $x'$. Let $x''$ be the nearest common ancestor of $x'$ and $c''$. Then $d(c'', x'') \le R - 2$ and

$d(x, x'') \leq R - 2$, yielding $d(x, c'') \leq 2R - 4 < 2R - 1$, contrary to the fact that $x$ is at distance $\geq 2R - 1$ from any of its descendants from $C$. This contradiction completes the proof of the case $x \in V - S$. If $x \in S - \{r\} \cup L_0$, we can take as $a$ and $a'$ the two occurrences of $x$ in $A$. □

**Lemma 5.** *For any vertex $x \in V - \{r\}$, there exists $a \in A$ such that $d(x, a) \leq R - 2$.*

*Proof.* Let $x \notin L_0 \cup S$, otherwise we can set $a = x$ by definition of $A$. Suppose by way of contradiction that all vertices of $A$ are at distance $\geq R - 1$ from $x$ and pick a vertex $c' \in C \subseteq A$ at distance $R - 1$ to $x$. If $c'$ is a descendant of $x$, then $x$ will be inserted in $A$ when $c'$ will be analyzed in Phase 2 of the algorithm, a contradiction. So, all descendants of $x$ from $C$ are at distance $\geq R$ to $x$. Let $x'$ be the son of $x$ and let $c''$ be any closest to $x'$ vertex of $C$. Since $d(x, c'') \geq R - 1$, $c''$ must be a descendant of $x'$. Thus $d(x', c'') = R - 1$ by what has been shown about $x$. As $P(c'', x) \cap C = \{c'\}$, the vertex $x'$ will be inserted in $A$ when $c''$ will be analyzed. Since $R \geq 3$, we get a contradiction with the choice of $x$. □

**Lemma 6.** *For any edge $e \in E \cup E'$, we have $diam(H - e) \leq D$.*

*Proof.* Pick two arbitrary vertices $u, v \in V$. Combining Lemmata 5.1 and 5.2 we can conclude that there exist four elements $a, a', b, b' \in A$ such that $a \neq a', b \neq b'$, $u \in P(a, a'), v \in P(b, b')$, $d(u, a) \leq R - 2, d(u, a') \leq R - 1$, and $d(v, b) \leq R - 2, d(v, b') \leq R - 1$. If $r \in \{u, v\}$ then obviously we have two disjoint paths of length at most $R$ between $u$ or $v$ in $H$. Now, suppose that $r \notin \{u, v\}$. For any edge $e$, the distances in $H - e$ between $u$ and $r$ and between $v$ and $r$ are at most $R$. If one of these two distances is at most $R - 1$, we are done. Otherwise, we must have either $a = b$ and $e = ar \in E'$ or $e$ belong to $P(u, a) \cap P(v, b)$. In the first case, $e$ does not belong to the path $P(u, v)$ and the length of this path is at most $D$, because $d(u, v) \leq d(u, a) + d(v, b) \leq 2R - 4 \leq D$. Now, suppose that $e$ belong to $P(u, a) \cap P(v, b)$. Then $u$ and $v$ belong to the same tree component $T$ of $F$. If $e$ does not belong to $P(u, v)$, then $d(u, v) \leq d(u, a) + d(v, b) \leq 2R - 4$. Otherwise, if $e \in P(u, v) \cap P(u, a) \cap P(v, b)$ then the pairs $u, b$ and $v, a$ lie in different connected components of $T - e$. Therefore $d(v, a) + d(u, b) \leq d(u, a) + d(v, b) \leq 2R - 4$ and the smallest of the distances $d(v, a)$ and $d(u, b)$, say the first, is at most $R - 2$. Thus $d_{H-e}(u, v) \leq d_{H-e}(u, r) + d_{H-e}(r, a) + d_{H-e}(a, v) \leq R + 1 + R - 2 = 2R - 1$, yielding $diam(H - e) \leq 2R - 1 = D$ for every edge $e \in E \cup E'$. □

The next lemma follows immediately from the analysis of the algorithm:

**Lemma 7.** $|A| \leq 2|C|$.

From Proposition 3.1 and previous lemmata, we obtain the following result (the algorithm for the case $D = 2R$ is similar, the correctness proof follows the same lines, however this proof is technically more involved, and is not given here due to space limitation):

**Theorem 3.** *Algorithm ADCE is a factor 4 approximation algorithm for the problem ADCE on forests $F = (V, E)$ for any $D \geq 5$.*

The algorithms A2VDBP and ADCE can be implemented in $O(|V|log|V|)$ time. Phase 1 of both algorithms takes linear time, while Phase 2 of A2VDBP can be easily performed in $O(|V|log|V|)$ time. In order to find the vertices of type $c_1$ and $c_2$, one can use mergeable heaps (supporting each of the operations INSERT, MINIMUM, EXTRACT-MIN, and UNION in $O(log|V|)$ time), i.e. in overall $O(|V|log|V|)$ time. Finally, the vertices $c''$ occurring in third case of Phase 3 of A2VDBP can be computed in total linear time.

In most of the cases occurring in the proof of Lemma 5.3, the algorithm ADCE returns two edge-disjoint paths of length $\leq D$. Nevertheless, the solutions returned by this algorithm are not always feasible augmentations for the problem A2EDBP. This leads to the question of finding a factor 4 approximation algorithm for the problem A2EDBP.

# References

1. N. Alon, A. Gyarfas, M. Ruszinko, Decreasing the diameter of bounded degree graphs, *J. Graph Theory*, **35** (2000), 161–172.
2. G.J. Chang, Labeling algorithms for domination problems in sun-free chordal graphs, *Discrete Appl. Math.*, **22** (1988/1989), 21–34.
3. V. Chepoi, B. Estellon, K. Nouioua, Y. Vaxès, Mixed covering of trees and the augmentation problem with mixted integer constraints (submitted).
4. V. Chepoi, Y. Vaxès, Augmenting trees to meet biconnectivity and diameter constraints, *Algorithmica*, **33** (2002), 243–262.
5. F. R. K. Chung, M. R. Garey, Diameter bounds for altered graphs, *J. Graph Theory*, **8** (1984), 511–534.
6. D. Dolev, J. Halpern, B. Simons, H.R. Strong, A new look at fault tolerant network routing, *Information and Computation*, **72** (1987), 180–196.
7. K. P. Eswaran, R. E. Tarjan, Augmentation problems, *SIAM J. Computing*, **5** (1976), 653–665.
8. A. M. Farley, A. Proskurowski, Self-repairing networks, *Parallel Processing Letters*, **3** (1993) 381–391.
9. M.R. Garey, D.S. Johnson, *Computers and Intractability: A Guide to the Theory of NP-Completeness*, Freeman, San Francisco, CA, 1979.
10. T. Ishii, S. Yamamoto, H. Nagamochi, Augmenting forests to meet odd diameter requirements, *International Symposium on Algorithms and Computation* (ISAAC'03), Lecture Notes in Computer Science, **2906** (2003), pp. 434–443.
11. G. Kant, H.L. Bodlaender, Planar graph augmentation problems, *Workshop on Algorithms and Data Structures* (WADS'91), Lecture Notes in Computer Science **519** (1991), pp. 286–298.
12. Ch.-L. Li, S.Th. McCormick, D. Simchi–Levi, On the minimum-cardinality-bounded-diameter and the bounded-cardinality-minimum-diameter edge addition problems, *Operations Research Letters* **11** (1992), 303–308.
13. A.A. Schoone, H.L. Bodlaender, J. van Leeuwen, Diameter increase caused by edge deletion, *J. Graph Theory* **11** (1987), 409–427.
14. V.V. Vazirani, *Approximation Algorithms*, Springer-Verlag, Berlin, 2001.
15. D. B. West, *Introduction to Graph Theory*, Prentice Hall, London, 2001.

# A Dynamic Implicit Adjacency Labelling Scheme for Line Graphs

David Morgan

University of Alberta, Edmonton AB T6G 2E8, Canada
davidm@cs.ualberta.ca

**Abstract.** As defined by Muller (Muller, Ph.D. thesis, Georgia Tech, 1988) and Kannan, Naor, and Rudich (Kannan et al., SIAM J Disc Math, 1992), an *adjacency labelling scheme* labels the vertices of a graph so the adjacency of two vertices can be deduced implicitly from their labels. In general, the labels used in adjacency labelling schemes cannot be tweaked to reflect small changes in the graph.

Motivated by the necessity for further exploration of dynamic (implicit) adjacency labelling schemes we introduce the concept of error detection, discuss metrics for judging the quality of such dynamic schemes, and develop a dynamic scheme for line graphs that allows the addition and deletion of vertices and edges. The labels used in this scheme require $O(\log n)$ bits and updates can be performed in $O(e)$ time, where $e$ is the number of edges added to or deleted from the line graph. This compares to the best known (static) adjacency labelling scheme for line graphs which uses $O(\log n)$ bit labels and requires $\Theta(n)$ time to generate a labelling even when provided with the line graph representation.

## 1  Introduction

Consider a finite simple undirected graph $G = (V_G, E_G)$ with $n$ vertices and $m$ edges; typically, we represent $G$ using an adjacency matrix, labelling the vertices from 1 to $n$. These labels distinguish between the vertices but tell us nothing about $G$. Moreover, the matrix is usually maintained as a global resource.

What if we could determine the adjacency of two vertices of $G$ in a more local manner, that is, by using only their labels? One such way is to use an adjacency labelling scheme as defined by Muller [1] and Kannan et al. [2]. An *adjacency labelling scheme* of a family $\mathcal{G}$ of finite graphs is a pair $(M, D)$ for which

- $M$ is a vertex labelling algorithm (marker) whose input is a member of $\mathcal{G}$.
- $D$ is a polynomial time evaluation algorithm (decoder) which correctly determines the adjacency of two vertices using only their labels (we will say that $D$ is adjacency-correct).

In essence, an adjacency labelling scheme is a distributed data structure that allows us to quickly determine adjacency from local information. Allowing sufficiently large labels we can create an adjacency labelling scheme for any

F. Dehne, A. López-Ortiz, and J.-R. Sack (Eds.): WADS 2005, LNCS 3608, pp. 294–305, 2005.

family of graphs; for instance, we can use adjacency lists or the rows of adjacency matrices to generate adjacency labelling schemes with $O(n \log n)$ and $\Theta(n)$ bit labels, respectively, for any graph class. Typically, however, properties of the graph class allow us to use smaller labels. To date, space-optimal adjacency labelling schemes have been developed for a variety of graph classes, such as bounded arboricity graphs, line graphs, and interval graphs [1, 2].

For example, consider the following adjacency labelling scheme for interval graphs [1]. Each interval graph on $n$ vertices has an interval representation in which each vertex is assigned a closed interval of reals with unique endpoints in $\{1, \ldots, 2n\}$, with adjacency determined according to the intersection of the intervals. The marker labels each vertex with the two endpoints of its associated interval while the decoder determines adjacency in $O(1)$ time by comparing these integers just as it would two intervals (throughout this work we assume a word-level RAM computation model for the marker and decoder). Given that there are $2^{\Omega(n \log n)}$ unlabelled interval graphs on $n$ vertices [3], any adjacency labelling scheme requires $\Omega(n \log n)$ bits to represent a graph (as the adjacency function uniquely determines the graph). Having used $O(\log n)$ bits per vertex, the aforementioned adjacency labelling scheme is space optimal.

Adjacency can be replaced by any function $f$ defined on sets of vertices; in turn, for any set $S$ of vertices on which $f$ is defined, $D$ must output the correct value of $f$ on $S$ using only the labels of the vertices in $S$. By setting adjacency labelling schemes in the larger context of *informative labelling schemes*, Peleg [4] rejuvenated interest in the idea of space efficient distributed data structures as introduced by Muller [1] and Kannan, Naor, and Rudich [2]. To date, informative labelling schemes have been developed for a variety of functions including distance, routing, center of three vertices, ancestor, and nearest common ancestor. A survey of such labelling schemes can be found in [5], and detailed discussions of their applications to XML search engines and communication networks can be found in [6], [7], [8], [9], and [10].

In many applications the underlying topology is constantly changing and we desire algorithms which can accommodate these changes. At present, algorithms for finding informative labelling schemes are static, that is, if a graph is changed then the algorithm must devise a labelling of the new graph from scratch. The dynamic version of adjacency labelling schemes was mentioned in [2], however, Kannan et al. did not consider the problem in any detail. The first paper to address this dynamic problem is that of Brodal and Fagerberg [11] who develop a dynamic adjacency labelling scheme for graphs of bounded arboricity, providing the graph operations do not cause the arboricity bound to be violated. More recently, the papers of Korman and Peleg [12] and Korman, Peleg, and Rodeh [13] have considered the dynamic problem for trees in the context of distributed computing. Cohen, Kaplan, and Milo [14] consider dynamic ancestor labellings of XML trees with persistent labels, that is, the label of a vertex cannot be changed once it has been assigned; in contrast, our labels can change over time. By not using persistent labels it is possible to reduce label size as we can change the labels as required, or, as desired.

As a continuation of [11], [12], [13], and [14] there is a need for further discussion on, and development of, dynamic schemes. In particular, algorithms developed for dynamic schemes should incorporate some form of error detection; that is, the algorithms should recognize when the modified graph is no longer a member of the family under consideration. In Section 2 we discuss dynamic (implicit) adjacency labelling schemes and in Section 3 we present a dynamic (implicit) adjacency labelling scheme for line graphs, a class of graphs fundamental in the study of intersection graph theory [15].

## 2    Dynamic Implicit Adjacency Labelling Schemes

A dynamic implicit adjacency labelling scheme is the natural dynamization of an adjacency labelling scheme (note the inclusion of "implicit" which underlines the fact that adjacency is *implicitly* deduced from the vertex labels). Let $f$ be a function defined on sets of vertices. A *dynamic implicit adjacency labelling scheme* of a family $\mathcal{G}$ of finite graphs is a tuple $(M, D, \Delta, R)$ for which

- $(M, D)$ is an adjacency labelling scheme of $\mathcal{G}$.
- $\Delta$ is a set of functions which map graphs in $\mathcal{G}$ to other graphs.
- $R$ is a polynomial time relabelling algorithm (relabeller) which, using only a vertex labelling, maintains an adjacency-correct labelling while a dynamic graph operation in $\Delta$ acts on a member of $\mathcal{G}$, providing the operation produces another graph in $\mathcal{G}$. Moreover, we say that the dynamic scheme is error-detecting if $R$ can determine when a dynamic graph operation produces a graph that does not belong to $\mathcal{G}$.

In a less formal context, $R$ can be considered as the composition of algorithms required by the graph operations found in $\Delta$. For instance, if $\Delta$ permitted the addition or deletion of any edge from a graph, we might consider $R$ to be comprised of two algorithms, ADDEDGE($e,L_G$) and DELETEEDGE($e,L_G$), which use a labelling $L_G$ to relabel $G + e$ and $G - e$, respectively. Again, note that the algorithms ADDEDGE and DELETEEDGE do not directly receive $G$ as input, rather, they are provided with $L_G$ (if we maintained an adjacency matrix or adjacency lists to represent $G$ then there would be no need for the adjacency labelling scheme!). Moreover, in practice we are not interested in maintaining a labelling for every graph in the family, rather, we use the labelling of a graph to determine a labelling of a slightly modified graph, discarding the labelling of the original graph in the process. That is, the above algorithms might be presented as ADDEDGE($e$) and DELETEEDGE($e$).

Just as an adjacency labelling scheme can be created for any graph class when we allow sufficiently large labels, sufficiently weak choices of $M$, $\Delta$, and $R$ will result in a dynamic implicit adjacency labelling scheme for any class; again, consider using adjacency lists or the rows of an adjacency matrix. As a result, there are several ways in which one might judge the quality of a dynamic implicit adjacency labelling scheme. First of all, we might judge a dynamic scheme according to the time taken by $R$ relative to the time taken by the fastest marker of a (static) adjacency labelling scheme. Secondly, since a dynamic implicit adjacency labelling scheme includes an adjacency labelling scheme, we might also judge a dynamic scheme according to the size of the labels generated by $M$ and $R$. Finally, we might judge a dynamic scheme according to the range of operations contained in $\Delta$. The importance of each of these metrics is dependent on the domain in which the scheme is required; for our purposes we will try to generate space-optimal labels using $M$ and $R$ while allowing the addition and deletion of vertices and edges in an error-detecting scheme.

One might also be interested in how the effects of a dynamic change permeate through the graph; perhaps, depending on the domain, this could be a more important metric than those described above. To measure this change we define a quantity called the *modification locality*. The modification locality of $R$ is the maximum value over all operations in $\Delta$ and all possible labellings produced in the dynamic implicit adjacency labelling scheme, of the maximum distance to the set of vertices with changed neighbourhoods from a vertex whose label has changed, but whose neighbourhood has not. In essence, the modification locality measures the distance between the vertices whose labels we expected to change and the vertices whose labels we did not expect to change.

To illustrate how modification locality is calculated, consider the relabelling depicted in Figure 1. The neighbourhoods of $b$, $c$, and $e$ change whereas the labels of $d$ and $h$ are modified, but their neighbourhoods do not change. Therefore, the modification locality of this relabelling is,

$$\max_{x \in \{d,h\}} \{dist_G(x, \{b,c\}), dist_{\delta(G)}(x, \{b,c,e\})\} = dist_G(h, \{b,c\}) = 2.$$

Having considered the relabeller to be the composition of several smaller relabelling algorithms, we can consider its modification locality in terms of these smaller algorithms. Not only will this help us calculate the modification locality, this will also help us better understand the effect of specific dynamic changes on the labelling of the graph.

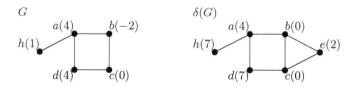

**Fig. 1.** Adding a vertex to a graph (*labels in brackets*)

Like their static counterparts, we can consider dynamic implicit adjacency labelling schemes in the larger context of dynamic implicit informative labelling schemes. Replacing adjacency with any function $f$ defined on sets of vertices, we can develop analogous measures of quality, including modification locality.

## 3    Error-Detecting Dynamic Implicit Adjacency Labelling Schemes for Line Graphs

In the remainder of this work we consider error-detecting dynamic implicit adjacency labelling schemes for line graphs. Line graphs are defined as follows [15].

**Definition 1.** *Given a graph $G = (V_G, E_G)$, its line graph is the graph $L(G) = (E_G, E_{L(G)})$ for which $\{u, v\} \in E_{L(G)}$ if and only if $u$ and $v$ are incident in $G$.*

We observe that by adding isolated vertices to $G$ we obtain infinitely many graphs with the same line graph. As such, if a graph $G$ has no isolated vertices we will refer to it as a base of $L(G)$. In [16] Whitney has shown that every connected line graph has a unique base, up to isomorphism, except for $K_3$ which has two bases, namely, $K_3$ and $K_{1,3}$. Just as a graph "generates" a line graph, we can can say that an edge labelled graph "generates" a vertex labelled line graph. For this reason, we will also use the term "base" to refer to an edge labelled graph which "generates" a particular vertex labelled line graph.

Our work on line graphs requires a concept similar to that of isomorphism, but involving edge labellings of base graphs. Given an edge labelling $\psi$ of a graph (in which each label is unique), for each edge label $\alpha$ we let $P_\alpha^\psi$ denote the partition of the labels incident with $\alpha$ that is determined by the endpoints of $\alpha$. We define two bases of a vertex labelled line graph $L(G)$, having edge labellings $\psi_1$ and $\psi_2$, to be *partition isomorphic* if the bases are isomorphic and $P_\alpha^{\psi_1} = P_\alpha^{\psi_2}$, for all edge labels $\alpha$. For example, the two bases shown in Figure 2(b) are not partition isomorphic. When we consider the theorem of Whitney in the context of labelled line graphs we arrive at the following theorem which we present without proof.

**Theorem 1.** *Every vertex labelled connected line graph, except those shown in Figure 2(a), has a unique (edge labelled) base, up to partition isomorphism. For each of the four exceptions, a vertex labelled graph has two bases that are not partition isomorphic.*

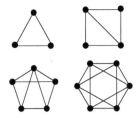
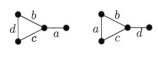
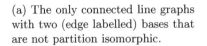

(a) The only connected line graphs with two (edge labelled) bases that are not partition isomorphic.

(b) Two bases of a vertex labelled line graph, namely $K_4 - e$, that are not partition isomorphic.

**Fig. 2.** Partition isomorphism of graphs

In [1], Muller presents a space-optimal implicit adjacency labelling scheme for the class of line graphs. Consider a line graph $L(G)$ on $n$ vertices, with base graph $G$. To each vertex in $G$ the marker assigns a unique prelabel from $\{1, \ldots, |V_G|\}$ then labels each vertex $v$ of $L(G)$ as $(ep_0, ep_1)$, where $ep_0$ and $ep_1$ are the prelabels of the endpoints of the edge of $G$ corresponding to $v$. Since the base graph has no isolated vertices, $|V_G| \le 2|E_G| = 2|V_{L(G)}| = 2n$, so each vertex label uses $O(\log n)$ bits. Presumably, the marker knows the structure of $G$, as well as the correspondence between edges in $G$ and vertices in $L(G)$, so it can generate this labelling in $\Theta(n)$ time. If the marker does not have this information then it must determine $G$ from $L(G)$ using an algorithm like those found in [17] or [18] which have running time at best $\Theta(m+n)$. The decoder can determine the adjacency of two vertices, with labels $(ep_0, ep_1)$ and $(ep_2, ep_3)$, in $O(1)$ time by checking if $\{ep_0, ep_1\} \cap \{ep_2, ep_3\} = \emptyset$.

Our dynamic labelling is similar to that of Muller, but we also incorporate circular doubly linked lists of the edges incident with each vertex in the base graph. Specifically, the labels used in our dynamic scheme will be comprised of the following information.

*pre*: Each vertex of the line graph will be assigned a unique prelabel from $\{1, \ldots, |V_{L(G)}|\}$; *pre* is the prelabel of the vertex. For simplicity, we refer to a vertex by its prelabel. Since each vertex in the line graph corresponds to an edge in the base, *pre* will also be used to refer to the edge of the base corresponding to *pre*.

$pre.ep_0, pre.ep_1$: Each vertex of the base will be assigned a unique prelabel from $\{1, \ldots, |V_G|\}$. Considered as an edge in the base, *pre* has two endpoints; $pre.ep_0$ and $pre.ep_1$ are the prelabels of these endpoints.

$pre.nn_0, pre.nn_1$: The values of $|N(pre.ep_0)|$ and $|N(pre.ep_1)|$ (in the base), respectively, where $N(x)$ denotes the open neighbourhood of the vertex $x$.

$pre.prev_0, pre.prev_1, pre.nx_0, pre.nx_1$: With *pre* as the current edge in the circular doubly linked lists about $pre.ep_i$, the prelabels of the previous and next edges are $pre.prev_i$ and $pre.nx_i$, respectively.

In particular, the label of a vertex is ($pre$: $pre.ep_0$; $pre.ep_1$; $pre.nn_0$; $pre.nn_1$; $pre.prev_0$; $pre.nx_0$; $pre.prev_1$; $pre.nx_1$); an example of this labelling is presented in Figure 3. Like the static scheme of Muller, we assume that the marker knows the structure of $G$, so it can generate an initial labelling in $\Theta(n)$ time; otherwise, it must use an algorithm like that found in [17] or [18]. Again, the decoder can determine the adjacency $v_1$ and $v_2$ in $O(1)$ time by checking if $\{v_1.ep_0, v_1.ep_1\} \cap \{v_2.ep_0, v_2.ep_1\} = \emptyset$.

Consider a line graph with $n$ vertices. If $l(string)$ denotes the number of bits required to represent *string* then the number of bits used in the label of *pre* is

$$l(pre) + \sum_{i=0}^{1} \Big( l(pre.ep_i) + l(pre.nn_i) + l(pre.prev_i) + l(pre.nx_i) \Big).$$

Observe that if the graph had been obtained by the deletion of vertices then it is possible that the largest prelabel of a vertex might be larger than $n$; as such, let

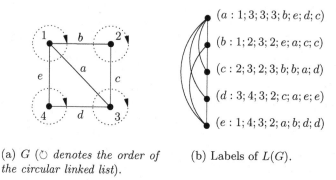

(a) $G$ ($\circlearrowleft$ denotes the order of     (b) Labels of $L(G)$.
the circular linked list).

**Fig. 3.** Our dynamic implicit adjacency labelling scheme for line graphs

the largest prelabel of a vertex in the line graph be $L$ and let the largest prelabel of a vertex in the base graph be $B$. Thereby, $l(pre), l(pre.prev_i), l(pre.nx_i) \in O(\log L)$ and $l(pre.ep_i) \in O(\log B)$. Moreover, $|V_G| \leq 2n$, so $l(pre.nn_i) \in O(\log n)$ and the label of $pre$ uses $O(\log L + \log B + \log n)$ bits. If $L, B \in O(n)$, which we assume hereafter, then this reduces to $O(\log n)$. That is, the graph is represented using $O(n \log n)$ bits. Using an argument found in Spinrad [19] (p. 18), we can show that there are $2^{\Omega(n \log n)}$ labelled line graphs on $n$ vertices, thereby proving the dynamic scheme space-optimal for labelled line graphs. This scheme may also be space-optimal for unlabelled line graphs, however, this lower bound has not yet been established in the unlabelled case.

## 3.1     Algorithms Used in the Dynamic Scheme

In the remainder of this work we discuss the graph operations included in our dynamic implicit adjacency labelling scheme for line graphs. In particular, we permit the addition or deletion of a single vertex (along with its incident edges) and the addition or deletion of a single edge. In this work, we present pseudocode only for the algorithm which deals with the deletion of a vertex; the pseudocode for the other algorithms has been omitted due to their length.

**Deleting a Vertex from the Line Graph.** One change we can make is to delete a vertex along with its incident edges. As with all of our graph modifications, it is imperative to understand how a change in the line graph causes a change in the base. By deleting a vertex from the line graph we delete the corresponding edge in the base. Letting $pre$ be the vertex of the line graph to be deleted, DELETEVERTEX, the algorithm presented in Figure 4, traverses the circular linked lists at each of the endpoints of $pre$ (in the base) so as to decrement by one the the number of edges incident with these endpoints, then removes $pre$ from the circular linked lists of its endpoints, and frees the prelabel $pre$ for future use. DELETEVERTEX runs in $O(|N(pre)|)$ time, where $|N(pre)| \in O(n)$ (we also assume a word-level RAM computation model for the relabeller). Recall that,

DELETEVERTEX($pre$)
1    **for** $i \leftarrow 0$ **to** 1 **do**
2        **if** $pre.nn_i = 1$ **then**
3            FREEBASE($pre.ep_i$)
4        **else** DECREMENTLIST($pre, i$)
5            REMOVEFROMLIST($pre, i$)
6    FREELINE($pre$)

DECREMENTLIST($x, xend$)
1    $x.nn_{xend} \leftarrow x.nn_{xend} - 1$
2    $u \leftarrow x$
3    $uend \leftarrow xend$
4    **while** $u.nx_{uend} \neq x$ **do**
5        $u \leftarrow u.nx_{uend}$
6        $uend \leftarrow$ END($u, x.ep_{xend}$)
7        $u.nn_{uend} \leftarrow u.nn_{uend} - 1$

REMOVEFROMLIST($x, xend$)
1    $y \leftarrow x.nx_{xend}$
2    $z \leftarrow x.prev_{xend}$
3    $yend \leftarrow$ END($y, x.ep_{xend}$)
4    $zend \leftarrow$ END($y, x.ep_{xend}$)
5    $y.nx_{yend} \leftarrow z$
6    $z.nx_{yend} \leftarrow y$

END($e, v$)
1    **if** $e.ep_0 = v$ **then**
2        **return** 0
3    **else return** 1

FREEBASE($bpre$)
1    free prelabel $bpre$ for future

FREELINE($lpre$)
1    free prelabel $lpre$ for future

**Fig. 4.** The algorithm DELETEVERTEX used to delete a vertex from a line graph

by the definition of a dynamic implicit informative labelling scheme, DELETEV-ERTEX is input with the entire labelling of the graph, which uses $\Theta(n \log n)$ bits, thereby, the running time of DELETEVERTEX is polynomial in the size of its inputs (this would have also been true even if we had assumed a log-cost RAM computation model in determining the running time of the relabeller). Moreover, DELETEVERTEX is error-detecting because any vertex induced subgraph of a line graph is also a line graph (hereditary).

**Proposition 1.** *The modification locality of* DELETEVERTEX *is zero.*

*Proof.* First, observe that the set of vertices whose neighbourhoods change is $N_{L(G)}[pre]$, the closed neighbourhood of $pre$ in $L(G)$. Vertex labels are only modified during a call of REMOVEFROMLIST where, in particular, if the label of a vertex $v$ is modified, then its corresponding edge in the base had been in one of the circular linked lists about an endpoint of $pre$ (in the base, $G$). That is, $v \in N_{L(G)}[pre]$. Therefore the set of vertices with modified labels is a subset of the set of vertices whose neighbourhoods change, giving the desired result.    □

**Adding a Vertex to the Line Graph.** Adding a vertex, along with its incident edges, to the line graph is equivalent to adding an edge to the base graph. In adding the vertex $pre$ to the line graph we must specify the set of neighbours of $pre$. The endpoints of $pre$ (in the base) must cover $N(pre)$ (as edges in the base), moreover, these endpoints must be incident only with these edges. If $N(pre) = \emptyset$, then ADDVERTEX creates two new vertices in the base and puts $pre$ between them. If $N(pre) \neq \emptyset$ then we are looking for a set $S$ of vertices in the base for which each of the following conditions hold.

– $1 \leq |S| \leq 2$.
– each edge of $N(pre)$ (in the base) has exactly one endpoint in $S$.
– no edge of the base not in $N(pre)$ has an endpoint in $S$.

We will call such a set $S$ *valid*.

To find a valid set ADDVERTEX calls FINDVALID. FINDVALID selects an edge of the base, $edge_0$, from $N(pre)$ and tries to include $edge_0.ep_0$ in a valid set. Letting $X$ be the subset of edges in $N(pre)$ that are not incident with $edge_0.ep_0$, we observe that if we require another vertex in the valid set then it must come from an edge in $X$. We initially set $X$ to $N(pre)$, then traverse the circular linked list structure about $edge_0.ep_0$ to eliminate edges from $X$. If at any point we find an edge which does not belong to $N(pre)$ then $edge_0.ep_0$ cannot be in the valid set, so we backtrack and try endpoint 1. If endpoint 1 is similarly problematic then the base will not yield a valid set. However, before concluding that $pre$ cannot be added to the line graph we must determine if the component of the line graph containing $edge_0$ has another base which is not partition isomorphic. If so, we repeat our efforts on $edge_0$ using this new base.

Providing some endpoint of $edge_0$ can be added to the valid set, FINDVALID now selects an edge, $edge_1$, from $X$ and tries to include $edge_1.ep_0$ in the valid set. Letting $X_1$ be the subset of edges of $X$ that are not incident with $edge_1.ep_0$, we observe that $edge_1.ep_0$ can be added to complete the valid set if and only if all of the edges found in the circular linked list about $edge_1.ep_0$ belong to $X$ and $X_1 = \emptyset$. We determine $X_1$ in a manner similar to that described for finding $X$ above, then backtrack if necessary. By backtracking, FINDVALID exhausts all combinations of bases and endpoints in finding a valid set. In particular, backtracking first tries a new endpoint then, if necessary, a new base.

If a valid set is found then we add $pre$ to the base graph using the vertices in the valid set. ADDVERTEX then inserts $pre$ into the circular linked lists about its endpoints and traverses these circular linked lists in order to adjust the $.nn$ fields of the labels to reflect the addition of $pre$. Since $N[pre]$ contains all the vertices whose labels change, the modification locality of ADDVERTEX is zero.

From Theorem 1 we see that any component of the line graph with two bases that are not partition isomorphic has $O(1)$ vertices. Therefore, FINDVALID requires at most $O(1)$ backtracks, where each base change takes $O(1)$ time, so the running time of FINDVALID is dominated by the time taken to traverse the circular linked lists when eliminating edges from $X$ and $X_1$. Consequently, FINDVALID runs in $O(|N(pre)|)$ time, as does ADDVERTEX. As per our comments earlier in this section, the running time of ADDVERTEX is polynomial in the size of the input. Moreover, ADDVERTEX is error-detecting since our use of backtracking guarantees that a valid set will be found, providing one exists.

**Deleting an Edge from the Line Graph.** Consider the act of deleting an edge from a line graph; this is equivalent to "pulling apart" two incident edges in the base. If there are additional edges incident with the vertex of the base at

**Table 1.** Possible cases for deleting an edge from (or adding an edge to) a line graph. In each case the edge $\{a, b\}$ is deleted from the line graph. The use of ellipses indicates that the line graph extends arbitrarily from the indicated vertex

| Case | A | B | C |
|------|---|---|---|
| $G$ | | | |
| $G'$ | | | |

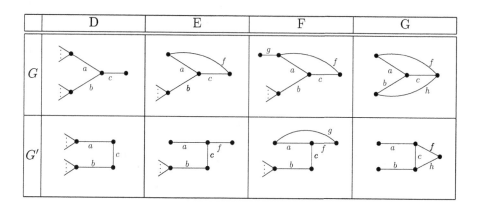

| | D | E | F | G |
|---|---|---|---|---|
| $G$ | | | | |
| $G'$ | | | | |

| | H | I | J |
|---|---|---|---|
| $G$ | | | |
| $G'$ | | | |

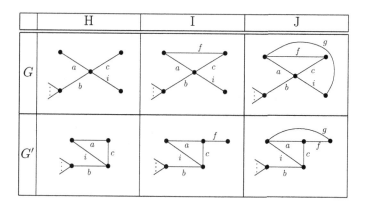

which these two incident edges were joined, then it becomes increasingly difficult to determine the new base graph. Fortunately, there are a finite number of cases to be considered; we classify these cases in the following theorem whose lengthy proof has been omitted.

**Theorem 2.** *Table 1 classifies all of the possible base graphs, up to symmetry, for which the edge $\{a, b\}$ can be deleted from the corresponding line graph to produce a new line graph. For each base graph $G$ the base graph of the new line graph is given by $G'$.*

Since the circular linked list structure distributes the information about the neighbourhood of a vertex across the labels of its neighbourhood, the vertex labels are sufficient to determine the local structures depicted in Table 1 (in fact, we can perform both depth first and breadth first search on the line graph and its base). Consequently, DELETEEDGE needs only identify the structure of the base graph then alter the labels to represent the new base graph.

From Table 1 we see that only edges of the base whose endpoints change are those which are incident with $a$ or $b$; which is to say that the only vertices of the line graph whose labels change are those which are adjacent to $a$ or $b$ (as vertices in the line graph). As such, the modification locality of DELETEEDGE is one. Moreover, given that the degrees of all the modified vertices is $O(1)$, the running time of DELETEEDGE is $O(1)$, which is sub-linear in the size of its inputs (we don't actually need access to all the labels).

**Adding an Edge to the Line Graph.** Since the process of adding an edge is exactly the reverse of deleting an edge, Table 1 classifies all the possibilities. Again, the labels of the vertices in the line graph are sufficient to determine the local structures depicted in Table 1 so the algorithm ADDVERTEX needs only identify the structure of the base graph, then alter the labels to represent the new structure. Like DELETEEDGE, ADDVERTEX runs in $O(1)$ time and has a modification locality of one.

## 4    Conclusion

In this work we have discussed the concept of a dynamic implicit adjacency labelling scheme and presented a scheme for line graphs that allows the addition and deletion of a vertex or edge. In developing this dynamic scheme we have used circular doubly linked lists to distribute information about the neighbourhood of a vertex across the labels of the vertices in that neighbourhood.

Future research will reveal dynamic schemes for additional classes of graphs, as well as for functions other than adjacency. Preliminary research indicates that the technique of using circular linked lists to store information about the vertices can also be used to design dynamic schemes for other classes defined by maximal cliques [20] and maximal vertex induced complete bipartite subgraphs.

# References

1. Muller, J.H.: Local structure in graph classes. PhD thesis, Georgia Institute of Technology (1988)
2. Kannan, S., Naor, M., Rudich, S.: Implicit representation of graphs. SIAM J. on Discrete Mathematics 5 (1992) 596–603
3. Gavoille, C., Peleg, D.: Compact and localized distributed data structures. J. of Distributed Computing 16 (2003) 111–120
4. Peleg, D.: Informative labeling schemes for graphs. In: Mathematical Foundations of Computer Science 2000, Proc. $25^{th}$ Int'l. Symp., LNCS 1893, Springer-Verlag (2000) 579–588
5. Gavoille, C., Paul, C.: Optimal distance labeling for interval and circular-arc graphs. In: Proc. $11^{th}$ Eur. Symp. on Algorithms, LNCS 2832. (2003) 254–265
6. Peleg, D.: Proximity-preserving labeling schemes and their applications. In: Graph Theoretic Concepts in Computer Science, Proc. of the $25^{th}$ Int'l. Workshop, LNCS 1665, Springer-Verlag (1999) 30–41
7. Kaplan, H., Milo, T.: Short and simple labels for small distances and other functions. In: Algorithms and Data Structures, Proc. $7^{th}$ Int'l. Workshop, LNCS 2125, Springer-Verlag (2001) 246–257
8. Abiteboul, S., Kaplan, H., Milo, T.: Compact labeling schemes for ancestor queries. In: Proc. $12^{th}$ ACM-SIAM Symp. on Discrete Algorithms, ACM (2001) 547–556
9. Alstrup, S., Gavoille, C., Kaplan, H., Rauhe, T.: Nearest common ancestors: A survey and a new distributed algorithm. In: Proc. $14^{th}$ ACM Symp. on Parallel Algorithms and Architectures, ACM (2002) 258–264
10. Thorup, M., Zwick, U.: Compact routing schemes. In: Proc. $13^{th}$ ACM Symp. on Parallel Algorithms and Architectures, ACM (2001) 1–10
11. Brodal, G.S., Fagerberg, R.: Dynamic representation of sparse graphs. In: Algorithms and Data Structures, Proc. $6^{th}$ Int'l. Workshop, LNCS 1663, Springer-Verlag (1999) 342–351
12. Korman, A., Peleg, D.: Labelling schemes for weighted dynamic trees. In: Automata, Languages and Programming, Proc. $30^{th}$ Int'l. Coll., LNCS 2719, Springer-Verlag (2003) 369–383
13. Korman, A., Peleg, D., Rodeh, Y.: Labeling schemes for dynamic tree networks. Theory of Computing Systems 37 (2004) 49–75
14. Cohen, E., Kaplan, H., Milo, T.: Labeling dynamic XML trees. In: Proc. $21^{st}$ ACM Symp. on Principles of Database Systems. (2002) 271–281
15. Branstädt, A., Le, V.B., Spinrad, J.P.: Graph Classes: A Survey. SIAM Monographs on Discrete Mathematics and Applications. SIAM, Philadelphia (1999)
16. Whitney, H.: Congruent graphs and the connectivity of graphs. American J. of Mathematics 54 (1932) 150–168
17. Lehot, P.G.H.: An optimal algorithm to detect a line graph and output its root graph. J. of the ACM 21 (1974) 569–575
18. Roussopoulos, N.D.: A max$\{m, n\}$ algorithm for determining the graph $H$ from its line graph $G$. Information Processing Letters 2 (1973) 108–112
19. Spinrad, J.: Efficient Graph Representation. Fields Institute Monographs. AMS, Providence (2003)
20. Morgan, D.: Useful names for vertices: An introduction to dynamic implicit informative labelling schemes. Technical Report TR05-04, Dept. of Computing Science, University of Alberta (2005)
21. Morgan, D.: A dynamic implicit adjacency labelling scheme for line graphs. Technical Report TR05-03, Dept. of Computing Science, University of Alberta (2005)

# The On-line Asymmetric Traveling Salesman Problem

Giorgio Ausiello, Vincenzo Bonifaci, and Luigi Laura

University of Rome "La Sapienza",
Department of Computer and System Sciences,
Via Salaria, 113 - 00198 Rome, Italy
{ausiello, bonifaci, laura}@dis.uniroma1.it

**Abstract.** We consider two on-line versions of the asymmetric traveling salesman problem with triangle inequality. For the *homing* version, in which the salesman is required to return in the city where it started from, we give a $\frac{3+\sqrt{5}}{2}$-competitive algorithm and prove that this is best possible. For the *nomadic* version, the on-line analogue of the shortest asymmetric hamiltonian path problem, we show that the competitive ratio of any on-line algorithm has to depend on the amount of asymmetry of the space in which the salesman moves. We also give bounds on the competitive ratio of on-line algorithms that are *zealous*, that is, in which the salesman cannot stay idle when some city can be served.

## 1 Introduction

In the classical traveling salesman problem, a set of cities has to be visited in a single tour with the objective of minimizing the total length of the tour. This is one of the most studied problems in combinatorial optimization, together with its dozens of variations [9, 14]. In the asymmetric version of the problem, the distance from one point to another in a given space can be different from the inverse distance. This variation, known as the Asymmetric Traveling Salesman Problem (ATSP) arises in many applications; for example, one can think of a delivery vehicle traveling through one-way streets in a city, or of gasoline costs when traveling through mountain roads.

The ATSP has been much studied from the point of view of approximation algorithms. However, if the condition is that every city or place has to be visited *exactly* once, the problem is **NPO**-complete and thus no approximation is at all possible in polynomial time, unless **P**=**NP** [17]. Instead, in the case where every city or place given in the input has to be visited *at least* once or, equivalently, the distance function satisfies the triangular inequality, approximation algorithms exist. In particular, the best algorithms known have an approximation ratio of $O(\log n)$ [8, 11]. The problem is also known to be **APX**-hard [16]. The question of the existence of an algorithm with a constant approximation ratio for the asymmetric case is still open after more than two decades.

Here we are interested in the on-line version of the ATSP, named OL-ATSP. The on-line versions of a number of vehicle routing problems, including the

F. Dehne, A. López-Ortiz, and J.-R. Sack (Eds.): WADS 2005, LNCS 3608, pp. 306–317, 2005.
© Springer-Verlag Berlin Heidelberg 2005

standard TSP, the traveling repairman problem, the quota TSP and dial-a-ride problems have been studied recently [2, 3, 4, 6, 12, 13, 15]. In the on-line TSP and ATSP, the places to visit in the space are requested over time and a server (the salesman or vehicle) has to decide in what order to serve them, without knowing the entire sequence of requests beforehand. The objective is to minimize the completion time of the server. We use the established framework of *competitive analysis* [5, 7, 18], where the cost of the algorithm being studied is compared to that of an ideal optimum off-line server, knowing in advance the entire sequence of requests (notice, however, that even the off-line server cannot serve a request before it is released). The ratio between the on-line and the off-line costs is called the *competitive ratio* of the algorithm and is a measure of the loss of efficacy due to the absence of information on the future.

As we will see, the asymmetric TSP is substantially harder than the normal TSP when considered from an on-line point of view; in other words, OL-ATSP is not a trivial extension of OL-TSP. In fact, as Table 2 shows, most bounds on the competitive ratio are strictly higher than the corresponding bounds for OL-TSP, and in particular in the nomadic case there cannot be on-line algorithms with a constant competitive ratio.

The rest of this paper is organized as follows. After the necessary definitions and the discussion of the model, we study in Section 3 the homing case of the problem, in which the server is required to finish its tour in the same place where it started; we give a $\frac{3+\sqrt{5}}{2}$-competitive algorithm and show that this is also best possible. In Section 4, we address the nomadic version, also known as the wandering traveling salesman problem [10], in which the server is not required to finish its tour in the origin. For this case we show that in general an on-line algorithm with a competitive ratio independent of the space cannot exist; indeed, we show that the competitive ratio is a precise function of the amount of asymmetry of the space. In Section 5 we explain how our algorithms can be combined with polynomial time approximation algorithms. In the last section, we give our conclusions and discuss some open problems.

## 2    The Model

An input for the OL-ATSP consists of a space $M$ from the class $\mathcal{M}$ defined below, a distinguished point $O \in M$, called the origin, and a sequence of requests $r_i = (t_i, x_i)$ where $x_i$ is a point of $M$ and $t_i \in \mathbb{R}^+$ is the time when the request is presented. The sequence is ordered so that $i < j$ implies $t_i \leq t_j$.

The server is located at the origin $O$ at time 0 and the distances are scaled so that, without loss of generality, the server can move at most at unit speed.

We will consider two versions of the problem. In the *nomadic* version, the server can end its route anywhere in the space; the objective is just to minimize the completion time required to serve all presented requests. In the *homing* version, the objective is to minimize the completion time required to serve all presented requests and return to the origin.

An *on-line* algorithm for the OL-ATSP has to determine the behavior of the server at a certain moment $t$ as a function only of the requests $(t_i, x_i)$ such that $t_i \leq t$. Thus, an on-line algorithm does not have knowledge about the number of requests or about the time when the last request is released. We will use $p^{\mathrm{OL}}(t)$ to denote the position of the on-line server at time $t$. Sometimes we will let $T$ be some tour or route over a subset of the requests; in this case, $|T|$ will be the length of that tour.

We will use $Z^{\mathrm{OL}}$ to denote the completion time of the solution produced by a generic on-line algorithm OL, while $Z^*$ will be the completion time of the optimal off-line solution. An on-line algorithm OL is *c-competitive* if, for any sequence of requests, $Z^{\mathrm{OL}} \leq cZ^*$.

A word is in order about the spaces that we are going to allow for our problems. Usually, in the context of the on-line TSP, path-metric spaces are considered [3]. However, here the main issue is precisely asymmetry, so we have to drop the requisite that for every $x$ and $y$, $d(x,y) = d(y,x)$. Thus we obtain *path-quasi-metric spaces*. We review here the definitions. A set $M$, equipped with a distance function $d : M^2 \to \mathbb{R}^+$, is called a *quasi-metric space* if, for all $x, y, z \in M$:

(i)  $d(x,y) = 0$ if and only if $x = y$;[1]
(ii) $d(x,y) \leq d(x,z) + d(z,y)$.

We call a space $M$ *path-metric* if, for any $x, y \in M$, there is a function $f : [0,1] \to M$ such that $f(0) = x$, $f(1) = y$ and $f$ is continuous, in the following sense: $d(f(a), f(b)) = (b - a)d(x,y)$ for any $0 \leq a \leq b \leq 1$. This function represents a *shortest path* from $x$ to $y$. Notice that the path-metric property implies connectivity.

We will use $\mathcal{M}$ to denote the class of path-quasi-metric spaces. Notice that discrete metrics (i.e., those with a finite number of points) are not path-metric. However, we can always make a space path-metric by adding (an infinity of) extra points between pairs of vertices.

In particular, to see how a directed graph with positive weights on the arcs can define a path-quasi-metric space, consider the all-pairs shortest paths matrix of the graph. This defines a finite quasi-metric. Now we add, for every ordered pair of nodes $x$ and $y$, an infinity of points $\pi_\gamma$, indexed by a parameter $\gamma \in (0,1)$. Let $\pi_0$ and $\pi_1$ denote $x$ and $y$ respectively. We extend the distance function $d$ so that:

$$d(\pi_\gamma, \pi_{\gamma'}) = (\gamma' - \gamma)d(x,y) \quad \text{for all } 0 \leq \gamma < \gamma' \leq 1.$$

It can be verified that $\pi$ represents a shortest path from $x$ to $y$. For $\gamma \notin \{0,1\}$, the distance from a point $\pi_\gamma$ to a point $z$ not in $\pi$ is defined as $d(\pi_\gamma, z) = d(\pi_\gamma, y) + d(y, z)$; that is, the shortest path from $\pi_\gamma$ to $z$ passes through $y$. Viceversa, the distance from $z$ to $\pi_\gamma$ is defined as $d(z, \pi_\gamma) = d(z, x) + d(x, \pi_\gamma)$. Finally,

$$d(\pi_{\gamma'}, \pi_\gamma) = (1 - (\gamma' - \gamma))d(x,y) + d(y,x) \quad \text{for all } 0 \leq \gamma < \gamma' \leq 1.$$

---

[1] This condition is not strictly essential; we could consider *quasi-pseudo-metric* spaces, for which this condition is relaxed to $d(x,x) = 0$ for all $x \in M$.

We say that such a space is *induced* by the original directed weighted graph. We remark that our model, differently from the originally proposed one [3], allows to model the case in which the server is not allowed to do U-turns.

It will be useful to have a measure of the amount of asymmetry of a space. Define as the *maximum asymmetry* of a space $M \in \mathcal{M}$ the value

$$\Psi(M) = \sup_{x,y \in M} \frac{d(x,y)}{d(y,x)}.$$

We will say that a space $M$ has *bounded asymmetry* when $\Psi(M) < \infty$.

## 3    Homing OL-ATSP

In this section we consider the homing version of the on-line ATSP, in which the objective is to minimize the completion time required to serve all presented requests and return to the origin. We establish a lower bound of about 2.618 and a matching upper bound. Note that in the case of symmetric on-line TSP, the corresponding bounds are both equal to 2 [3, 12].

Let $\phi$ denote the golden ratio, that is, the unique positive solution to $x = 1 + 1/x$. In closed form, $\phi = \frac{1+\sqrt{5}}{2} \simeq 1.618$.

**Theorem 1.** *The competitive ratio of any on-line algorithm for homing OL-ATSP is at least $1 + \phi$.*

*Proof.* Fix any $\epsilon > 0$. The space used in the proof is the one induced by the graph depicted in Figure 1. The graph has $7 + 4n$ nodes, where $n = 1 + \lceil \frac{\phi-1}{\epsilon} \rceil$, and the length of every arc is $\epsilon$, except for those labeled otherwise. Observe that the space is symmetric with respect to an imaginary vertical axis passing through $O$. Thus, we can assume without loss of generality that, at time 1, the on-line server is in the left half of the space. Then at time 1 a request is given in point $A$, in the other half. Now let $t$ be the first time at which the on-line server reaches point $D$ or $E$.

If $t \geq \phi$, no further request is given. In this case $Z^{OL} \geq t + 1 + 2\epsilon$ while $Z^* \leq 1 + 3\epsilon$ so that, when $\epsilon$ approaches zero, $Z^{OL}/Z^*$ approaches $1 + t \geq 1 + \phi$.

Otherwise, if $t \in [1, \phi]$, at time $t$, we can assume that the on-line server has just reached $E$ (again, by symmetry). At this time, the adversary gives a request in $B_i$, where $i = \lceil \frac{t-1}{\epsilon} \rceil$. Now the on-line server has to traverse the entire arc $EC$ before it can go serve $B_i$, thus

$$Z^{OL} \geq t + 1 + 3\epsilon + 1 + \epsilon \left\lceil \frac{t-1}{\epsilon} \right\rceil + 2\epsilon \geq 2t + 1 + 5\epsilon.$$

Instead, the adversary server will have moved from $O$ to $B_i$ in time at most $t + 2\epsilon$ and then served $B_i$ and $A$, achieving the optimal cost $Z^* \leq t + 4\epsilon$. Thus, when $\epsilon$ approaches zero, $Z^{OL}/Z^*$ approaches $2 + \frac{1}{t} \geq 1 + \phi$.    □

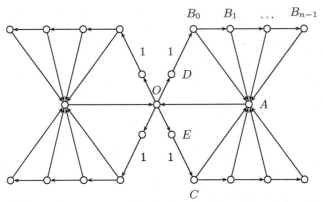

**Fig. 1.** The space used in the homing lower bound proof

To prove a matching upper bound on the competitive ratio, we use a variation of algorithm SMARTSTART, first introduced by Ascheuer et al. [1].

**Algorithm.** SMARTSTART($\alpha$)
The algorithm keeps track, at every time $t$, of the length of an optimal tour $T^*(t)$ over the unserved requests. At the first instant $t$ such that $t \geq \alpha|T^*(t)|$, the server starts following at full speed the currently optimal tour, ignoring temporarily every new request. When the server is back in the origin, it stops and returns monitoring the value $|T^*(t)|$, starting as before when necessary. As we will soon see, the best value of $\alpha$ is $\alpha^* = \phi$.

**Theorem 2.** SMARTSTART($\phi$) *is* $(1 + \phi)$-*competitive for homing* OL-ATSP.

*Proof.* We distinguish two cases depending if the last request arrives while the server is waiting in the origin or not.

In the first case, let $t$ be the release time of the last request. If the server starts immediately at time $t$, it will follow a tour of length $|T^*(t)| \leq t/\alpha$, ending at time at most $(1+1/\alpha)t$, while the adversary pays at least $t$, so the competitive ratio is at most $1 + 1/\alpha$. Otherwise, the server will start at a time $t' > t$ such that $t' = \alpha|T^*(t)|$ (since $T^*$ does not change after time $t$) and pay $(1+\alpha)|T^*(t)|$, so the competitive ratio is at most $1 + \alpha$.

In the second case, let $T^*(t)$ be the tour that the server is following while the last request arrives; that is, we take $t$ to be the starting time of that tour. Let $T'(t)$ be an optimal tour over the requests released *after* time $t$. If the server has time to wait in the origin when it finishes following $T^*(t)$, the analysis is the same as in the first case. Otherwise, the completion time of SMARTSTART is $t + |T^*(t)| + |T'(t)|$. Since SMARTSTART has started following $T^*(t)$ at time $t$, we have $t \geq \alpha|T^*(t)|$. Then

$$t + |T^*(t)| \leq (1 + 1/\alpha)t.$$

Also, if $r_f = (t_f, x_f)$ is the first request served by the adversary having release time at least $t$, we have that $|T'(t)| \leq d(O, x_f) + Z^* - t$ since a possibility for

$T'$ is to go to $x_f$ and then do the same as the adversary (subtracting $t$ from the cost since we are computing a length, not a completion time, and on the other hand the adversary will not serve $r_f$ at a time earlier than $t$).

By putting everything together, we have that SMARTSTART pays at most

$$(1 + 1/\alpha)t + d(O, x_f) + Z^* - t$$

and since two obvious lower bounds on $Z^*$ are $t$ and $d(O, x_f)$, this is easily seen to be at most $(2 + 1/\alpha)Z^*$.

Now $\max\{1 + \alpha, 2 + \frac{1}{\alpha}\}$ is minimum when $\alpha = \alpha^* = \phi$. For this value of the parameter the competitive ratio is $1 + \phi$.     □

## 3.1   Zealous Algorithms

In the previous section we have seen that the optimum performance is achieved by an algorithm that before starting to serve requests, waits until a convenient starting time is reached. In this section we consider instead the performance that can be achieved by *zealous* algorithms [4]. A zealous algorithm does not change the direction of its server unless a new request becomes known, or the server is in the origin or at a request that has just been served; furthermore, a zealous algorithm moves its server always at full (that is, unit) speed.

We show that, for zealous algorithms, the competitive ratio has to be at least 3 and, on the other hand, we give a matching upper bound.

**Theorem 3.** *The competitive ratio of any zealous on-line algorithm for homing OL-ATSP is at least 3.*

*Proof.* We use the same space used in the lower bound for general algorithms (Figure 1). At time 1, the server has to be in the origin and the adversary gives a request in $A$. Thus, at time $1 + \epsilon$ the server will have reached w.l.o.g. $E$ (by symmetry) and the adversary gives a request in $B_0$. The completion time of the on-line algorithm is at least $3 + 6\epsilon$, while $Z^* \leq 1 + 3\epsilon$. The result follows by taking a sufficiently small $\epsilon$.     □

The following algorithm is best possible among the zealous algorithms for homing OL-ATSP.

**Algorithm.** PLAN AT HOME
When the server is in the origin and there are unserved requests, the algorithm computes an optimal tour over the set of unserved requests and the server starts following it, ignoring temporarily every new request, until it finishes its tour in the origin. Then it waits in the origin as before.

**Theorem 4.** PLAN AT HOME *is zealous and* 3-*competitive for homing* OL-ATSP.

*Proof.* Let $t$ be the release time of the last request. If $p(t)$ is the position of PLAN AT HOME at time $t$ and $T$ is the tour it was following at that time, we have that PLAN AT HOME finishes following $T$ at time $t' \leq t + |T|$. At that time, it will eventually start again following a tour over the requests which remain unserved at time $t'$. Let us call $T'$ this other tour. The total cost payed by PLAN AT HOME will be then at most $t + |T| + |T'|$. But $t \leq Z^*$, since even the off-line adversary cannot serve the last request before it is released, and on the other hand both $T$ and $T'$ have length at most $Z^*$, since the off-line adversary has to serve all of the requests served in $T$ and $T'$. Thus, $t + |T| + |T'| \leq 3Z^*$. □

## 4   Nomadic OL-ATSP

In this section we consider the nomadic version of the on-line ATSP, in which the server can end its route anywhere in the space. We show that no on-line algorithm can have a constant competitive ratio (that is, independent of the underlying space). Then we show, for spaces with a maximum asymmetry $K$, a lower bound $\sqrt{K}$ and an upper bound $1 + \sqrt{K+1}$. Note that in the case of symmetric nomadic on-line TSP, the best lower and upper bounds are 2.03 and $1 + \sqrt{2}$, respectively [15].

**Theorem 5.** *For every $L > 0$, there is a space $M \in \mathcal{M}$ such that the competitive ratio of any on-line algorithm for nomadic OL-ATSP on $M$ is at least $L$.*

*Proof.* For a fixed $\epsilon > 0$, consider the space induced by a directed cycle on $n = \lceil \frac{L}{\epsilon} \rceil$ nodes, where every arc has length $\epsilon$ (Figure 2). At time 0 a request is given in node $A_3$. Let $t$ be the first time the on-line algorithm reaches node $A_2$.

Now if $t \geq 1$, the adversary does not release any other request so that $Z^* = 2\epsilon$, $Z^{OL} \geq 1 + \epsilon$ and $Z^{OL}/Z^* \geq \frac{1}{2\epsilon} + \frac{1}{2}$.

Otherwise, if $t \leq 1$, at time $t$ the adversary releases a request in the origin. It is easily seen that $Z^* \leq t + 2\epsilon$ and $Z^{OL} \geq t + \epsilon(\lceil \frac{L}{\epsilon} \rceil - 1) \geq t + 2\epsilon + L - 3\epsilon$ so that

$$Z^{OL}/Z^* \geq 1 + \frac{L - 3\epsilon}{t + 2\epsilon} \geq 1 + \frac{L - 3\epsilon}{1 + 2\epsilon}.$$

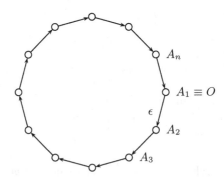

**Fig. 2.** The space used in the nomadic lower bound proof

By taking $\epsilon$ close to zero we see that in the first case the competitive ratio grows indefinitely while in the second case it approaches $L$. □

**Corollary 1.** *There is no on-line algorithm for nomadic* OL-ATSP *on all spaces $M \in \mathcal{M}$ with a constant competitive ratio.*

We also observe that the same lower bound can be used when the objective function is the sum of completion times.

Thus, we cannot hope for an on-line algorithm which is competitive for all spaces in $\mathcal{M}$. Indeed, we will now show that the amount of asymmetry of a space is related to the competitive ratio of any on-line algorithm for that space.

**Theorem 6.** *For every $K \geq 1$, there is a space $M \in \mathcal{M}$ with maximum asymmetry $K$ such that any on-line algorithm for nomadic* OL-ATSP *on $M$ has competitive ratio at least $\sqrt{K}$.*

*Proof.* Consider a set of points $M = \{x_\gamma : \gamma \in [0,1]\}$ with a distance function

$$d(x_\gamma, x_{\gamma'}) = \begin{cases} \gamma' - \gamma & \text{if } \gamma \leq \gamma' \\ K(\gamma - \gamma') & \text{if } \gamma \geq \gamma'. \end{cases}$$

The origin is $x_0$. The adversary releases a request at time 1 in point $x_1$. Let $t$ be the time the on-line algorithm serves this request. If $t \geq \sqrt{K}$, no more requests are released and $Z^{\text{OL}} \geq \sqrt{K}$, $Z^* = 1$, $Z^{\text{OL}}/Z^* \geq \sqrt{K}$.

Otherwise, if $t \leq \sqrt{K}$, at time $t$ a request is given in the origin. Now $Z^{\text{OL}} \geq t + K$, $Z^* \leq t + 1$ and

$$Z^{\text{OL}}/Z^* \geq \frac{t+K}{t+1} = 1 + \frac{K-1}{t+1} \geq 1 + \frac{K-1}{\sqrt{K}+1} = \sqrt{K}. □$$

A natural algorithm, on the lines of the best known algorithm for the symmetric version of the problem [15], gives a competitive ratio which is asymptotically the same as that of this lower bound.

**Algorithm.** RETURN HOME$(\alpha)$
At any moment at which a new request is released, the server returns to the origin via the shortest path. Once in the origin at time $t$, it computes an optimal route $T$ over all requests presented up to time $t$ and then starts following this route, staying within distance $\alpha t'$ of the origin at any time $t'$, by reducing the speed at the latest possible time.

**Theorem 7.** *For every space $M \in \mathcal{M}$ with maximum asymmetry $K$, there is a value of $\alpha$ such that* RETURN HOME$(\alpha)$ *is $(1 + \sqrt{K+1})$-competitive on $M$.*

*Proof.* There are two cases to consider. In the first case RETURN HOME does not need to reduce its speed after the last request is released. In this case, if $t$ is the release time of the last request, we have

$$Z^{\text{RH}} \leq t + K\alpha t + |T| \leq Z^* + K\alpha Z^* + Z^* = (2 + K\alpha)Z^*.$$

In the second case, let $t$ be the last time RETURN HOME is moving at reduced speed. At that time, RETURN HOME has to be serving some request; let $x$ be the location of that request. Since RETURN HOME is moving at reduced speed we must have $d(O, x) = \alpha t$; afterwards RETURN HOME will follow the remaining part $T_x$ of the route at full speed. Thus

$$Z^{\mathrm{RH}} \leq t + |T_x| = (1/\alpha)d(O, x) + |T_x|.$$

On the other hand, $Z^* \geq |T| \geq d(O, x) + |T_x|$. Thus, in this case, the competitive ratio is at most $1/\alpha$.

Obviously, we can choose $\alpha$ in order to minimize $\max\{2 + K\alpha, 1/\alpha\}$. This gives a value of $\alpha^* = \frac{\sqrt{K+1}-1}{K}$, for which we obtain the competitive ratio of the theorem. $\qquad \square$

### 4.1    Zealous Algorithms

Also in the case of the nomadic version of the on-line ATSP, we wish to consider the performance of zealous algorithms. Of course, no zealous algorithm will be competitive for spaces with unbounded asymmetry. Here we show that the gap between non-zealous and zealous algorithms is much higher than in the homing case, increasing from $\Theta(\sqrt{K})$ to $\Theta(K)$.

**Theorem 8.** *For every $K \geq 1$, there is a space $M \in \mathcal{M}$ with maximum asymmetry $K$ such that the competitive ratio of any zealous on-line algorithm for nomadic OL-ATSP on $M$ is at least $\frac{1}{2}(K + 1)$.*

*Proof.* We use the same space used in the proof of Theorem 6 (Figure 2). At time 0, the adversary releases a request in point $x_1$. The on-line server will be at point $x_1$ exactly at time 1. Then, at time 1, the adversary releases a request in point $x_0$. It is easy to see that $Z^{\mathrm{OL}} \geq 1 + K$, while $Z^* = 2$. $\qquad \square$

We finally observe that RETURN HOME(1) is a zealous algorithm for nomadic OL-ATSP and, by the proof of Theorem 7, it has competitive ratio $K + 2$.

## 5    Polynomial Time Algorithms

None of the algorithms that we have proposed in the previous sections runs in polynomial time, since all of them need to compute optimal tours on some subsets of the requests. On the other hand, a polynomial time on-line algorithm with a constant competitive ratio could be used as an approximation algorithm for the ATSP, and thus we do not expect to find one easily. However, our algorithms use off-line optimization as a black box and thus can use approximation algorithms as subroutines in order to give polynomial time on-line algorithms, the competitive ratio depending of course on the approximation ratio.

The basic problem that has to be solved in the homing version is the off-line ATSP. The best polynomial time algorithm for this problem has an approximation ratio of $\rho \simeq 0.842 \log n$ [11]. For the nomadic version, the corresponding off-line problem is the shortest asymmetric hamiltonian path. We are not

**Table 1.** The competitive ratio as a function of $\rho$ and $K$

| Problem | Algorithm | Competitive ratio |
|---------|-----------|-------------------|
| Homing OL-ATSP | SMARTSTART($\alpha^*$) | $(1 + 2\rho + \sqrt{1 + 4\rho})/2$ if $\rho \leq 2$ <br> $2\rho$ if $\rho \geq 2$ |
| Homing OL-ATSP | PLAN AT HOME | $1 + 2\rho$ |
| Nomadic OL-ATSP | RETURN HOME($\alpha^*$) | $2K(\sqrt{(1+\rho)^2 + 4K} - (1+\rho))^{-1}$ |
| Nomadic OL-ATSP | RETURN HOME(1) | $1 + \rho + K$ |

aware of studies about algorithms with a guaranteed approximation ratio for this problem; however, for spaces with bounded asymmetry, an algorithm with approximation ratio $\rho = (1 + K)\rho_T$ can be easily obtained by using the solution found by any algorithm for ATSP with an approximation ratio of $\rho_T$.

We do not repeat here the proofs of our theorems taking into account the approximation ratio of the off-line solvers, since they are quite straightforward. However, we give the competitive ratio of our algorithms as a function of $\rho$, the approximation ratio, and $K$, the maximum asymmetry of the space, in Table 1.

# 6   Conclusions

We have examined some of the on-line variations of the asymmetric traveling salesman problem. Table 2 compares the bounds on the competitive ratio of the problems considered in this paper with those of the corresponding symmetric variations. The table confirms that the asymmetric problems are indeed harder and not simply extensions than their symmetric counterparts.

The main conclusion is that, as usual in on-line vehicle routing when minimizing the completion time, waiting can improve the competitive ratio a lot. This is particularly evident in the case of nomadic ATSP on spaces with bounded

**Table 2.** The competitive ratio of symmetric and asymmetric routing problems

| Problem | Best Lower Bound | Best Upper Bound | References |
|---------|------------------|------------------|------------|
| Homing OL-TSP | 2 | 2 | [3, 12] |
| Homing OL-ATSP | 2.618 | 2.618 | |
| Homing OL-TSP (zealous) | 2 | 2 | [3] |
| Homing OL-ATSP (zealous) | 3 | 3 | |
| Nomadic OL-TSP | 2.03 | 2.414 | [15] |
| Nomadic OL-ATSP | $\sqrt{K}$ | $1 + \sqrt{K + 1}$ | |
| Nomadic OL-TSP (zealous) | 2.05 | 2.5 | [3, 15] |
| Nomadic OL-ATSP (zealous) | $\frac{1}{2}(K + 1)$ | $2 + K$ | |

asymmetry, where zealous algorithms have competitive ratio $\Omega(K)$ while RE-TURN HOME is $O(\sqrt{K})$-competitive.

We expect the competitive ratio of the homing OL-ATSP to be somewhat lower than $1 + \phi$ when the space has bounded asymmetry. Also, since the proof that no on-line algorithm can have a constant competitive ratio in the nomadic case also applies to the on-line asymmetric traveling repairman problem, it would be interesting to investigate this problem in spaces with bounded asymmetry.

Finally, we remark that the existence of polynomial time $O(1)$-competitive algorithms is indissolubly tied to the existence of an $O(1)$-approximation algorithm for the off-line ATSP.

## Acknowledgments

The authors would like to thank Luca Allulli for some helpful discussion.

## References

1. N. Ascheuer, S. O. Krumke, and J. Rambau. Online dial-a-ride problems: Minimizing the completion time. In *Proc. 17th Symp. on Theoretical Aspects of Computer Science*, volume 1770 of *Lecture Notes in Computer Science*, pages 639–650. Springer-Verlag, 2000.
2. G. Ausiello, M. Demange, L. Laura, and V. Paschos. Algorithms for the on-line quota traveling salesman problem. In K.-Y. Chwa and J. I. Munro, editors, *Proc. 10th Annual International Computing and Combinatorics Conference*, volume 3106 of *Lecture Notes in Computer Science*, pages 290–299. Springer-Verlag, 2004.
3. G. Ausiello, E. Feuerstein, S. Leonardi, L. Stougie, and M. Talamo. Algorithms for the on-line travelling salesman. *Algorithmica*, 29(4):560–581, 2001.
4. M. Blom, S. O. Krumke, W. E. de Paepe, and L. Stougie. The online-TSP against fair adversaries. *INFORMS Journal on Computing*, 13:138–148, 2001.
5. A. Borodin and R. El-Yaniv. *Online Computation and Competitive Analysis*. Cambridge University Press, 1998.
6. E. Feuerstein and L. Stougie. On-line single-server dial-a-ride problems. *Theoretical Computer Science*, 268:91–105, 2001.
7. A. Fiat and G. J. Woeginger, editors. *Online Algorithms: The State of the Art*. Springer-Verlag, 1998.
8. A. M. Frieze, G. Galbiati, and F. Maffioli. On the worst-case performance of some algorithms for the asymmetric traveling salesman problem. *Networks*, 12(1):23–39, 1982.
9. G. Gutin and A. P. Punnen, editors. *The Traveling Salesman Problem and its Variations*. Kluwer, Dordrecht, The Nederlands, 2002.
10. M. Jünger, G. Reinelt, and G. Rinaldi. The traveling salesman problem. In M. O. Ball, T. Magnanti, C. L. Monma, and G. Nemhauser, editors, *Network Models, Handbook on Operations Research and Management Science*, volume 7, pages 225–230. Elsevier, 1995.
11. H. Kaplan, M. Lewenstein, N. Shafrir, and M. Sviridenko. Approximation algorithms for asymmetric TSP by decomposing directed regular multigraphs. In *Proc. 44th Symp. Foundations of Computer Science*, pages 56–66, 2003.

12. S. O. Krumke. Online optimization: Competitive analysis and beyond. Habilitation Thesis, Technical University of Berlin, 2001.

13. S. O. Krumke, W. E. de Paepe, D. Poensgen, and L. Stougie. News from the online traveling repairman. In J. Sgall, A. Pultr, and P. Kolman, editors, *Proc. 26th Symp. on Mathematical Foundations of Computer Science*, volume 2136 of *Lecture Notes in Computer Science*, pages 487–499. Springer-Verlag, 2001.

14. E. L. Lawler, J. K. Lenstra, A. R. Kan, and D. B. Shmoys. *The Traveling Salesman Problem: A Guided Tour of Combinatorial Optimization.* Wiley, Chichester, England, 1985.

15. M. Lipmann. *On-Line Routing.* PhD thesis, Technical University of Eindhoven, 2003.

16. C. H. Papadimitriou and M. Yannakakis. The traveling salesman problem with distances one and two. *Mathematics of Operations Research*, 18(1):1–11, 1993.

17. S. Sahni and T. F. Gonzalez. P-complete approximation problems. *Journal of the ACM*, 23(3):555–565, 1976.

18. D. Sleator and R. E. Tarjan. Amortized efficiency of list update and paging rules. *Communications of the ACM*, 28(2):202–208, 1985.

# All-Pairs Shortest Paths with Real Weights in $O(n^3/\log n)$ Time

Timothy M. Chan

School of Computer Science, University of Waterloo,
Waterloo, Ontario N2L 3G1, Canada
tmchan@uwaterloo.ca

**Abstract.** We describe an $O(n^3/\log n)$-time algorithm for the all-pairs-shortest-paths problem for a real-weighted directed graph with $n$ vertices. This slightly improves a series of previous, slightly subcubic algorithms by Fredman (1976), Takaoka (1992), Dobosiewicz (1990), Han (2004), Takaoka (2004), and Zwick (2004). The new algorithm is surprisingly simple and different from previous ones.

## 1   Introduction

The *all-pairs-shortest-paths* problem (APSP) is of course one of the most well-studied problems in algorithm design. We consider here the general case where the input is a weighted directed graph and edge weights are arbitrary real numbers. The problem is to compute the shortest-path distance between every pair of vertices, together with a representation of these shortest paths (so that the shortest path for any given vertex pair can be retrieved in time linear in its length).

The classical Floyd–Warshall algorithm [6] solves the APSP problem in $O(n^3)$ time for a graph with $n$ vertices. Fredman [10] was the first to realize that subcubic running time is attainable: he gave an algorithm with an impressive-looking time bound of $O(n^3(\log\log n/\log n)^{1/3})$. Later, Takaoka [19] and Dobosiewicz [7] refined Fredman's approach and reduced the bound to $O(n^3\sqrt{\log\log n/\log n})$ and $O(n^3/\sqrt{\log n})$ respectively. Just last year, several interesting, independent developments have occurred: first Han [12] announced an improved $O(n^3(\log\log n/\log n)^{5/7})$-time algorithm, then Takaoka [20] announced an even better $O(n^3(\log\log n)^2/\log n)$-time algorithm, and finally Zwick [23] found the algorithm with the current record of $O(n^3\sqrt{\log\log n}/\log n)$ time. The record turns out to be short-lived—in this note, we obtain yet another algorithm with a further improved running time of $O(n^3/\log n)$.

**Related work.** For *sparse* graphs, a more efficient solution to APSP is to apply Dijkstra's single-source algorithm $n$ times, as described in any decent algorithm textbook [6]. Using a Fibonacci-heap implementation (with Johnson's preprocessing step if negative weights are allowed), the running time is $O(n^2\log n + mn)$, where $m$ denotes the number of edges. For a long time, this

F. Dehne, A. López-Ortiz, and J.-R. Sack (Eds.): WADS 2005, LNCS 3608, pp. 318–324, 2005.
© Springer-Verlag Berlin Heidelberg 2005

was the best result known, until recently Pettie and Ramachandran [14] and Pettie [13] have managed to bring the time bounds down to $O(mn \log \alpha(m, n))$ and $O(n^2 \log \log n + mn)$ for undirected and directed graphs, respectively, using rather complicated techniques.

A flurry of activities in the last decade has concentrated on the case of graphs with *small integer weights* (and, in particular, unweighted graphs), where a number of genuinely subcubic algorithms [2, 11, 16, 17, 22] have been developed using known methods for matrix multiplication over rings [5, 18]. Currently, the best such APSP algorithms for undirected and directed graphs run in $O(n^{2.376}M)$ and $O(n^{2.575}M^{0.681})$ time respectively [17, 22], where $M$ denotes the maximum edge weight (in absolute value). Note that these running times are subcubic only when $M \ll n^{0.634}$. It is not known whether such matrix multiplication methods can help for APSP in the case of real weights, or for that matter, integer weights from the range $\{0, 1, \ldots, n\}$. (Even if the answer is affirmative, algorithms that involve so-called "fast" matrix multiplication are not necessarily attractive from a practical point of view.) Feder and Motwani [9] described an $O(n^3/\log n)$-time algorithm that avoids fast matrix multiplication but their algorithm works only for unweighted, undirected graphs.

**About the new algorithm.** We confess that our result represents only a minute improvement over previous slightly subcubic algorithms in the general real-weight case—a mere $\sqrt{\log \log n}$-factor speedup over the previous result by Zwick! However, we believe that our algorithm is interesting, because (i) it is conceptually very simple (note the length of the paper) and (ii) it is markedly different from previous approaches:

The approach originated by Fredman [10] (and later continued by Takaoka [19]) broke the $O(n^3)$ barrier by relying on *table-lookup* tricks (storing solutions to all small-sized subproblems in an array for later retrieval in constant time). Dobosiewicz's approach [7] avoided explicit table lookups by exploiting *word-RAM operations* (specifically, performing bitwise-logical operations on $(\log n)$-bit words in unit time). The recent algorithms by Han [12], Takaoka [20], and Zwick [23] all involved even more complicated combinations of approaches. For example, Zwick [23] nontrivially combined Dobosiewicz's approach with a known table-lookup technique for Boolean matrix multiplication [3], resulting in an algorithm that uses both table lookups and word-RAM operations. In contrast, our approach uses *neither* table lookups nor word operations! In fact, our algorithm is readily implementable within the pointer-machine model.[1] Curiously, our approach is *geometrically* inspired (based on a multidimensional divide-and-conquer technique commonly seen in computational geometry). Considering the long history of the APSP problem, it is amusing that this little idea alone can beat all previous algorithms for arbitrary, real-weighted, dense graphs.

---

[1]   To be fair, we should mention that some algorithms based on table lookups can be modified to run on pointer machines, for example, by using variants of radix sort [4]. The same could be true for some of the previous APSP algorithms, but such a modification would seem to require much additional effort.

## 2     A Geometric Subproblem

We begin with what may at first appear to be a complete digression: a problem in computational geometry, concerning a special case of off-line orthogonal range searching (also similar to the "maxima" problem) [15]. The problem is to find all *dominating pairs* between a red point set and a blue point set in $d$-dimensional space, where a red point $p = (p_1, \ldots, p_d) \in \mathbb{R}^d$ and a blue point $q = (q_1, \ldots, q_d) \in \mathbb{R}^d$ are said to form a dominating pair iff $p_k \leq q_k$ for all $k = 1, \ldots, d$. The algorithm in the lemma below is standard [15], but unlike traditional analysis in computational geometry, we are interested in the case where the dimension is not a constant.

**Lemma 1.** *Given $n$ red/blue points in $\mathbb{R}^d$, we can report all $K$ dominating pairs in $O(c_\varepsilon^d n^{1+\varepsilon} + K)$ time for any constant $\varepsilon \in (0,1)$, where $c_\varepsilon := 2^\varepsilon/(2^\varepsilon - 1)$.*

*Proof.* We describe a simple divide-and-conquer algorithm. If $n = 1$, we stop. If $d = 0$, we just output all pairs of red and blue points. Otherwise, we compute the median $z$ of the $d$-th coordinates of all points and let $P_{\text{left},\gamma}$ (resp. $P_{\text{right},\gamma}$) denote the subset of all points of color $\gamma$ with $d$-th coordinates at most $z$ (resp. at least $z$). (Note that we can avoid a linear-time median-finding algorithm if we pre-sort all the $d$-th coordinates.) We then recursively solve the problem for $P_{\text{left},\text{red}} \cup P_{\text{left},\text{blue}}$, for $P_{\text{right},\text{red}} \cup P_{\text{right},\text{blue}}$, and finally for the projection of $P_{\text{left},\text{red}} \cup P_{\text{right},\text{blue}}$ to the first $d - 1$ coordinates. (Note that we can avoid actually projecting the points, by just ignoring the $d$-th coordinates.) Correctness is immediate.

Excluding the output cost, the running time obeys the recurrence

$$T_d(n) \leq 2T_d(n/2) + T_{d-1}(n) + O(n),$$

with $T_d(1) = O(1)$ and $T_0(n) = O(n)$. The additional output cost is bounded by $O(K)$, since each pair is reported once.

Naively, one can establish by induction on $d$ that $T_d(n) = O(n \log^d n)$, yielding an $O(n \log^d n + K)$-time algorithm. This result is already known. (In fact, it is known that one can save one or two logarithmic factors by handling the base cases $d = 1$ and $d = 2$ directly.)

We offer an alternative analysis of the recurrence that is better for certain non-constant values of $d$ and is thus slightly more effective for the application in the next section. We make a change of variable: fixing a parameter $b$ and letting $T'(N) := \max_{b^d n \leq N} T_d(n)$, we have

$$T'(N) \leq 2T'(N/2) + T'(N/b) + cN,$$

for some constant $c$. This single-variable recurrence can be solved by standard techniques. For example, by induction, the bound $T'(N) \leq c'[N^{1+\varepsilon} - N]$ follows from $T'(N) \leq 2c'[(N/2)^{1+\varepsilon} - N/2] + c'[(N/b)^{1+\varepsilon} - N/b] + cN$, as long as the constant $c'$ is sufficiently large, and

$$2/2^{1+\varepsilon} + 1/b^{1+\varepsilon} = 1,$$

which holds by setting $b := c_\varepsilon^{1/(1+\varepsilon)}$. Thus, $T'(N) = O(N^{1+\varepsilon})$, implying $T_d(n) = O((b^d n)^{1+\varepsilon}) = O(c_\varepsilon^d n^{1+\varepsilon})$, and the lemma follows. □

(We note in passing that the above two-variable recurrence can also be recast to fit the type studied by Eppstein [8], by considering the exponential function $T''(r,d) := T_d(2^r)$.)

## 3   The APSP Algorithm

We are now ready to present our new APSP algorithm. Like previous algorithms, we employ a well-known reduction from APSP to the computation of the *distance product* (also known as the *min-plus product*) of two $n \times n$ matrices: given matrices $A = \langle a_{ik}\rangle_{i,k=1,\dots,n}$ and $B = \langle b_{kj}\rangle_{k,j=1,\dots,n}$, the result of this multiplication is defined as the matrix $C = \langle c_{ij}\rangle_{i,j=1,\dots,n}$ with $c_{ij} := \min_k(a_{ik} + b_{kj})$. Given an algorithm for the distance product, we can solve the APSP problem by repeated squaring [6], but this would increase the running time by a logarithmic factor (which we obviously cannot afford); instead, we apply the reduction described in the text by Aho et al. [1–Section 5.9, Corollary 2], which avoids the extra logarithmic factor. It is thus sufficient to upper-bound the complexity of the distance product problem. We emphasize that Strassen's matrix multiplication method and its relatives cannot be applied directly, because in the min-plus case, elements only form a semi-ring.

For notational simplicity, we assume that the minimum term in the expression $\min_k(a_{ik} + b_{kj})$ is unique. (General perturbation techniques can ensure this but are not necessary if we break ties in a consistent manner.) We note that our algorithm can automatically identify the index $k$ attaining the minimum for each $c_{ij}$. (This property is required so that one can not only determine each shortest path distance but retrieve each shortest path.)

In the following lemma, we reveal the key connection between our earlier geometric problem and distance products of rectangular matrices:

**Lemma 2.** *We can compute the distance product of an $n \times d$ matrix $A$ and a $d \times n$ matrix $B$ in $O(dc_\varepsilon^d n^{1+\varepsilon} + n^2)$ time.*

*Proof.* The outline of the algorithm is simple: For each $k = 1, \dots, d$, we compute the set of pairs $X_k = \{(i,j) \mid \forall k' = 1,\dots,d, \ a_{ik} + b_{kj} \le a_{ik'} + b_{k'j}\}$; we then set $c_{ij} = a_{ik} + b_{kj}$ for every $(i,j) \in X_k$.

We first make an obvious observation (used also in previous approaches): $a_{ik} + b_{kj} \le a_{ik'} + b_{k'j}$ is equivalent to $a_{ik} - a_{ik'} \le b_{k'j} - b_{kj}$. We now make the next observation (which was missed in previous approaches): computing $X_k$ for a fixed $k$ corresponds exactly to computing all dominating pairs between the two $d$-dimensional point sets

$$\{(a_{ik} - a_{i1}, a_{ik} - a_{i2}, \dots, a_{ik} - a_{id})\}_{i=1,\dots,n} \text{ and}$$
$$\{(b_{1j} - b_{kj}, b_{2j} - b_{kj}, \dots, b_{dj} - b_{kj})\}_{j=1,\dots,n}.$$

(The dimension is actually $d - 1$, since the $k$-th coordinates are all 0's.) By Lemma 1, this computation takes $O(c_\varepsilon^d n^{1+\varepsilon} + |X_k|)$ time for each $k$. Since $\sum_{k=1}^d |X_k| = n^2$, the lemma follows. □

The highlight of our approach is already over. To get a subcubic algorithm for the distance product of square matrices, and consequently for APSP, all that remains is to choose an appropriate value for the dimensional parameter $d$:

**Theorem 1.** *We can compute the distance product of two $n \times n$ matrices in $O(n^3/\log n)$ time.*

*Proof.* We split the first matrix into $n/d$ matrices $A_1, \ldots, A_{n/d}$ of dimension $n \times d$, and the second matrix into $n/d$ matrices $B_1, \ldots, B_{n/d}$ of dimension $d \times n$. We compute the distance product of $A_\ell$ and $B_\ell$ for each $\ell = 1, \ldots, n/d$, by Lemma 2, and return the element-wise minimum of these $n/d$ matrices of dimension $n \times n$. The total time is at most the time bound of Lemma 2 multiplied by $n/d$, i.e.,

$$O\left(c_\varepsilon^d n^{2+\varepsilon} + \frac{n^3}{d}\right).$$

The theorem follows by choosing $d$ to be $\log n$ times a sufficiently small constant (depending on $\varepsilon \in (0, 1)$). For example, we can set $\varepsilon \approx 0.38$ (with $c_\varepsilon \approx 4.32$) and $d \approx 0.42 \ln n$, to minimize the constant factor in the dominant term. □

**Corollary 1.** *We can solve the APSP problem in $O(n^3/\log n)$ time.*

We conclude by mentioning how easy it is to adapt our algorithm to run on pointer machines: Lemma 1 poses no problem by using linked lists. In Lemma 2, for each pair $(i, j) \in X_k$, we cannot directly set the value of $c_{ij}$ since random access is forbidden; instead, we insert the pair $(j, k)$ into $i$'s "bucket". Afterwards, for each $i$, we sort its bucket according to the $j$ value (by scanning through all pairs $(j, k)$ in the bucket, putting the index $k$ into $j$'s "slot", and collecting all slots at the end). We can then set $c_{ij}$ for every $i$ and $j$, all within $O(n^2)$ time.

## 4   Discussion

We have demonstrated that a slightly subcubic time bound for the general APSP problem with real weights can be obtained without "cheating" on the RAM via table lookups or word operations, and without algebraic techniques for fast matrix multiplication. Although we have taken a geometric approach, the resulting algorithm shares some similarities with previous algorithms: for example, like in our proof of Lemma 2, Fredman's algorithm and its successors use the same primitive operation on the weights (comparing values each of which is the difference of two entries from a common row or column); in addition, Dobosiewicz's algorithm also goes through each index $k$ and compute the same set $X_k$ of index pairs (but in a different way, of course).

Among the series of slightly sub-cubic upper bounds obtained, $O(n^3/\log n)$ looks the most "natural", and it is interesting to contemplate whether we have reached the limit, at least as far as nonalgebraic algorithms are concerned. In any case, reducing the running time further by more than a logarithmic factor would be difficult: even for the simpler problem of *Boolean matrix multiplication*, the best known algorithm without algebraic techniques [3] runs in $O(n^3/\log^2 n)$ time and has not be improved for over three decades.

Although our algorithm is simple enough for implementation, it is primarily of theoretical interest. Some preliminary experiments seem to indicate that even for $n$ about 1000 (where the size of the graph is on the order of a million), the best choice of $d$ is still 2 (i.e., the dimension for the geometric subproblems is 1). Compared to the naive cubic method for computing distance products, the $\log n$-factor speedup can only be "felt" when the input size is very large, but in such cases, caching and other issues become more important.

An interesting theoretical question is whether a similar log-factor-type speedup is possible for sparse graphs. For example, for the simpler problem of computing the *transitive closure* of an unweighted directed graph, Yuster and Zwick in a recent paper [21] asked for an $o(mn)$-time algorithm, but an $O(mn/\log n + n^2)$ time bound is actually easy to get on the word RAM.[2] Can a similar $o(mn)$ running time be obtained for APSP for real-weighted graphs? (The author has recently made progress on this question for the case of unweighted, undirected graphs.)

Finally, we remark that in his original paper [10], Fredman was concerned with decision-tree complexities and found a (nonalgorithmic) way to solve the general APSP problem using $O(n^{2.5})$ comparisons of sums of edge weights (which then led to his slightly subcubic algorithmic result). It remains an open problem to find improved upper bounds or nontrivial lower bounds on the number of comparisons required.

**Acknowledgement.** I thank a reviewer for bringing up references [12, 20] to my attention. This work has been supported by NSERC.

# References

1. A. V. Aho, J. E. Hopcroft, and J. D. Ullman. *The Design and Analysis of Computer Algorithms.* Addison-Wesley, Reading, MA, 1974.
2. N. Alon, Z. Galil, and O. Margalit. On the exponent of the all pairs shortest path problem. *J. Comput. Sys. Sci.*, 54:255–262, 1997.

---

[2]   Proof: Assume that the graph is acyclic, since we can precompute the strongly connected components in linear time and contract each component. We want to find the set $S_u$ of all vertices reachable from each vertex $u$. For each vertex $u$ in reverse topological order, we can compute $S_u$ by taking the union of $S_v$ over all vertices $v$ incident from $u$. Each of these $O(m)$ set-union operations can be carried out in $O(n/\log n)$ time by representing a set as an $(n/\log n)$-word vector and by using the bitwise-or operation.

3. V. L. Arlazarov, E. C. Dinic, M. A. Kronrod, and I. A. Faradzev. On economical construction of the transitive closure of a directed graph. *Soviet Math. Dokl.*, 11:1209–1210, 1970.

4. A. L. Buchsbaum, H. Kaplan, A. Rogers, and J. R. Westbrook. Linear-time pointer-machine algorithms for least common ancestors, MST verification, and dominators. In *Proc. 30th ACM Sympos. Theory Comput.*, pages 279–288, 1998.

5. D. Coppersmith and S. Winograd. Matrix multiplication via arithmetic progressions. *J. Symbolic Comput.*, 9:251–280, 1990.

6. T. H. Cormen, C. E. Leiserson, R. L. Rivest, and C. Stein. *Introduction to Algorithms.* McGraw-Hill, 2nd ed., 2001.

7. W. Dobosiewicz. A more efficient algorithm for the min-plus multiplication. *Int. J. Computer Math.*, 32:49–60, 1990.

8. D. Eppstein. Quasiconvex analysis of backtracking algorithms. In *Proc. 15th ACM-SIAM Sympos. Discrete Algorithms*, pages 788–797, 2004.

9. T. Feder and R. Motwani. Clique partitions, graph compression and speeding-up algorithms. *J. Comput. Sys. Sci.*, 51:261–272, 1995.

10. M. L. Fredman. New bounds on the complexity of the shortest path problem. *SIAM J. Comput.*, 5:49–60, 1976.

11. Z. Galil and O. Margalit. All pairs shortest paths for graphs with small integer length edges. *J. Comput. Sys. Sci.*, 54:243–254, 1997.

12. Y. Han. Improved algorithm for all pairs shortest paths. *Inform. Process. Lett.*, 91:245–250, 2004.

13. S. Pettie. A new approach to all-pairs shortest paths on real-weighted graphs. *Theoret. Comput. Sci.*, 312:47–74, 2004.

14. S. Pettie and V. Ramachandran. A shortest path algorithm for real-weighted undirected graphs. *SIAM J. Comput.*, to appear.

15. F. P. Preparata and M. I. Shamos. *Computational Geometry: An Introduction.* Springer-Verlag, New York, 1985.

16. R. Seidel. On the all-pairs-shortest-path problem in unweighted undirected graphs. *J. Comput. Sys. Sci.*, 51:400–403, 1995.

17. A. Shoshan and U. Zwick. All pairs shortest paths in undirected graphs with integer weights. In *Proc. 40th IEEE Sympos. Found. Comput. Sci.*, pages 605–614, 1999.

18. V. Strassen. Gaussian elimination is not optimal. *Numerische Mathematik*, 13:354–356, 1969.

19. T. Takaoka. A new upper bound on the complexity of the all pairs shortest path problem. *Inform. Process. Lett.*, 43:195–199, 1992.

20. T. Takaoka. A faster algorithm for the all-pairs shortest path problem and its application. In *Proc. 10th Int. Conf. Comput. Comb.*, Lect. Notes Comput. Sci., vol. 3106, Springer-Verlag, pages 278–289, 2004.

21. R. Yuster and U. Zwick. Fast sparse matrix multiplication. In *Proc. 12th European Sympos. Algorithms*, Lect. Notes Comput. Sci., vol. 3221, Springer-Verlag, pages 604–615, 2004.

22. U. Zwick. All-pairs shortest paths using bridging sets and rectangular matrix multiplication. *J. ACM*, 49:289–317, 2002.

23. U. Zwick. A slightly improved sub-cubic algorithm for the all pairs shortest paths problem with real edge lengths. In *Proc. 15th Int. Sympos. Algorithms and Computation*, Lect. Notes Comput. Sci., vol. 3341, Springer-Verlag, pages 921–932, 2004.

# $k$-Link Shortest Paths in Weighted Subdivisions

Ovidiu Daescu[1,*], Joseph S.B. Mitchell[2,**], Simeon Ntafos[1],
James D. Palmer[1], and Chee K. Yap[3,* * *]

[1] Department of Computer Science, University of Texas at Dallas,
Richardson, TX 75080, USA
{daescu, ntafos, jdp011100}@utdallas.edu
[2] Department of Applied Mathematics and Statistics,
Stony Brook University, Stony Brook, NY 11794, USA
jsbm@ams.sunysb.edu
[3] Department of Computer Science,
New York University, New York, NY 10012, USA
yap@cs.nyu.edu

**Abstract.** We study the shortest path problem in weighted polygonal
subdivisions of the plane, with the additional constraint of an upper
bound, $k$, on the number of links (segments) in the path. We prove
structural properties of optimal paths and utilize these results to obtain approximation algorithms that yield a path having $O(k)$ links and
weighted length at most $(1 + \epsilon)$ times the weighted length of an optimal
$k$-link path, for any fixed $\epsilon > 0$. Some of our results make use of a new
solution for the 1-link case, based on computing optimal solutions for a
special sum-of-fractionals (SOF) problem. We have implemented a system, based on the CORE library, for computing optimal 1-link paths;
we experimentally compare our new solution with a previous method for
1-link optimal paths based on a prune-and-search scheme.

## 1   Introduction

A weighted subdivision $R$ in the plane is a decomposition of the plane into
polygonal regions, each with an associated nonnegative weight. The weighted
length of a line segment, $\overline{ab}$, joining two points $a$ and $b$ within the same region
$R_i \in R$ is defined as the product of the weight $w_i$ of region $R_i$ and the Euclidean
length $|\overline{ab}|$ of the line segment $\overline{ab}$. For a polygonal path $p$, the weighted length
is given by a finite sum of subsegment (Euclidean) lengths, each multiplied by
the weight of the region containing the subsegment.

We are motivated by applications that require paths that are optimal with
respect to more than one criterion. For example, in emergency and medical

---

* Corresponding author. Daescu's work is supported by NSF grant CCF-0430366.
** J. Mitchell is partially supported by the U.S.-Israel Binational Science Foundation
(2000160), NASA (NAG2-1620), NSF (CCR-0098172, ACI-0328930, CCF-0431030),
and Metron Aviation.
* * * Yap's work is supported by NSF Grant CCF-0430836.

F. Dehne, A. López-Ortiz, and J.-R. Sack (Eds.): WADS 2005, LNCS 3608, pp. 325–337, 2005.

interventions, in military route planning, and in air traffic applications, one may desire polygonal paths having only a few links (turns), while also having a small (weighted) length. A minimum-weight path may have unacceptably many turns.

In this paper we study the *k-link shortest path problem* in weighted regions, in which we place an upper bound, $k$, on the number of links (edges) in the polygonal path, while minimizing the weighted length of the path. We compute paths from a given source region $R_s$ to a target region $R_t$ in a weighted subdivision $R$. An important special case, which arises as a subproblem in our approach, is that of computing a *1-link shortest path* from a source to a target.

**Related Work.** In the unweighted setting, approximation algorithms are known for $k$-link shortest paths inside simple polygons and polygons with holes [16]. In weighted subdivisions, turn-constrained optimal paths have been studied in the context of air traffic routing (using a grid-based dynamic programming algorithm) [10] and, very recently, in the context of mine avoidance routing (using a genetic algorithm) [13]; neither of these approaches gives approximation algorithm guarantees. The 1-link shortest path problem to compute an optimal "link" (or "penetration") in weighted subdivisions has been studied in [2, 3, 6], where it is shown that the special structure of the optimal solution allows for efficient search for solutions.

Without a bound on the number of links, several results are known for computing shortest paths in weighted regions [1, 9, 11, 12, 14, 15, 19], beginning with the first polynomial-time results of Mitchell and Papadimitriou [15], who compute $(1 + \epsilon)$-approximate geodesic shortest paths on weighted terrains.

**Our Results.** We present the following results. (1) We prove there exists a $(2k - 1)$-link path $p$ from source $R_s$ to target $R_t$ that turns only on the edges of $R$, such that the weighted length of $p$ is at most that of an optimal $k$-link path $p^*$ from $R_s$ to $R_t$. (2) We give two approximation algorithms for computing $k$-link shortest paths in weighted regions. The first one requires the computation of 1-link shortest paths and produces a $(5k - 2)$-link path whose weight is within factor $(1 + \epsilon)$ of optimal. The second algorithm relies only on computation of $(1 + \epsilon/6)$-approximate 1-link shortest paths and produces a $14k$-link path whose weight is within factor $(1 + \epsilon)$ of optimal. (3) We give a new (in this context) algorithm for computing 1-link shortest paths, based on solving a variant of the sum-of-linear-fractionals (SOLF) problem [8]. (4) We have implemented a system, based on the CORE library [5], for computing 1-link shortest paths. We compare experimentally two algorithms for 1-link shortest paths: one based on our variant of the SOLF problem, and one based on a prune-and-search scheme [6].

**Preliminaries.** Let $R$ be a planar weighted subdivision with a total of $n$ vertices and a set $\mathcal{E}$ of $O(n)$ edges. Without loss of generality, we assume $R$ is triangulated, the source and target regions can be separated by a vertical line, and the vertices of $R$ are in general position (no three collinear). For a path $p$, we let $|p|$ denote the Euclidean length of $p$ and $\|p\|$ the weighted length of $p$. A polygonal path $p$ whose turn points all lie on the edge set $\mathcal{E}$ is said to be *edge-restricted*.

Consider a link (line segment) $l$ between two edges $e_s$ and $e_t$ of $R$, with $e_s$ and $e_t$ not bounding the same (triangular) face (otherwise, the problem is trivial). The weighted length of the link $l$ is $d(l) = \sum_{i:l\cap R_i \neq \emptyset} w_i d_i(l)$, where $w_i$ is the weight of $R_i$ and $d_i(l)$ is the Euclidean length of $l$ within the region $R_i \in R$.

Let the equation of the line supporting $l$ be $y = mx + b$. Let $R_i$ be a region intersected by $l$, with $s_i^1$ and $s_i^2$ the two sides of $R_i$ that intersect $l$, at points $v_i^1(x_i^1, y_i^1)$ and $v_i^2(x_i^2, y_i^2)$, respectively. From $d_i(l) = \sqrt{1 + m^2}|x_i^2 - x_i^1|$, we have that

$$d(l) = \sqrt{1 + m^2} \sum_{i:l\cap R_i \neq \emptyset} w_i |x_i^2 - x_i^1| = \sqrt{1 + m^2} \sum_i \sigma_i w_i \frac{b_i - b}{m - m_i} \qquad (1)$$

where $\sigma_i$ is +1 or -1, $m_i$ and $b_i$ are constants, and the number of terms in the summation is $O(n)$. If $l$ is rotated and translated, while keeping its endpoints on $e_s$ and $e_t$, the expression for $d(l)$ does not change as long as no vertex of $R$ is crossed by $l$. The corresponding set of pairs $(m, b)$ defines a two-dimensional convex domain $D$ whose edges correspond to $l$ being tangent to a vertex of $R$ and whose vertices correspond to $l$ passing through two vertices of $R$ [2]. The region swept by $l$, while maintaining its combinatorial type, is called an *hourglass* (Fig. 1). For a fixed slope $m$, we see from (1) that $d(l)$ is linear in $b$ as $l$ varies within the hourglass; thus, there exists a segment $l$ minimizing $d(l)$, over the hourglass, passing through a vertex $v$ of $R$ [6]. For a fixed choice of the vertex $v$ of an hourglass, the expression for $d(l)$ depends only on the slope $m$ of the line through $l$:

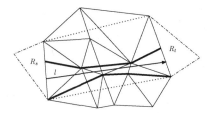

**Fig. 1.** An hourglass for which the formula for $d(l)$ does not change

$$d(l) = \sqrt{1 + m^2}\left(d_0 + \sum_i \frac{a_i}{m + b_i}\right), \qquad (2)$$

where $d_0, a_i$ and $b_i$ are constants. Note that $d(l)$ is bounded and positive.

## 2   Approximating $k$-Link Shortest Paths

**Lemma 1.** *Let $p$ be a shortest $k$-link path between edges $e_s$ and $e_t$ of $R$. Then, each link of $p$ has an endpoint on an edge of $R$ or goes through a vertex of $R$.*

*Proof.* The proof is by contradiction. Let $p$ be a $k$-link path between $e_s$ and $e_t$, let $l_1, l_2$ and $l_3$ be three consecutive links of $p$ and refer to Figure 2. Assume that $p$ makes two consecutive region interior turns, one at the common endpoint of the links $l_1$ and $l_2$ and the other at the common endpoint of the links $l_2$ and $l_3$ (Figure 2 (a)). Assume also that the turn for $l_1$ and $l_2$ is inside the region $R_1$, the turn for $l_2$ and $l_3$ is inside the region $R_2$, and extend $l_1$ and $l_3$ to intersect

**Fig. 2.** (a) Three consecutive links $l_1, l_2$, and $l_3$; (b) the middle link ends on edges; (c) an inside region turn with a link ending on an edge and (d) an inside turn with a link *stopped* at a vertex of the subdivision

the boundaries of $R_1$ and $R_2$, respectively. If we slide $l_2$ parallel to itself (i.e., the slope of $l_2$ is fixed) then the length of $p$ changes only locally, corresponding to the changes in length for $l_1$, $l_2$, and $l_3$. The change in length for $l_2$ is a linear function of the intercept of the line supporting $l_2$. It can also be shown that the changes for $l_1$ and $l_3$ are linear functions of the same variable. Thus, $p$ can be improved locally by sliding $l_2$ until either it hits a vertex of $R$ or an intersection point of $l_1 \cap R_1$ or $l_3 \cap R_2$. Figure 2 (b), (c) and (d) shows three possible cases for consecutive links on an optimal path $p$. The other cases can be obtained by symmetry. The rest follows by induction on the number of links on the path. □

**Theorem 1.** *There exists a path $p$ between $e_s$ and $e_t$ with at most $(2k-1)$-links that turns only on the edges of $R$ and such that the weighted length of $p$ is at most that of a $k$-link shortest path $p^*$ from $e_s$ to $e_t$.*

We now show how to modify previous discretization schemes (e.g., [1, 19]) in order to approximate edge-restricted $k$-link shortest paths. We say a path $p$ from source point $s$ to destination point $t$ is an $\epsilon$-*good approximate shortest path* if $\|p\| \leq (1 + \epsilon)\|p_k(s,t)\|$, where $p_k(s,t)$ is an edge-restricted $k$-link shortest path from $s$ to $t$.

Let $\mathcal{E}$ be the set of boundary edges of $R$. Let $\mathcal{E}(v)$ be the set of subdivision edges having endpoint $v$, and let $d(v)$ be the minimum distance between $v$ and the edges in $\mathcal{E} \setminus \mathcal{E}(v)$. (Note that if $v$ is not a vertex of $R$, then $\mathcal{E}(v) = \emptyset$.) For each edge $e \in \mathcal{E}$, let $d(e) = \sup_{x \in e} d(x)$. Let $v(e)$ be a point on $e$ such that $d(v(e)) = d(e)$. For each vertex $v \in R$, let $r(v) = \epsilon \frac{d(v)}{5}$. We refer to $r(v)$ as the vertex radius for $v$. The disk of radius $r(v)$ centered at $v$ defines the *vertex-vicinity* $S(v)$ of the vertex $v$.

We now describe how the Steiner points on an edge $e = \overline{v_1 v_2}$ are chosen. Each vertex $v_i$, where $i = 1, 2$, has a vertex-vicinity $S(v_i)$ of radius $r(v_i)$ and the Steiner points $v_{i,1}, \ldots, v_{i,j_i}$ are placed on $e$ such that $|v_i v_{i,1}| = r(v_1)$ and $|v_{i,m} v_{i,m+1}| = \epsilon d(v_{i,m})$, $m = 1, 2, \ldots, j_i - 1$. The value of $j_i$ is such that $v_{i,j_i} = v_i(e)$. We call the line segment formed by two adjacent Steiner points $v_{i,m}$ and $v_{i,m+1}$ a *Steiner edge*. The pairing of any two Steiner edges forms a quadrilateral shape called a *Steiner strip*. The shape could be degenerate if the Steiner edges are on the same boundary edge. In [19], it has been shown such a discretization scheme can be used to guarantee a $3\epsilon$-good approximate shortest path. With $\delta$ the maximum number of Steiner points placed on an edge, this discretization scheme gives $\delta = O(\frac{1}{\epsilon} \log \frac{1}{\epsilon})$. (It is important to note that $\delta$ also depends on some geometric parameters of $R$, such as the longest edge; see [1].)

Let $(e_i, e_j)$ be a pair of edges of the subdivision and assume a $k$-link edge-restricted shortest path has a turn on $e_i$ and the next turn on $e_j$. Then, the shortest path link $l^*$ between $e_i$ and $e_j$ is contained in an hourglass corresponding to one of the $O(n^2)$ 1-link shortest path subproblems defined by the pair $(e_i, e_j)$ (see Fig. 1). Each of the two endpoints of the shortest path link $l^*$ lies between either two Steiner points or a vertex and a Steiner point. Assume each endpoint is between two Steiner points. Let $l$ be one of the four line segments between $e_i$ and $e_j$ that are defined by the four Steiner points and further assume that $l$ is fully contained in the same hourglass as $l^*$. Then, $d(l)$ and $d(l^*)$ have similar description and $d(l) - d(l^*) = \sum_{i=1}^{m} w_i d_i(l) - \sum_{i=1}^{m} w_i d_i(l^*) = \sum_{i=1}^{m} w_i(d_i(l) - d_i(l^*))$ where $m = O(n)$ and, without loss of generality, we assume that $l$ and $l^*$ intersect the regions $R_1, R_2, \ldots, R_m$ of subdivision $R$. Asking that $\sum_{i=1}^{m} w_i(d_i(l) - d_i(l^*)) \leq \epsilon d(l^*)$ implies $\sum_{i=1}^{m} w_i(d_i(l) - d_i(l^*)) \leq \epsilon \sum_{i=1}^{m} w_i d_i(l^*)$ and thus $\sum_{i=1}^{m} w_i d_i(l) \leq \sum_{i=1}^{m} w_i((1 + \epsilon)d_i(l^*))$. Thus, the Steiner points on the edges $e_i$ and $e_j$ should be such that $d_i(l) \leq (1 + \epsilon)d_i(l^*)$, $i = 1, 2, \ldots, m$.

Clearly, if $l^*$ passes very close to a vertex $v$ of $R$ and intersects two or more edges incident to $v$, a discretization scheme cannot guarantee that $d_i(l) \leq (1 + \epsilon)d_i(l^*)$, for the corresponding distances. More generally, a similar situation appears when the optimal path has multiple turns very close to $v$, e.g., with link endpoints between a vertex and a Steiner point (Fig. 3 (a)). Another potential problem is that it is possible that none of the four line segments joining the points that define the Steiner strip between $e_i$ and $e_j$ is fully contained in the same hourglass as $l^*$, implying that $l$ and $l^*$ intersect different subsets of regions of $R$. The challenge, then, is to formulate approximation schemes that either avoid or address these problems.

We address the first problem using normalization. A path $p$ is said to be *normalized* if it does not turn on edges within a vertex-vicinity. In Lemma 1 of [19], Sun and Reif show that for any path $p$ from $s$ to $t$, there is a normalized path $\hat{p}$ from $s$ to $t$ such that $||\hat{p}|| = (1 + \frac{\epsilon}{2})||p||$. For $k$-link edge-restricted paths we have the following related lemma.

**Lemma 2.** *For any $k$-link edge-restricted path $p_k$ from $s$ to $t$, there is a normalized path $\hat{p}$ from $s$ to $t$ such that $||\hat{p}|| = (1 + \epsilon/2)||p_k||$.*

*Proof.* We observe that $p$ need not be an optimal path for Sun and Reif's Lemma 1 to hold. Thus, the same proof holds for a $k$-link path, $p_k$. However, there is no guarantee that $\hat{p}$ has only $k$ links. □

We would like to bound the number of links that may be "added" to $\hat{p}$ relative to $p_k$. Let $u_1''$ be the boundary point where $p_k$ first enters a region adjacent to $v$ and let $u_2''$ be the boundary point where $p_k$ leaves a region adjacent to $v$. Let $u_1' \in p_k$ be the boundary point on the cheapest region intersected by $p_k$ before entering the vicinity of $v$ and let $u_2' \in p_k$ be the boundary point on the cheapest region intersected by the last link of $p_k$ that has nonempty intersection with the vicinity of $v$ (see Fig. 3 (a)). In constructing $\hat{p}$, we may remove zero or more links in $p_k$ completely contained in the vertex-vicinity (link $\overline{u_1 u_2}$ in Fig. 3 (a))

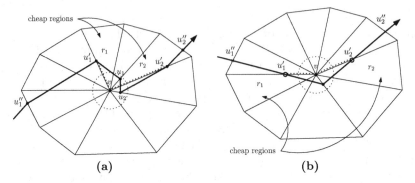

**Fig. 3.** (a) The solid line path is part of an optimal $k$-link path. The dotted path represents a normalized path. (b) The solid line path represents a single turn in an optimal $k$-link path. The dotted path represents a normalized path

and then add up to two additional links that begin outside the vertex-vicinity and pass through $v$ (links $\overline{u_1'v}$ and $\overline{vu_2'}$ in Fig. 3 (a)). Fig. 3 (a) represents a situation in which the number of links in the subpath $p[u_1'', u_2'']$ in the original path $p_k$ is one more than in the normalized path, $\hat{p}$. Fig. 3 (b) is representative of the worst case, where two links are added and none are removed.

**Lemma 3.** *For any $k$-link path $p_k$ from $s$ to $t$, there is a normalized edge-restricted path $\hat{p}$ such that $||\hat{p}|| = (1 + \epsilon/2)||p_k||$ and $\hat{p}$ has at most $3k - 2$ links.*

*Proof.* We observe that at most two links need to be added for each link of $p_k$ when constructing a normalized path $\hat{p}$ from a path $p_k$. Let $k_1$ be the number of links that need normalization. Then, they are replaced by $3k_1 - 2$ links. Let $k_2$ be the remaining links (i.e., $k = k_1 + k_2$). These links may have an endpoint that is not on a vertex or edge of $R$. From Theorem 1, at most $2k_2 - 1$ links are required to create an approximating path restricted to edges. Note that each of the at most $k_2 - 1$ links that are within a triangle of $R$ may also require normalization, which would add $k_2 - 1$ extra links. Thus, in a normalized edge restricted path the $k_2$ links are replaced by at most $3k_2 - 2$ links. □

**An approximation using exact optimal links.** Recall that we refer to the line segment formed by two adjacent Steiner points $v_{i,j}$ and $v_{i,j+1}$ on an edge incident to vertex $v_i \in R$ as a *Steiner edge*. The pairing of any two Steiner edges forms a *Steiner strip*. The pairing of a Steiner edge with a vertex of $R$ forms a *Steiner-vertex strip*. The pairing of any two vertices of $R$ forms a *vertex-pair*.

Consider a shortest normalized $k$-link path, $p_k$, that only turns on edges. Each link $l_i$ in $p_k$ is "captured" either by a Steiner strip, a Steiner-vertex strip, or a vertex-pair.

If $l_i$ is captured by a vertex-pair $(u, v)$ the weighted length of $l_i$ can be easily computed as $||\overline{uv}||$. Then, the difficulty in approximating the weighted length of $l_i$ is when $l_i$ is captured by either a Steiner strip or a Steiner-vertex strip, where the Steiner strip is the more general case.

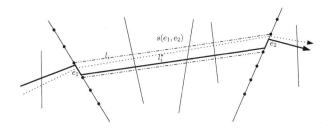

**Fig. 4.** The dotted line path represents an optimal $k$-link path. The solid line path represents a $2k$ approximation made up of optimal links and "small" connecting, edge-crawling links

Let $s(e_1, e_2)$ be a Steiner strip that captures $l_i$, where $e_1$ and $e_2$ are the Steiner edges at which $l_i$ originates and terminates, respectively. Although $l_i$ forms part of an optimal $k$-link path, $l_i$ is not necessarily an optimal link between $e_1$ and $e_2$. This suggests one could try to replace $l_i$ with the optimal link $l_i^*$ between $e_1$ and $e_2$. We show in the next lemma (proof omitted here) that using $k$ optimal links and $k$ "small" connecting links we can construct an approximating path with $2k$ links that provides an $\epsilon$-good approximation of $p_k$. We define an edge-crawling link as a link along an edge of $R$ and contained within a Steiner edge (see Fig. 4).

**Lemma 4.** *A normalized $k$-link path, $p_k$, that turns on edges can be approximated by an $\epsilon$-good $2k$-link path made up of $k$ optimal links connected by $k$ edge-crawling links.*

**Theorem 2.** *For sufficiently small positive $\epsilon$, a $k$-link shortest path can be approximated with a normalized $\epsilon$-good $(5k - 2)$-link edge restricted path.*

*Proof.* It follows from Lemma 3 that there is a normalized edge-restricted path with $3k - 2$ links. When applying Lemma 4, only $2k$ optimal links on this path need small edge crawling links. The approximation factor is $(1 + \epsilon/2)(1 + \epsilon) = (1 + 3\epsilon/2 + \epsilon^2/2) \leq (1 + 2\epsilon)$, assuming $\epsilon \leq 1/2$. Then, we can use $\epsilon = \epsilon/2$ when generating the Steiner points to get the claimed result (we will no longer mention this in the remaining proofs). $\square$

We now show how to use this discretization scheme to construct a weighted graph $G_\epsilon(V, E)$ that captures approximate paths. Each node $v \in V$ corresponds to a vertex of $R$ or a Steiner edge in our discretization scheme. Each edge $e \in E$ corresponds to either a Steiner strip, a Steiner-vertex strip or a vertex-pair. The weight of $e$ is the weighted length of a shortest 1-link path through the respective Steiner strip, Steiner-vertex strip or vertex-pair. (Some other geometric informations are associated with $G_\epsilon$; we defer this to the full paper.) This differs from the discretization graph in [1, 19], where $V$ is formed of Steiner points and $E$ is made up of links between Steiner points.

The number of vertices in $V$ is $O(\delta n)$ and the number of edges in $E$ is $O((\delta n)^2)$. Computing a single edge in $G_\epsilon$ corresponds to solving a 1-link shortest path problem for a specific hourglass. Let this time be $T_h(n)$. Thus, the time to compute $G_\epsilon$ is $O((\delta n)^2 T_h(n))$. Once $G_\epsilon$ is constructed, we can use dynamic programming to find a $k$-link shortest path in $G_\epsilon$ in $O(k(\delta n)^4)$ time.

**Theorem 3.** *There exists a path approximation graph $G_\epsilon$ of size $O((\delta n)^2)$, that can be constructed in $O((\delta n)^2 T_h(n))$ time, and can be used to report in $O(k(\delta n)^4)$ time a $(5k-2)$-link path that $(1+\epsilon)$-approximates a $k$-link shortest path.*

**An approximation using approximate optimal links.** Finding optimal 1-link paths can be computationally expensive (e.g., see Section 3). We now describe a technique for computing approximate optimal 1-link paths for the subproblems that arise in building the path approximation graph.

Observe that a link $l$ between two Steiner edges $e_1$ and $e_2$ may intersect several Steiner edges placed on the edges of each region $l$ crosses. Each region that $l$ intersects captures part of $l$ in a Steiner strip with one exception. If $l$ passes within the vertex-vicinity of a vertex in $R$ then part of $l$ is not captured. Furthermore, $l$ could intersect $O(n)$ vertex-vicinities.

We would like to find an approximating path that has no vertex-vicinity intersections. However, as we have seen earlier, avoiding a vertex vicinity could add two extra links on the approximating path. To reduce the increase in the number of links on the approximating path we then need to reduce the number of vertex-vicinities that can be intersected by a single link.

We accomplish this by changing the discretization scheme slightly. Let $T$ be the set of all possible vertex triplets formed by the vertices in $R$. $O(n^3)$ such triplets exist that correspond to $O(n^3)$ triangles. For each vertex in each triangle we can compute the minimum distance to the opposite edge. Let $\gamma_i$ be the minimum such distance obtained from a triplet containing the vertex $v_i$, and let $\gamma = \min\{\gamma_i/2 \mid i = 1, 2, \ldots, n\}$. By a recent result in [7], $\gamma_i$ can be found in $O(n \log n)$ time and thus $\gamma$ can be computed in $O(n^2 \log n)$ time.

Let the new vertex-vicinity radius, $r'(v_i)$, be $\min(r(v_i), \gamma)$. The first Steiner point after a vertex $v_i$ is placed such that $|v_i v_{i,1}| = r'(v_i)$. A path $p$ is said to be $\gamma$-*normalized* if each link in $p$ does not turn within a vertex-vicinity or pass through a vertex-vicinity without also passing through the vertex.

The proofs of Lemmas 5-8 are deferred to the full paper.

**Lemma 5.** *For any $k$-link path $p_k$ from $s$ to $t$, there is a $\gamma$-normalized edge-restricted path $\hat{p}$ from $s$ to $t$ such that (1) $\|\hat{p}\| = (1 + \frac{\epsilon}{2})\|p_k\|$ and (2) $\hat{p}$ has no more than eight times as many links as $p_k$.*

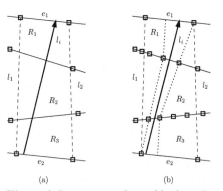

(a)                    (b)

**Fig. 5.** A Steiner strip formed by lines $l_1$ and $l_2$ may intersect edges that are more coarsely (a) or more finely (b) sampled

Next, we consider the computation of a single link. Fig. 5 (a) illustrates a situation in which a Steiner strip intersects one or more edges with sparsely placed

Steiner points. Fig. 5 (b) illustrates a situation where a Steiner strip intersects one or more edges with densely placed Steiner points. In Fig. 5 (a), the line segments $l_1$ and $l_2$ provide an $\epsilon$-good approximation for all line segments captured between $l_1$ and $l_2$. In Fig. 5 (b), $l_1$ and $l_2$ only provide $\epsilon$-good approximations for the lines that pass between the same Steiner points as $l_1$ and $l_2$.

The number of possible approximating 1-link paths for a pair of edges is $O((\delta n)^2)$. The complexity to store both every possible approximation and the ranges over which those approximations are valid for the entire subdivision is then $O((\delta n)^4)$.

**Lemma 6.** *Let $\hat{l}$ be an optimal $\gamma$-normalized link between two edges $e_1$ and $e_2$. A single $(1+\epsilon)$-factor approximating link can be computed in $O(n(\delta n)^2)$ time.*

**Lemma 7.** *For sufficiently small positive $\epsilon$, a $k$-link shortest path can be approximated with a $\gamma$-normalized $\epsilon$-good $(14k)$-link edge restricted path.*

**Lemma 8.** *The path approximation graph $G_\epsilon$ has size $O((\delta n)^2)$ and can be constructed in $O(n(\delta n)^4)$ time.*

Using a dynamic programming algorithm for computing $k$-link shortest paths in weighted graphs, one can find an approximate solution for the $k$-link shortest path using $G_\epsilon$ in $O(k(\delta n)^4)$ time.

**Theorem 4.** *There exists a path approximation graph $G_\epsilon$ of size $O((\delta n)^2)$, that can be constructed in $O(n(\delta n)^4)$ time, and can be used to report in $O(k(\delta n)^4)$ time a $14k$-link path that $(1+\epsilon)$-approximates a $k$-link shortest path.*

# 3    Optimal 1-Links: A Sum of Fractionals Approach

To compute 1-link shortest paths, we adapt an algorithm for minimizing a sum of linear fractional functions (SOLF) [8], to the sum of fractional functions (SOF) problem that describes an optimal 1-link path. In order to find a 1-link shortest path one needs to solve a number of optimization problems of the form $\min_{x \in S}\{\sum_{i=1}^m \sqrt{1+x^2}\frac{a_i}{b_ix+c_i}\} = \min_{x \in S}\{\sum_{i=1}^m r_i(x)\}$, where $b_1 = 0$, $b_i = 1$, $i = 2, 3, \ldots, m$, $a_i, c_i$ are constants and $b_ix + c_i > 0$ over $S$, $i = 1, 2, \ldots m$. Thus, the functions we try to optimize are 1-dimensional SOFs with generic term $r_i(x) = \sqrt{1+x^2}(a_i/(b_ix + c_i))$ rather than 1-dimensional SOLFs, where the generic term would have the form $r_i(x) = a_i/(x + c_i)$. While in general one may not be able to apply the $d$-dimensional SOLF algorithm in [8] for a SOF problem, we will show below that this is possible for our objective function. Our choice of method is based on the results in [4], which show that the one-dimensional SOLF algorithm is very fast in practice. Since our function is a one-dimensional SOF, adapting the SOLF method for SOF functions may lead to similar results.

The only place in the SOLF algorithm where the expression of the ratio $r_i(x)$ is important is in the optimization subproblems that require to minimize (or maximize) $r_i(x)$ over a convex domain (an interval in our case). For a 1-dimensional SOLF, this reduces to minimizing a linear function over an interval and thus takes constant time.

**Lemma 9.** *The function* $r(x) = \sqrt{1 + x^2} \frac{a_i}{b_i x + c_i}$ *is unimodal if* $b_i x + c_i > 0$ *(or* $b_i x + c_i < 0$), *with extremal value obtained at* $x^* = 1/c_i$.

From Lemma 9 it follows that for the SOF problems associated with the 1-link shortest path, the optimization subproblems for $r_i(x)$ can be solved exactly in constant time each and thus we can apply the SOLF algorithm for these SOFs. As shown in [4], an iteration of the algorithm can be implemented to run in $O(m)$ time for the 1-dimensional case, while some special steps (executed in case of a stall) require altogether $O(m^2)$ time.

One way to speed up the computation is to process each of the SOF problems in turn, temporarily suspending the processing of the current SOF before $k$ has been incremented (i.e. before the execution of Step 5). Each time an upper or lower bound is updated, we can use the new bound to remove or cull SOF problems from the problem space. Experimental results suggest this culling process very quickly removes many subproblems that will not lead to an optimal solution. A hybrid implementation where the subdivision algorithm in [6] is applied to stalled regions would also fit into this framework.

## 4    Implementation and Experiments

We have implemented two algorithms for solving the weighted region 1-link shortest path problem. The first one is based on the prune-and-search scheme in [6] and the second one is based on the SOF algorithm in Section 3. Our results show that both algorithms are fast on random generated subdivisions. To ensure robustness for the special cases when a source-to-target link is close to an edge or vertex of $R$ our implementation uses the CORE library [5].

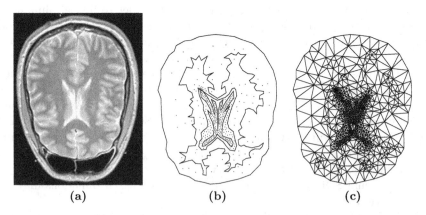

|       (a)       |       (b)       |       (c)       |

**Fig. 6.** (a) The original transverse CT scan, (b) a trace of structural elements in the scan and (c) the triangulation of that structure

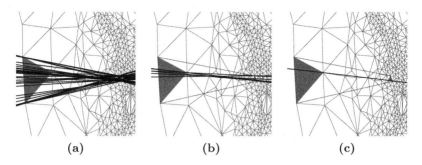

**Fig. 7.** Prune-and-search progress after (a) zero, (b) one, and (c) twenty one steps

We exemplify our software package on a problem inspired by biomedicine. Consider the CT scan in Fig. 6 (a) which was taken from the Visible Human Project [17]. The source image is made up of pixels which are samples from the CT process. Fig. 6 (b) shows one possible "trace" of this data which emphasizes certain structures. Using Shewchuk's triangulator each region can be tessellated into triangles using different constraints [18]. This is illustrated in Fig. 6 (c) where we have chosen to make the area constraints on the interior regions greater than those of the exterior region.

Fig. 7 shows the source and target regions we chose for this example. There are over 1500 triangles in this mesh, but far fewer are of interest after the bounding box is applied. The prune-and-search technique begins by enumerating possible optimization subproblems, with each subproblem corresponding to a double wedge through a vertex of $R$. The initial double wedges are represented as dotted lines in Fig. 7 (a). 37 wedges are found initially for the example in Fig. 7 but after only one round of culling based on the upper and lower bounds for each subproblem, there are only four wedges at step two (see Fig. 7 (b)). The number of wedges continues to grow and shrink in next steps, and for our example the algorithm terminates in twenty one steps (see Fig. 7 (c)).

While in general this algorithm performs well, the difficulty in assessing it is that it has very good best-case behavior and very bad worst-case behavior. In the worst case, the algorithm may continue subdividing a double wedge region over which the value of the objective function changes extremely slowly, such that subdivided double wedges cannot be culled. Fig. 8 shows an example of the best, worst and average number of subproblems competing in each step.

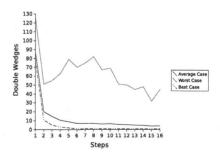

**Fig. 8.** The wedge count at each step represents the number of active problems

In applying the SOF technique to the same problem we find similar variabilities in performance. In some instances the SOF technique approaches a solution faster than the subdivision approach and in other instances the SOF technique stalls and fails to progress. If the SOF algorithm stalls, it is often necessary to run the SOF algorithm recursively on a subdivided problem. Unfortunately it is possible the SOF algorithm will then continue to stall. When the SOF algorithm was applied directly to each of the subproblems in the series, we found that 61.71% of the subproblems stalled. An average of 23.9 iteration steps and 7.64 recursive calls were required to solve each subproblem. If we consider only the subproblems which did not stall, we find that on average only 2.4 iterations were required. One improvement that has shown some success is the adoption of a hybrid solution where a SOF approach is used until a stall is detected and then in that case we revert to the subdivision technique in order to avoid a recursive stall. However, more experiments are needed to decide which of the two algorithms performs better in practice.

# References

1. L. Aleksandrov, A. Maheshwari, and J.-R. Sack. Determining approximate shortest paths on weighted polyhedral surfaces. *Journal of the ACM*, 52(1):25–53, 2005.
2. D. Z. Chen, O. Daescu, X. Hu, X. Wu, and J. Xu. Determining an optimal penetration among weighted regions in two and three dimensions. *Journal of Combinatorial Optimization*, 5(1):59–79, 2001.
3. D. Z. Chen, X. Hu, and J. Xu. Optimal beam penetration in two and three dimensions. *Journal of Combinatorial Optimization*, 7(2):111–136, 2003.
4. D. Z. Chen, J. Xu, N. Katoh, O. Daescu, X. Wu, and Y. Dai. Efficient algorithms and implementations for optimizing the sum of linear fractional functions, with applications. In *Journal of Combinatorial Optimization*, 9(1):69–90, 2005.
5. CORE library project. http://www.cs.nyu.edu/exact/core.
6. O. Daescu. Improved optimal weighted links algorithms. In *Proc. 2nd Internat. Workshop on Computational Geometry and Applications*, pages 65–74, 2002.
7. O. Daescu and J. Luo. Proximity problems on line segments spanned by points. In *14th Annual Fall Workshop on Computational Geometry*, Nov 2004.
8. J. E. Falk and S. W. Palocsay. Optimizing the sum of linear fractional functions. In *Recent advances in global optimization*, pages 221–258, 1992.
9. L. Gewali, A. Meng, J. S. B. Mitchell, and S. Ntafos. Path planning in $0/1/\infty$ weighted regions with applications. *ORSA J. on Computing*, 2(3):253–272, 1990.
10. J. Krozel, C. Lee, and J. S. B. Mitchell. Estimating time of arrival in heavy weather conditions. In *Proc. AIAA Guidance, Navig., and Control*, pages 1481–1495, 1999.
11. M. Lanthier, A. Maheshwari, and J.-R. Sack. Approximating shortest paths on weighted polyhedral surfaces. *Algorithmica*, 30(4):527–562, 2001.
12. C. Mata and J. Mitchell. A new algorithm for computing shortest paths in weighted planar subdivisions. *Proc. 13th Ann. Symp. Comput. Geom.*, pages 264–273, 1997.
13. C. B. McCubbin, C. D. Piatko, A. V. Peterson, and C. R. Donnald. Cooperative organic mine avoidance path planning. In *Proc. of the SPIE*, vol. 5794, 2005.
14. J. S. B. Mitchell. Geometric shortest paths and network optimization. In J.-R. Sack and J. Urrutia, eds, *Handbook of Computational Geometry*. Elsevier, 2000.

15. J. S. B. Mitchell and C. H. Papadimitriou. The weighted region problem: Finding shortest paths through a weighted planar subdivision. *Journal of the ACM*, 38(1):18–73, Jan. 1991.
16. J. S. B. Mitchell, C. Piatko, and E. Arkin. Computing a shortest *k*-link path in a polygon. In *Proc. 33rd IEEE Sympos. Found. Comput. Sci.*, pages 573–582, 1992.
17. National Library of Medicine. The visible human project. `http://www.nlm.nih.gov/research/visible`.
18. J. R. Shewchuk. Triangle: Engineering a 2D Quality Mesh Generator and Delaunay Triangulator. In M. C. Lin and D. Manocha, eds., *Appl. Computat. Geom.: Towards Geometric Engineering*, LNCS vol. 1148, pages 203–222. Springer-Verlag, 1996.
19. Z. Sun and J. H. Reif. Adaptive and compact discretization for weighted region optimal path finding. In *Proc. 14th Sympos. on Fundamentals of Computation Theory*, LNCS, Springer-Verlag, 2003.

# Power-Saving Scheduling for Weakly Dynamic Voltage Scaling Devices*

Jian-Jia Chen[1], Tei-Wei Kuo[2], and Hsueh-I Lu[2]

[1] Department of Computer Science and Information Engineering,
National Taiwan University, Taiwan, Republic of China
[2] Department of Computer Science and Information Engineering,
Graduate Institute of Networking and Multimedia,
National Taiwan University, Taiwan, Republic of China
{r90079, ktw, hil}@csie.ntu.edu.tw

**Abstract.** We study the problem of non-preemptive scheduling to minimize energy consumption for devices that allow dynamic voltage scaling. Specifically, consider a device that can process jobs in a non-preemptive manner. The input consists of (i) the set $R$ of available speeds of the device, (ii) a set $J$ of jobs, and (iii) a precedence constraint $\Pi$ among $J$. Each job $j$ in $J$, defined by its arrival time $a_j$, deadline $d_j$, and amount of computation $c_j$, is supposed to be processed by the device at a speed in $R$. Under the assumption that a higher speed means higher energy consumption, the power-saving scheduling problem is to compute a feasible schedule with speed assignment for the jobs in $J$ such that the required energy consumption is minimized.

This paper focuses on the setting of *weakly* dynamic voltage scaling, i.e., speed change is not allowed in the middle of processing a job. To demonstrate that this restriction on many portable power-aware devices introduces hardness to the power-saving scheduling problem, we prove that the problem is NP-hard even if $a_j = a_{j'}$ and $d_j = d_{j'}$ hold for all $j, j' \in J$ *and* $|R| = 2$. If $|R| < \infty$, we also give fully polynomial-time approximation schemes for two cases of the general NP-hard problem: (a) all jobs share a common arrival time, and (b) $\Pi = \emptyset$ and for any $j, j' \in J$, $a_j \leq a_{j'}$ implies $d_j \leq d_{j'}$. To the best of our knowledge, there is no previously known approximation algorithm for any special case of the NP-hard problem.

## 1 Introduction

With the increasing popularity of portable systems, energy efficiency has become a major design issue in hardware and software implementations [4, 9, 16, 21, 22, 27]. Power-aware resource management for portable devices is a critical design factor

---

* Support in parts by research grants from ROC National Science Council NSC-93-2752-E-002-008-PAE.

F. Dehne, A. López-Ortiz, and J.-R. Sack (Eds.): WADS 2005, LNCS 3608, pp. 338–349, 2005.

since most of them are driven by their own power sources (e.g., batteries). Energy-efficient electronic circuit designs, e.g., [1, 34], were proposed in the past decade, and various vendors have provided processors, memory chips, storage devices, or even motherboards equipped with the voltage scaling technology. For example, the flash-memory chip designed by the Intel Corp. [12] supports several voltage levels for operations. Because of the characteristics of many similar storage devices, the supply voltage for an I/O operation remains unchanged for the entire duration of the operation. Non-preemptivity in operation scheduling is also the inherent nature of many I/O devices. Besides, for performance-sensitive devices, the balance between performance and energy consumption must be taken into considerations. We consider power savings on a device capable of supporting several levels of supply voltages with predictable execution times and energy consumption to process jobs.

Let $X_R$ be a device that processes jobs one at a time in a non-preemptive manner, where $R$ consists of the available speeds of $X_R$. As restricted for most portable devices, speed change is not allowed in the middle of processing a job, e.g., in the flash-memory chip [12]. The *energy-consumption rate* of $X_R$ is a function $\phi$ over $R$ such that the energy required for $X_R$ to process a job at speed $r \in R$ for $t$ time units is $\phi(r) \cdot t$. The rest of the paper makes the physically reasonable assumption that the function $\phi(r)/r$ is monotonically increasing, i.e., $r > r'$ implies $\frac{\phi(r)}{r} > \frac{\phi(r')}{r'}$.

For any set $S$, let $|S|$ denote the cardinality of $S$. All numbers throughout the paper are rational. Let $J$ consist of jobs $1, 2, \ldots, |J|$ to be processed by $X_R$. For each $j \in J$, let $c_j$ be the time required for $X_R$ to process job $j$ at speed 1, let $d_j$ be the deadline for completing job $j$, and let $a_j$ be the arrival time for job $j$. If the precedence constraint $\Pi$ is present on $J$, a schedule cannot execute job $j'$ before $j$ when $j'$ is a successor to $j$. It is reasonable to assume that $d_j \leq d_{j'}$ if $j'$ is a successor to $j$. For notational brevity, let $d_1 \leq d_2 \leq \cdots \leq d_{|J|}$. If $d_j = d_{j'}$ and $a_j < a_{j'}$, then $j < j'$. Without loss of generality, we assume $\min_{j \in J} a_j = 0$. A schedule $s$ for $J$ is *feasible* if each job $j \in J$ is assigned a speed $s_j \in R$ and processed after $a_j$ and before $d_j$ without speed change, preemption, or violating the precedence constraints. We say that $s$ is an *earliest-deadline-first schedule* for $J$ (with respect to $X_R$) if each job $j \in J$ is processed by $X_R$ before jobs $j + 1, \ldots, |J|$ as early as possible. It is well known (e.g., [8–§A5.1]) that if the jobs in $J$ have arbitrary arrival times and deadlines, determining whether the jobs in $J$ can be scheduled to meet all deadlines is strongly NP-complete even if $|R| = 1$. The time required for $X_R$ to process job $j$ at speed $r \in R$ is assumed being $c_j/r$. The *energy consumption* $\Phi(s)$ of a schedule $s$ is $\sum_{j \in J} \phi(s_j)c_j/s_j$. The POWER-SAVING SCHEDULING problem is to find a feasible schedule $s$ for $J$ with minimum $\Phi(s)$. Clearly, if the energy-consumption rate of $X_R$ is linear, which is highly unlikely in practice, then any schedule for $J$ has the same energy consumption.

Regardless of the property of the energy consumption function, it is not hard to show that the POWER-SAVING SCHEDULING problem does not admit any polynomial-time approximation algorithm unless $P = NP$, when jobs have arbitrary arrival times and deadlines. Assuming that there exists a polynomial-time approximation algorithm $\mathcal{K}$ for the POWER-SAVING SCHEDULING problem, algorithm $\mathcal{K}$ can be used to solve the following NP-complete 3-PARTITION problem [8] in polynomial time: Given a set $A$ of $3M$ elements, a bound $B \in Z^+$, and a size $w(a) \in Z^+$ for each $a \in A$, where $B/4 < w(a) < B/2$ and $\sum_{a \in A} w(a) = MB$, the 3-PARTITION problem is to find a partition of $A$ into $M$ disjoint sets $A_1, A_2, \cdots, A_M$ such that $\sum_{a \in A_j} w(a) = B$ for $1 \leq j \leq M$. For each element $a \in A$, a unique job $j$ is created by setting $a_j = 0$, $d_j = (M+1)B - 1$, and $c_j = s_{|R|} \cdot w(a)$. Another job set $J'$ is also constructed by creating $M - 1$ jobs, where $a_j = (j+1)B - 1$, $d_j = (j+1)B$, and $c_j = s_{|R|}$ for the $j$-th job in $J'$. It is clear that there exists a feasible schedule for the resulting job set above if and only if there exists a partition for the 3-PARTITION problem. Since $\mathcal{K}$ is a polynomial-time approximation algorithm the POWER-SAVING SCHEDULING problem, $\mathcal{K}$ can be applied to determine the 3-PARTITION problem in polynomial time by examining the feasibility of the derived schedule of $\mathcal{K}$. This contradicts the assumption that $P \neq NP$.

*Our contribution* We investigate the intractability of the problem. Moreover, we give a fully polynomial-time approximation scheme for two practically important cases: (a) all jobs share a common arrival time, and (b) $\Pi = \emptyset$ and for any $j, j' \in J$, $a_j \leq a_{j'}$ implies $d_j \leq d_{j'}$ ($a_j \leq a_{j'}$ if $j < j'$).

Let *basic case* stand for the situation that $|R| = 2$ and that all jobs in $J$ have a common deadline $d$ and a common arrival time 0. We show that the POWER-SAVING SCHEDULING problem is NP-complete even for the basic case. The hardness comes from the dis-allowance of speed change on devices with a finite set of speeds to select from. Moreover, we give a fully polynomial-time approximation scheme for the POWER-SAVING SCHEDULING problem when (a) all jobs share a common arrival time, and (b) $\Pi = \emptyset$ and for any $j, j' \in J$, $a_j \leq a_{j'}$ implies $d_j \leq d_{j'}$, based upon standard techniques of dynamic programming and rounding. Specifically, we first show that a special case of the problem can be solved in pseudopolynomial-time algorithm by a dynamic program. (We comment that the dynamic programs studied by Chen, Lu, and Tang [5] would have sufficed if their schedules were not allowed to violate deadlines.) To turn a pseudopolynomial-time algorithm into a fully polynomial-time approximation scheme via rounding, we need the following lemma to obtain a good estimate of the minimum energy consumption.

**Lemma 1 (Yao, Demers, and Shenker [35]).** *Let $R$ be the set of non-negative rational numbers. Then, given a set of jobs with arbitrary arrival times and deadlines, a feasible schedule for $J$ on $X_R$ allowing preemption with minimum energy consumption can be solved in $O(|J| \log^2 |J|)$ time.*

Comment: the result stated in Lemma 1 requires that the energy consumption rate function $\phi$ is convex and increasing, which is implied by the global assumption of the present paper that $\phi(r)/r$ is monotonically increasing.

*Related work* In [14], a 3-approximation algorithm is proposed for off-line job scheduling, when processors have a special state *sleep*. If a processor has multiple special states beside *running*, (e.g., *idle*, *sleep*, and *standby*), processors could consume much less energy while no job is processing. Although special states provide more flexibility for energy-aware job scheduling, overheads on energy consumption and time latencies for state transitions must be considered. The on-line competitive algorithm proposed by Yao, et al. [35] is extended to handle processors with a single *sleep* state in [14] or multiple special states in [13].

Ishihara and Yasuura [15] show that the minimum energy consumption problem can be formulated as an integer linear program, when $|R|$ is finite, speed change is allowed, and all jobs have the same arrival time and deadline. They show that there exists an optimal schedule with at most two processor speeds and at most one job is processed at two different processor speeds. However the proofs in [15] consider only a specific processor model which is proposed in [1, 34]. In [3], the results in [15] are extended for any convex function. Energy-efficient scheduling has been extensively studied in off-line, e.g., [10, 19, 26], or on-line, e.g., [17, 23–25, 28–31], fashions. Algorithms considering time-cost trade-offs on scheduling are proposed in [6, 7, 32, 33].

The results on minimization of energy consumption on processors could not be applied directly to I/O devices. Generally, processors can process jobs in a preemptive manner. In contrast, I/O devices might perform operations in a non-preemptive manner, and no speed change is allowed in the middle of processing an operation. Chang, Kuo, and Lo [2] propose an adjustment mechanism to dynamically adjust the supply voltage for a flash memory to reduce the energy consumption. The proposed heuristic algorithm is efficient but the optimality is not proved. Hong, Kirovski, Qu, Potkonjak, and Srivastava [10] propose a heuristic algorithm for minimization of the energy consumption on a non-preemptive device, where the supply voltages of the device are available between two given positive thresholds. Besides, Manzak and Chakrabarti [20] consider a minimum energy consumption problem on a system equipped with non-preemptive I/O devices capable of supporting multiple levels of voltages, where voltage switch introduces overheads in energy and time. On-line algorithms are proposed to minimize the energy consumption for executions of real-time tasks, provided that there is a given feasible schedule as an input. Although the derived schedules are feasible and power savings were achieved in the experimental results, the optimality is not shown. To the best of our knowledge, no previous scheduling algorithm to minimize the energy consumption for $X_R$ is known with theoretical analysis of optimality for energy consumption or power savings.

For the remainder of the paper, define $R = \{r_1, r_2, \ldots, r_{|R|}\}$ with $r_1 < r_2 < \cdots < r_{|R|}$. The rest of the paper is organized as follows. Section 2 addresses the basic case. Section 3 shows our approximation scheme. Section 4 concludes this paper.

# 2   Basic Case

In this section, we show that finding a schedule with minimum energy consumption is NP-complete for the basic case. We then present a fully polynomial-time approximation scheme for the basic case as a warm-up. For any subset $I$ of $J$, let $s(I)$ be the schedule $s$ for $J$ with $s_i = r_1$ for each $i \in I$ and $s_i = r_2$ for each $i \in J - I$. Clearly, we have

$$\Phi(s(I)) = \sum_{i \in I} \phi(r_1)c_i/r_1 + \sum_{i \in J-I} \phi(r_2)c_i/r_2.$$

**Theorem 1.** *The* POWER-SAVING SCHEDULING *problem for* $X_R$ *is NP-complete even for the basic case.*

*Proof.* It is clear that the POWER-SAVING SCHEDULING problem is in NP. It suffices to show the NP-hardness by a reduction from the following NP-complete SUBSET SUM problem [8]: Given a number $w_j$ for each index $j \in J$ and another arbitrary number $w$, the problem is to determine whether there is a subset $I$ of $J$ with $\sum_{i \in I} w_i = w$.

An instance for the basic case of the POWER-SAVING SCHEDULING problem is constructed as follows: For each index $j \in J$, we create a job $j$ with $c_j = r_1 r_2 w_j$. Let the common deadline $d$ be $w(r_2 - r_1) + r_1 \sum_{j \in J} w_j$. Clearly, the set $J$ of jobs is feasible. Also, for any subset $I$ of $J$, we have

$$\Phi(s(I)) = r_2\phi(r_1) \sum_{i \in I} w_i + r_1\phi(r_2) \sum_{i \in J-I} w_i = (r_2\phi(r_1) - r_1\phi(r_2)) \sum_{i \in I} w_i + r_1\phi(r_2) \sum_{j \in J} w_j. \tag{1}$$

We show that there is a set $I \subseteq J$ with $\sum_{i \in I} w_i = w$ if and only if the set $J$ of jobs admits a feasible schedule $s$ with $\Phi(s) = w(r_2\phi(r_1) - r_1\phi(r_2)) + r_1\phi(r_2) \sum_{j \in J} w_j$. As for the if-part, let $I$ consist of the jobs $i$ with $s_i = r_1$, which implies $s(I) = s$. By Equation (1) and $\Phi(s) = w(r_2\phi(r_1) - r_1\phi(r_2)) + r_1\phi(r_2) \sum_{j \in J} w_j$, we know $(r_2\phi(r_1) - r_1\phi(r_2)) \sum_{i \in I} w_i = w(r_2\phi(r_1) - r_1\phi(r_2))$. Since $r_2\phi(r_1) < r_1\phi(r_2)$, we have $\sum_{i \in I} w_i = w$. As for the only-if-part, let $s = s(I)$. One can easily see the feasibility of $s$ by verifying that $\sum_{i \in I} w_i = w$ implies $\sum_{i \in I} c_i/r_1 + \sum_{i \in J-I} c_i/r_2 = r_2 \sum_{i \in I} w_i + r_1 \sum_{i \in J-I} w_i = d$. By $\sum_{i \in I} w_i = w$ and Equation (1), we have $\Phi(s) = w(r_2\phi(r_1) - r_1\phi(r_2)) + r_1\phi(1) \sum_{j \in J} w_j$. □

Given a number $w_j$ for each index $j \in J$ and another arbitrary number $w$, the MAXIMUM SUBSET SUM problem is to find a subset $I$ of $J$ with $\sum_{i \in I} w_i \leq w$ such that $\sum_{i \in I} w_i$ is maximized. For the rest of the section, we show how to obtain a fully polynomial-time approximation scheme for the basic case of the POWER-SAVING SCHEDULING problem based upon the following lemma.

**Lemma 2 (Ibarra and Kim [11]).** *The* MAXIMUM SUBSET SUM *problem admits a fully polynomial-time $\frac{1}{(1-\delta)}$-approximation algorithm* SUBSET$(w_1, w_2, \ldots, w_{|J|}, w, \delta)$ *for any $0 < \delta < 1$.*

Given the algorithm shown in Algorithm 1, we have the following lemma.

**Lemma 3.** *Algorithm 1 is a $(1 + \epsilon)$-approximation for the basic case of the* POWER-SAVING SCHEDULING *problem for any $\epsilon > 0$.*

*Proof.* Let $I^*$ be a subset of $J$ with $\sum_{i \in I^*} c_i \leq c$ such that $\sum_{i \in I^*} c_i$ is maximized. By the choice of $c$, one can verify that both $s(I)$ and $s(I^*)$ are feasible schedules

---

**Algorithm 1**

---

**Input:** $J, r_1, r_2, d, \epsilon$;
**Output:** A feasible schedule $s$ with almost minimum energy consumption;
1: let $c = r_1(r_2 d - \sum_{j \in J} c_j)/(r_2 - r_1)$;
2: let $I$ be the subset returned by SUBSET$(c_1, c_2, \ldots, c_{|J|}, c, \delta)$, where $\delta = \frac{\epsilon r_2 \phi(r_1)}{r_1 \phi(r_2)}$;
3: output $s(I)$;

---

for $J$. Moreover, $s(I^*)$ is an optimal schedule. We show $\Phi(s(I)) \leq (1+\epsilon)\Phi(s(I^*))$ as follows. By Lemma 2, we have $\sum_{i \in I} c_i \leq \sum_{i \in I^*} c_i \leq \sum_{i \in I} c_i/(1-\delta)$. It follows that

$$\Phi(s(I)) - \Phi(s(I^*)) = \phi(r_1)(\sum_{i \in I} c_i - \sum_{i \in I^*} c_i)/r_1 + \phi(r_2)(\sum_{i \in I^*} c_i - \sum_{i \in I} c_i)/r_2$$

$$\leq \phi(r_2)(\sum_{i \in I^*} c_i - \sum_{i \in I} c_i)/r_2 \leq \delta \sum_{i \in I^*} \phi(r_2) \cdot c_i/r_2$$

$$= \epsilon \sum_{i \in I^*} \phi(r_1) \cdot c_i/r_1 \leq \epsilon \cdot \Phi(s(I^*)).$$

The lemma is proved.                                                              □

## 3    Our Approximation Scheme

Recall that $R = \{r_1, r_2, \ldots, r_{|R|}\}$, where $r_1 < r_2 < \cdots < r_{|R|}$. Define

$$\gamma = \max_{2 \leq i \leq |R|} \frac{r_{i-1} \cdot \phi(r_i)}{r_i \cdot \phi(r_{i-1})}.$$

An execution sequence is said to be optimal if any feasible schedule can be translated into such a sequence without increasing the energy consumption. Let $s^*$ be a feasible schedule $s$ for the input job set $J$ with minimum $\Phi(s)$. We propose our approximation scheme based on the following lemma which ignores the precedence constraint $\Pi$ first.

**Lemma 4.** *Suppose that we are given a schedule $\hat{s}$ satisfying $\Phi(s^*) \leq \Phi(\hat{s}) \leq \gamma\Phi(s^*)$, the earliest-deadline-first execution sequence is optimal, and $\Pi = \emptyset$, then it takes $O(|R||J|^2(\epsilon^{-1} + \log \gamma))$ time and $O(\epsilon^{-1}|R||J|^2)$ space to compute a $(1 + \epsilon)$-optimal solution for the POWER-SAVING SCHEDULING problem for any parameter $0 < \epsilon \leq 1$.*

*Proof.* Our approximation scheme is based upon the standard rounding technique. For each $j \in J$ and each $r \in R$, let $\psi(j, r)$ denote the energy consumption $\phi(r)c_j/r$ required by $X_R$ for processing job $j$ at speed $r$. That is, $\Phi(s) = \sum_{j \in J} \psi(j, s_j)$ for any schedule $s$. For any positive number $q$, define

$$\psi_q(j, r) = \frac{\lceil q \cdot \psi(j, r) \rceil}{q} \qquad \text{for any } j \in J \text{ and } r \in R;$$

$$\Phi_q(s) = \sum_{j \in J} \psi_q(j, s_j) \qquad \text{for any schedule } s \text{ for } J.$$

Clearly, $q \cdot \psi_q(j, r)$ is an integer and $\psi(j, r) \le \psi_q(j, r) \le \psi(j, r) + \frac{1}{q}$ holds for each $j \in J$ and $r \in R$. Therefore,

$$\Phi(s) \le \Phi_q(s) \le \Phi(s) + \frac{|J|}{q} \tag{2}$$

holds for any schedule $s$ of $J$. In other words, $\Phi_q(s)$ is the "rounded-up" energy consumption, which can be a good estimate for $\Phi(s)$ as long as $q$ is sufficiently large. Finding $s^*$ is NP-hard, but a feasible schedule $s^q$ for $J$ with minimum $\Phi_q(s^q)$ can be computed via the standard technique of dynamic programming as follows.

For any index $j \in J$ and any nonnegative $k$, let $\tau(j, k) = \infty$ signify that $\Phi_q(s) > k$ holds for any feasible schedule $s$ for the job subset $\{1, 2, \ldots, j\}$. If $\tau(j, k) \ne \infty$, let $\tau(j, k)$ be the minimum *completion* time required by any feasible schedule $s$ for the job subset $\{1, 2, \ldots, j\}$ with $\Phi_q(s) \le k/q$. For notational brevity, define

$$\tau(0, k) = \begin{cases} 0 & \text{if } k \ge 0; \\ \infty & \text{otherwise} \end{cases} \tag{3}$$

for any integer $k$. By the optimality of the earliest-deadline-first execution sequence, it is not difficult to verify that the following recurrence relation holds for any $j \in J$ and any positive integer $k$:

$$\tau(j, k) = \min_{r \in R} \begin{cases} \max(\tau(j-1, k - q \cdot \psi_q(j, r)), a_j) & \text{if } \max(\tau(j-1, k - q \cdot \psi_q(j, r)), a_j) \\ \quad + c_j/r & \quad + c_j/r \le d_j; \\ \infty & \text{otherwise.} \end{cases} \tag{4}$$

Let $k_q$ be the minimum $k$ with $\tau(|J|, k) < \infty$. Clearly, $\Phi_q(s) = k_q/q$. Since each $q \cdot \psi_q(j, r)$ is an integer, a feasible schedule $s^q$ for $J$ with minimum $\Phi_q(s^q)$ can be obtained by a standard dynamic-programming algorithm, based upon Equations (3) and (4), in

$$O(|R||J| \cdot k_q) = O(|R||J| \cdot \Phi_q(s^q) \cdot q) \tag{5}$$

time and space. By Equation (2) and the optimality of $s^*$ (respectively, $s_q$) with respect to $\Phi$ (respectively, $\Phi_q$), we have

$$\Phi(s^*) \le \Phi(s^q) \le \Phi_q(s^q) \le \Phi_q(s^*) \le \Phi(s^*) + \frac{|J|}{q}. \tag{6}$$

It remains to determine $q$. Clearly, we want $q$ to be sufficiently large so that $\Phi(s^q)$ can be close enough to $\Phi(s^*)$. For example, by $\epsilon \le 1$, we can prove that

$$q \ge \frac{2|J|}{\epsilon \cdot \Phi_q(s^q)} \tag{7}$$

implies $\Phi(s^q) \leq (1+\epsilon) \cdot \Phi(s^*)$ as follows. By Equation (7) and the last inequality in Equation (6), we have

$$\frac{|J|}{q} \leq \frac{\epsilon \cdot \Phi_q(s^q)}{2} \leq \frac{\epsilon \cdot \Phi(s^*)}{2} + \frac{\epsilon \cdot |J|}{2q}.$$

It follows that

$$\frac{|J|}{q} \leq \frac{\epsilon \cdot \Phi(s^*)}{2 - \epsilon} \leq \epsilon \cdot \Phi(s^*),$$

which by Equation (6) implies $\Phi(s^q) \leq (1+\epsilon) \cdot \Phi(s^*)$.

Of course the immediate problem is that schedule $s^q$ depends on the value of $q$. That is, we need to know the value of $q$ in order to compute $s^q$, so it seems difficult to enforce $q \geq \frac{2|J|}{\epsilon \cdot \Phi_q(s^q)}$ at one shot. Fortunately, we can use the following trick of "doubling the value of $q$ in each iteration": Initially, we let $q$ be

$$q' = \frac{|J|}{\epsilon \cdot \Phi(\hat{s})}. \tag{8}$$

In each iteration, we first compute $s^q$. If Equation (7) holds, we output the current $s^q$ as a $(1 + \epsilon)$-optimal solution. Otherwise, we double the value of $q$ and then proceed to the next iteration. Let $q''$ be the value of $q$ in the last iteration.

What is the required running time? By Equations (5) and (6) we know that the first iteration runs in $O(|J|^2|R|/\epsilon)$ time and space. Therefore, we focus on the case that the above procedure runs for more than one iteration. By Equations (5) and (6), we know that each iteration runs in $O(|R||J|(q \cdot \Phi(s^*) + |J|))$ time and space. Since the value of $q$ is doubled in each iteration, the overall running time is

$$O\left(|R||J|\left(q'' \cdot \Phi(s^*) + |J| \log \frac{q''}{q'}\right)\right).$$

Since Equation (7) does not hold in the second-to-last iteration, we have

$$\frac{q''}{2} < \frac{2|J|}{\epsilon \cdot \Phi_{q''/2}(s^{q''/2})}. \tag{9}$$

Besides, by Equation (6), we know

$$\Phi(s^*) \leq \Phi_{q''/2}(s^{q''/2}). \tag{10}$$

Combining Equations (6), (9), and (10), we know $q'' \cdot \Phi(s^*) = O(|J|/\epsilon)$. By Equations (9) and (10) we have

$$q'' < \frac{4|J|}{\epsilon \cdot \Phi(s^*)}. \tag{11}$$

Combining the given relation of $\Phi(s^*)$ and $\Phi(\hat{s})$, Equations (8) and (11), we have

$$\log_2 \frac{q''}{q'} < \log_2 \frac{4\Phi(\hat{s})}{\Phi(s^*)} = O(\log \gamma).$$

The theorem is proved.                                                          □

Since $\log \gamma$ is polynomial in the number of bits required to encode the input $r$ and $\phi(r)$ for all $r \in R$, Lemma 4 provides a fully polynomial-time approximation scheme for the POWER-SAVING SCHEDULING problem when we are given a schedule $\hat{s}$ satisfying $\Phi(s^*) \le \Phi(\hat{s}) \le \Phi(s^*) \cdot \gamma$ and the earliest-deadline-first sequence is known to be optimal, if there are no precedence constraints on $J$. Let $X_{R'}$ be the (imaginary) device with $R' = \{r \mid r_1 \le r \le r_{|R|}\}$ for any energy consumption function $\phi'$ that coincides with $\phi$ at all speeds in $R$. We need the following lemma to derive $\hat{s}$.

**Lemma 5.** *If preemption is allowed for the* POWER-SAVING SCHEDULING *problem, then there exists a polynomial-time algorithm to derive an optimal schedule on $X_{R'}$.*

*Proof.* Let $s'$ be the schedule obtained by applying Lemma 1. That is, job $j$ is to be executed at speed $s'_j$ according to schedule $s'$. We now transform $s'$ into $\bar{s}$ so that $\bar{s}$ is optimal and feasible on $X_{R'}$. We define $J_l$ as $\{j \mid s'_j \ge r_1\}$. For each job $j \in J_l$, $\bar{s}$ just copies the schedule of $j$ on $s'$, including the speed setting and processing time intervals. For each job $j \in J - J_l$, if $s'$ executes $j$ in the interval $[z_1, z_2]$ (there could be more than one interval), then $\bar{s}$ executes $j$ in the interval $[z_1, z_1 + \frac{s'_j(z_2 - z_1)}{r_1}]$. It is clear that only one job in $\bar{s}$ is processed at one time. Therefore, $\bar{s}$ is feasible. We adopt the terminologies used in [35] to prove the optimality of $\bar{s}$: $g(I)$ for a time interval $I = [z, z']$ is defined as $g(I) = \frac{\sum_{j \in R_I} c_i}{z' - z}$, where $R_I$ is the set of jobs satisfying $a_j \ge z$ and $d_j \le z'$. A critical interval $I^*$ satisfies $g(I^*) \ge g(I)$ for any interval $I$. Theorem 1 in [35] shows that there exists an optimal schedule $S$, which executes every job in $R_{I^*}$ at the speed $g(I^*)$ completely within $I^*$ and executes no other jobs during $I^*$. The algorithm for achieving Lemma 1 is obtained by computing a sequence of critical intervals iteratively. Therefore, it is clear that if the input job set $J$ is feasible on $X_R$, $g(I^*)$ must be no larger than $r_{|R|}$. We know that $\bar{s}$ is a feasible schedule on $X_{R'}$ (allow preemption). Besides, the optimality of Theorem 1 in [35] fails on $X_{R'}$ only when $g(I^*) < r_1$. However, executing jobs at the speed $r_1$ results in an optimal solution in this situation. Therefore, $\bar{s}$ is optimal on $X_{R'}$ and obtainable in $O(|J| \log^2 |J|)$ time if preemption is allowed.           □

**Theorem 2.** *The* POWER-SAVING SCHEDULING *problem admits a fully polynomial-time approximation scheme when (a) all jobs share a common arrival time, and (b) $\Pi = \emptyset$ and for any $j, j' \in J$, $a_j \le a_{j'}$ implies $d_j \le d_{j'}$.*

*Proof.* We first consider only the timing constraints of $J$ in this paragraph. Based on Lemma 4, we just have to show that we can derive a schedule $\hat{s}$ satisfying $\Phi(s^*) \le \Phi(\hat{s}) \le \Phi(s^*) \cdot \gamma$ efficiently and prove the optimality of the earliest-deadline-first execution sequence for these job sets. The *preemptive earliest deadline first rule* is defined in [18]: If a job arrives or is completed at time

$t$, execute an arrived (ready) job with the earliest deadline at time $t$. Given a feasible schedule $s$, it is also feasible to schedule jobs according to the preemptive earliest-deadline-first rule by setting the processing speed of job $j$ as $s_j$ [18] (if preemption is allowed). Since the job sets under considerations satisfy the condition $a_j \leq a_{j'}$ and $d_j \leq d_{j'}$ if $j' > j$, it is clear that each job in the resulting schedule following the preemptive earliest-deadline-first rule is non-preempted. The energy consumption for the resulting schedule remains, because $s_j$ does not be increased or decreased. Therefore, the earliest-deadline-first sequence is known to be optimal for the POWER-SAVING SCHEDULING problem. Let $\bar{s}$ be the resulting schedule from Lemma 5. Let $s'$ be the schedule for $J$ defined by letting $s'_j$ be the smallest $r \in R$ with $\bar{s}_j \leq r$ for each $j \in J$ and schedule jobs according to the earliest-deadline-first execution sequence. It is clear that $\Phi(\bar{s}) \leq \Phi(s^*) \leq \Phi(s') \leq \Phi(\bar{s}) \cdot \gamma \leq \Phi(s^*) \cdot \gamma$. If $\Pi = \emptyset$, the fully polynomial-time approximation scheme is obtained by setting $\hat{s} = s'$.

In the following, we shall show how to deal with $\Pi$ ($a_j = 0$ for all $j \in J$). Given an earliest-deadline-first feasible schedule $s$ for $J$, we claim that we can transform $s$ into another schedule $s''$ in $O(|J|^2)$ time such that $\Phi(s'') = \Phi(s)$ and $s''$ satisfies the precedence and timing constraints. If this claim stands, we can transform the derived schedule $s^q$ in Lemma 4 (respectively, $s'$ and $s^*$ in the previous paragraph) into a feasible schedule without increasing $\Phi(s^q)$ (respectively, $\Phi(s')$ and $\Phi(s^*)$). This implies the existence of a fully polynomial-time approximation scheme. Initially, we let $s''$ be $s$. We consider a job $j$ from job 2 to $|J|$. While considering job $j$, we look backward if there is a successor $k$ to $j$ is executed before $j$ in $s''$ such that the ordered jobs $J_{kj}$ executed between $k$ and $j$ are not successors to $j$. If such a job $k$ exists, the execution sequence is modified by delaying $k$ to be executed immediately after $j$ finishes. Because all jobs are ready at time 0, job $j$ and jobs in $J_{jk}$ complete earlier. Since $k$ is a successor to $j$ ($d_k \geq d_j$) and all jobs are ready at time 0, the resulting schedule is feasible while considering jobs $j$, $k$, and $J_{kj}$. We repeat the previous procedure until no such a job $k$ exists, and let $s''$ be the final schedule. It is clear that $s''$ is feasible for the job sets $\{1, 2, \ldots, j\}$ after we perform the above rescheduling by considering job $j$. After we consider job $|J|$, $s''$ satisfies the precedence and timing constraints. The time complexity for the above rescheduling is $O(|J|^2)$.                                                                                    □

## 4    Conclusion

This paper targets non-preemptive scheduling for minimization of energy consumption on devices that allow weakly dynamic voltage scaling. The problem is shown to be NP-hard even if the device has only two speeds and all jobs share the same arrival time and deadline. Moreover, we provide a fully polynomial-time approximation scheme of the NP-hard problem for two cases: (a) all jobs share a common arrival time, and (b) $\Pi = \emptyset$ and for any $j, j' \in J$, $a_j \leq a_{j'}$ implies $d_j \leq d_{j'}$.

An interesting direction for future research is to extend our approximation scheme to handle the overheads on voltage/speed switches.

# References

1. A. Chandrakasan, S. Sheng, and R. Broderson. Lower-power CMOS digital design. *IEEE Journal of Solid-State Circuit*, 27(4):473–484, 1992.
2. L.-P. Chang, T.-W. Kuo, and S.-W. Lo. A dynamic-voltage-adjustment mechanism in reducing the power consumption of flash memory for portable devices. In *Proceedings of IEEE International Conference on Consumer Electronics*, pages 218–219, 2001.
3. J.-J. Chen, T.-W. Kuo, and C.-L. Yang. Profit-driven uniprocessor scheduling with energy and timing constraints. In *ACM Symposium on Applied Computing*, pages 834–840. ACM Press, 2004.
4. J. Y. Chen, W. B. Jone, J. S. Wang, H.-I. Lu, and T. F. Chen. Segmented bus design for low-power systems. *IEEE Transactions on VLSI Systems*, 7(1):25–29, 1999.
5. Z. Chen, Q. Lu, and G. Tang. Single machine scheduling with discretely controllable processing times. *Operations Research Letters*, 21(2):69–76, 1997.
6. P. De, J. E. Dunne, J. B. Ghosh, and C. E. Wells. Complexity of the discrete time-cost tradeoff problem for project networks. *Operations Research*, 45(2):302–306, 1997.
7. V. G. Deĭneko and G. J. Woeginger. Hardness of approximation of the discrete time-cost tradeoff problem. *Operations Research Letters*, 29(5):207–210, 2001.
8. M. R. Garey and D. S. Johnson. *Computers and intractability: A guide to the theory of NP-completeness*. W. H. Freeman and Co., 1979.
9. V. Gutnik and A. P. Chandrakasan. Embedded power supply for low-power DSP. *IEEE Transactions on VLSI Systems*, 5(4):425–435, 1997.
10. I. Hong, D. Kirovski, G. Qu, M. Potkonjak, and M. B. Srivastava. Power optimization of variable voltage core-based systems. In *Proceedings of the 35th Annual Conference on Design Automation Conference*, pages 176–181. ACM Press, 1998.
11. O. H. Ibarra and C. E. Kim. Fast approximation algorithms for the knapsack and sum of subsets problems. *Journal of the ACM*, 22(4):463–468, 1975.
12. Intel Corporation. *28F016S5 5-Volt FlashFile Flash Memory Datasheet*, 1999.
13. S. Irani, S. Shukla, and R. Gupta. Competitive analysis of dynamic power management strategies for systems with multiple saving states. In *Proceedings of the Design Automation and Test Europe Conference*, 2002.
14. S. Irani, S. Shukla, and R. Gupta. Algorithms for power savings. In *Proceedings of the Fourteenth Annual ACM-SIAM Symposium on Discrete Algorithms*, pages 37–46. Society for Industrial and Applied Mathematics, 2003.
15. T. Ishihara and H. Yasuura. Voltage scheduling problems for dynamically variable voltage processors. In *Proceedings of the International Symposium on Low Power Electroncs and Design*, pages 197–202, 1998.
16. W.-B. Jone, J. S. Wang, H.-I. Lu, I. P. Hsu, and J.-Y. Chen. Design theory and implementation for low-power segmented bus systems. *ACM Transactions on Design Automation of Electronic Systems*, 8(1):38–54, 2003.
17. S. Lee and T. Sakurai. Run-time voltage hopping for low-power real-time systems. In *Proceedings of the 37th Conference on Design Automation*, pages 806–809. ACM Press, 2000.
18. C. L. Liu and J. W. Layland. Scheduling algorithms for multiprogramming in a hard-real-time environment. *Journal of the ACM*, 20(1):46–61, 1973.
19. A. Manzak and C. Chakrabarti. Variable voltage task scheduling algorithms for minimizing energy. In *Proceedings of the 2001 International Symposium on Low Power Electronics and Design*, pages 279–282. ACM Press, 2001.

20. A. Manzak and C. Chakrabarti. Energy-conscious, deterministic I/O device scheduling in hard real-time systems. *IEEE Transactions on Computer-Aided Design of Integrated Circuits and Systems*, 22(7):847–858, 2003.
21. A. Manzak and C. Chakrabarti. Variable voltage task scheduling algorithms for minimizing energy/power. *IEEE Transactions on VLSI Systems*, 11(2):270–276, 2003.
22. M. Pedram and J. M. Rabaey. *Power Aware Design Methodologies*. Kluwer Academic Publishers, 2002.
23. T. Pering, T. Burd, and R. Brodersen. The simulation and evaluation of dynamic voltage scaling algorithms. In *Proceedings of the 1998 International Symposium on Low Power Electronics and Design*, pages 76–81. ACM Press, 1998.
24. T. Pering, T. Burd, and R. Brodersen. Voltage scheduling in the iparm microprocessor system. In *Proceedings of the 2000 International Symposium on Low Power Electronics and Design*, pages 96–101. ACM Press, 2000.
25. J. Pouwelse, K. Langendoen, and H. Sips. Energy priority scheduling for variable voltage processors. In *Proceedings of the 2001 International Symposium on Low Power Electronics and Design*, pages 28–33. ACM Press, 2001.
26. G. Quan and X. Hu. Energy efficient fixed-priority scheduling for real-time systems on variable voltage processors. In *Proceedings of the 38th Conference on Design Automation*, pages 828–833. ACM Press, 2001.
27. V. Raghunathan, M. B. Srivastava, and R. K. Gupta. A survey of techniques for energy efficient on-chip communication. In *Proceedings of the 40th Conference on Design Automation*, pages 900–905. ACM Press, 2003.
28. D. Shin and J. Kim. A profile-based energy-efficient intra-task voltage scheduling algorithm for real-time applications. In *Proceedings of the 2001 International Symposium on Low Power Electronics and Design*, pages 271–274. ACM Press, 2001.
29. D. Shin, J. Kim, and S. Lee. Low-energy intra-task voltage scheduling using static timing analysis. In *Proceedings of the 38th Conference on Design Automation*, pages 438–443. ACM Press, 2001.
30. Y. Shin and K. Choi. Power conscious fixed priority scheduling for hard real-time systems. In *Proceedings of the 36th ACM/IEEE Conference on Design Automation Conference*, pages 134–139. ACM Press, 1999.
31. Y. Shin, K. Choi, and T. Sakurai. Power optimization of real-time embedded systems on variable speed processors. In *Proceedings of the 2000 IEEE/ACM International Conference on Computer-Aided Design*, pages 365–368. IEEE Press, 2000.
32. M. Skutella. Approximation algorithms for the discrete time-cost tradeoff problem. In *Proceedings of the Eighth Annual ACM-SIAM Symposium on Discrete Algorithms*, pages 501–508. Society for Industrial and Applied Mathematics, 1997.
33. M. Skutella. Approximation algorithms for the discrete time-cost tradeoff problem. *Mathematics of Operations Research*, 23(4):909–929, 1998.
34. M. Weiser, B. Welch, A. Demers, and S. Shenker. Scheduling for reduced CPU energy. In *Proceedings of Symposium on Operating Systems Design and Implementation*, pages 13–23, 1994.
35. F. Yao, A. Demers, and S. Shenker. A scheduling model for reduced CPU energy. In *Proceedings of the 36th Annual Symposium on Foundations of Computer Science*, pages 374–382. IEEE, 1995.

# Improved Approximation Algorithms for Metric Maximum ATSP and Maximum 3-Cycle Cover Problems

Markus Bläser[1], L. Shankar Ram[1], and Maxim Sviridenko[2]

[1] Institut für Theoretische Informatik, ETH Zürich,
CH-8092 Zürich, Switzerland
{mblaeser, lshankar}@inf.ethz.ch
[2] IBM T.J. Watson Research Center
sviri@us.ibm.com

**Abstract.** We consider an APX-hard variant ($\Delta$-Max-ATSP) and an APX-hard relaxation (Max-3-DCC) of the classical traveling salesman problem. $\Delta$-Max-ATSP is the following problem: Given an edge-weighted complete loopless directed graph $G$ such that the edge weights fulfill the triangle inequality, find a maximum weight Hamiltonian tour of $G$. We present a $\frac{31}{40}$-approximation algorithm for $\Delta$-Max-ATSP with polynomial running time. Max-3-DCC is the following problem: Given an edge-weighted complete loopless directed graph, compute a spanning collection of node-disjoint cycles, each of length at least three, whose weight is maximum among all such collections. We present a $\frac{3}{4}$-approximation algorithm for this problem with polynomial running time. In both cases, we improve on the previous best approximation performances. The results are obtained via a new decomposition technique for the fractional solution of an LP formulation of Max-3-DCC.

## 1 Introduction

Travelling salesman problems have been studied for many decades. Classically, one deals with minimization variants, that is, one wants to compute a shortest (i.e., minimum weight) Hamiltonian tour. But also the corresponding maximization variants have been investigated. At a first glance, computing a tour of maximum weight seems to be unnatural, but this problems has its applications for instance in maximum latency delivery problems [5] or in the computation of shortest common superstrings [4].

### 1.1 Notations and Definitions

Let $G = (V, E)$ be a complete loopless directed graph and $w : E \to \mathbb{Q}_{\geq 0}$ be a weight function that assigns each edge a nonnegative weight. A cycle of $G$ is a (strongly) connected subgraph such that each node has indegree and outdegree one. (Since $G$ has no loops, every cycle has length at least two.) The weight $w(c)$

F. Dehne, A. López-Ortiz, and J.-R. Sack (Eds.): WADS 2005, LNCS 3608, pp. 350–359, 2005.

of a cycle $c$ is the sum of weigths of the edges contained in it. A *Hamiltonian tour* of $G$ is a cycle that contains all nodes of $G$. The problem of finding a Hamiltonian tour of minimum weight is the well-studied asymmetric traveling salesman problem (ATSP). The problem is called asymmetric, since $G$ is directed. The special case that $G$ is undirected or, equivalently, that $w$ is symmetric has received even more attention. But also the maximization variant—given $G$, find a Hamiltonian tour of maximum weight—has been studied. This problem is for instance used for maximum latency delivery problems [5] and as a blackbox in shortest common superstring computations [4]. We here study the variant of Maximum ATSP where $w$ in addition fulfills the triangle inequality, that is,

$$w(u, v) + w(v, x) \geq w(u, x) \qquad \text{for all nodes } u, v, x.$$

We call this problem $\Delta$-Max-ATSP.

A *cycle cover* of $G$ is a collection of node-disjoint cycles such that each node is part of exactly one cycle. Every Hamiltonian tour is obviously a cycle cover. We call a cycle a *k-cycle* if it has *exactly* $k$ edges (and nodes). A cycle cover is a *k-cycle cover* if each cycle in the cover has *at least* $k$ edges. We call the problem of finding a maximum weight 3-cycle cover Max-3-DCC. Note that we here do not require $w$ to fulfill the triangle inequality.

## 1.2     Previous Results

For the general Maximum ATSP, Nemhauser, Fisher, and Wolsey [6] present a $\frac{1}{2}$-approximation algorithm with polynomial time. Kosaraju, Park, and Stein [8], Bläser [1], Levenstein and Sviridenko [10], and Kaplan et al. [7] improve on this by giving polynomial time approximation algorithm with performances $\frac{38}{63}$, $\frac{8}{13}$, $\frac{5}{8}$, and $\frac{2}{3}$, respectively. For $\Delta$-Max-ATSP, Kostochka and Serdyukov [9] provide a $\frac{3}{4}$-approximation algorithm with polynomial running time. Kaplan et al. [7] improve on this by giving a polynomial time $\frac{10}{13}$-approximation algorithm.

$\Delta$-Max-ATSP is APX-hard, even if the weight function is $\{1, 2\}$-valued. This follows basically from the hardness proof of the corresponding minization variant given by Papadimitriou and Yannakakis [12]

The problem of computing a maximum weight 2-cycle cover is solvable in polynomial time, see Section 2.1. But already the problem of computing maximum weight 3-cycle covers is APX-hard, even if $w$ attains only two different values [3]. Bläser and Manthey [2] give a $\frac{3}{5}$-approximation algorithm with polynomial running time for Max-3-DCC. Kaplan et al. [7] improve on this by giving a $\frac{2}{3}$-approximation algorithm with polynomial running time.

## 1.3     New Results

As a main technical contribution, we present a new decomposition technique for the fractional solution of an LP for computing maximum weight 3-cycle covers. The new idea is to ignore the directions of the edges and decompose the fractional solution into undirected cycle covers, that means, that after ignoring directions, the subgraph is a cycle cover. This has of course the drawback that viewed as a directed graph, the edges of the cycles might not point into the same direction. The advantage, on the other hand is, that all cycles in the cycle covers obtained have length at least three. In previous approaches such a fractional solution always was decomposed into directed cycle covers in which every cycle could have length two (but one had some additional knowledge about the distribution of the 2-cycles in the covers.) We apply this method to $\Delta$-Max-ATSP and Max-3-DCC.

For $\Delta$-Max-ATSP, this results in a $\frac{31}{40}$-approximation algorithm improving on the previous best algorithm which has approximation performance $\frac{10}{13}$. Note that $\frac{31}{40} = 0.775$ and $\frac{10}{13} \approx 0.769$.

For Max-3-DCC, we get a $\frac{3}{4}$-approximation algorithm. This improves the previous best algorithm which has performance $\frac{2}{3}$.

# 2     Computing Undirected 3-Cycle Covers

Let $G$ be a complete directed graph without loops with $n$ nodes and let $w$ be a weight function on the edges of $G$.

## 2.1     LP for 3-Cycle Covers

Maximum weight cycle covers can be computed by solving the following well-known LP:

$$\text{Maximize} \sum_{(u,v)} w(u,v)x_{u,v} \text{ subject to}$$

$$\sum_{u \in V} x_{u,v} = 1 \text{ for all } v \in V, \quad \text{(indegree constraints)}$$

$$\sum_{v \in V} x_{u,v} = 1 \text{ for all } u \in V, \quad \text{(outdegree constraints)} \tag{1}$$

$$x_{u,v} \geq 0 \quad \text{for all } (u,v).$$

The variable $x_{u,v}$ is the indicator variable of the edge $(u,v)$. The matrix corresponding to (1) is totally unimodular (see e.g. [11]), thus any optimum basic solution of (1) is integer valued (indeed $\{0,1\}$ valued). When one wants to use cycle covers as a relaxation for approximating Hamiltonian tours, the worst case is a cycle cover that consists solely of cycles of length two, so-called *2-cycles*. To avoid this, one can add 2-cycle elimination constraints to the LP:

$$x_{u,v} + x_{v,u} \leq 1 \quad \text{for all } (u,v) \quad \text{(2-cycle elimination)} \tag{2}$$

These constraints are a subset of the so-called *subtour elimination constraints*.

If we consider the LP above as an integer LP, then the 2-cycle elimination constraints ensure that there are no 2-cycles in a feasible solution. However, after adding the 2-cycle elimination constraints, the basic feasible solutions of the relaxed LP may not be integral anymore.

## 2.2    Decomposition Into Directed 2-Cycle Covers

Let $(x_{u,v}^{\star})$ denote an optimal fractional solution of the relaxed LP (1) together with (2). Let $W^{\star}$ be its weight. Let $N$ be the smallest common multiple of all denominators of the $x_{u,v}^{\star}$. From $(x_{u,v}^{\star})$, we build a directed multigraph $M^{\star}$: Each edge $(u, v)$ in $M^{\star}$ has multiplicity $x_{u,v}^{\star} \cdot N$. By using standard scaling and rounding we may also assume that $N$ is a power of two, i.e. $N = 2^{\nu}$ for some integer $\nu$ polynomially bounded in the input length (see the journal version of [7] for details).

We change the solution $(x_{u,v}^{\star})$ and corresponding multigraph $M^{\star}$ in the following way. Construct undirected graph $H$ by defining an undirected edge $\{u, v\}$ if $M^{\star}$ contains edges between vertices $u$ and $v$ in both directions. If $H$ contains a cycle $C$ then $M^{\star}$ contains two corresponding oppositely oriented cycles $C_1$ and $C_2$ of length $\geq 3$. The multiplicity of those cycles is $\min\{x_{u,v}^{\star} : (u, v) \in C_1\} \cdot N$ and $\min\{x_{u,v}^{\star} : (u, v) \in C_2\} \cdot N$. W.l.o.g assume that the weight of $C_1$ is no more than the weight of $C_2$. We delete $\min\{x_{u,v}^{\star} : (u, v) \in C_1\} \cdot N$ copies of $C_1$ from $M^{\star}$ and add $\min\{x_{u,v}^{\star} : (u, v) \in C_1\} \cdot N$ copies of $C_2$. We also change the current solution $(x_{u,v}^{\star})$ to reflect the change in $M^{\star}$. The new solution is also an optimal solution of the LP (1) together with (2) since we did not decrease the value of the solution and did not violate the LP constraints during the transformation. Repeating the procedure $O(n^2)$ times we could guarantee that we have an optimal solution $(x_{u,v}^{\star})$ such that graph $M^{\star}$ does not have oppositely oriented cycles of length larger than two.

Lewenstein and Sviridenko [10] showed how to compute a collection of cycle covers $C_1, \ldots, C_N$ from $M^{\star}$ with the following properties:

(P1)  $M^{\star}$ is the union of $C_1, \ldots, C_N$, considered as a multigraph. Thus the total weight of $C_1, \ldots, C_N$ equals $N \cdot W^{\star}$.

(P2)  Between any pair of nodes $u$ and $v$, the total number of edges in $C_1, \ldots, C_N$ between $u$ and $v$ is at most $N$.

The number $N$ might be exponential, however, Lewenstein and Sviridenko also gave a succinct representation consisting of at most $n^2$ cycle covers with appropriate multiplicities. This will be sufficient for our algorithms.

The discussion above implies that $C_1, \ldots, C_N$, fulfill the additional property:

(P3)  Let $H$ be the undirected graph that contains an edge $\{u, v\}$ iff $u$ and $v$ are contained in a 2-cycle in at least one of $C_1, \ldots, C_N$. Then $H$ is acyclic.

## 2.3    Decomposition Into Undirected 3-Cycle Covers

For our algorithms, we now redistribute the edges of $C_1, \ldots, C_N$. Each copy of an edge $(u, v)$ in $M^{\star}$ appears in exactly one of the $C_1, \ldots, C_N$. Color an edge of $M^{\star}$ red, if it occurs in a 2-cycle in the particular $C_i$. Otherwise color it blue. Note that by (P3), red edges cannot form a cycle of length strictly larger than two.

**Lemma 1.** *Let $U$ be an undirected $2N$-regular multigraph that has at most $N$ copies of any edge where $N = 2^{\nu}$ is a power of two. Then there are undirected*

*3-cycle covers* $D_1, \ldots, D_N$ *such that* $U$ *is the union of* $D_1, \ldots, D_N$. *This decomposition can be performed in polynomial time.*

*Proof.* The proof is by induction on $N$: If $N = 1$, then $U$ is a 3-cycle cover.

Assume that $N > 1$. We will now decompose $U$ into two $N$-regular multigraph $U_1$ and $U_2$, each of them containing at most $N/2$ copies of each edge. By the induction hypothesis, these multigraphs can be decomposed into $N/2$ 3-cycle covers each. This proves the lemma.

Let $e$ be an edge of $U$ and let $m$ be its multiplicity. We move $\lfloor m/2 \rfloor$ copies to $U_1$ and $\lfloor m/2 \rfloor$ to $U_2$. If $m$ is even, this distributes all copies of $e$ between $U_1$ and $U_2$. If $m$ is odd, then one copy remains, that is, the multigraph $U'$ that remains after treating all edges in this way is indeed a graph. Since $U$ is $2N$-regular and we remove an even number of copies of each edge, the degree of each node in $U'$ is even. Therefore, each connected component of $U'$ is Eulerian. For each such component, we compute an Eulerian tour. We take the edges of each component of $U'$ in the order induced by the Eulerian tour and move them in an alternating fashion to $U_1$ and $U_2$. In this way, the degree at each node in $U'$ is "halved", therefore both $U_1$ and $U_2$ are $N$-regular.

It remains to show that every edge in $U_1$ and $U_2$, respectively, has multiplicity at most $N/2$. This is clearly true if the multiplicity $m$ of an edge $e$ in $U$ was even. If $m$ is odd, then $m < N$, since $N$ is even. Thus $U_1$ and $U_2$ get $\lfloor m/2 \rfloor < N/2$ copies. The last copy is then either moved to $U_1$ or $U_2$, but in both cases, the multiplicity is thereafter $\leq N/2$. □

Let $W_2$ be the average weight of all 2-cycles and $W_3$ be the average weight of all cycles of length at least three in $C_1, \ldots, C_N$.

We now consider $M^\star$ as an undirected graph. It is $2N$-regular and by (P2), there are at most $N$ edges between any pair of nodes. By Lemma 1, we can decompose $M^\star$ into undirected 3-cycle covers $D_1, \ldots, D_N$. As already mentioned, none of the cycles in $D_1, \ldots, D_N$ solely consists of red edges. Now we view $D_1, \ldots, D_n$ as directed graphs again. They may not be directed cycle covers anymore, since the cycles may not be directed cycles. For all $i$ and for any red edge in $D_i$, we add the other edge of the corresponding 2-cycle to $D_i$. Let $\hat{D}_1, \ldots, \hat{D}_N$ be the resulting graphs. The average weight of $\hat{D}_1, \ldots, \hat{D}_N$ is

$$\frac{w(\hat{D}_1) + \cdots + w(\hat{D}_N)}{N} = 2W_2 + W_3, \tag{3}$$

because every edge in a 2-cycle was added a second time.

## 3   Metric Maximum ATSP

Throughout this section, we assume that $w$ fulfills the triangle inequality, i.e.,

$$w(u, v) + w(v, x) \geq w(u, x) \qquad \text{for all nodes } u, v, x.$$

Our goal is to find a Hamiltonian tour of maximum weight.

## 3.1  First Algorithm

By exploiting an algorithm due to Kostochka and Serdyukov [9], Kaplan et al. [7] show how to compute a Hamiltonian tour of weight at least $\frac{3}{4}W_2 + \frac{5}{6}W_3$.

**Theorem 1.** *There is an algorithm with polynomial running time that given* $C_1, \ldots, C_N$, *computes a Hamiltonian tour of weight* $\frac{3}{4}W_2 + \frac{5}{6}W_3$.     □

This will be the first algorithm that we use. It is favorable if $W_2$ is small. Next we design an algorithm that works well if $W_2$ is large.

## 3.2  Second Algorithm

**Lemma 2.** *Let $K$ be a connected component of $\hat{D}_i$. After deleting one blue edge of $K$, we can construct in polynomial time two node-disjoint directed paths $P_1$ and $P_2$ such that*

1. *$P_1$ and $P_2$ span the same nodes as $K$,*
2. *$P_1$ can be transformed into $P_2$ by reversing all directions of its edges and vice versa,*
3. *except the discarded blue edge, all edges of $K$ are in $P_1 \cup P_2$, and*
4. *the discarded edge connects the two end-points of $P_1$ and $P_2$, respectively.*

*Proof.* The component corresponding to $K$ in $D_i$ is an undirected cycle. At least one of the edges on this cycle is blue by (P3). Discard one blue edge. Let $v_1, \ldots, v_\ell$ be the nodes of $K$ (in this order) and assume that the edge between $v_\ell$ and $v_1$ was discarded. Between any two nodes $v_\lambda$ and $v_{\lambda+1}$, there are at most two edges and if there are two edges, then these edges are red and point into different directions, since we added the other edge of the 2-cycles. Therefore, the paths $v_1, v_2, \ldots, v_\ell$ and $v_\ell, v_{\ell-1}, \ldots, v_1$ contain all edges of $K$ except the one that we discarded.     □

By applying Lemma 2 to each component of $\hat{D}_i$, we obtain two collections of node-disjoint paths $P_{i,1}$ and $P_{i,2}$ such that each connected component of $\hat{D}_i$ corresponds to two oppositely directed paths in $P_{i,1}$ and $P_{i,2}$ respectively. Next we are going to construct Hamiltonian tours $H_{i,1}$ and $H_{i,2}$ out of $P_{i,1}$ and $P_{i,2}$.

**Lemma 3.** *Given $P_{i,1}$ and $P_{i,2}$, we can construct in polynomial time, two Hamiltonian tours $H_{i,1}$ and $H_{i,2}$ such that $H_{i,1}$ and $H_{i,2}$ contain all the weight of the red edges and $1/2$ of the weight of the blue edges of $\hat{D}_i$.*

*Proof.* Let $p_{j,1}, \ldots, p_{j,k}$ be the paths of $P_{i,j}$ for $j = 1, 2$. Assume that $p_{1,\kappa}$ and $p_{2,\kappa}$ span the same nodes but have opposite directions. Let $p_{1,\kappa}$ be the path that forms a cycle with the discarded blue edge.

We first describe a randomized algorithm. We first select paths $q_1, \ldots, q_k$: $q_\kappa$ is $p_{1,\kappa}$ or $p_{2,\kappa}$, both with probability $1/2$. All coin flips are independent. The cycle $H_{i,1}$ is obtained by patching the paths together in the order $q_1, \ldots, q_k$, and the cycle $H_{i,2}$ by patching the paths together in the opposite order $q_k, \ldots, q_1$.

With the exception of the discarded blue edges in Lemma 2, an edge of $\hat{D}_i$ is included twice with probability $1/2$, namely once in $H_{i,1}$ and once in $H_{i,2}$. Thus we get all the weight of these edges in expectation.

We now show that we can get some weight of the discarded blue edges back (in expectation) during the patching process. Figure 3.2 shows two discarded blue edges $e$ and $f$. There are four possibilities how the corresponding paths can be directed. Each occurs with probability $1/4$. The edges introduced by the patching are $x_j$ and $y_j$, respectively. The expected weight we get is

$$\tfrac{1}{4}(w(x_1) + w(y_1) + w(x_2) + w(y_2) + w(x_3) + w(y_3) + w(x_4) + w(y_4)).$$

By the triangle inequality, we have

$$w(e) \leq w(x_1) + w(y_2),$$
$$w(e) \leq w(x_2) + w(y_1),$$
$$w(f) \leq w(x_3) + w(y_1),$$
$$w(f) \leq w(x_1) + w(y_3).$$

Thus we recover $\tfrac{1}{4}(w(e) + w(f))$ in expectation.[1] But we will recover $\tfrac{1}{4}w(e)$ on the lefthand side of the path of $e$ and $\tfrac{1}{4}w(f)$ on the righthand side of the path of $f$. Thus the total weight is $1/2$ of the weight of the discarded blue edges.

The above randomized procedure can be easily derandomized by exploiting the method of conditional expectations.                                    □

**Theorem 2.** *There is an algorithm with polynomial running time that given $C_1, \ldots, C_N$, computes a Hamiltonian tour of weight at least $W_2 + \tfrac{1}{4}W_3$.*

*Proof.* The algorithm computes the graphs $\hat{D}_1, \ldots, \hat{D}_N$ and decomposes them into $2N$ collections of paths $P_{1,1}, P_{1,2}, \ldots, P_{N,1}, P_{N,2}$ as in Lemma 2. From each pair $P_{i,1}, P_{i,2}$, it computes Hamiltonian tours $H_{i,1}$ and $H_{i,2}$ as in Lemma 3. It then outputs the tour with the largest weight. By (3), $\hat{D}_1, \ldots, \hat{D}_N$ have weight $N \cdot (2W_2 + W_3)$. When constructing the collections of paths, we might loose up to weight $N \cdot W_3$. But when forming the tours, we get half of it back by Lemma 3. Altogether, there are $2N$ Hamiltonian tours. The heaviest of them has weight at least $W_2 + \tfrac{1}{4}W_3$.                                    □

### 3.3     Final Algorithm

**Theorem 3.** *There is a $\tfrac{31}{40}$-approximation algorithm for $\Delta$-Max-ATSP with polynomial running time.*

*Proof.* The algorithm runs the algorithms of Theorems 1 and 2 and outputs the heavier tour. Balancing the approximation performances of both algorithms

---

[1] Note that we need $w(x_1)$ and $w(y_1)$ twice, but $w(x_4)$ and $w(y_4)$ is not of any use for us.

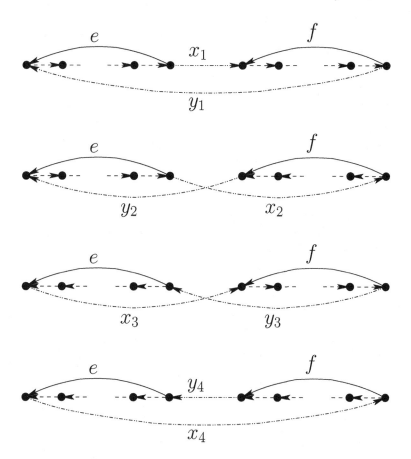

**Fig. 1.** Two discarded blue edges $e$ and $f$ (drawn solid) of two consecutive path (drawn dashed). There are four possibilities how the paths can be chosen. $x_j$ and $y_j$ are the edges used for the patching, the $x_j$ are used when patching from left to right, the $y_j$ when patching from right to left

yields the desired result. This is easiliy seen by the following probabilistic argument: Choose the output of the first algorithm with probability 9/10 and the output of the second one with probability 1/10. An easy calculation shows that the expected weight is $\frac{31}{40}(W_2 + W_3)$. The weight of the heavier tour is certainly at least as large as the expected weight. $\qquad\square$

## 4   Maximum 3-Cycle Cover

In this section, we only assume that $w$ is nonnegative. In particular, $w$ is *not* required to fulfill the triangle inequality. Our goal is to compute a directed 3-cycle cover of maximum weight.

## 4.1    First Algorithm

Bläser and Manthey [2] show how to compute a 3-cycle cover of weight $\frac{1}{2}W_2 + W_3$.

**Theorem 4.** *There is an algorithm with polynomial running time that given* $C_1, \ldots, C_N$, *computes a 3-cycle cover of weight* $\frac{1}{2}W_2 + W_3$.     □

This will be our first algorithm. In the next subsection, we design an algorithm that is favorable if $W_2$ is large.

## 4.2    Second Algorithm

**Lemma 4.** *Let* $K$ *be a connected component of* $\hat{D}_i$. *We can construct in polynomial time two node-disjoint directed cycles* $Z_1$ *and* $Z_2$ *such that*

1. $Z_1$ *and* $Z_2$ *span the same nodes as* $K$,
2. $Z_1$ *can be transformed into* $Z_2$ *by reversing all directions of its edges and vice versa,*
3. *all edges of* $K$ *are in* $Z_1 \cup Z_2$,
4. *and the length of* $Z_1$ *and* $Z_2$ *is at least three.*

*Proof.* The component corresponding to $K$ in $D_i$ is an undirected cycle of length at least three. After possibly adding some edges to $K$, $K$ consists of two oppositely oriented directed cycles of length at least three.     □

**Theorem 5.** *There is an algorithm with polynomial running time that given* $C_1, \ldots, C_N$, *computes a 3-cycle cover of weight* $W_2 + \frac{1}{2}W_3$.

*Proof.* The algorithm computes the graphs $\hat{D}_1, \ldots, \hat{D}_N$ and decomposes them into $2N$ collections of 3-cycle covers by treating each component as in Lemma 4. It then outputs the 3-cycle cover with the largest weight. By (3), $\hat{D}_1, \ldots, \hat{D}_N$ have weight $N \cdot (2W_2 + W_3)$. There are $2N$ 3-cycle covers. The heaviest of them has weight at least $W_2 + \frac{1}{2}W_3$.     □

## 4.3    Final Algorithm

**Theorem 6.** *There is a $\frac{3}{4}$-approximation algorithm for Max-3-DCC with polynomial running time.*

*Proof.* The algorithm runs the algorithms of Theorems 4 and 5 and outputs the heavier 3-cycle cover. Balancing the approximation performances of both algorithms yields the desired result.     □

# References

1. Markus Bläser. An $\frac{8}{13}$-approximation algorithm for the asymmetric maximum tsp. *J. Algorithms*, 50(1):23–48, 2004.
2. Markus Bläser and Bodo Manthey. Two approximation algorithms for 3-cycle covers. In *Proc. 5th Int. Workshop on Approximation Algorithms for Combinatorial Optimization (APPROX)*, volume 2462 of *Lecture Notes in Comput. Sci.*, pages 40–50, 2002.

3. Markus Bläser and Bodo Manthey. Approximating maximum weight cycle covers in directed graphs with edge weights zero and one. *Algorithmica*, 2005.
4. Dany Breslauer, Tao Jiang, and Zhigen Jiang. Rotations of periodic strings and short superstrings. *J. Algorithms*, 24:340–353, 1997.
5. P. Chalasani and R. Motwani. Approximating capacitated routing and delivery problems. *SIAM J. Comput*, 28:2133–2149, 1999.
6. M. L. Fisher, L. Nemhauser, and L. A. Wolsey. An analysis of approximations for finding a maximum weight Hamiltonian circuit. *Networks*, 12(1):799–809, 1979.
7. H. Kaplan, M. Lewenstein, N. Shafrir, and M. Sviridenko. Approximation algorithms for asymmetric tsp by decomposing directed regular multigraphs. In *Proc. 44th Ann. IEEE Symp. on Foundations of Comput. Sci. (FOCS)*, pages 56–65, 2003.
8. S. Rao Kosaraju, James K. Park, and Clifford Stein. Long tours and short superstrings. In *Proc. 35th Ann. IEEE Symp. on Foundations of Comput. Sci. (FOCS)*, 1994.
9. A. V. Kostochka and A. I. Serdyukov. Polynomial algorithms with the estimates $\frac{3}{4}$ and $\frac{5}{6}$ for the traveling salesman problem of the maximum. *Upravlyaemye Sistemy*, 26:55–59, 1985. (in Russian).
10. Moshe Lewenstein and Maxim Sviridenko. A 5/8 approximation algorithm for the maximum asymmetric TSP. *SIAM J. Disc. Math.*, 17(2):237–248, 2003.
11. C. H. Papadimitriou and K. Steiglitz. *Combinatorial Optimization: Algorithms and Complexity*. Prentice-Hall, 1982.
12. C. H. Papadimitriou and M. Yannakakis. The traveling salesman problem with distances one and two. *Math. Operations Research*, 18:1–11, 1993.

# On the Vehicle Routing Problem

Piotr Berman[1] and Surajit K. Das[2]

[1] The Pennsylvania State University
berman@cse.psu.edu
[2] Penske Logistics
surajit.das@penske.com

**Abstract.** In the Vehicle Routing Problem (VRP), as in the Traveling Salesman Problem (TSP), we have a metric space of customer points, and we have to visits all customers. Additionally, every customer has a *demand*, a quantity of a commodity that has to be delivered in our vehicle from a single point called *the depot*. Because the vehicle capacity is bounded, we need to return to the depot each time we run out of the commodity to distribute. We describe a fully polynomial time algorithm with approximation 2.5, we also modify this algorithm for the on-line version of VRP, the randomized version has competitive ratio of 2.5 on the average, and the deterministic version has ratio 4. We also describe 2 approximation for a restricted version of the problem.

## 1  Introduction

In general, the VRP seeks to determine the minimum cost (length) route for a vehicle to deliver goods to (or pick up goods from) a set of $n$ customers. The route is divided into trips that start and end at the supply depot,[1] with the constraint that the cumulative weight of goods delivered to customers (alternatively, the cumulative weight of collected goods) on a single trip does not exceed a certain bound, which we set to 1 w.l.o.g. In literature it is often assumed that demands are integer and that vehicle capacity is some $Q$ — in our setting, this means that the demands are integer multiples of some $1/Q$.

An instance of VRP is defined by the distance matrix $C$ for the set of points $\{0, \ldots, n\}$ where 0 is the location of the depot and the other points are the customer locations, and the vector $d$ for the customer demands. We assume that the distances satisfy the triangle inequality. The objective is to minimize the total distance traveled by the vehicle while satisfying demand at all the customers. (Note that fulfillment of a customer demand might require visiting the custumer on more than one trip).

Haimovich et al. [9] gave the first approximation algorithm for the VRP. Their Optimal Tour Partition (OTP) heuristic approximates the solution with a ratio

---

[1] Charikar *et al.* [5] consider a version of VRP with multiple sources of goods rather than a single depot.

F. Dehne, A. López-Ortiz, and J.-R. Sack (Eds.): WADS 2005, LNCS 3608, pp. 360–371, 2005.

of $\tau + \frac{\lceil n/Q \rceil}{n/(Q-\tau)}$, assuming a $\tau$-optimal[2] solution to the TSP. In later work, they described improved OTP algorithm that is $(\tau + 1 - \frac{3}{2Q})$-optimal. The running time of OTP was not discussed, but it can be seen to be $O(nQ)$.

Note that the best known values of $\tau$ are 1.5 for the general case (see Christofides [6]) and $1 + \varepsilon$ if distances represent $L_p$ metric in $R^k$ for a finite $k$ (see Arora [1]).

A VRP where one or more of the parameters are random variables is referred to as a stochastic VRP (SVRP). Some examples are VRPSD (VRP with Stochasic Demands) and VRPSCD (VRP with Stochastic Customers and Demands). The application that gave birth to SVRP in the academic literature was refuelling of terminal tankage facilities from a refinery, with demand at the tanks a random variable (Charnes and Cooper [4]). Many other applications are described by Bertsimas [3], Secomandi [11] and Gendreau et al. [7]. One can see that it is very typical in logistics to serve many customers from a single depot, when the actual demands of the customers are either unknown or are subject to random fluctuations.

With his *Cyclic Heuristic (CH)* for the VRPSD, Bertsimas [3] obtains an approximation ratio of $\tau + 1 + O(1/n)$ for the case of i.i.d. demands and $Q + 1$ for the case of non-i.i.d. demands. CH, the stochastic counterpart of OTP, determines the best tour on the basis of the *expected lengths* of the tours. Assuming that the maximum demand $K$ is not larger than $Q$, CH requires $O(n^3 K^2)$ time, a bound later improved to $O(n^2 KQ)$ by Gendreau et al. [7].

We make the following contributions to this problem. In the general offline case we obtain the same approximation ratio of $\tau + 1$ as in [10] but our approximation algorithm is fully polynomial. In the on-line case in which the demand of a customer is known only when the vehicle visits it, we get the same ratio of $\tau + 1$ and this subsumes any possible random distribution of the demands, thus improving upon [3]. We also show competitive ratio $2\tau + 1$ against an *adaptive* adversary. All of these results are in Sec. 3. Finally, in Sec. 4 we describe a 2-optimal algorithm for a restricted version of the problem. While this version is a practical problem in its own right we hope that the primal-dual technique that we describe will lead to an improvement for the the general version as well.

## 2    Linear Program for VRP

The multiset of edges used by all trips can be viewed as a vector $x$ in a linear space with one base vector for each edge. In this section we will describe a linear program with the necessary conditions for a valid solution and consequently, the corresponding dual linear program that provides valid lower bounds for the cost of an optimal solution.

We use notation $V = \{0, 1, \ldots, n\}$ (set of nodes), $\mathbf{P}(V) = \{S : S \subseteq V\}$, $E = \{e \in \mathbf{P}(V) : |e| = 2\}$ (set of edges); for $S \in \mathbf{P}(V)$ we have $d_S = \sum_{i \in S} d_i$;

---

[2] We use "$\tau$-optimal" as a synonim for "with approximation ratio $\tau$".

we also define a 0-1 matrix $\Delta$ with a column $\beta(e)$ for every edge and a row $\delta(S)$ for every $S \in \mathbf{P}(V)$; $\Delta_{S,e} = \delta(S)_e = \beta(e)_S = 1$ iff $|e \cap S| = 1$. The following linear program is a fractional relaxation of VRP:

$$\text{Minimize } cx$$
$$\text{subject to}$$

| | | | |
|---|---|---|---|
| $x^T \geq 0$ | or | $x_e \geq 0$ | for every $e \in E$ | (2.1) |
| $\Delta x^T \geq 2\lceil d \rceil^T$ | or | $\delta(S)x_e \geq 2\lceil d_S \rceil$ | for every $S \in \mathbf{P}(V)$ | (2.2) |

Even an integral solution to this linear program may fail to correspond to a valid solution. However, to prove approximation (or competitive) ratio of an algorithm we can use solutions of the following dual linear program:

$$\text{Maximize } 2\lceil d \rceil y$$
$$\text{subject to}$$

| | | | |
|---|---|---|---|
| $y^T \geq 0$ | or | $y_S \geq 0$ | for every $S \in \mathbf{P}(V)$ | (2.3) |
| $\Delta^T y^T \leq c^T$ | or | $\beta(e)y \leq c_e$ | for every $e \in E$ | (2.4) |

Assume that $c_{0,1} \leq c_{0,2} \leq \ldots c_{0,n}$ and let $S(i) = \{i, \ldots, n\}$. Define $y_{S(i)}^{RB} = c_{0,i} - c_{0,i-1}$, and for every other $S$ define $y_S^{RB} = 0$. The triangle inequality implies that $\beta(i,j)y^{RB} = |c_{0,i} - c_{0,j}| \leq c_{i,j}$, hence $y^{RB}$ satisfies (2.3) and (2.4), which proves the validity of the *radial lower bound* $RB = 2dy^{RB} = 2\sum_i d_i c_{0,i}$. Here, $RB$ is a lower bound for the solution cost that is based solely on "radial" distances from the depot to the customers.

## 3    From TSP Solution to VRP Solution

We simplify OTP of [9, 10] so its running time becomes strongly polynomial and it can be easily adapted to the on-line cases. First, we add a customer with a demand $d_0$ at the depot location to ensure that (a) the tour of the customer sites includes the depot and (b) the sum of demands is integer. Our algorithm will choose some $\theta$ such that $0 < \theta \leq 1$ and load the vehicle with $\theta$ units of the commodity for the first trip. The vehicle returns to the depot only if it runs out of the commodity or if all customer demands are satisfied. Any subsequent trip after the first commences with a load of 1 unit of the commodity and starts with a visit to the last customer on the immediately preceding trip if that customer was not completely served, and starts with the next customer otherwise.

Let $D_i = \sum_{j=0}^{i-1} d_j$. We can work out the condition that within the solution that was determined by $\theta$ we make a roundtrip between node $i$ and node 0: for some integer $k$, $D_i < k+\theta$ and $D_{i+1} \geq k+\theta$; equivalently, $\lfloor D_i - \theta \rfloor < \lfloor D_i + d_i - \theta \rfloor$. In other words, we make $f(i, \theta) = \lfloor D_i + d_i - \theta \rfloor - \lfloor D_i - \theta \rfloor$ roundtrips between $i$ and 0. Therefore the cost of this solution is the TSP cost plus

$$F(\theta) = 2 \sum_{i=1}^{n} f(i, \theta)c_{0,i}.$$

**Observation 1.** As a function of $\theta$, $f(i, \theta)$ is piecewise linear, with one or two non-zero pieces. E.g. if $D_i = 5.7$ and $d_i = 0.4$ then $f(i, \theta) = 1$ if $0 \le \theta \le 0.1$ or $0.7 < \theta < 1$.

Obs. 1 allows us to calculate the minimum of $F(\theta)$ in time $n \log n$. If we add this calculation as a post-processing step to a $\tau$-optimal TSP algorithm, we get a strongly polynomial $(\tau + 1)$-optimal VRP algorithm. This proves

**Theorem 1.** *There exists a deterministic polynomial time algorithm for the VRP with approximation ratio 2.5.*

**Observation 2.** If we pick $\theta$ uniformly at random, $E[f(i, \theta)] = d_i$; consequently, $E[F(\theta)] = RB$. Thus, with no prior knowledge of $d$, we can fill the vehicle with random amount $\theta$ and start following the TSP tour making roundtrips to the depot whenever the vehicle runs out of the commodity that is being distributed; this results in average (or *expected*) additional cost being equal to $RB$.

Because in this randomized algorithm we do not need to know the demands, this proves

**Theorem 2.** *There exists a randomized polynomial time algorithm for on-line VRP problem with competitive ratio 2.5.*

This result is similar to the one by Bertsimas [3], but stronger and simpler.

**Observation 3.** If $t$ is the cost of the TSP tour that is the basis of our algorithm, then $F(\theta') - F(\theta) \le t$ for every $\theta$ and $\theta'$.

To see that Obs. 3 is true, compare solutions determined by $\theta$ and $\theta'$. The former splits the tour of cost $t$ into pieces and given such a path, say, from node $i$ to $j$, it adds a traversal from depot node to $i$ and from $j$ to the depot node. The latter picks a node, say $k$, in every path created by the former, say from $i$ to $j$. We could extend the path from $k$ to the depot node by the path fragment from $i$ to $k$, and the path to $k$ can be extended to $j$. By the triangle inequality, the cost of transformation is not greater than the path length; by adding over all paths, we transform solution determined by $\theta$ to the one determined by $\theta'$ with the extra cost no greater than $t$. Because we can pick $\theta' = 0$ we can see that

**Theorem 3.** *There exists a deterministic polynomial time algorithm for on-line VRP problem with competitive ratio $2\tau + 1$, which equals 4 if we use the Christofides algorithm.*

## 4   Restricted VRP

In this section, we obtain an approximation ratio 2 for a restricted version of VRP. We will assume that every demand is either *large* or *small*, and that in an optimum solution at most one customer with large demand is visited on each trip, while there is no restriction on the number of small demands that can be satisfied on a single trip.

This situation arises in practice, for instance, for a representative of a company that sells and maintains X-ray machines. One vehicle can carry no more than one such machine; however, any number of service visits can be undertaken by a given vehicle.

We can model this by picking a very small number $z$, setting each large demand to $1 - nz$ and each small demand to $z$. As a result, if $S \neq \varnothing$, $\lceil d_S \rceil = \max(|S \cap L|, 1)$ where $L$ is the set of nodes with large demands. With these assumptions, we can use linear program (2.1-2) and the dual linear program (2.3-4). We will refer to small demands as *zero demands* and to large demands as *one demands*.

## 4.1    Zero Demands

We now consider the case with zero-demands only, so that $\lceil d \rceil$ is a vector of all 1's. In this case the primal program (2.1-2) and the dual program (2.3-4) are very similar to those used by Agrawal *et al.* [2] (see also [8]) for the Steiner Tree Problem, and we will adapt their *Colliding Trees* method for our purposes. We describe the solution in some detail because later we incorporate it into the more general case.

For simplicity we present the algorithm in a form that is rather inefficient but that still runs in polynomial time (the simple algorithm explicitly maintains the values of all dual variables; the efficient version described by [2] has a data structure that allows us to update many such values in a single operation). The algorithm uses a data structure $\Pi$ that represents a partition of $\{0, \ldots, n\}$ and it considers only those subsets[3] $S \subset \{0, \ldots, n\}$ that are or were present in $\Pi$. For any such set $S$ we define

$y_S$ — the dual variable of $S$;
$LB_S = \sum_{U \subset S} y_U$ — the part of the lower bound attributable to set $S$;
$a_S$ — boolean variable indicating the status of $S$ in our algorithm, we identify **true** with 1 and **false** with 0, if $a_S$ is **true** we say that $S$ is *active*;
$p_S$ — the "anchor point" of $S$, the tour traversing the minimum cost spanning tree of $S \cup \{p_S\}$ is considered a possible part of our solution.

The algorithm is shown in Fig. 1.

When the algorithm terminates, every set $S$ has $a_S = $ **false**, and $p_S$ is defined. Moreover, by following the method of [2] one can prove that $\overline{T}_S$ is a spanning tree for $S \cup \{p_S\}$ with cost at most $2LB_S$.

In particular, we can define a *radius* $r_i$ of point $i$ as the sum of $y_S$'s such that $i \in S$. Iterations of the algorithm preserve the following invariants:

(a)   all elements of active sets ($S$ with $a_S$ true) have the same radius $r_{active}$,
(b)   all elements of a set $S$ (active or not) have the same radius $r_S \leq r_{active}$,
(c)   tree $T_S$ spans $S$ with cost at most $2(LB_S - r_S)$,
(d)   if $S$ is not active, tree $\overline{T}_S$ spans $S \cup \{p_S\}$ with cost at most $2LB_S$.

---

[3]   We use $\subset$ to denote a *proper* subset, *i.e.* one that is not equal.

initialize $\Pi$ with sets $\{0\}, \ldots, \{n\}$
for $0 \le u \le n$
    $\mathbf{a}_{\{u\}} \leftarrow \mathbf{true}, T_{\{u\}} \leftarrow \emptyset$
$\mathbf{a}_{\{0\}} \leftarrow \mathbf{false}$
while $\mathbf{a} \ne 0$ do
    find maximal $\varepsilon$ such that $\beta(e)(y + \varepsilon \mathbf{a}) \le c_e$ for every edge $e$
    $y \leftarrow y + \varepsilon \mathbf{a}$
    for every $e = \{u, v\}$ such that $\beta(e)y = c_e$
        $U \leftarrow Find_\Pi(u), V \leftarrow Find_\Pi(v)$
        if $\mathbf{a}_U$ and $\mathbf{a}_V$
            $W \leftarrow Union_\Pi(U, V)$
            $\mathbf{a}_W \leftarrow \mathbf{true}$
            $T_W \leftarrow T_U \cup T_V \cup \{e\}$
        if $\mathbf{a}_U$
            $\overline{T}_U \leftarrow T_U \cup \{e\}$
            $p_U \leftarrow v$
            $\mathbf{a}_U \leftarrow \mathbf{false}$
        if $\mathbf{a}_V$
            $\overline{T}_V \leftarrow T_V \cup \{e\}$
            $p_V \leftarrow u$
            $\mathbf{a}_V \leftarrow \mathbf{false}$

**Fig. 1.** Algorithm for zero demands

When the algorithm terminates the union of $\overline{T}_S$ trees forms a spanning tree of all the points, and the constructed vector $y$ proves that the cost of this tree is within factor 2 of the minimal cost of a Steiner tree. The lower bound for TSP implied by the dual program is twice that for a Steiner tree, and we obtain a valid tour by traversing each edge of each $\overline{T}_S$ twice. Thus we have a factor two approximation. Importantly, trees of the form $\overline{T}_S$ will be used as parts of the (primal) solution and $LB_S$'s will be used as terms in the lower bound (dual solution).

### 4.2   One Demands

If we have only one-demands we make a separate trip to every customer and the resulting cost is $2 \sum_{i=1}^{n} c_{0,i} = RB$ (see Sec. 2). Because $RB$ is a lower bound, in this case we have an optimum solution.

If we are content with approximation ratio 2, we can relax the restriction on the size of demands so that each is at least $1/2$. The solution will remain the same, while $RB$ will decrease by a factor not larger than 2.

### 4.3   Zero-One Demands

We now combine the ideas of the previous two subsections. A valid solution is a set of tours originating at 0 that collectively visit all the customer points, but with at most one node from $L$ per trip.

We proceed in two phases. Phase 1 is similar to our algorithm for zero de-mands. The difference is that if $i \in L$, we treat $i$ in the same way as the depot, *i.e.* we initialize $\mathbf{a}_{\{i\}} = \mathbf{false}$ and $p_{\{i\}} = i$. When Phase 1 terminates, the union of $\overline{T}_S$ trees forms a spanning forest with one tree containing the depot and each of the other trees containing at most one node from $L$.

Vector $y$ defines for every edge $e$ its *residual length* $c_e - \mathbf{b}_e y$. The notion of residual length can be naturally extended to paths. The minimum residual length of a path from a point $p$ to the depot will be called the *residual distance* of $p$, denoted $\mathbf{r}_y(p)$.

In phase 2 we reduce some of the coefficients $y_S$; if $y_S$ is reduced and positive we say that $S$ is *reduced*, if $y_S$ is reduced to zero we say that $S$ is *nixed*, otherwise $S$ is *not reduced*. We will maintain two properties:

① if $S \subset U$ and $S$ is reduced then $U$ is nixed;
② if $S \subset U$ and $U$ is reduced then $S$ is not reduced.

The goal of Phase 2 is to find a set $\mathcal{P}$ of $2|L|$ paths such that

③ for each $i \in L$ the set $\mathcal{P}$ has two paths from 0 to $i$ with the minimum residual length;
④ if $S$ is not nixed then at most one path from $\mathcal{P}$ goes through a point in $S$;
⑤ if no path from $\mathcal{P}$ goes through a point in $S$ then $S$ is not reduced.

**Lemma 1.** *If conditions ①-⑤ hold then we can efficiently find a solution to VRP that has cost within factor 2 of the optimum.*

**Proof.** Let $\mathcal{A}(y)$ be the set of maximal non-nixed sets, and let $S(\mathcal{P})$ be the sets of points in the paths from $\mathcal{P}$.

The sum $2 \sum_{S \in \mathcal{A}(y)} LB_S$ plus the sum of residual lengths of paths in $\mathcal{P}$ forms a lower bound $TB(y)$ for the cost of a solution to our VRP instance. To show this, we extend the vector $y$ to form the corresponding dual solution; assume that points in $L$ have the following nondecreasing set of residual distances: $a_1, ..., a_k$, while $a_0 = 0$. We define $S(i) = \{p \in V : \mathbf{r}_y(p) \geq a_i\}$ and we set $y_{S(i)} = a_i - a_{i-1}$. Because we use residual distances, the inequalities (2.4) are still satisfied.

Note that $\frac{1}{2} TB(y) = \sum_{S \in \mathcal{A}(y)} LB(S) + \sum_{p \in L} \mathbf{r}_y(p)$. From now on, we will consider only dual solutions in which sets have coefficients not exceeding those from Phase 1, and use the last expression for the implied lower bound.

We now modify the dual solution and the set of paths $\mathcal{P}$ so that the sum

$$\sum_{P \in \mathcal{P}} \text{residual length of } P + 4 \sum_{S \in \mathcal{A}(y)} LB(S)$$

does not increase, and eventually every $S$ is nixed. Consider first $S \in \mathcal{A}(y)$ that intersects $S(\mathcal{P})$. Let $e_1$ and $e_2$ be the first and last edges of $P$ with a point in $S$ and $z = y_S$. Suppose that between $e_1$ and $e_2$ path $P$ departs from $S$ and re-enters; because $P$ has the minimal residual length, this is possible only if $S$ and all sets visited in that detour are non-reduced. In this case we replace this portion of $P$ with a traversal in $T_S$ properties ①-⑤ are preserved. Now we nix $S$; $4LB_S$ is decreased by $4z$ and the residual length of $P$ increases by $2z$.

Now every set $S$ that intersects $S(\mathcal{P})$ is nixed and every $S$ without such a point is not reduced. This implies that every point in $V - S(\mathcal{P})$ belongs to a non-reduced set. For each non-reduced set $S$ we add to our solution the cyclic path that traverses edges of $\overline{T}_S$ twice. The cost of these paths is bounded by $4\sum_{S \in \mathcal{A}(y)} LB(S)$ so it suffices to show that the graph formed by the paths of our solution is connected.

Since every point $p \in V - S(\mathcal{P})$ is on a cyclic path, it suffices to show that each $\overline{T}_S$ defining such a cyclic path is connected, perhaps indirectly, to $S(\mathcal{P})$. We can show it by induction on the time when $\mathbf{a}_S$ becomes false.

At the time when $\mathbf{a}_S$ became false the anchor point $p_S$ was defined. If $p_S$ belongs to some $U$ such that $\mathbf{a}_U$ became false earlier, we use the inductive hypothesis. Suppose not, then $\mathbf{a}_U$ became false at the same time as $\mathbf{a}_S$, which means that these two sets are siblings. Because the parent of $S$ and $U$ is nixed, $S \cup U$ intersects $S(\mathcal{P})$, and because $S$ does not intersect $S(\mathcal{P})$, $U$ does. This means that $U$ is nixed. Now, either $p_S \in S(\mathcal{P})$ and we are done, or $p_S$ belongs to a subset of $U$, say $W$, that is not nixed, and for which $\mathbf{a}_W$ became false earlier. ❐

### Network flow formulation of properties ①-④

Below we define a network $NW_1(y)$ such that if it has a flow with value $2|L|$ we can satisfy properties ①-④, and if it has a cut of capacity below $2|L|$ we can decrease $y_S$'s of some sets from $\mathcal{A}(y)$ and increase the lower bound $TB(y)$.

1. The nodes are: source node $s$, sink node $t$, node $p$ for each $p \in L$, and for each $S \in \mathcal{A}(y)$ we have $S_{in}$ and $S_{out}$.
2. For each $p \in L$ we have edge $(s, p)$ with capacity 2;
3. For each $S \in \mathcal{A}(y)$ we have edge $(S_{in}, S_{out})$ with capacity 1;
4. If a path from $p \in L$ to 0 with the minimum residual length has an edge $(q, r)$ then we have a network edge $(\overline{q}, \overline{r})$ with capacity $\infty$ where
   - if $q = p \in L$ then $\overline{q} = p$,
   - if $q \in S \in \mathcal{A}(y)$ then $\overline{q} = S_{out}$,
   - if $r = 0$ then $\overline{r} = t$,
   - if $r \in S \in \mathcal{A}(y)$ then $\overline{r} = S_{in}$.

**Lemma 2.** *Assume that $y$ satisfies properties ①-② and in $NW_1(y)$ there exists a flow with value $2|L|$. Then there exists a set of paths $\mathcal{P}$ with properties ③-④.*

**Proof.** Because all capacities in $NW_1(y)$ are integer, we can decompose the flow into $2|L|$ unit flows, each following a simple path.

Suppose that one of these paths is $(s, p, t)$. This means that $p \in L$ and the residual length of $(p, 0)$ is the same as the original length. Thus we can place in $\mathcal{P}$ two copies of the 1-edge path $(p, 0)$.

Now consider a path $(s, p, S_{in}^1, S_{out}^1, \ldots, S_{in}^k, S_{out}^k, t)$. Assume that $(p, q_1)$ is the shortest edge from $p$ to $S^1$, $(r_{i-1}, q_i)$ is the shortest edge from $S^{i-1}$ to $S^i$ and $(r_k, 0)$ is the shortest edge from $S^k$ to 0. We get a path prototype

$(p, q_1, r_1, \ldots, q_k, r_k, 0)$. We convert this prototype to a path for our set $\mathcal{P}$ by replacing each $(q_i, r_i)$ with the unique path from $q_i$ to $r_i$ in the tree $T_{S^i}$.

Our paths use only the edges from paths of smallest residual length and edges of residual length 0, hence they satisfy ③. Property ④ is ensured by the unit capacity of $(S_{in}, S_{out})$ edges.    ❒

**Lemma 3.** *Assume that $y$ satisfies properties ①-② and in $NW_1(y)$ there exists a cut with capacity lower than $2|L|$. Then there exists $\mathcal{B} \subset \mathcal{A}(y)$ such that after subtracting sufficiently small $\varepsilon$ from $y_S$ for each $S \in \mathcal{B}$ we increase $TB(y)$ by at least $2\varepsilon$, while $y$ still satisfies ①-②.*

**Proof.** Consider a cut $C$ that has capacity lower than $2|L|$. Let $K = \{p \in L : (s, p) \in C\}$ and $\mathcal{B} = \{S \in \mathcal{A}(y) : (S_{in}, S_{out}) \in C\}$. Because $2|K| + |\mathcal{B}| < 2|L|$, $|\mathcal{B}| < 2|L - K|$. Subtract a small $\varepsilon$ from $y_S$ for each $S \in \mathcal{B}$. The sum of $LB_S$'s decreases by $|\mathcal{B}|\varepsilon$, $\mathbf{r}_p$ increases by $2\varepsilon$ for each $p \in L - K$ while $\mathbf{r}_p$ cannot decrease for $p \in L$. Thus $\frac{1}{2}TB(y)$ increases by $(2|L - K| - |\mathcal{B}|)\varepsilon \geq \varepsilon$.    ❒

There are two considerations that limit $\varepsilon$ in the last lemma. First, we cannot decrease any $y_S$ below zero. Second, as we make some of the current shortest paths longer, some competing paths may become shortest as well. In either case, the network $NW_1(y)$ would have to be altered.

To limit the number of repetitions of this process we can restrict all distances to a small set of integers, say from 0 to $n^2$; this involves a very small increase of the approximation ratio. This way in polynomially many steps we get $y$ and $\mathcal{P}$ that satisfy conditions ①-④.

### Network flow formulation of properties ①-⑤

We now assume that we have $y$ such that $NW_1(y)$ has flow with value $2|L|$, which by Lem. 2 can be converted into a set $\mathcal{P}$ of $2|L|$ paths that satisfy properties ①-④. $\mathcal{P}$ fails to satisfy property ⑤ if there is a reduced $y_S$ such that no path goes through the flow does not use edge $(S_{in}, S_{out})$. This means that in the network $NW_1$ we not only bound the flow through some edges from above but also in some $(S_{in}, S_{out})$ edges we request exactly unit flow. Another way we may fail property ⑤ is if the flow does not use any edge through some singleton set $\{i\}$ that is nixed.

Note that if $i$ does not belong to any shortest path from $L$ to 0 we can increase $y_{\{i\}}$. Otherwise we insert $\{i\}$ to $\mathcal{A}(y)$ and we create $\{i\}_{in}, \{i\}_{out}$ as for other sets in $\mathcal{A}(y)$, except the capacity of $(\{i\}_{in}, \{i\}_{out})$ is $\infty$.

Now we request that the minimum flow through the edges of reduced or nixed sets is 1. The question of whether such a flow exists can be translated into a normal maximum-flow question in another network $NW_2(y)$ as follows:

Let $\mathcal{R}(y)$ be the set of $S$'s such that $y_S$ is reduced. To create $NW_2(y)$ from $NW_1(y)$, we (a) introduce a new sink node $t'$ and an edge $(t, t')$ of capacity $2|L|$, and (b) for each $S \in \mathcal{R}(y)$, subtract 1 from the capacity of $(S_{in}, S_{out})$ and add two edges $(s, S_{out})$ and $(S_{in}, t')$, each of capacity 1. Note that in (b) the capacity can be reduced from 1 to 0 or from $\infty$ to $\infty$ (if $S$ is a nixed singleton).

**Lemma 4.** *Assume that $y$ satisfies ①-② and in $NW_2(y)$ there exists a flow with value $2|L| + |\mathcal{R}(y)|$. Then there exists a flow in $NW_1(y)$ with value $2|L|$ which has value 1 for each edge $(S_{in}, S_{out})$ such that $S \in \mathcal{R}(y)$.*

**Proof.** Note that the total capacity of edges that leave $s$ is $2|L| + |\mathcal{R}(y)|$, and that the total capacity of edges that enter $t'$ is the same. Hence in our flow all these edges have flow equal to the capacity.

Now, step by step we can transform the network back to $NW_1(y)$. For $S \in \mathcal{R}(y)$ we (a) subtract 1 from the flow in $(s, S_{out})$ and $(S_{in}, t')$, (b) add 1 to the flow in edge $(S_{in}, S_{out})$. After completion of all such steps, we get the desired flow in $NW_1(y)$.    ◻

**Lemma 5.** *Assume that $y$ satisfies ①-② and in $NW_2(y)$ there exists a cut with capacity lower than $2|L| + |\mathcal{R}(y)|$. Then there exist sets $\mathcal{B} \subset \mathcal{A}(y)$ and $\mathcal{D} \subset \mathcal{R}(y)$ such that for a sufficiently small $\varepsilon$ we can increase $y_S$ by $\varepsilon$ for every $S \in \mathcal{D}$ and decrease $y_S$ by $\varepsilon$ for every $S \in \mathcal{B}$, and we will obtain a valid solution to the dual program that satisfies ①-②, while $TB(y)$ increases by at least $2\varepsilon$.*

**Proof.** Assume that $C$ is the cut in $NW_2(y)$ with capacity lower than $2|L| + |\mathcal{R}(y)|$. For the sake of brevity, in the extended abstract we consider only the case then $(t, t') \notin C$. We use $C$ to define the following sets:

$$
\begin{aligned}
K &= \{p \in L : (s, p) \in C\} \\
\mathcal{B}_0 &= \{S \in \mathcal{R}(y) : (s, S_{out}) \in C \text{ and } (S_{in}, t') \in C\} \\
\mathcal{B}_1 &= \{S \in \mathcal{A}(y) - \mathcal{R}(y) : (S_{in}, S_{out}) \in C\} \\
\mathcal{D} &= \{S \in \mathcal{R}(y) : (s, S_{out}) \notin C \text{ and } (S_{in}, t') \notin C\} \\
\mathcal{N}_{\leftarrow} &= \{S \in \mathcal{R}(y) : (s, S_{out}) \in C \text{ and } (S_{in}, t') \notin C\} \\
\mathcal{N}_{\rightarrow} &= \{S \in \mathcal{R}(y) : (s, S_{out}) \notin C \text{ and } (S_{in}, t') \in C\} \\
\mathcal{B} &= \mathcal{B}_0 \cup \mathcal{B}_1 \text{ and } \mathcal{N} = \mathcal{N}_{\leftarrow} \cup \mathcal{N}_{\rightarrow}
\end{aligned}
$$

Note that we never increase $y_S$ beyond its value from Phase 1, *i.e.* we never increase $y_S$ if $S \in \mathcal{A}(y) - \mathcal{R}(y)$. This is possible because $\mathcal{D} \subseteq \mathcal{R}(y)$ and we can choose a sufficiently small $\varepsilon$.

Changing $y_S$ as described in the lemma decreases $\sum y_S/\varepsilon$ by $|\mathcal{B}| - |\mathcal{D}|$. We need to show that the increase of $\sum_{p \in L} \mathbf{r}_p/\varepsilon$ is larger. We consider two cases.
**Case:** $(t, t') \in C$. This case is depicted in the upper right diagram of Fig. 2. From a node in $p \in L - K$ we cannot go to $\mathcal{D}$ directly; instead, before each visit to $\mathcal{D}$ we must traverse a set from $\mathcal{B}$; as a result $\mathbf{r}_p$ cannot become smaller — traversals through $\mathcal{D}$ decrease the residual distance by $2\varepsilon$, and the traversals through $\mathcal{B}$ increase it by $2\varepsilon$, and we must have at least as many of the latter as we have of the former.

For $p \in K$, we can go directly to $\mathcal{B}$, but later the same reasoning holds, so we may decrease $\mathbf{r}_p$ by at most $2\varepsilon$. Therefore $\sum_{p \in L} \mathbf{r}_p/\varepsilon$ decreases by at most $2|K|$. It suffices to show that $|\mathcal{D}| - |\mathcal{B}| - 2|K| > 0$, or that $|\mathcal{B}| + 2|K| < |\mathcal{D}|$.

The capacity of the cut equals $2|K| + 2|\mathcal{B}_0| + |\mathcal{B}_1| + |\mathcal{N}| + 2|L|$, where the first four terms can be seen in the left diagram of Fig. 2 and the last term is the capacity of $(t, t')$. According to our assumption, this capacity is lower than $|L| + |\mathcal{B}_0| + |\mathcal{D}| + |\mathcal{N}|$, hence $2|K| + |\mathcal{B}_0| + |\mathcal{B}_1| + |L| < |\mathcal{D}|$.

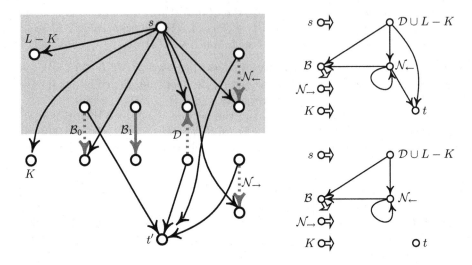

**Fig. 2.** On the left we depict the position of nodes in $NW_2(y)$ in respect to cut $C$. The cut consists of edges that go out from the gray rectangle. For each class of sets the gray edge indicates the position of $(S_{in}, S_{out})$ edges; they are dashed if their capacity is decreased. Note that we cannot have infinite capacity edges in the cut, which limits possible shortest paths from sets that have $S_{out}$ in the gray rectangle. These limits are depicted in the diagrams on the right. In these diagrams wide short arrows indicates that a shortest path from the respective sets can go anywhere, whereas longer thin arrows indicate all possibilities for the respective sets. The upper right diagram is valid if $(t, t')$ is in the cut (so $t$ is inside the gray rectangle), and the lower right diagram is valid if $(t, t')$ is not in the cut

**Case:** $(t, t') \notin C$. This case is depicted in the lower right diagram of Fig. 2, and the reasoning for that case differs from the case above in only two aspects. We can increase the estimate of each new $\mathbf{r}_p$ by $2\varepsilon$ because on a path from $L$ to $t$ we must visit $\mathcal{B}$, and after each visit to $\mathcal{D}$ we must have another visit to $\mathcal{D}$, as we cannot go from $\mathcal{D} \cup \mathcal{N}_\leftarrow \cup L - K$ to $\{t\} \cup \mathcal{N}_\rightarrow$ without traversing $\mathcal{B}$ first. Thus it suffices to show that $|\mathcal{D}| - |\mathcal{B}| + 2|L - K| > 0$. The second difference is that the capacity of the cut equals $2|K| + 2|B_0| + |B_1| + |\mathcal{N}|$. The effects of these two differences cancel each other.                                                                    □

### Conclusion for zero-one demands.

**Theorem 4.** *For every* $\varepsilon > 0$ *there exists a polynomial time approximation algorithm for VRP with demands zero or one with approximation ratio* $2 + \varepsilon$.

**Proof.** We can start by finding an approximate solution with cost $x$ that is at most 2.5 times the optimum. Then we can rescale the edge lengths so that $x = 5n^2/\varepsilon$. Next we round down up each edge cost to the nearest integer; because we have at most $n$ trips, we use edges at most $n^2$ times, so we increase the optimum solution cost by less than $n^2$; because the optimum is at least $2n^2/\varepsilon$, this is an increase by factor smaller than $1 + \frac{1}{2}\varepsilon$. Now it suffices to provide in polynomial time a solution within factor 2 of the optimum.

We start from Phase 1 and keep on increasing the lower bound $TB(y)$ by finding a small cut in $NW_1(y)$ or in $NW_2(y)$ and using Lem. 3 or Lem. 5. If the lower bound cannot be increased we obtain a flow in $NW_2(y)$ that satisfies the assumptions of Lem. 4. This flow can be converted into a flow in $NW_1(y)$ that satisfies ①-⑤, and by Lem. 2 it can in turn be converted into a valid solution within factor 2 of the optimum. Restriction of the edge lengths to integers ensures that the increases of the lower bound are integer, and the small range of these integers ensures that we will have a polynomial number of iterations.      ⊐

# References

1. Arora, S. Polynomial Time Approximation Schemes for Euclidean Traveling Salesman and other Geometric Problems. J. ACM 45(5): 753-782 (1998).
2. Agrawal, A., P. Klein and R. Ravi. 1991. When Trees Collide: An approximation algorithm for the generalized Steiner problem on networks. *Proceedings of 23rd STOC* (1991): 134-144.
3. Bertsimas, Dimitris J. A Vehicle Routing Problem with Stochastic Demand. *Operations Research* 40 (3): 574-585 (May-June 1992).
4. Charnes, A. and W. Cooper. 1959. Chance Constrained Programming. *Management Science* (6): 73-79. *Transportation Science* (29): 2.
5. Charikar, M., S. Khuller and B. Raghavachari. Algorithms for Capacitated Vehicle Routing. *SIAM Journal of Computing.* 31(3): 665-682 (2001)
6. Christofides, N. Worst Case Analysis of a new heuristic for the TSP. Report 388, Graduate School of Industrial Administration, Carnegie Mellon University, Pittsburgh, PA, (1976).
7. Gendreau, M., G. Laporte and Rene Seguin. An Exact Algorithm for the VRP with Stochastic Demands and Customers. 1995. *Transportation Science* 29, 2.
8. Goemans, M. X. and D. P. Williamson. A General Approximation Technique For Constrained Forest Problems. *SIAM Journal of Computing* (24): 296-317, (1995).
9. Haimovich, M. and A. Rinnooy Kan. Bounds and Heuristics for the Capacitated Routing Problem *Mathematics of Operations Research* (10): 527-542, 1985.
10. Haimovich, M., A. Rinnooy Kan and L.Stougie. Analysis of heuristics for Vehicle Routing Problems. *Vehicle Routing: Methods and Studies.* B.L. Golden and A. A. Assad (Eds.). North-Holland, Amsterdam, The Netherlands, 1988.
11. Secomandi, N. Exact and Heuristic DP algorithms for the VRP with Stochastic Demands. Ph.D. Dissertation, Dept of Decision and Info Sciences, University of Houston, Texas, 1998.

# The Structure of Optimal Prefix-Free Codes in Restricted Languages: The Uniform Probability Case

## (Extended Abstract)*

Mordecai J. Golin and Zhenming Liu

**Abstract.** In this paper we discuss the problem of constructing minimum-cost, prefix-free codes for equiprobable words under the assumption that all codewords are restricted to belonging to an arbitrary language $\mathcal{L}$ and extend the classes of languages to which $\mathcal{L}$ can belong.

*Varn Codes* are minimum-cost prefix-free codes for equiprobable words when the encoding alphabet has *unequal-cost letters*. They can be modelled by the leaf-set of minimum external-path length *lopsided trees*, which are trees in which different edges have different lengths, corresponding to the costs of the different letters of the encoding alphabet. There is a very large literature in the information theory and algorithmic literature devoted to analyzing the cost [24] [18] [10] [11] [3] [16] [21] [1] [7] [22] [7] [22] of such codes/trees and designing efficient algorithms for building them [16] [8] [26] [9] [20] [15] [7].

It was recently shown [13] that the Varn coding problem can be rewritten as the problem of constructing a minimum-cost prefix-free code for equiprobable words, under the assumption that all codewords are restricted to belonging to an arbitrary language $\mathcal{L}$, where $\mathcal{L}$ is a special type of language, specifically a regular language accepted by a DFA with only one accepting state. Furthermore, [13] showed that the techniques developed for constructing Varn Codes could then be used to construct optimal codes restricted to *any* regular $\mathcal{L}$ that is accepted by a DFA with only *one* accepting state. Examples of such languages are where $\mathcal{L}$ is "all words in $\Sigma^*$ ending with a particular given string $P \in \Sigma^*$," i.e., $\mathcal{L} = \Sigma^* P$ (the simplest case of such a language are the 1-ended codes, $\mathcal{L} = (0+1)^*1$ [4, 5]). A major question left open was how to construct minimum-cost prefix-free codes for equiprobable words restricted to $\mathcal{L}$ when $\mathcal{L}$ does not fit this criterion.

In this paper we solve this open problem for *all* regular $\mathcal{L}$, i.e., languages accepted by Deterministic Finite Automaton, as long as the language satisfies a very general non-degeneracy criterion. Examples of such languages are $\mathcal{L}$ of the type, $\mathcal{L}$ is all words in $\Sigma^*$ ending with *one* of the given strings $P_1, P_2, \ldots, P_n \in \Sigma^*$. More generally our technique will work when $\mathcal{L}$ is a language accepted by

---

* Department of Computer Science, HKUST, Clear Water Bay, Kowloon, Hong Kong. email:{golin,cs_lzm}@cs.ust.hk. The work of both authors was partially supported by HK RGC CERG grants HKUST6312/04E and DAG03/04.EG06.

F. Dehne, A. López-Ortiz, and J.-R. Sack (Eds.): WADS 2005, LNCS 3608, pp. 372–384, 2005.

$$A_1 = \{aabca, babca, cabca, dabca\}, \quad A_2 = \{01, 0011, 1100, 1010, 1001\}$$
$$B_1 = \{aabca, babca, cabca, aaabca\}, \quad B_2 = \{01, 10, 0011, 101010, 000111\}$$

**Fig. 1.** Examples of optimal $(A_1, A_2)$ and non-optimal $(B_1, B_2)$ codes in $\mathcal{L}_1$ and $\mathcal{L}_2$. $\mathrm{cost}(A_1) = 20$; $\mathrm{cost}(B_1) = 21$; $\mathrm{cost}(A_2) = 18$; $\mathrm{cost}(B_2) = 20$

any Deterministic Automaton, even automaton with a countably *infinite* number of states, as long as the number of accepting states in the automaton is finite.

Our major result is a combinatorial theorem that, given language $\mathcal{L}$ accepted by a Deterministic Automaton, exactly describes the general *structure* of all optimal prefix-free codes restricted to $\mathcal{L}$. This theorem immediately leads to a simple algorithm for *constructing* such codes given the restriction language $\mathcal{L}$ and the number of leaves $n$.

## 0.1    Formal Statement of the Problem

We start with a quick review of basic definitions. Let $\Sigma$ be a finite alphabet, e.g., $\Sigma = \{0, 1\}$, or $\Sigma = \{a, b, c\}$. A *code* is a set of words $C = \{w_1, w_2, \ldots, w_n\} \subset \Sigma^*$. A word $w = \sigma_1 \sigma_2 \ldots \sigma_l$ is a *prefix* of another word $w' = \sigma_1' \sigma_2' \ldots \sigma_{l'}'$ if $w$ is the start of $w'$. For example 01 is a prefix of 010011. Finally, a code is said to be *prefix-free* if for all pairs $w, w' \in C$, $w$ is not a prefix of $w'$.

Let $P = \{p_1, p_2, p_3, \ldots, p_n\}$ be a discrete probability distribution, that is, $\forall i, 0 \le p_i \le 1$ and $\sum_i p_i = 1$. The *cost* of code $C$ with distribution $P$ is $\mathrm{cost}(C, P) = \sum_i |w_i| \cdot p_i$ where $|w|$ is the length of word $w$; $\mathrm{cost}(C, P)$ is therefore the average length of a word under probability distribution $P$. The *prefix-coding problem* is, given $P$, to find a prefix-free code $C$ that minimizes $\mathrm{cost}(C, P)$. This problem is well-studied and can easily and efficiently be solved by the well-known Huffman-coding algorithm When the codewords are *equiprobable*, i.e., $\forall i$, $p_i = 1/n$, then $\mathrm{cost}(C, P) = \frac{1}{n}\sum_i |w_i| = \frac{1}{n}\mathrm{cost}(C)$ where $\mathrm{cost}(C) = \sum_i |w_i|$. $\mathrm{cost}(C, P)$ is then minimized when $\mathrm{cost}(C)$ is minimized. We will call such a code an *optimal uniform-cost code*.

In this paper we are interested in what happens to the uniform-cost code problem when it is restricted so that all of the words in $C$ must be contained in some language $\mathcal{L} \subseteq \Sigma^*$,. As examples consider $\mathcal{L} = \mathcal{L}_1$, the set of all words in $\{a, b, c\}^*$ that end with the pattern $abca$ and $\mathcal{L} = \mathcal{L}_2$, the set of all words in $\{0, 1\}^*$ in which the number of '0's is equal to the number of '1's.

In Figure 1, the codes $A_i$ are optimal prefix-free codes (for 4/5) words in $\mathcal{L}_i$ (i=1,2). That is, no codes with the same number of words in $\mathcal{L}_i$ have smaller cost than the $A_i$. The $B_i$ are non-optimal codes in the same languages.

Let language $\mathcal{L}$ be fixed. We would like to answer the questions:

- What is the optimal (min-cost) prefix-free code $C_n$ containing $n$ words in $\mathcal{L}$?
- How does $C_n$ change with $n$?

We call this the *$\mathcal{L}$-restricted prefix-coding problem*. Our major tools for attacking this problem are *generalized lopsided trees*.

*Note: In this extended abstract we only state our main results and provide intuition as to why they are correct. The full proofs are omitted.*

*A version of this paper with more diagrams and worked examples can be found as 2005 HKUST Theoretical Computer Science Group research report HKUST-TCSC-2005-04 at* `http://www.cs.ust.hk/tcsc/RR`.

# 1   Generalized Lopsided Trees

**Definition 1.** *See Figures 2 and 3.*
*We are given a finite set $T = \{t_1, t_2, ..., t_k\}$ and two functions*

$$cost(\cdot, \cdot) : T \times N^+ \to N^+ \quad and \quad type(\cdot, \cdot) : T \times N^+ \to T$$

*where $N^+$ is the set of nonnegative integers; $T$, cost() and type() are the tree parameters.*

- *A **generalized lopsided tree** for $T$, $cost(\cdot, \cdot)$ and $type(\cdot, \cdot)$ is a tree (of possibly unbounded node-degree) in which every node is labelled with one element $T$.*
- *The label of a node is its **type**; equivalently, a node of type $t_i$ is a $t_i$-node.*
- *By convention, unless otherwise explicitly stated, the root of a generalized lopsided tree must be a $t_1$-node.*
- *The jth child of a $t_i$ node, if it exists, will have type $type(t_i, j)$. The **length** (weight) of the edge from a $t_i$-node to its jth child, will be $cost(t_i, j)$. By convention, we will assume that if $j \leq j'$, then $cost(t_i, j) \leq cost(t_i, j')$ .*

*Note that it is possible that a type $t_i \in T$ node could be restricted to have at most a finite number $k$ of possible defined children. In this case, $cost(t_i, j)$ and $type(t_i, j)$ are undefined for $j > k$.*

*Note too that it is possible for a node to be "missing" its middle children, e.g, the 1st and 3rd child of a node might be in the tree, but the 2nd child might not.*

*When designing an algorithm for constructing optimal trees we will assume that the values $cost(t_i, j)$, $type(t_i, j)$ and $Num(i, m, h) = |\{j : cost(t_i, j) = h \text{ and } type(t_i, j) = t_m\}|$, can all be returned in $O(1)$ time by some oracle.*

*Finally, we point out that our definition restricts $cost(\cdot, \cdot)$ to be nonnegative integers. If $cost(\cdot, \cdot)$ were arbitrary nonnegative rationals they could be scaled to be integers and the problem would not change. Allowing $cost(\cdot, \cdot)$ to be nonnegative irrationals would change the problem and require modifying many of the lemmas and theorems in this paper. In this extended abstract we restrict ourselves to the simpler integer case since, as we will soon see, restricted languages can be modelled using integer costs.*

**Definition 2.** *See Figures 2 and 3. Let $u$ be a node and $Tr$ be a generalized lopsided tree.*

- *depth(u) is the sum of the lengths of the edges on the path connecting the root to $u$.*
- *The **height** of $Tr$ is $H(Tr) = \max_{u \in Tr} depth(u)$.*

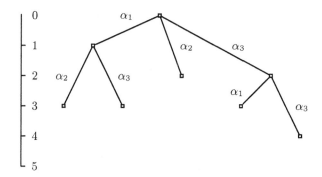

**Fig. 2.** Example: Using a lopsided tree (with only one type of node) to model a Varn code with letter costs $c_1 = 1$, $c_2 = c_3 = 2$. Edge costs are represented by vertical distances in the diagram. Let $T = \{t\}$. Then $\text{cost}(t, 1) = 1$, and $\text{cost}(t, 2) = \text{cost}(t, 3) = 2$. The code represented by the tree is the set of all external paths, which is $\alpha_1\alpha_2, \alpha_1\alpha_3, \alpha_2, \alpha_3\alpha_1, \alpha_3\alpha_3$. The cost of the tree is $2 + 3 + 3 + 3 + 4 = 15$; its height is 4

- The **leaf set** of $Tr$ is $\text{leaf}(Tr)$, the set of leaves of $Tr$.
- The **cost** of $Tr$ is its **external path length** or $C(Tr) = \sum_{v \in \text{leaf}(Tr)} \text{depth}(v)$

*Tree $Tr$ is optimal if it has minimum external path length over all trees with $|\text{leaf}(Tr)|$ leaves, i.e.,*

$$\text{cost}(Tr) = \min\{\text{cost}(Tr') : Tr' \text{ a tree with } |\text{leaf}(Tr')| = |leaf(Tr)|\}$$

**Definition 3.** $\text{opt}(n)$ *denotes an arbitrary generalized lopsided tree that has minimum cost among all generalized lopsided trees with $n$ leaves.*

For given $T$, $cost$, and $type$ the problem in which we are interested is: **Given $n$, characterize the combinatorial structure of $\text{opt}(n)$ and propose an algorithm for the construction of $\text{opt}(n)$.**

Figure 2 illustrates a case in which $|T| = 1$ and Figure 3 a case in which $|T| = 2$.

The $|T| = 1$ case has been extensively studied in the literature under the name *lopsided trees* (hence, *generalized* lopsided trees for the extension studied here). The name *lopsided trees* was introduced in 1989 by Kapoor and Reingold [16] but the trees themselves have been implicitly present in the literature at least since 1961 when Karp [17] used them to model minimum-cost prefix-free (Huffman) codes in which the length of the edge of the letters in the encoding alphabet were unequal; $c_i$ represented the length of the $i^{\text{th}}$ letter in the encoding alphabet (the *idea* of such codes was already present in Shannon [24]).

A major motivation for analyzing lopsided trees was the study of Varn-codes [26] [21]. Suppose that we wish to construct a prefix-free encoding of $n$ symbols using an encoding alphabet of $r$ letters, $\Sigma = \{\alpha_1, \ldots, \alpha_r\}$ in which the length of character $\alpha_i$ is $c_i$, where the $c_i$s may all be different.

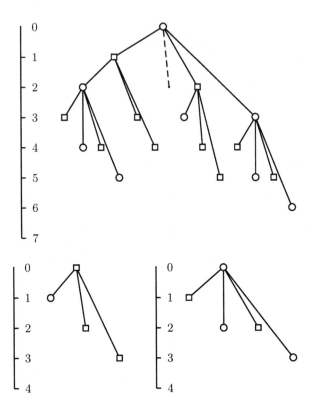

$$\text{type}(sq, 1) = circ \qquad \text{cost}(sq, 1) = 1$$
$$\text{type}(sq, 2) = sq \qquad \text{cost}(sq, 1) = 2$$
$$\text{type}(sq, 3) = sq \qquad \text{cost}(sq, 1) = 3$$

$$\text{type}(circ, 1) = sq \qquad \text{cost}(circ, 1) = 1$$
$$\text{type}(circ, 2) = circ \qquad \text{cost}(circ, 2) = 2$$
$$\text{type}(circ, 3) = sq \qquad \text{cost}(circ, 2) = 2$$
$$\text{type}(circ, 4) = circ \qquad \text{cost}(circ, 3) = 4$$

**Fig. 3.** A Generalized Lopsided tree (on the top) with $T = \{circle(circ), square(sq)\}$. Cost of the tree is $3 \cdot 3 + 5 \cdot 4 + 4 \cdot 5 + 6 = 51$; height is 6. The two trees on the bottom describe the functions cost and type on the two types of nodes, *(sq)* and *(circ)*. For comparison's sake, the functions are also explicitly written out. **Note that the second child of the root is missing**

If a symbol is encoded using string $\omega = \alpha_{i_1}\alpha_{i_2}\ldots\alpha_{i_l}$, then $cost(\omega) = \sum_{j \leq l} c_{i_j}$ is the length of the string. For example if $r = 2$, $\Sigma = \{0, 1\}$ and $c_1 = c_2 = 1$ then the cost of the string is just the number of bits it contains. This last case is the basic one encountered in regular Huffman encoding.

Now suppose that the $n$ symbols to be encoded are known to occur with equal frequency. The *cost* of the code is then defined to be $\sum_{i \leq n} cost(\omega_i)$ (which divided by $n$ is the average cost of transmitting or *length* of a symbol). Given $c_1 \leq c_2 \leq \cdots \leq c_r$, a *Varn-code* for $n$ symbols is a minimum-cost code. Varn codes have been extensively studied in the compression and coding literature([21] [2] both contain large bibliographies).

Such codes can be naturally modelled by lopsided trees in which the length of the edge from a node to its $i^{\text{th}}$ child is $c_i$. See Figure 2. Suppose that $v$ is a leaf in a lopsided tree and the unique path from the tree's root to $v$ first traverses an $i_1^{\text{st}}$ edge then an $i_2^{\text{nd}}$ edge and so on up to an $i_l^{\text{th}}$ edge. We can then associate with this leaf the codeword $\omega = \alpha_{i_1}\alpha_{i_2}\ldots\alpha_{i_l}$. The cost of this codeword is exactly the same as the depth of $v$ in the tree, i.e., $\sum_{j\leq l}c_{i_j}$. Using this correspondence, every tree with $n$ leaves corresponds to a prefix-free set of $n$ codewords and vice-versa; the cost of the code is exactly equal to the external path length of the tree which we will henceforth call the *cost* of the tree. This correspondence is extensively used, for example, in the analysis of Huffman codes.

A lopsided tree with minimal cost for $n$ leaves will be called an *optimal (lopsided) tree.*

With this correspondence and notation we see that the problems of constructing a Varn code and calculating its cost are equivalent to those of constructing an optimal (lopsided) tree and calculating its cost. This is what was studied by most of the papers listed in the first paragraph of this note and this problem is now essentially fully understood.

[13] noted that if $\Sigma = \{\alpha_1,\ldots,\alpha_r\}$ and the $c_i$ are all integral then the Varn coding problem can be modelled by introducing new alphabet $\Sigma' = \{x_1,x_2,\ldots,x_r\}$ and *Varn language* $\mathcal{L} = (x_1^{c_1} + x_2^{c_2} + \ldots + x_r^{c_r})^* \subseteq \Sigma'^*$. A 1-1 correspondence between character $\alpha_i$ and string $x_i^{c_i}$ shows that there is a 1-1 correspondence between Varn codes and prefix-codes restricted to $\mathcal{L}$ and, similarly, between lopsided trees and prefix-codes restricted to $\mathcal{L}$. Thus, the problem of finding the smallest cost prefix-code restricted to $\mathcal{L}$ is equivalent to finding a min-cost lopsided tree. [13] then noted that this was true not just for codes restricted to Varn-languages but that codes restricted to *any* regular $\mathcal{L}$ accepted by a DFA with one accepting state (like Varn Languages) could also (almost) be modelled by lopsided trees, thus permitting using the same techniques to find the cost of such codes. For example let $\mathcal{L} = \Sigma^* P$ for some fixed $P \in \Sigma^*$, i.e., $\mathcal{L}$ is all words that end in $P$. Such a $\mathcal{L}$ is always accepted by some DFA with one accepting state, so the results in [13] permit finding optimal codes of $n$ words restricted to such $\mathcal{L}$.

A problem that was left open in [13] was how to solve this problem if $\mathcal{L}$ is not in this restricted form. For example, let $\Sigma = \{0,1\}$. The simple regular language $\mathcal{L} = \Sigma^*(000 + 111)\Sigma^*$ of all words containing at least one occurance of 000 or 111 is not accepted by any DFA with only one accepting state. Another example of a regular language not accepted by any DFA with only one accepting state is $\mathcal{L} = 0^*1(0000^*1)^*0^*$ the language containing all words in which every two consecutive ones are separated by at least 3 zeros.

We now see that the $\mathcal{L}$-*restricted prefix-coding problem* can be modelled using *generalized lopsided trees* for regular languages $\mathcal{L}$. Let $\mathcal{L}$ be accepted by some Deterministic Automaton $\mathcal{M}$ with accepting states $A = \{a_1, a_2, ..., a_n\}$. Without loss of generality we may assume that the empty string $\epsilon \in \mathcal{L}$, so the *start state* of $\mathcal{M}$ is in $A$. Now define the parameters of the lopsided tree as follows: $T = \{t_1, ..., t_k\}$ where $t_i$ corresponds to state $a_i$. For any fixed $i$ enumerate, by increasing length (breaking ties arbitrarily) all paths in $\mathcal{M}$ that start at $a_i$ and end at some node $a_j \in A$, *without passing through any other node in $A$ in the interior of the path*. Let these paths be $p_i^{(1)}, p_i^{(2)}, ....$ Set $\mathbf{end}(p) = j$, where $a_j$ terminates path $p$. We complete the remaining parameters of the generalized trees by defining the functions

$$\text{type}(t_i, j) = t_{\mathbf{end}\left(p_i^{(j)}\right)}, \qquad \text{cost}(t_i, j) = \mathbf{length}\left(p_i^{(j)}\right). \tag{1}$$

There is then a simple one-one correspondence between prefix-free codes restricted to $\mathcal{L}$ and the leaves of the defined generalized lopsided tree with the cost of the code being equal to the cost (external path length) of the tree. Thus, finding the min-cost prefix free code with $n$ words restricted to $\mathcal{L}$ is exactly equivalent to finding the min-cost generalized lopsided tree with $n$ leaves. The remainder of this paper will therefore be devoted to analyzing generalized lopsided trees and how they change as $n$ grows.

As mentioned, the case of regular lopsided trees, i.e., when $|T| = 1$, is well-understood. The difficulty in extending the results on the growth of lopsided trees to that of generalized lopsided trees is that there is a fundamental difference between $|T| = 1$ and $|T| > 1$. Let opt$(n)$ be the optimal lopsided tree with $n$ nodes and $I_n$ the set of *internal (non-leaf) nodes* in opt$(n)$. In [7] it was shown that, even though it is not necessarily true that opt$(n) \subset$ opt$(n + 1)$, i.e., the trees can not be grown greedily, it is always true that $I_n \subseteq I_{n+1}$. So, with a little more analysis, one can "incrementally" construct the trees by greedily growing the set of internal nodes. Because of the interactions between the various types of nodes, this last property is *not* true for *generalized* lopsided trees. We therefore have to develop a new set of tools to analyze these trees, which is the purpose of this paper.

*Note: our correspondence only required that $\mathcal{L}$ be accepted by a Deterministic Automaton with a finite set of accepting states. Since all regular languages are accepted by Deterministic Finite Automatons our technique will suffice to analyze all restrictions to regular languages.*

*We point out that there are many non-regular languages accepted by Non-finite Deterministic Automata (automaton that can have countable infinite states) with a finite set of accepting states. For example, the language $\mathcal{L}_2$, the set of all words in $\{0, 1\}^*$ in which the number of "0"s is equal to the number of "1"s, has this property. Since these can also be modelled by generalized lopsided trees, our technique will work for restrictions to those languages as well.*

## 2    Definitions

In this section, we introduce definitions that will be used in the sequel. In what follows $T$, cost and type will be assumed fixed and given.

**Definition 4.** *Let $Tr$ be a generalized lopsided tree and $v$ a node in $Tr$.*

- *$internal(Tr)$ is the set of internal nodes of $Tr$.*
- *$type(v)$ is the type of $v$*
- *$parent(v)$ is the parent of $v$;note that the parent of the root is undefined*

Our main technique will involve building a larger tree $Tr'$ out of smaller tree $Tr$ by replacing some leaf $v \in leaf(Tr)$ with some new tree $T_2$ rooted at a type($v$)-node. The increase in the number of leaves from $Tr$ to $Tr'$ is $|leaf(T_2)| - 1$. The average cost $cost(T_2)/(|leaf(T_2)| - 1)$ of the new leaves will be crucial to our analysis and we therefore define

**Definition 5.** *The **average replacement cost** of tree $Tr$ is*

$$\mathrm{ravg}(Tr) = \mathrm{cost}(Tr)/(|\mathrm{leaf}(Tr) - 1).$$

Intuitively, we prefer to use the subtree with smallest ravg to expand the existing lopsided tree. This motivates us to study the trees with minimum ravg. Recall that the set $T$ represents the collection of types.

**Definition 6.** *Let $t_k \in T$. Set*

$$MinS(t_k) = \min\{\mathrm{ravg}(Tr), : type(root(Tr)) = t_k\}. \tag{2}$$

*The corresponding tree attaining $MinS(t_k)$ is denoted by* **MinS**$(t_k)$.

Note that this definition *does not depend upon $n$*, but only upon $T$, cost(), and type(). There might be more than one tree that attains[1] the minimum cost. In such a case, we select an arbitrary tree attaining the minimum that contains the least number of nodes.

We can now define certain essential quantities regulating the growth of lopsided trees.

**Definition 7.** *Let $l$ be an integer. Set $bottom(l) = l + \min_i\{\lfloor MinS(t_i) \rfloor\}$ and*

$$\mathrm{lev}_{t_k}(l) = \begin{cases} bottom(l) - MinS(t_k) & \text{if } MinS(t_k) \text{ is an integer;} \\ \lfloor bottom(l) + 1 - MinS(t_k) \rfloor & \text{if } MinS(t_k) \text{ is not an integer.} \end{cases} \tag{3}$$

In our analysis we often manipulate unused or *free* nodes. In order to do so, we must first introduce a reference tree containing all nodes.

**Definition 8.** *The **Infinite Generalized Lopsided Tree (ILT)** is the rooted infinite tree such that for each node $v$ in the tree, the $i^{th}$ child's type is $type(v, i)$; the length of the edge connecting $v$ and its $i^{th}$ child is $cost(v, i)$, i.e., every node contains all of its legally defined children.*

We can now define

---

[1] It is easy to prove that the minimum *is* attained but we do not do so in this extended abstract.

**Definition 9.** *A leaf $v$ in the infinite lopsided tree is* free with respect to *tree $Tr$, if $v \notin Tr$ and* parent$(v) \in Tr$; *the* free set *of $Tr$ is*

$$free(Tr) = \{v : v \text{ is free with respect to } Tr\}.$$

In our study of lopsided trees we need to somehow avoid repeated paths that do not contribute any benefit to the tree. We therefore define:

**Definition 10.** *An* improper lopsided tree *is a tree containing a path $p_1 p_2 .... p_k$, where $k > |T|$ in which each $p_i$ has only one child (that is, $p_1$ has only child $p_2$, $p_2$ has only child $p_3$, ...). A* proper lopsided tree *is a tree which is not improper.*

It is not difficult to see that improper trees can not be optimal. We may therefore restrict ourselves to studying proper trees. Note that a proper tree with $n$ leaves can only contain $O(n)$ nodes in total (where the constant in the $O()$ depends upon the tree parameters); we will need this fact in the sequel.

We need one more definition:

**Definition 11.** *Lopsided Tree parameters $T$*, cost()*, and* type() *are non-degenerate if they satisfy the following condition:*
*There exists $N > 0$ such that, $\forall l \geq N$; if the number of nodes on level $l$ in* **ILT** *is $\neq 0$ then the number of nodes on level* (bottom$(l) + 1$) *in* **ILT** *is $\geq$* $\max_i \{|\text{leaf}(\mathbf{MinS}(t_i))|\}$.

Essentially, the parameters are non-degenerate if deep enough into the infinite tree, the number of nodes per level can't get too small. A technicality occurs because it is quite easy to construct languages in which many levels of the infinite tree have *no* nodes, e.g., the language of all words in $w \in \{0,1\}^*$ in which # of 0's in $w$ equals # of 1's in $w$. In this language all words have even length, so all odd levels are empty. The condition is stated to handle such cases as well. While the non-degeneracy definition is quite technical, it is usually quite easy to show that most interesting classes of languages satisfy it. For example, $\mathcal{L}$ of the type, $\mathcal{L}$ is "all words in $\Sigma^*$ containing at least one of the specified patterns $P_1, P_2, \ldots, P_k \in \Sigma^*$" always satisfy this condition.

## 3     The Structure of Optimal Generalized Trees

**Theorem 1.** *Let $Tr$ be any optimal tree, $v_1, v_2$ two nodes in in $Tr$ with* type$(v_1)$ = type$(v_2)$. *Then if $v_1$ is internal in $Tr$ and $v_2$ is a leaf then* depth$(v_1) \leq$ depth$(v_2)$. *Furthermore, there exists a constant $N$, dependent only upon the tree parameters, such that if $Tr$ has $n \geq N$ leaves then*

1. *if $v$ is a leaf in $Tr$, then $H(Tr) - depth(v) \leq \lceil MinS(type(v)) \rceil$ and*
2. *if $v$ is internal in $Tr$, then $H(Tr) - depth(v) \geq \lceil MinS(type(v)) \rceil - 1$*

This lemma can be read as saying that opt$(n)$ always has a layered structure, i.e., there exists integers $l_1, \ldots l_{|T|}$, such that (i) all $t_i$ nodes *on or above* level $l_i$ are internal (ii) all $t_i$ nodes *below* level $l_i + 1$ are leaves and (ii) $t_i$ nodes *on* level

$l_i+1$ could be either internal or leaves. Furthermore, $H(Tr)-(l_i+\lceil MinS(t_i)\rceil) \in \{0,1\}$ so (up to an additive factor of 1), it is *independent* of $n$.

The proof of this theorem is a quite technical case-by-case one and is omitted from this extended abstract. The basic intuition behind it is quite simple, though. First, it is easy to see that, for fixed type $t_i$, there must be *some* level $l_i$ above which all $t_i$-nodes are internal and below which all $t_i$-nodes are leaves; otherwise, we can swap a higher leaf with a lower internal to get a cheaper tree with the same number of leaves. The actual *location* of $l_i$ is derived by (i) calculations noting that if a leaf $v$ is higher than the given level, then the tree can be improved by turning $v$ into an internal node by rooting a $\mathbf{MinST}(t_i)$ tree at it and removing $|\mathrm{leaf}(\mathbf{MinST}(t_i))|$ leaves from the bottom level of the tree; and (ii) calculations noting that if an internal node $v$ is lower than the specified level then it and all of its descendents can be removed and replaced by new free leaves located at the bottom level or one level below the bottom. The existence of the nodes in (i) to remove and nodes in (ii) to add follows from the non-degeneracy condition.

**Definition 12.** *Set*

$$V(l) = \{v \in \mathbf{ILT} \mid \mathrm{depth}(v) \leq \mathrm{lev}_{\mathrm{type}(v)}(l)\},$$

*that is, for each $i$, $V(l)$ contains exactly all of the $t_i$ nodes with depth $\leq \mathrm{lev}_{t_i}(l)$. Now set*

$$\mathbf{TreeA}(l) = V(l) \cup \{v \mid v \in \mathbf{ILT} \text{ and } \mathrm{parent}(v) \in V(l) \text{ and } \mathrm{depth}(v) \leq \mathrm{bottom}(l)\}$$

*and*

$$\mathbf{TreeB}(l) = V(l) \cup \{v \mid v \in \mathbf{ILT} \text{ and } \mathrm{parent}(v) \in V(l) \text{ and } \mathrm{depth}(v) \leq \mathrm{bottom}(l)+1\}$$
$$= \mathbf{TreeA}(l) \cup \{v \mid v \in \mathbf{ILT} \text{ and } \mathrm{parent}(v) \in V(l) \text{ and } \mathrm{depth}(v) = \mathrm{bottom}(l)+1\}$$

Note that $V(l)$ is the set of internal nodes of $\mathbf{TreeA}(l)$ and also the set of internal nodes of $\mathbf{TreeB}(l)$.

**Lemma 1.** *Let $l$ be an integer, then*

$$|\mathrm{leaf}(\mathbf{TreeA}(l))| \leq |\mathrm{leaf}(\mathbf{TreeB}(l))| \leq |\mathrm{leaf}(\mathbf{TreeA}(l+1))|.$$

Even though it is possible that, for some $l$, $|\mathrm{leaf}(\mathbf{TreeA}(l))| = |\mathrm{leaf}(\mathbf{TreeA}(l+1))|$ it is not difficult to see that, if the non-degeneracy condition is satisfied, $\lim_{l \to \infty} |\mathrm{leaf}(\mathbf{TreeA}(l))| = \infty$ so, for every $n$ we can find an $l$ such that $|\mathrm{leaf}(\mathbf{TreeA}(l))| \leq n < |\mathrm{leaf}(\mathbf{TreeA}(l+1))|$.

We can now state our main theorem:

**Theorem 2.** *Suppose parameters $T$, cost(), and type() are non-degenerate. For a given integer $l$, set $A(l) = \mathrm{leaf}(\mathbf{TreeA}(l))$ and $B(l) = \mathrm{leaf}(\mathbf{TreeB}(l))$ Then*

1. *If $n = |A(l)|$, then the tree $\mathbf{TreeA}(l)$ is optimal.*
2. *If $|A(l)| < n \leq |B(l)|$, then the tree obtained by appending the $n - |A(l)|$ highest free (with respect to $\mathbf{TreeA}(l)$) leaves to $\mathbf{TreeA}(l)$ is optimal.*

3. *If* $|B(l)| < n < |A(l+1)|,$
   - *All nodes in $V(l)$ are internal in* opt$(n)$.
   - *No $t_i$-node whose depth is greater than* lev$_{t_i}(l) + 1$ *is internal in* opt$(n)$.

This suggests how to find opt$(n)$ given $n$. First, find $l$ such that $A(l) \leq n < A(l+1)$. Then, calculate $B(l)$. If $A(l) \leq n \leq B(l)$ then opt$(n)$ is just $TreeA(l)$ with the highest $n - A(l)$ free leaves in $TreeA(l)$ added to it. The complicated part is when $B(l) < n < A(l+1)$. In this case Theorem 2 tells us that the set of $t_i$ internal nodes in opt$(n)$ is all of the $t_i$ nodes on or above depth lev$_{t_i}(l) + 1$ plus some $t_i$ nodes at depth lev$_{t_i}(l) + 1$. If we exactly knew the set of all internal nodes we can easily construct the tree by appending the $n$ highest leaves. So, our problem reduces down to finding exactly how many $t_i$ internal nodes there are on lev$_{t_i}(l) + 1$. We therefore define a vector that represents these numbers:

**Definition 13.** *Let $n$ and $l$ be such that $B(l) < n < A(l+1)$ and Let* opt$(n)$ *be an optimal tree for $n$ leaves and $v_i$ be the number of $t_i$-internal nodes exactly at depth* lev$_{t_i}(l) + 1$. *The* feature vector *for* opt$(n)$ *is* $\mathbf{v} = (v_1, v_2, ..., v_{|T|})$.

Theorem 2, our combinatorial structure theorem, now immediately yields a straightforward algorithm for constructing opt$(n)$. The first stage of the algorithm is to find $l$ such that $A(l) \leq n < A(l+1)$. Note that this can be done in $O(|T|^2 l^2)$ time by iteratively building $A(1), A(2), \ldots, A(l+1)$ ($l$ is the first integer such that $n < A(l+1)$). This is done not by building the actual tree but by constructing an *encoding* of the tree that, on each level, keeps track of how many $t_i$-leaves and $t_i$ internals there are on each level. So, an encoding of a height $i$ tree uses $O(|T|i)$ space. From the definition of **TreeA**$(i)$ it is easy to see that its encoding can be built from the encoding of **TreeA**$(i-1)$ in $O(|T|^2 i)$ time so $l$ can be found in $\sum_{i \leq l+1} O(|T|^2 i) = O(|T|^2 l^2)$.

Now note that, because the tree is *proper*, the total number of nodes in opt$(n)$ is $O(n)$ (where the constants in the $O()$ depend upon the parameters of the lopsided tree) so all of the $v_i = O(n)$. In particular, this means that there are at most $O(n^{|T|})$ possible feature vectors.

Given $n$, $l$ and some vector $\mathbf{v}$ it is easy to check, in $O(|T|^2 l)$ time, whether a tree with feature vector $\mathbf{v}$ actually exists. This can be done by starting with the encoding of **TreeA**$(l)$ and then, working from level $l_{|T|}$ down, using the given $\mathbf{v}$ to decide whether there are enough type-$t_i$ leaves available on level $l_i$ to transform into internals and, if there are, then transforming them. While doing this, we always remember how many leaves $L$ exist *above* the current level. After finishing processing level $l_1$, we then add the highest available $n - L$ leaves below $l_1$ if they exist, or find that no such tree exists. If such a tree can be built, then, in $O(|T|l)$ time, its cost can be calculated from the encoding.

Combining the above then gives an $O\left(|T|^2 l^2 n^{|T|}\right)$ algorithm for constructing opt$(n)$. Simply try every possible feature vector and return the one that gives the minimal cost. The fact that the tree is proper implies that $l = O(n)$ so, in the worst case, this is an $O\left(|T|^2 n^{|T|+2}\right)$ algorithm. In many interesting cases, e.g., when all nodes have a bounded number of defined children, $l = O(\log n)$ so this beomes an $O\left(\log^2 n \, |T|^2 \, n^{|T|}\right)$ algorithm.

# References

1. Julia Abrahams, "Varn codes and generalized Fibonacci trees," *The Fibonacci Quarterly,* **33**(1) (1995) 21-25.
2. Julia Abrahams, "Code and Parse Trees for Lossless Source Encoding," *Communications in Information and Systems,* **1** (2) (April 2001) 113-146.
3. Doris Altenkamp and Kurt Mehlhorn, "Codes: Unequal Probabilities, Unequal Letter Costs," *Journal of the ACM,* **27**(3) (July 1980) 412-427.
4. T. Berger and R. W. Yeung, "Optimum "1"-ended Binary Prefix codes," *IEEE Transactions on Information Theory,* **36** (6) (Nov. 1990), 1435-1441.
5. R. M. Capocelli, A. De Santis, and G. Persiano, "Binary Prefix Codes Ending in a "1"," *IEEE Transactions on Information Theory,* **40** (1) (Jul. 1994), 1296-1302.
6. Chan Sze-Lok and M. Golin, "A Dynamic Programming Algorithm for Constructing Optimal "1"-ended Binary Prefix-Free Codes," *IEEE Transactions on Information Theory,* **46** (4) (July 2000) 1637-44.
7. V. S.-N. Choi and Mordecai J. Golin, "Lopsided Trees I: Analyses," *Algorithmica,* **31**, 240-290, 2001.
8. D.M. Choy and C.K. Wong, "Construction of Optimal Alpha-Beta Leaf Trees with Applications to Prefix Codes and Information Retrieval," *SIAM Journal on Computing,* **12**(3) (August 1983) pp. 426-446.
9. N. Cot, "Complexity of the Variable-length Encoding Problem," *Proceedings of the 6th Southeast Conference on Combinatorics, Graph Theory and Computing,* (1975) 211-224.
10. I. Csiszár, "Simple Proofs of Some Theorems on Noiseless Channels," *Inform. Contr.,* bf 14 (1969) pp. 285-298
11. I. Csiszár, G. Katona and G. Tsunády, "Information Sources with Different Cost Scales and the Principle of Conservation of Energy," *Z. Wahrscheinlichkeitstheorie verw,* **12**, (1969) pp. 185-222
12. M. Golin and G. Rote, "A Dynamic Programming Algorithm for Constructing Optimal Prefix-Free Codes for Unequal Letter Costs," *IEEE Transactions on Information Theory,* **44**(5) (September 1998) 1770-1781.
13. M. Golin and HyeonSuk Na, "Optimal prefix-free codes that end in a specified pattern and similar problems: the uniform probability case.," *Data Compression Conference, DCC'2001,* (March 2001) 143-152.
14. M. Golin and Assaf Schuster "Optimal Point-to-Point Broadcast Algorithms via Lopsided Trees," *Discrete Applied Mathematics,* **93** (1999) 233-263.
15. M. Golin and N. Young, "Prefix Codes: Equiprobable Words, Unequal Letter Costs," *SIAM Journal on Computing,* **25**(6) (December 1996) 1281-1292.
16. Sanjiv Kapoor and Edward Reingold, "Optimum Lopsided Binary Trees," *Journal of the Association for Computing Machinery,* **36** (3) (July 1989) 573-590.
17. R. M. Karp, "Minimum-redundancy coding for the discrete noiseless channel," *IRE Transactions on Information Theory,* **7**, pp. 27-39, 1961
18. R. M. Krause, "Channels Which Transmit Letters of Unequal Duration," *Inform. Contr.,* **5** (1962) pp. 13-24,
19. Harry R. Lewis and Christos H. Papadimitriou, *Elements of the Theory of Computation (2nd ed.),* Prentice Hall. (1998).
20. Y. Perl, M. R. Garey, and S. Even. "Efficient Generation of Optimal Prefix Code: Equiprobable Words Using Unequal Cost Letters," *Journal of the Association for Computing Machinery,* 22(2):202–214, April 1975.
21. Serap A. Savari, "Some Notes on Varn Coding," *IEEE Transactions on Information Theory,* **40**(1) (Jan. 1994) 181-186.

22. Serap A. Savari, "A Probabilistic Approach to Some Asymptotics in Noiseless Communications," *IEEE Transactions on Information Theory,* **46**(4) (July 2000) 1246-1262.
23. Raymond Yeung, *A First Course in Information Theory,* Kluwer Academic/Plenum Publishers, New York. (2002).
24. C.E. Shannon "A Mathematical Theory of Communication," *Bell System Technical Journal* **27** (1948) 379-423, 623-656.
25. L. E. Stanfel, "Tree Structures for Optimal Searching," *Journal of the Association for Computing Machinery,* **17**(3) (July 1970) 508-517.
26. B.F. Varn, "Optimal Variable Length Codes (Arbitrary Symbol Costs and Equal Code Word Probabilities)," *Informat. Contr.,* **19** (1971) 289-301

# Tradeoffs Between Branch Mispredictions and Comparisons for Sorting Algorithms

Gerth Stølting Brodal[1,*] and Gabriel Moruz[1]

BRICS[**], Department of Computer Science, University of Aarhus,
IT Parken, Åbogade 34, DK-8200 Århus N, Denmark
{gerth, gabi}@daimi.au.dk

**Abstract.** Branch mispredictions is an important factor affecting the running time in practice. In this paper we consider tradeoffs between the number of branch mispredictions and the number of comparisons for sorting algorithms in the comparison model. We prove that a sorting algorithm using $O(dn \log n)$ comparisons performs $\Omega(n \log_d n)$ branch mispredictions. We show that Multiway MergeSort achieves this tradeoff by adopting a multiway merger with a low number of branch mispredictions. For adaptive sorting algorithms we similarly obtain that an algorithm performing $O(dn(1 + \log(1 + \mathrm{Inv}/n)))$ comparisons must perform $\Omega(n \log_d(1 + \mathrm{Inv}/n))$ branch mispredictions, where Inv is the number of inversions in the input. This tradeoff can be achieved by GenericSort by Estivill-Castro and Wood by adopting a multiway division protocol and a multiway merging algorithm with a low number of branch mispredictions.

## 1 Introduction

Modern CPUs include branch predictors in their architecture. Increased CPU pipelines enforce the prediction of conditional branches that enter the execution pipeline. Incorrect predictions determine the pipeline to be flushed with the consequence of a significant performance loss (more details on branch prediction schemes can be found in Section 2).

In this paper we consider comparison based sorting algorithms, where we assume that all element comparisons are followed by a conditional branch on the outcome of the comparison. Most sorting algorithms satisfy this property. Our contributions consist of tradeoffs between the number of comparisons required and the number of branch mispredictions performed by deterministic comparison based sorting and adaptive sorting algorithms.

We prove that a comparison based sorting algorithm performing $O(dn \log n)$ comparisons uses $\Omega(n \log_d n)$ branch mispredictions. We show that a variant

---

[*] Supported by the Carlsberg Foundation (contract number ANS-0257/20) and the Danish Natural Science Foundation (SNF).
[**] Basic Research in Computer Science, www.brics.dk, funded by the Danish National Research Foundation.

F. Dehne, A. López-Ortiz, and J.-R. Sack (Eds.): WADS 2005, LNCS 3608, pp. 385–395, 2005.

of Multiway MergeSort adopting a $d$-way merger with a low number of branch mispredictions can achieve this tradeoff.

A well known result concerning sorting is that an optimal comparison based sorting algorithm performs $\Theta(n \log n)$ comparisons [4–Section 9.1]. However, in practice, there is often the case that the input sequence is nearly sorted. In such cases, one would expect a sorting algorithm to be faster than on random input inputs. To quantify the presortedness of a given sequence, several *measures of presortedness* have been proposed. A common measure of presortedness is the number of inversions in the input, Inv, formally defined by $\text{Inv}(X) = |\{(i,j) \mid i < j \ \land x_i > x_j\}|$ for a sequence $X = (x_1, \ldots, x_n)$.

A sorting algorithm is denoted *adaptive* if its time complexity is a function that depends both on the size of the input sequence and the presortedness existent in the input [14]. For a survey concerning adaptive sorting algorithms and definitions of different measures of presortedness refer to [6].

For comparison based adaptive sorting algorithms we prove that an algorithm that uses $O(dn(1 + \log(1 + \text{Inv}/n)))$ comparisons performs $\Omega(n \log_d(1 + \text{Inv}/n))$ branch mispredictions. This tradeoff is achieved by GenericSort introduced by Estivill-Castro and Wood [5] by adopting a $d$-way division protocol and $d$-way merging that performs a low number of branch mispredictions. The division protocol is a $d$-way generalization of the binary greedy division protocol considered in [1].

In [2] it was shown that the number of mispredictions performed by standard binary MergeSort is adaptive with respect to the measure Inv. The number of comparisons and branches performed is $O(n \log n)$ but the number of branch mispredictions is $O(n \log(\text{Inv}/n))$, assuming a dynamic prediction scheme that predicts the next outcome of a branch based on the previous outcomes of the same branch.

Sanders and Winkel [15] presented a version of distribution sort that exploited special machine instructions to circumvent the assumption that each comparison is followed by a conditional branch. E.g. does the Intel Itanium 2 have a wide variety of *predicated instructions*, i.e. instructions that are executed even if its predicate is false, but the results of that instruction are not committed into program state if the predicate is false. Using predicated instructions Heap-Sort [7, 16] can be implemented to perform $O(n \log n)$ comparisons, $O(n \log n)$ predicated increment operations, and $O(n)$ branch mispredictions (assuming a static prediction scheme, see Section 2), by simply using a predicated increment operation for choosing the right child of a node during the bubble-down phase of a deletemin operation.

The rest of the paper is structured as follows. In Section 2 we give an overview of the different branch predictions schemes implemented in the nowadays CPUs. In Section 3 we prove lower bound tradeoffs between the number of comparisons and the number of branch mispredictions for comparison based sorting and adaptive sorting algorithms. Matching upper bounds are provided in Sections 4 and 5, where we show how variants of multiway MergeSort and Generic-Sort, respectively, achieve the optimal tradeoffs between comparisons and branch mispredictions.

## 2    Branch Prediction Schemes

Branch mispredictions are an important factor affecting the running time in practice [9]. Nowadays CPUs have high memory bandwidth and increased pipelines, e.g. Intel Pentium IV Prescott has a 31 stage pipeline. The high memory bandwidth severely lowers the effect of caching over the actual running time when computation takes place in the internal memory.

When a conditional branch enters the execution pipeline of the CPU, its outcome is not known and therefore must be predicted. If the prediction is incorrect, the pipeline is flushed as it contains instructions corresponding to a wrong execution path. Obviously, each branch misprediction results in performance losses, which increase with the length of the pipeline.

Several branch prediction schemes have been proposed. A classification of the branch prediction schemes is given in Figure 1.

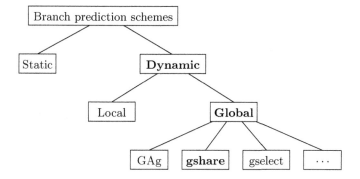

**Fig. 1.** A classification of the branch prediction schemes. The most popular branch predictors in each category are emphasized

In a static prediction scheme, every branch is predicted in the same direction every time according to some simple heuristics, e.g. all forward branches taken, all backward branches not taken. Although simple to implement, their accuracy is low and therefore they are not widely used in practice.

The dynamic schemes use the execution history when predicting a given branch. In the local branch prediction scheme (see Figure 2, left) the direction of a branch is predicted using its past outputs. It uses a *pattern history table* (*PHT*) to store the last branch outcomes, indexed after the lower $n$ bytes of the address of the branch instruction. However, the direction of a branch might depend on the output of other previous branch instructions and the local prediction schemes do not take advantage of it. To deal with this issue global branch prediction schemes were introduced [17]. They use a *branch history register* (BHR) that stores the outcome of the most recent branches. The different global prediction schemes vary only in the way the prediction table is looked up.

Three global branch prediction schemes proved very effective and are widely implemented in practice [13]. The *GAg* (Figure 2, middle) uses only the last $m$

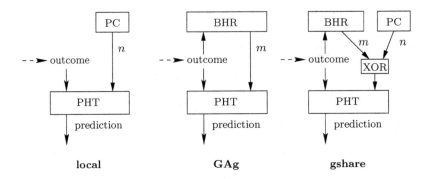

**Fig. 2.** Branch misprediction schemes

bits of the *BHR* to index the pattern history table, while *gshare* address the *PHT* by xor-ing the last bits $n$ of the branch address with the last $m$ bits of the *BHR*. Finally *gselect* concatenates the *BHR* with the lower bits of the branch address to obtain the index for the *PHT*.

The predictions corresponding to the entries in the *PHT* are usually obtained by the means of *two-bit saturating counters*. A two-bit saturating counter is an automaton consisting of four states, as shown in Figure 3.

Note that for the dynamic branch prediction schemes the same index in the *PHT* might correspond to several branches which would affect each other's predictions, constructively or destructively. This is known as the *aliasing effect* and reducing its negative effects is one of the main research areas in branch prediction schemes design.

Much research has been done on modeling branch mispredictions, especially in static analysis for upper bounding the worst case execution time (also known as WCET) [3, 11]. However, the techniques proposed involve too many hardware details and are too complicated to be used for giving branch misprediction complexities for algorithms. For the algorithms introduced in this paper, we show that even using a static branch prediction scheme, we can yield algorithms that achieve the lower bound tradeoffs between the number of comparisons and the number of branch mispredictions performed.

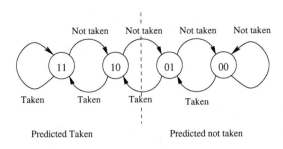

**Fig. 3.** Two-bit saturating counter

# 3    Lower Bounds for Sorting

In this section we consider deterministic comparison based sorting algorithms and prove lower bound tradeoffs between the number of comparisons and the number of branch mispredictions performed, under the assumption that each comparison between two elements in the input is immediately followed by a conditional branch that might be predicted or mispredicted. This property is satisfied by most sorting algorithms.

Theorem 1 introduces a worst case tradeoff between the number of comparisons and the number of branch mispredictions performed by sorting algorithms.

**Theorem 1.** *Consider a deterministic comparison based sorting algorithm $A$ that sorts input sequences of size $n$ using $O(dn \log n)$ comparisons, $d > 1$. The number of branch mispredictions performed by $A$ is $\Omega(n \log_d n)$.*

*Proof.* Let $T$ be the decision tree corresponding to $A$ (for a definition of decision trees see e.g. [4–Section 9.1]). By assumption, each node in the tree corresponds to a branch that can be either predicted or mispredicted. We label the edges corresponding to mispredicted branches with 1 and the edges corresponding to correctly predicted branches with 0. Each leaf is uniquely labeled with the labels on the path from the root to the given leaf. Assuming the depth of the decision tree is at most $D$ and the number of branch mispredictions allowed is $k$, each leaf is labeled by a sequence of at most $D$ 0's and 1's, containing at most $k$ 1's. By padding the label with 0's and 1's we can assume all leaf labels have length exactly $D + k$ and contain exactly $k$ 1's. It follows that the number of labelings is at most the binomial coefficient $\binom{D+k}{k}$ and therefore the number of leaves is at most $\binom{D+k}{k}$.

Denoting the number of leaves by $N \geq n!$, we obtain that $\binom{D+k}{k} \geq N$, which implies $\log \binom{D+k}{k} \geq \log N$. Using $\log \binom{D+k}{k} \leq k(O(1) + \log \frac{D}{k})$ we obtain that:

$$k \left( O(1) + \log \frac{D}{k} \right) \geq \log N . \tag{1}$$

Consider $D = \delta \log N$ and $k = \varepsilon \log N$, where $\delta \geq 1$ and $\varepsilon \geq 0$. We obtain:

$$\varepsilon \log N \left( O(1) + \log \frac{\delta}{\varepsilon} \right) \geq \log N ,$$

and therefore $\varepsilon \left( O(1) + \log \frac{\delta}{\varepsilon} \right) \geq 1$. Using $\delta = O(d)$ we obtain $\varepsilon = \Omega(1/\log d)$. Taking into account that $\log N \geq \log(n!) = n \log n - O(n)$ we obtain $k = \Omega(n \log_d n)$. □

Manilla [12] introduced the concept of optimal adaptive sorting algorithms. Given an input sequence $X$ and some measure of presortedness $M$, consider the set below$(X, M)$ of all permutations $Y$ of $X$ such that $M(Y) \leq M(X)$. Considering only inputs in below$(X, M)$, a comparison based sorting algorithm

| Measure | Comparisons | Branch mispredictions |
|---|---|---|
| Dis | $O(dn(1 + \log(1 + \text{Dis})))$ | $\Omega(n \log_d(1 + \text{Dis}))$ |
| Exc | $O(dn(1 + \text{Exc} \log(1 + \text{Exc})))$ | $\Omega(n\text{Exc} \log_d(1 + \text{Exc}))$ |
| Enc | $O(dn(1 + \log(1 + \text{Enc})))$ | $\Omega(n \log_d(1 + \text{Enc}))$ |
| Inv | $O(dn(1 + \log(1 + \text{Inv}/n)))$ | $\Omega(n \log_d(1 + \text{Inv}/n))$ |
| Max | $O(dn(1 + \log(1 + \text{Max})))$ | $\Omega(n \log_d(1 + \text{Max}))$ |
| Osc | $O(dn(1 + \log(1 + \text{Osc}/n)))$ | $\Omega(n \log_d(1 + \text{Osc}/n))$ |
| Reg | $O(dn(1 + \log(1 + \text{Reg})))$ | $\Omega(n \log_d(1 + \text{Reg}))$ |
| Rem | $O(dn(1 + \text{Rem} \log(1 + \text{Rem})))$ | $\Omega(n\text{Rem} \log_d(1 + \text{Rem}))$ |
| Runs | $O(dn(1 + \log(1 + \text{Runs})))$ | $\Omega(n \log_d(1 + \text{Runs}))$ |
| SMS | $O(dn(1 + \log(1 + \text{SMS})))$ | $\Omega(n \log_d(1 + \text{SMS}))$ |
| SUS | $O(dn(1 + \log(1 + \text{SUS})))$ | $\Omega(n \log_d(1 + \text{SUS}))$ |

**Fig. 4.** Lower bounds on the number of branch mispredictions for deterministic comparison based adaptive sorting algorithms for different measures of presortedness, given the upper bounds on the number of comparisons

performs at least $\log |\text{below}(X, M)|$ comparisons in the worst case. In particular, an adaptive sorting algorithm that is optimal with respect to measure Inv performs $O(n(1 + \log(1 + \text{Inv}/n)))$ comparisons [6].

Theorem 2 introduces a worst case tradeoff between the number of comparisons and the number of branch mispredictions for comparison based sorting algorithms that are adaptive with respect to measure Inv.

**Theorem 2.** *Consider a deterministic comparison based sorting algorithm $A$ that sorts an input sequence of size $n$ using $O(dn(1 + \log(1 + \text{Inv}/n)))$ comparisons, where* Inv *denotes the number of inversions in the input. The number of branch mispredictions performed by $A$ is $\Omega(n \log_d(1 + \text{Inv}/n))$.*

*Proof.* We reuse the proof of Theorem 1 by letting $N = |\text{below}(X, M)|$, for an input sequence $X$.

Using (1), with the decision tree depth $D = \delta n(1 + \log(1 + \text{Inv}/n))$ when restricted to inputs in $\text{below}(X, M)$, $k = \varepsilon n(1 + \log(1 + \text{Inv}/n))$ branch mispredictions, and $\log N = \Omega(n(1 + \log(1 + \text{Inv}/n)))$ [8], we obtain:

$$\varepsilon n \left(1 + \log\left(1 + \frac{\text{Inv}}{n}\right)\right)\left(O(1) + \log\frac{\delta}{\varepsilon}\right) = \Omega\left(n\left(1 + \log\left(1 + \frac{\text{Inv}}{n}\right)\right)\right) .$$

This leads to:

$$\varepsilon\left(O(1) + \log\frac{\delta}{\varepsilon}\right) = \Omega(1) ,$$

and therefore $\varepsilon = \Omega(1/\log\delta)$. Taking into account that $\delta = O(d)$ we obtain that $\varepsilon = \Omega(1/\log d)$, which leads to $k = \Omega(n \log_d(1 + \text{Inv}/n))$. $\qquad\square$

Using a similar technique, lower bounds for other measures of presortedness can be obtained. For comparison based adaptive sorting algorithms, Figure 4 states lower bounds on the number of branch mispredictions performed in the worst case, assuming the given upper bounds on the number of comparisons. For definitions of different measures of presortedness, refer to [6].

## 4    An Optimal Sorting Algorithm

In this section we introduce *Insertion d-way MergeSort*. It is a variant of $d$-way MergeSort that achieves the tradeoff stated in Theorem 1 by using an insertion sort like procedure for implementing the $d$-way merger. The merger is proven to perform a linear number of branch mispredictions.

We maintain two auxiliary vectors of size $d$. One of them stores a permutation $\pi = (\pi_1, \ldots, \pi_d)$ of $(1, \ldots, d)$ and the other one stores the indices in the input of the current element in each subsequence $i = (i_{\pi_1}, \ldots, i_{\pi_d})$, such that the sequence $(x_{i_{\pi_1}}, \ldots, x_{i_{\pi_d}})$ is sorted. During the merging, $x_{i_{\pi_1}}$ is appended to the output sequence and $i_{\pi_1}$ is incremented by 1 and then inserted in the vector $i$ in a manner that resembles insertion sort: in a scan the value $y = x_{i_{\pi_1}}$ to be inserted is compared against the smallest elements of the sorted sequence until an element larger than $y$ is encountered. This way, the property that the elements in the input sequence having indices $i_{\pi_1}, \ldots, i_{\pi_d}$ are in sorted order holds at all times. We also note that for each insertion the merger performs $O(1)$ branch mispredictions, even using a static branch prediction scheme.

**Theorem 3.** *Insertion d-way MergeSort performs $O(dn \log n)$ comparisons and $O(n \log_d n)$ branch mispredictions.*

*Proof.* For the simplicity of the proof, we consider a static prediction scheme where for the merging phase the element to be inserted is predicted to be larger than the minimum in the indices vector.

The number of comparisons performed at each level of recursion is $O(dn)$, since in the worst case each element is in the worst case compared against $d - 1$ elements at each level. Taking into account that the number of recursion levels is $\lceil \log_d n \rceil$, the total number of comparisons is $O(dn \log_d n) = O(dn \log n)$.

In what concerns the number of branch mispredictions, for each element Insertion $d$-way MergeSort performs $O(1)$ branch mispredictions for each recursion level. That is because each element is inserted at most once in the indices array $i$ at a given recursion level and for insertion sort each insertion is performed by using a constant number of branch mispredictions. Therefore we conclude that Insertion $d$-way MergeSort performs $O(n \log_d n)$ branch mispredictions.    □

We stress that Theorem 3 states an optimal tradeoff between the number of comparisons and the number of branch mispredictions. This allows tuning the parameter $d$, such that Insertion $d$-way Mergesort can achieve the best running time on different architectures depending on the CPU characteristics, i.e. the clock speed and the pipeline length.

## 5    Optimal Adaptive Sorting

In this section we describe how $d$-way merging introduced in Section 4 can be integrated within *GenericSort* by Estivill-Castro and Wood [5], using a greedy-like division protocol. The resulting algorithm is proved to achieve the tradeoff between the number of comparisons and the number of branch mispredictions stated in Theorem 2.

*GenericSort* is based on MergeSort and works as follows: if the input is small, it is sorted using some alternate sorting algorithm; if the input is already sorted, the algorithm returns. Otherwise, it splits the input sequence into $d$ subsequences of roughly equal sizes according to some *division protocol*, after which the subsequences are recursively sorted and finally merged to provide the sorted output.

The division protocol that we use, *GreedySplit*, is a generalization of the binary division protocol introduced in [1]. It partitions the input in $d + 1$ subsequences $S_0, \ldots, S_d$, where $S_0$ is sorted and $S_1, \ldots, S_d$ have balanced sizes. In a single scan from left to right we build $S_0$ in a greedy manner while distributing the other elements to subsequences $S_1, \ldots, S_d$ as follows: each element is compared to the last element of $S_0$, if it is larger, it is appended to $S_0$; if not, it is distributed to an $S_j$ such that at all times the $i^{th}$ element in the input that is not in $S_0$ is distributed to $S_{1+i \bmod d}$. It is easy to see that $S_0$ is sorted and $S_1, \ldots, S_d$ have balanced sizes. For merging we use the insertion sort based merger introduced in Section 4.

Lemma 1 generalizes Lemma 3 in [1] to the case of $d$-way splitting.

**Lemma 1.** *If GreedySplit splits an input sequence $X$ in $d + 1$ subsequences $S_0, \ldots, S_d$, where $S_0$ is sorted and $d \geq 2$, then*

$$\mathrm{Inv}(X) \geq \mathrm{Inv}(S_1) + \cdots + \mathrm{Inv}(S_d) + \frac{d-1}{4}(\mathrm{Inv}(S_1) + \cdots + \mathrm{Inv}(S_d)) .$$

*Proof.* Let $X = (x_1, \ldots, x_n)$ and $S_i = (s_{i1}, \ldots, s_{it})$, for $1 \leq i \leq d$. For each $s_{ij}$ denote by $\delta_{ij}$ its index in the input. By construction, $S_i$ is a subsequence of $X$.

For some subsequence $S_i$ consider an inversion $s_{ii_1} > s_{ii_2}$, with $i_1 < i_2$. By construction we know that for each subsequence $S_k$, with $k \neq i$, there exists some $s_{k\ell} \in S_k$ such that in the input sequence we have $\delta_{ii_1} < \delta_{k\ell} < \delta_{ii_2}$, see Figure 5. We prove that there exists at least an inversion between $s_{k\ell}$ and $s_{ii_1}$ or $s_{ii_2}$ in $X$. If $s_{k\ell} < s_{ii_2} < s_{ii_1}$ then there is an inversion between $s_{k\ell}$ and $s_{ii_1}$; if $s_{ii_2} < s_{k\ell} < s_{ii_1}$ then there are inversions in the input between $s_{k\ell}$ and both $s_{ii_1}$ and $s_{ii_2}$; finally, if $s_{ii_2} < s_{ii_1} < s_{k\ell}$, there is an inversion between $s_{k\ell}$ and $s_{ii_2}$. Let $s_{k\ell_1}, \ldots, s_{k\ell_z}$ be all the elements in $S_k$ such that $i_1 < \delta_{k\ell_1} < \cdots < \delta_{k\ell_z} < i_2$, i.e. all the elements from $S_k$ that appear in the input between ranks $\delta_{ii_1}$ and $\delta_{ii_2}$.

We proved that there is an inversion between $s_{k\ell_{\lfloor (1+z)/2 \rfloor}}$ and at least one of $s_{ii_1}$ and $s_{ii_2}$. Therefore, for the inversion $(s_{ii_1}, s_{ii_2})$ in $S_i$ we have identified an inversion in $X$ between an element in $S_k$ and an element in $S_i$ that is not present in any of $S_1, \ldots, S_d$. But this inversion can be counted for at most two different pairs in $S_i$, namely $(s_{ii_1}, s_{ii_2})$ and $(s_{ii_1}, s_{i(i_2+1)})$ if there is an inversion between $s_{ii_1}$ and $s_{k\ell_{\lfloor (1+z)/2 \rfloor}}$ or $(s_{ii_1}, s_{ii_2})$ and $(s_{i(i_1-1)}, s_{ii_2})$ otherwise. In a similar manner in $S_k$ the same inversion can be counted two times. Therefore, we obtain that for each inversion in $S_i$ there is an inversion between $S_i$ and $S_k$ that can be counted four times. Taking into account that all the inversions in $S_1, \ldots, S_d$ are also in $X$, we obtain:

$$\mathrm{Inv}(X) \geq \mathrm{Inv}(S_1) + \cdots + \mathrm{Inv}(S_d) + \tfrac{d-1}{4}(\mathrm{Inv}(S_1) + \cdots + \mathrm{Inv}(S_d)) \, . \qquad \square$$

**Theorem 4.** *GreedySort performs* $O(dn(1 + \log(1 + \mathrm{Inv}/n)))$ *comparisons and* $O(n\log_d(1 + \mathrm{Inv}/n))$ *branch mispredictions.*

*Proof.* We assume a static branch prediction scheme. For the division protocol we assume that at all times the elements are smaller than the maximum of $S_0$, meaning that branch mispredictions occur when elements are appended to the sorted sequences. This leads to a total of $O(1)$ branch mispredictions per element for the division protocol, because the sorted sequences are not sorted recursively. For the merger, the element to be inserted is predicted to be larger than the minimum in the indices vector at all times. Following the proof of Theorem 3, we obtain that splitting and merging take $O(1)$ branch mispredictions per element for each level of recursion.

We follow the proof in [10]. First we show that at the first levels of recursion, until the number of inversions gets under $n/d$, GreedySort performs $O(dn(1 + \log(1+\mathrm{Inv}/n))$ comparisons and $O(n(1+\log_d(1+\mathrm{Inv}/n))$ branch mispredictions. Afterwards, we show that the remaining levels consume a linear number of branch mispredictions and comparisons.

We first find the level $\ell$ for which the number of inversions gets below $n/d$. Denote by $\mathrm{Inv}_i$ the total number of inversions in the subsequences at level $i$. Using the result in Lemma 1, we obtain $\mathrm{Inv}_i \leq \left(\frac{4}{d+3}\right)^i \mathrm{Inv}$. The level $\ell$ should therefore satisfy:

$$\left(\frac{4}{d+3}\right)^\ell \mathrm{Inv} \leq \frac{n}{d} \, ,$$

implying $\ell \geq \log_{\frac{d+3}{4}} \frac{\mathrm{Inv}\cdot d}{n}$.

Taking into account that at each level of recursion the algorithm performs $O(dn)$ comparisons and $O(n)$ branch mispredictions, we obtain that for the first $\ell = \lceil \log_{\frac{d+3}{4}} \frac{\mathrm{Inv}\cdot d}{n} \rceil$ levels we perform $O(dn\log_d(\mathrm{Inv}/n)) = O(dn\log(\mathrm{Inv}/n))$ comparisons and $O(n\log_d(\mathrm{Inv}/n))$ branch mispredictions.

We prove that for the remaining levels we perform a linear number of comparisons and branch mispredictions.

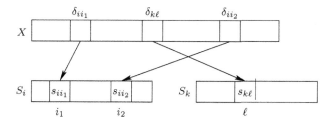

**Fig. 5.** Greedy division protocol. Between any two elements in $S_i$ there is at least one element in $S_k$ in the input sequence

Let $L(x)$ be the recursion level where some element $x$ is placed in a sorted sequence and $L(x) \geq \ell$. For each level of recursion $j$, where $\ell \leq j < L(x)$, $x$ is smaller than the maximum in the sorted subsequence $S_0$ and therefore there is an inversion between $x$ and the maximum in $S_0$ that does not exist in the recursive levels $j+1, j+2, \ldots..$ It follows that $L(x) - \ell$ is bounded by the number of inversions with $x$ at level $\ell$.

Taking into account that the total number of inversions at level $\ell$ is at most $n/d$ and that for each element at a level we perform $O(d)$ comparisons, we obtain that the total number of comparisons performed at the levels $\ell+1, \ell+2, \ldots$ is $O(n)$. Similarly, using the fact that for each element at each level $O(1)$ mispredictions are performed, we obtain that the total number of branch mispredictions performed for the levels below $\ell$ is $O(n/d)$.                                   □

## Acknowledgment

We would like to thank Peter Bro Miltersen for very helpful discussions.

## References

1. G. S. Brodal, R. Fagerberg, and G. Moruz. Cache-aware and cache-oblivious adaptive sorting. In *Proc. 32nd International Colloquium on Automata, Languages, and Programming*, Lecture Notes in Computer Science. Springer Verlag, 2005.
2. G. S. Brodal, R. Fagerberg, and G. Moruz. On the adaptiveness of quicksort. In *Proc. 7th Workshop on Algorithm Engineering and Experiments*. SIAM, 2005.
3. A. Colin and I. Puaut. Worst case execution time for a processor with branch prediction. *Real-Time Systems, Special issue on worst-case execution time analysis*, 18(2):249–274, april 2000.
4. T. H. Cormen, C. E. Leiserson, R. L. Rivest, and C. Stein. *Introduction to Algorithms, 2nd Edition*. MIT Press, 2001.
5. V. Estivill-Castro and D. Wood. Practical adaptive sorting. In *International Conference on Computing and Information - ICCI*, pages 47–54. Springer Verlag, 1991.
6. V. Estivill-Castro and D. Wood. A survey of adaptive sorting algorithms. *ACM Computing Surverys*, 24(4):441–475, 1992.
7. R. W. Floyd. Algorithm 245: Treesort3. *Communications of the ACM*, 7(12):701, 1964.
8. L. J. Guibas, E. M. McCreight, M. F. Plass, and J. R. Roberts. A new representation of linear lists. In *Proc. 9th Ann. ACM Symposium on Theory of Computing*, pages 49–60, 1977.
9. J. L. Hennesy and D. A. Patterson. *Computer Architecture: A Quantitative Approach*. Morgan Kauffman, 1996.
10. C. Levcopoulos and O. Petersson. Splitsort – an adaptive sorting algorithm. *Information Processing Letters*, 39(1):205–211, 1991.
11. X. Li, T. Mitra, and A. Roychoudhury. Modeling control speculation for timing analysis. *Real-Time Systems Journal*, 29(1), January 2005.
12. H. Manilla. Measures of presortedness and optimal sorting algorithms. *IEEE Trans. Comput.*, 34:318–325, 1985.

13. S. McFarling. Combining branch predictors. Technical report, Western Research Laboratory, 1993.

14. K. Mehlhorn. *Data structures and algorithms. Vol. 1, Sorting and searching.* Springer Verlag, 1984.

15. P. Sanders and S. Winkel. Super scalar sample sort. In *Proc. 12th European Symposium on Algorithms (ESA)*, volume 3221 of *Lecture Notes in Computer Science*, pages 784–796. Springer Verlag, 2004.

16. J. W. J. Williams. Algorithm 232: Heapsort. *Communications of the ACM*, 7(6):347–348, 1964.

17. Y.-Y. Yeh and Y. N. Patt. Alternative implementations of two-level adaptive branch prediction. In *ACM International Symposium on Computer Architecture (ISCA)*, 1992.

# Derandomization of Dimensionality Reduction and SDP Based Algorithms*

Ankur Bhargava and S. Rao Kosaraju

Dept. of Computer Science, Johns Hopkins University, Baltimore, MD-21218
{ankur, kosaraju}@cs.jhu.edu

**Abstract.** We present two results on derandomization of randomized algorithms. The first result is a derandomization of the Johnson-Lindenstrauss (JL) lemma based randomized dimensionality reduction algorithm. Our algorithm is simpler and faster than known algorithms. It is based on deriving a pessimistic estimator of the probability of failure. The second result is a general technique for derandomizing semidefinite programming (SDP) based approximation algorithms. We apply this technique to the randomized algorithm for Max Cut. Once again the algorithm is faster than known deterministic algorithms for the same approximation ratio. For this problem we present a technique to approximate probabilities with bounded error.

## 1    Introduction

We present two results in derandomization. The first result is a derandomization of the JL lemma based randomized dimensionality reduction algorithm. Derandomization is achieved with the help of a pessimistic estimator for the probability of failure. The second is a general technique to derandomize SDP based approximation algorithms. The derandomization here, is achieved by approximately computing probabilities with bounded error. The two techniques are independent of each other. Both the derandomizations use the method of conditional probabilities and these probabilities are upper-bounded or approximated rather than computed exactly.

First, we look at the JL lemma. Given a set of $n$ points in $d$ dimensional Euclidean space, the lemma guarantees an embedding of the $n$ points in $O(\log n)$ dimensional Euclidean space such that the Euclidean distances between all pairs of points is conserved to an arbitrarily small constant factor.

There are many proofs of the JL lemma [JL84, IM98, AV99, DG03, A01, FM88]. Indyk and Motwani [IM98] proposed using a random matrix to project vectors in high dimensions to lower dimensions, where each entry in the matrix is a standard normal r.v. Arriaga and Vempala [AV99] proposed using a projection matrix but with each entry chosen to be $\pm 1$ with equal probability. Achlioptas [A01] proved the lemma for $\pm 1$ valued r.v.'s. It is striking that $\pm 1$ valued r.v.'s of [AV99, A01] give the same error bounds as normal r.v.'s of [IM98].

---

* Supported by NSF Grant CCR-0311321.

F. Dehne, A. López-Ortiz, and J.-R. Sack (Eds.): WADS 2005, LNCS 3608, pp. 396–408, 2005.

There are two known derandomizations of randomized JL lemma constructions. Sivakumar [S02] gives a general derandomization technique, that can be applied to the randomized construction of [A01], but has a runtime in the order of $n^{100}$. Engebretsen, Indyk and O'Donnell [EIO02] give a derandomization for the version of [IM98], i.e. the one based on normal r.v.'s, and their algorithm runs in $O(dn^2 \log^{O(1)} n)$ time. It was generally felt that derandomization of the algorithm of [A01], which is based on $\pm 1$ valued r.v.'s is hard.

We present a derandomization of the algorithm of [A01]. Our algorithm runs in $O(dn^2 \log n)$ time. It is faster than the algorithm of [EIO02]. Our algorithm is extremely simple to program. We were able to program it in about 200 lines of C instructions. In section 2 we start with a general outline of this derandomization. The proofs may be found in the full version.

The second problem that we look at is the derandomization of SDP based approximation algorithms [GW94, KMS94, AK95, FJ95]. Such an algorithm first computes a set of vectors in a multi-dimensional space that is a solution to an SDP. Next, the algorithm generates a set of random vectors from a multivariate standard normal distribution and uses these random vectors to round the solution. There are analytical arguments that bound the expected performance of the algorithm. Our goal is to derandomize these algorithms based on such a statement of expectation, which has the following generic probability,

$$P[A \cdot X \geq 0 \text{ and } B \cdot X < 0], \tag{1}$$

where $A$ and $B$ are vectors in a Euclidean space and $X$ is a normally distributed r.v. in that space. The vectors $A$ and $B$ are computed by the underlying SDP. In some SDP formulations the statement of expectation is merely a sum of terms of the generic form in (1) while in others it may be a sum of products of such terms. Derandomization of randomized algorithms that use semidefinite programming boils down to computing (1) in polynomial time and the ability to maximize/minimize this expression when conditioned on some variable.

Engebretsen, Indyk and O'Donnell [EIO02] approximate semidefinite approximations and achieve derandomization for Max Cut [GW94] in $O(n^{3+o(1)} + n^2 2^{\log^{O(1)}(1/\epsilon)})$ time, and they achieve approximation of $(1-\epsilon)\Delta$, where $\Delta$ is the value guaranteed by the randomized algorithm. Their algorithm is based on projecting the vector solution of the SDP down to a constant number of dimensions while incurring a loss in the approximation factor. The derandomization runs in polynomial time only if $\epsilon$ is a constant. This means that their approximation ratio is always a constant factor poorer. We and [MR95] also achieve an approximation ratio $(1 - \epsilon)\Delta$ for this problem; however, we can make $\epsilon$ a small factor, such as $1/n^2$, at a small increase in speed. Our algorithm is faster and has a sharper approximation guarantee than that of [EIO02].

Mahajan and Ramesh [MR95] compute (1) by numerically evaluating integrals, as a result their algorithm is quite inefficient. In section 3 we show how the quantity in (1) can be approximated by low degree polynomials to sufficient accuracy, and then how to minimize or maximize it by finding the roots of polynomials. Using this procedure we have designed a deterministic $\tilde{O}(n^3)$ time algorithm for approximating Max Cut to a factor of 0.878.

## 2 Derandomization of Dimensionality Reduction

**Randomized Algorithm.** We project a given set of $m$ vectors, $\alpha_1, \ldots, \alpha_m \in \mathbb{R}^d$ into $k$ dimensions while preserving the $\ell_2$ norm of each vector approximately. Let $Y$ be a random matrix selected uniformly at random from $\{+1, -1\}^{d \times k}$. $Y$ is a linear projection matrix from $\mathbb{R}^d$ to $\mathbb{R}^k$. The *randomized dimensionality reduction algorithm* [A01] takes any vector $\alpha \in \mathbb{R}^d$ and maps it to $\alpha Y / \sqrt{k}$ in $\mathbb{R}^k$. The projection preserves the square of the $\ell_2$ norm of a vector.

**Lemma 1.** *If $Y$ is selected u.a.r. from $\{1, -1\}^{d \times k}$ and $k \geq \left( \frac{\epsilon^2}{4} - \frac{\epsilon^3}{6} \right)^{-1} \log \frac{2}{\delta}, \epsilon \leq 1$ then for any $\alpha \in \mathbb{R}^d$, where $k, d$ are intergers and $\delta, \epsilon$ are reals.*

$$\mathrm{P}\left[ \|\alpha\|_2^2 (1 - \epsilon) \leq \|\frac{1}{\sqrt{k}} \alpha Y\|_2^2 \leq \|\alpha\|_2^2 (1 + \epsilon) \right] > 1 - \delta.$$

We set the parameter $\delta$ (in lemma 1) to equal $1/m$. This means that the lemma will hold for $k = \frac{2}{\epsilon^2} \log 2m$ dimensions. Thus there exists a projection matrix $Y$ that preserves all $m$ vector lengths approximately with a distortion of at most $(1 + \epsilon)/(1 - \epsilon)$, while projecting the vectors into $\frac{2}{\epsilon^2} \log 2m$ dimensions.

**Conditional Probabilities.** The goal is to deterministically construct the matrix $Y$ in an efficient manner. For each vector $\alpha_i$, $1 \leq i \leq m$ let $B_i^+$ and $B_i^-$ be events defined as follows, $B_i^+ : \frac{1}{k} \|\alpha_i Y\|_2^2 > (1 + \epsilon) \|\alpha_i\|^2$; $B_i^- : \frac{1}{k} \|\alpha_i Y\|_2^2 < (1 - \epsilon) \|\alpha_i\|^2$. $B_i^+$ and $B_i^-$ denote the violation of the conditions in lemma 1, the upper and lower bounds respectively. In addition, we abbreviate $B_i^+ \cup B_i^-$ as $B_i$. Let $B$ be the event that any of the events, $B_1, \ldots, B_m$, occurs. Therefore, $\mathrm{P}[B] \leq \mathrm{P}[B_1] + \cdots + \mathrm{P}[B_m] = \mathrm{P}[B_1^+] + \mathrm{P}[B_1^-] + \cdots + \mathrm{P}[B_m^+] + \mathrm{P}[B_m^-]$. The reader can visualize the set of all possible outcomes of the randomized projection as a binary tree. The root of this tree is at depth $0$ and the leaves are at depth $kd$, i.e., one level for each random variable. There are $2^{kd}$ leaves in this tree, i.e., one leaf for each possible value of $Y \in \{+1, -1\}^{kd}$. At each internal node of the tree, there are two children. One child corresponds to setting the random variable corresponding to that level to $+1$ and the other to $-1$. We derandomize entries of matrix $Y$ one by one, by the conditional probability method [MR95b]. The sequence can be chosen arbitrarily. Here, we choose the following order, $Y_{1,1}, Y_{1,2}, \ldots, Y_{1,d}, Y_{2,1}, \ldots, Y_{2,d}, \ldots, Y_{k,1}, \ldots, Y_{k,d}$. The leaf nodes of this tree are of two types. There are leaves at which $B$ holds, i.e., at least one of the vectors violates the $1 \pm \epsilon$ bound, and the *good leaves*, which are of interest to us, where all the bounds hold.

Suppose that we have derandomized the following r.v.'s, $Y_{1,1} = y_{1,1}, \ldots, Y_{s,j-1} = y_{s,j-1}$. This corresponds to some path, $\pi$, starting at the root of the tree. We abbreviate the conditional probability $\mathrm{P}[B|Y_{1,1} = y_{1,1}, \ldots, Y_{s,j-1} = y_{s,j-1}]$ as $\mathrm{P}[B|\pi]$. We are at some internal node of the tree, which is succintly defined by $\pi$ and it is known that $\mathrm{P}[B|\pi] < 1$. If we could compute the quantities $\mathrm{P}[B|\pi, +1]$ and $\mathrm{P}[B|\pi, -1]$ then we merely need to take that path which corresponds to the lesser of the two because it will be less than 1. This follows from, $\mathrm{P}[B|\pi] < 1$ and $\mathrm{P}[B|\pi] = \mathrm{P}[B|\pi, +1]/2 + \mathrm{P}[B|\pi, -1]/2$. If it is possible to compute $\mathrm{P}[B|\pi, +1]$ and $\mathrm{P}[B|\pi, -1]$ then all we need to do is set $y_{s,j}$ to $+1$ if the former probability is the lesser of the two or to $-1$ if the latter probability is the lesser of the two.

**Pessimistic Estimator.** Our derandomization would be complete but for the fact that computing $P[B|\pi]$ is hard. In order to get around this we derive a pessimistic estimator for $P[B|\pi]$. We denote this estimator by $Q[\pi]$. The estimator, $Q[\pi]$ is a function that maps from the set of nodes in the tree to reals. In order that the estimate $Q$ is usable in place of the exact probability the function $Q$ must satisfy the following properties for all paths $\pi$ in the tree [R88],

(a) $Q[\varepsilon] < 1$;
(b) $P[B|\pi] \leq Q[\pi]$; and
(c) $Q[\pi, +1] + Q[\pi, -1] \leq 2\,Q[\pi]$,

where, $\varepsilon$ denotes a null path, i.e. the root of the tree. If there is a function $Q$ that satisfies the above conditions then it is called a *pessimistic estimator* [MR95b]. Condition (a) ensures that there exists at least one good leaf in the tree. Condition (b) states that at any node in the tree, denoted by $\pi$ which is also the path to that node from the root, if the estimator, $Q[\pi] < 1$ then at least one descendant leaf of that node is good. Condition (c) establishes that the estimator can be used to select which child is guaranteed to have a good descendant leaf. It states that the estimator of one of the children is bound to be at most the estimator of the parent. We now derive an estimator that satisfies the three conditions (a), (b), and (c). The estimator is an upper bound, so: $P[B|\pi] \leq \sum_{i=1}^{m} P[B_i^+|\pi] + P[B_i^-|\pi] \leq \sum_{i=1}^{m} Q_i^+[\pi] + Q_i^-[\pi] = Q[\pi]$. The estimator $Q$ is the sum of $2m$ terms that are upper bounds to the lower and upper tail probabilities. Using Markov's inequality and the moment generating function (see [AV99, A01]) it follows that,

$$P[B_i^+|\pi] \leq e^{-t_i\|\alpha_i\|_2^2(1+\epsilon)k} \prod_{l=1}^{k} E[e^{t_i(\alpha_i \cdot Y_l)^2}|\pi] \leq Q_i^+[\pi], \text{ and}$$

$$P[B_i^-|\pi] \leq e^{t_i\|\alpha_i\|_2^2(1-\epsilon)k} \prod_{l=1}^{k} E[e^{-t_i(\alpha_i \cdot Y_l)^2}|\pi] \leq Q_i^-[\pi],$$

(2)

where, $t_i = \epsilon/(2\|\alpha_i\|_2^2(1+\epsilon))$ and the estimators that we wish to derive, $Q_i^+$ and $Q_i^-$ must be upper bounds. We define the estimator functions $Q_i^+[\pi]$ and $Q_i^-[\pi]$ and show that $Q$ satisfies conditions (a), (b) and (c). We assume that $Y_{1,1}$ to $Y_{s,j-1}$ have been fixed corresponding to the path denoted by $\pi$. The upper bounds on the probability are a product of $k$ expectations: $Q_i^+[\pi] = \prod_{l=1}^{k} Q_{i,l}^+[\pi]$, $Q_i^-[\pi] = \prod_{l=1}^{k} Q_{i,l}^-[\pi]$. The overall structure of the estimator for $P[B|\pi]$ is as follows,

$$Q[\pi] = \sum_{i=1}^{m} \left( \prod_{l=1}^{k} Q_{i,l}^+[\pi] + \prod_{l=1}^{k} Q_{i,l}^-[\pi] \right) \geq \sum_{i=1}^{m} P[B_i^+|\pi] + P[B_i^-|\pi] \geq P[B|\pi].$$

Next, we define each of the $Q_{i,l}^+$ and $Q_{i,l}^-$'s in the following three cases.
**Case $l < s$:** Contribution from random vectors that have been covered by $\pi$,

$$E[e^{t_i(\alpha_i \cdot Y_l)^2}|\pi] = E[e^{t_i(\alpha_i \cdot Y_l)^2}|Y_l = y_l] = e^{t_i(\alpha_i \cdot y_l)^2},$$

$$Q_{i,l}^+[\pi] = e^{-t_i\|\alpha_i\|_2^2(1+\epsilon)} e^{t_i(\alpha_i \cdot y_l)^2},$$

$$Q_{i,l}^-[\pi] = e^{t_i\|\alpha_i\|_2^2(1-\epsilon)} e^{-t_i(\alpha_i \cdot y_l)^2}.$$

**Case $l > s$:** Contribution from random vectors that are not covered by $\pi$,

$$E[e^{t_i(\alpha_i \cdot Y_l)^2}|\pi] = E[e^{t_i(\alpha_i \cdot Y_l)^2}] \leq (1 - 2\|\alpha_i\|_2^2 t_i)^{-1/2}, \qquad \text{(from lem. 2)}$$

$$E[e^{-t_i(\alpha_i \cdot Y_l)^2}|\pi] = E[e^{-t_i(\alpha_i \cdot Y_l)^2}] \leq 1 - \|\alpha_i\|_2^2 t_i + 3\|\alpha_i\|_2^4 t_i^2/2, \quad \text{(from lem. 2)}$$

$$Q_{i,l}^+[\pi] = e^{-t_i\|\alpha_i\|_2^2(1+\epsilon)}(1 - 2\|\alpha_i\|_2^2 t_i)^{-1/2},$$

$$Q_{i,l}^-[\pi] = e^{t_i\|\alpha_i\|_2^2(1-\epsilon)}(1 - \|\alpha_i\|_2^2 t_i + 3\|\alpha_i\|_2^4 t_i^2/2).$$

**Case $l = s$:** Contribution from random vectors that have partly been covered by $\pi$, let $a_{i,j}^2 = \alpha_{i,j}^2 + \cdots + \alpha_{i,d}^2$ and let $c_{s,i,j} = \alpha_{i,1}y_{s,1} + \cdots + \alpha_{i,j-1}y_{s,j-1}$.

$$E[e^{t_i(\alpha_i \cdot Y_s)^2}|\pi] = E[e^{t_i(\alpha_i \cdot Y_s)^2}|Y_{s,1} = y_{s,1}, \ldots, Y_{s,j-1} = y_{s,j-1}]$$

$$\leq \exp(\frac{c_{s,i,j}^2 t_i}{1 - 2a_{i,j}^2 t_i})(1 - 2a_{i,j}^2 t_i)^{-1/2}, \qquad \text{(from lem. 2)}$$

$$E[e^{-t_i(\alpha_i \cdot Y_s)^2}|\pi] = E[e^{-t_i(\alpha_i \cdot Y_s)^2}|Y_{s,1} = y_{s,1}, \ldots, Y_{s,j-1} = y_{s,j-1}]$$

$$\leq 1 - t_i(c_{s,i,j}^2 + a_{i,j}^2) + t_i^2(3a_{i,j}^4 + 6a_{i,j}^2 c_{s,i,j}^2 + c_{s,i,j}^4)/2, \qquad \text{(from lem. 2)}$$

$$Q_{i,l}^+[\pi] = e^{-t_i\|\alpha_i\|_2^2(1+\epsilon)} \exp(\frac{c_{s,i,j}^2 t_i}{1 - 2a_{i,j}^2 t_i})(1 - 2a_{i,j}^2 t_i)^{-1/2},$$

$$Q_{i,l}^-[\pi] = e^{t_i\|\alpha_i\|_2^2(1-\epsilon)}\left(1 - t_i(c_{s,i,j}^2 + a_{i,j}^2) + t_i^2(3a_{i,j}^4 + 6a_{i,j}^2 c_{s,i,j}^2 + c_{s,i,j}^4)/2\right).$$

See lemma 2 for proofs of the upper-bounds used above. This completes the definition of the estimator Q. In order that property $(a)$ is satisfied, we select $k$ and $\epsilon$ appropriately at the outset. Property $(b)$ holds because we use upper bounds for the expectations in (2). To show that property $(c)$ holds we need to show that the case $l = s$ type of term of the estimator satisfies property $(c)$. This is sufficient because the other terms are constants when $Y_{s,j}$ is being derandomized. The estimators after we derandomize $Y_{s,j}$ to $y_{s,j}$ are,

$$Q_{i,l}^+[\pi, y_{s,j}] = e^{-t_i\|\alpha_i\|_2^2(1+\epsilon)} \exp(\frac{(c_{s,i,j} + \alpha_{i,j}y_{s,j})^2 t_i}{1 - 2(a_{i,j}^2 - \alpha_{i,j}^2)t_i})(1 - 2(a_{i,j}^2 - \alpha_{i,j}^2)t_i)^{-1/2},$$

$$Q_{i,l}^-[\pi, y_{s,j}] = e^{t_i\|\alpha_i\|_2^2(1-\epsilon)}\left(1 - t_i((c_{s,i,j} + \alpha_{i,j}y_{s,j})^2 + a_{i,j}^2 - \alpha_{i,j}^2) + \right.$$

$$\left. t_i^2(3(a_{i,j}^2 - \alpha_{i,j}^2)^2 + 6(a_{i,j}^2 - \alpha_{i,j}^2)(c_{s,i,j} + \alpha_{i,j}y_{s,j})^2 + (c_{s,i,j} + \alpha_{i,j}y_{s,j})^4)/2\right).$$

We can simplify the expressions by defining two functions, $q_1(c, a, t) = \exp\left(\frac{c^2 t}{1 - 2a^2 t}\right)$ $\frac{1}{\sqrt{1-2a^2t}}$, and $q_2(c, a, t) = 1 - (c^2 + a^2)t + (c^4 + 6c^2a^2 + 3a^4)\frac{t^2}{2}$. Then for $Q_{i,l}^+$,

$$Q_{i,l}^+[\pi] = e^{-t_i\|\alpha_i\|_2^2(1+\epsilon)} q_1(c_{s,i,j}, a_{i,j}, t_i),$$

$$Q_{i,l}^+[\pi, +1] = e^{-t_i\|\alpha_i\|_2^2(1+\epsilon)} q_1(c_{s,i,j} + \alpha_{i,j}, \sqrt{a_{i,j}^2 - \alpha_{i,j}^2}, t_i),$$

$$Q_{i,l}^+[\pi, -1] = e^{-t_i\|\alpha_i\|_2^2(1+\epsilon)} q_1(c_{s,i,j} - \alpha_{i,j}, \sqrt{a_{i,j}^2 - \alpha_{i,j}^2}, t_i).$$

There are a similar set of three expressions for $Q_{i,l}^-$, In order that property $(c)$ hold for $Q$ it is sufficient to show that the following two inequalities hold for all $i$, $l$, and $\pi$,

$$Q_{i,l}^+[\pi, +1] + Q_{i,l}^+[\pi, -1] \leq 2\, Q_{i,l}^+[\pi], Q_{i,l}^-[\pi, +1] + Q_{i,l}^-[\pi, -1] \leq 2\, Q_{i,l}^-[\pi].$$

This can be simplified further by requiring that $(c)$ holds for $Q$ provided that the following is true about functions $q_1$ and $q_2$,

$$q_1(c + x, \sqrt{a^2 - x^2}, t) + q_1(c - x, \sqrt{a^2 - x^2}, t) \leq 2q_1(c, a, t),$$
$$q_2(c + x, \sqrt{a^2 - x^2}, t) + q_2(c - x, \sqrt{a^2 - x^2}, t) \leq 2q_2(c, a, t),$$

where $c$ is any real, $0 \leq |x| \leq a$ and $2a^2t < 1$ and $t > 0$. Lemma 3 shows that this is true and therefore that property $(c)$ does indeed hold for $Q$.

Since $Q$ satisfies all three properties we can use it to derandomize the random projection matrix by the method of conditional probabilities.

There are $dk$ r.v.'s. In order to derandomize each r.v. we must compute two estimators which takes $O(m)$ time. Therefore, the derandomization algorithm runs in $O(mdk)$ time. As an example, consider some $n$ points in $d$ dimensional space. The JL lemma guarantees that they can be embedded into $O(\log n)$ dimensional space with constant distortion. Here, $m = O(n^2)$, if we wish to preserve all-pair distances. So, our derandomization will run in $O(dn^2 \log n)$.

**Technical Lemmas.**

**Lemma 2.** *If $Y$ is an r.v. which is selected u.a.r. from $\{+1, -1\}^d$ and $\alpha \in \mathbb{R}^d$ then for any reals $t \geq 0$ and $c$,*

$$\mathrm{E}[e^{t(\alpha \cdot Y + c)^2}] \leq \exp\left(\frac{c^2 t}{1 - 2\|\alpha\|_2^2 t}\right) \frac{1}{\sqrt{1 - 2\|\alpha\|_2^2 t}},$$
$$\mathrm{E}[e^{-t(\alpha \cdot Y + c)^2}] \leq 1 - (c^2 + \|\alpha\|_2^2)t + (c^4 + 6c^2\|\alpha\|_2^2 + 3\|\alpha\|_2^4)\frac{t^2}{2}.$$

**Lemma 3.** *If $c, a, t \geq 0$ are reals and $q_1(c, a) = \exp(\frac{c^2 t}{1 - 2a^2 t}) \frac{1}{\sqrt{1 - 2a^2 t}}$, $q_2(c, a) = 1 - (c^2 + a^2)t + (c^4 + 6c^2a^2 + 3a^4)\frac{t^2}{2}$, and $x$ is any real then*

$$q_1(c + x, \sqrt{a^2 - x^2}) + q_1(c - x, \sqrt{a^2 - x^2}) \leq 2q_1(c, a),$$
$$q_2(c + x, \sqrt{a^2 - x^2}) + q_2(c - x, \sqrt{a^2 - x^2}) \leq 2q_2(c, a).$$

## 3    Derandomization of SDP Based Approximation Algorithms

### Approximating Conditional Probabilities in Poly-log Time

The crux of derandomization by conditional probability lies in computing the probability that a particular choice leads to a favorable outcome. Here, the event of interest is when a random vector with a multi-dimensional standard normal distribution falls inside the 'wedge' formed by two given planes. Lemma 4 describes how the probability of this event can be computed in poly-logarithmic time provided that we know certain things about the planes before hand. The three necessary parameters are the angle between the two planes, and the distances of the two planes from the origin.

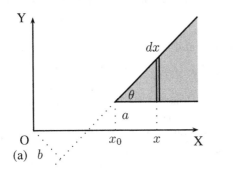
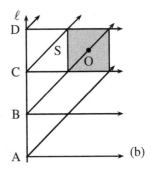

**Lemma 4.** *If $X_1, \ldots, X_n$ are i.i.d. standard normal r.v.'s and $a_1, \ldots, a_n, b_1, \ldots, b_n, c$ and $d$ are real numbers, then*

$$P\left[a_1 X_1 + \cdots + a_n X_n \geq c \text{ and } b_1 X_1 + \cdots + b_n X_n \geq d\right] \quad (3)$$

*can be approximated to within an additive error of $O(\epsilon)$ by a polynomial of degree $O(-\log \epsilon)$, and this polynomial can be constructed in $O(\log^2 \epsilon)$ time provided that we are given parameters $a, b$ and $\theta$, where, $a = \dfrac{c}{\sqrt{a_1^2 + \cdots + a_n^2}}, b = \dfrac{d}{\sqrt{b_1^2 + \cdots + b_n^2}}$, and $\cos \theta = \left(\dfrac{a_1 b_1 + \cdots + a_n b_n}{\sqrt{a_1^2 + \cdots + a_n^2}\sqrt{b_1^2 + \cdots + b_n^2}}\right)$.*

*Proof.* Each condition inside the probability describes a plane in $\mathbb{R}^n$. Due to the spherical symmetry of the distribution of $(X_1, \ldots, X_n)$ in $\mathbb{R}^n$ we can rotate the coordinate axes such that we are left with the problem of computing the probability inside a wedge in the first two dimensions (figure (a)), thereby reducing the problem to two dimensions, where we know the angle between the two lines is $\theta$ and the distances of the two lines from the origin are $a$ and $b$. Note that $\theta$, $a$ and $b$ are invariant under rotation. Since the wedge is convex, $0 \leq \theta \leq \pi$. Rotate the wedge such that one of the lines is parallel to the X-axis (figure (a)). The equations of the two lines are $y = a$ and $x \sin \theta - y \cos \theta = b$. There are four different pairs of lines possible, depending upon the signs of $a$ and $b$. There are four different wedges formed by each pair of lines, any one of these 16 choices could be the quantity of interest (in (3)). We proceed with one of these instances where the wedge, $W(a, b, \theta)$, is the set, $\{(x, y) : y \geq a \text{ and } x \sin \theta - y \cos \theta \geq b\}$. Analogous arguments can be recreated for the other 15 cases. Assume that the signs of $a$ and $b$ are correctly chosen and the shaded region in figure (a) is equal to the quantity in (3). Then the quantity in (3) is equal to $P[(X, Y) \in W(a, b, \theta)]$, where $X, Y$ are i.i.d. standard normal r.v.'s. We denote this quantity by $P(a, b, \theta) = \frac{1}{2\pi} \int_{W(a,b,\theta)} e^{-(x^2 + y^2)/2} dy \, dx$. Let $N(z) = 1/\sqrt{2\pi} \int_z^\infty e^{-z^2/2} dz$. Let $x_0 = b \csc \theta + a \cot \theta$. $(x_0, a)$ is the coordinate of the point at which the two lines that define the wedge meet. Observe that the following identities hold, $P(a, b, \theta) + P(a, -b, \pi - \theta) = N(a)$, $P(a, b, \frac{\pi}{2}) = N(a)N(b)$, and

$P(a, b, \frac{\pi}{2} - \theta) + P(x_0, -b, \theta) = P(a, x_0, \frac{\pi}{2})$. Therefore we can restrict the analysis to when $\theta \in [0, \pi/4]$. We use the Maclaurin series expansion of $e^z$ to approximate the integral. The rate of convergence of the series is dependent on the absolute value of $z$. For larger $z$ more terms are needed. However, the normal distribution rapidly drops for large $z$. We show that $O(-\log \epsilon)$ terms in the expansion of $e^z$, where $z^2 \le -2 \log \epsilon$, suffice for an accuracy of $O(\epsilon)$. The Maclaurin series expansion for $e^z$ is $\sum_{i=0}^{\infty} z^i/i!$. For brevity, let $\Delta^2 = -2 \log \epsilon$. $\Delta$ is a positive quantity. Consider the probability measure of the set $W \cap \{(x, y) : x^2 \ge \Delta^2\}$. This is at most the probability measure of the set $\{(x, y) : x^2 \ge \Delta^2\}$ which is

$$\frac{2}{\sqrt{2\pi}} \int_{\Delta}^{\infty} e^{-x^2/2} dx < \frac{2}{\sqrt{2\pi}} \int_{\Delta}^{\infty} x e^{-x^2/2} dx \le \frac{2}{\sqrt{2\pi}} \epsilon,$$

provided that, $\Delta^2 > 1$, which amounts to $\epsilon < e^{-1/2}$. Similarly, the probability measure of $W \cap \{(x, y) : y^2 \ge \Delta^2\}$ is less than $2\epsilon/\sqrt{2\pi}$. Let $S = \{(x, y) : x^2 \le \Delta^2$ and $y^2 \le \Delta^2\}$. There exists a $\delta_1$ where $|\delta_1| \le 4\epsilon/\sqrt{2\pi}$, such that,

$$P(a, b, \theta) = \frac{1}{2\pi} \int_{S \cap W(a,b,\theta)} e^{-x^2/2} e^{-y^2/2} dy \, dx + \delta_1.$$

It follows from the Maclaurin expansion that, $e^{-z^2/2} = \sum_{i=0}^{-c \log \epsilon} \frac{(-1)^i z^{2i}}{2^i i!} + \delta_2$, where $c$ is a constant. If we apply the ratio test to $i$th and the $i+1$th terms, we get $z^2/(2i+2) \le z^2/(2i) \le -\log \epsilon/i \le 1/c$, for all $i$ larger than $c \log n$ and all $z^2 \le \Delta^2$, and we get an error term $\delta_2 \le (c/(c+1))(\Delta^2)^{-c \log \epsilon}/(2^{-c \log \epsilon}(-c \log \epsilon)!)$. Then there exists a constant $c$ such that $|\delta_2|$ is $O(\epsilon)$. We can now substitute the expansions and evaluate the integrals,

$$P(a, b, \theta) = \frac{1}{2\pi} \int_{S \cap W(a,b,\theta)} \left( \sum_{i=0}^{-c \log \epsilon} \frac{(-1)^i y^{2i}}{2^i i!} \right) \left( \sum_{i=0}^{-c \log \epsilon} \frac{(-1)^i x^{2i}}{2^i i!} \right) dy \, dx + \delta_3,$$

where, $\delta_3$ is $O(\epsilon\sqrt{-\log \epsilon})$. The $\sqrt{-\log \epsilon}$ factor is due to the second integral, where we must integrate the error of the first integral. Depending on the geometric shape of $S \cap W(a, b, \theta)$ we may have to split the integration into two parts. On integrating and substituting the limits of the definite integrals we get a polynomial in $a$ and $b$. Some terms need to be expanded using the binomial expansion. In the end we are left with a polynomial in $a$ and $b$ of degree $-4c \log \epsilon + 2$. Therefore, $P(a, b, \theta)$ is approximable by a polynomial in $a$ and $b$ of degree $O(-\log \epsilon)$ to an accuracy of $O(\epsilon\sqrt{-\log \epsilon})$, when $\theta \in [0, \pi/4]$. It remains to show that $P(a, b, \pi/2)$ is expressible as a polynomial in $a$ and $b$, to a reasonable accuracy. We assume that both $a$ and $b$ are less than $\Delta$. If this is not the case then the integral is negligible anyway, so in that case it can be approximated by 0.

$$P(a, b, \pi/2) = \frac{1}{2\pi} \left( \int_a^{\infty} e^{-x^2/2} dx \right) \left( \int_b^{\infty} e^{-y^2/2} dy \right) =$$

$$\frac{1}{2\pi} \left( \int_{\max\{a, -\Delta\}}^{\Delta} e^{-x^2/2} dx \right) \left( \int_{\max\{b, -\Delta\}}^{\Delta} e^{-y^2/2} dy \right) + \delta_4,$$

where, $\delta_4$ is $O(\epsilon)$. Using the same Maclaurin series expansion as before we get,

$$P(a,b,\pi/2) = \frac{1}{2\pi}\left[\sum_{i=0}^{-c\log\epsilon}\frac{(-1)^i x^{2i+1}}{2^i i!(2i+1)}\right]_{\max\{a,-\Delta\}}^{\Delta}\left[\sum_{i=0}^{-c\log\epsilon}\frac{(-1)^i y^{2i+1}}{2^i i!(2i+1)}\right]_{\max\{b,-\Delta\}}^{\Delta} + \delta_5,$$

where, $\delta_5$ is $O(\epsilon\sqrt{-\log\epsilon})$. A subcase of the above argument shows that $N(a)$ is expressible as a polynomial in $a$ with degree $O(-\log\epsilon)$ and with error $O(\epsilon\sqrt{-\log\epsilon})$.

We have shown that with an error of $O(\epsilon\sqrt{-\log\epsilon})$ we can approximate the conditional probability with a polynomial of degree $O(-\log\epsilon)$ and in time $O(\log^2\epsilon)$. The time follows from the fact that we have to multiply two polynomials each of degree $O(-\log\epsilon)$. The statement can be reworded to say that with an error of $O(\epsilon)$ we can approximate the conditional probability with a polynomial of degree $O(-\log\epsilon)$ and in time $O(\log^2\epsilon)$. This is at the cost of a constant factor in the degree of the polynomial. Therefore, the lemma holds.

### Piecewise Approximation of Conditional Probability

In the process of derandomizing by conditional probability we are interested in the conditional probability,

$$P[a\cdot X\geq 0 \text{ and } b\cdot X\geq 0|X_1=x_1,\ldots,X_{j-1}=x_{j-1},X_j=x],$$

where, $X=(X_1,\ldots,X_n)$, is a multivariate standard normal r.v., $a$ and $b$ are vectors in $\mathbb{R}^n$, $x_1,\ldots,x_{j-1}$ are constants in $\mathbb{R}$ and $x$ is a variable. The probability is a function of $x$. Let it be $p(x)$. If we are given an $x$ then lemma 4 states that $p(x)$ is approximable by a polynomial $q(x)$ of degree $O(-\log\epsilon)$ that achieves bounded error, $p(x)=q(x)+\delta$, where $|\delta|$ is $O(\epsilon)$. The polynomial $q(x)$ is defined by a wedge formed in two dimensions. The distances of two lines (the lines that describe the wedge) from the origin and the angle between the two lines (the angle of the wedge) are resepctively: $\frac{a_1x_1+\cdots+a_{j-1}x_{j-1}+a_jx}{a_{j+1}^2+\cdots+a_n^2}$, $\frac{b_1x_1+\cdots+b_{j-1}x_{j-1}+b_jx}{b_{j+1}^2+\cdots+b_n^2}$, $\cos^{-1}\frac{a_{j+1}b_{j+1}+\cdots+a_nb_n}{\sqrt{a_{j+1}^2+\cdots+a_n^2}\sqrt{b_{j+1}^2+\cdots+b_n^2}}$. Note that the angle of the wedge does not depend upon $x$. However the distances from the origin are linear in $x$. Effectively, this implies that as $x$ changes the wedge moves along a line. See figure (b). The line $\ell$ is the path along which the wedge moves with changing $x$. Its orientation and angle are constant, i.e., independent of $x$. As the wedge travels along some arbitrary line $\ell$ in the 2d plane, its intersection with the square $S$ changes. Recall that the square $S$ is the region in which our Maclaurin expansion is convergent, and outside $S$ the probability density of the normal distribution is negligibly small. The example in figure (b) shows 5 intervals in which $p(x)$ is approximable by different polynomials. The intervals are $(-\infty, A)$, $(A, B)$, $(B, C)$, $(C, D)$ and $(D, \infty)$. Within each of these intervals, lemma 4 gives the same polynomial $q(x)$. Across intervals the polynomial is different. In general, $\ell$ may be any arbitrary line on the 2d plane, and the intervals may be different. Independent of the orientation of line $\ell$, the number of distinct intervals is bounded by a constant. Effectively, $p(x)$ is piecewise approximable with a constant sized set of polynomials. We state this formally in the following lemma,

**Lemma 5.** *If $X \in \mathbb{R}^n$ is a standard normal random vector, $a, b \in \mathbb{R}^n$, $c_1, c_2, d_1, d_2 \in \mathbb{R}$ and $x$ is a variable, then there exists a constant $k$ and the interval $(-\infty, \infty)$ can be partitioned into intervals $\{I_1, \ldots, I_k\}$, such that $\mathrm{P}[a \cdot X \geq c_1 + c_2 x$ and $b \cdot X \geq d_1 + d_2 x]$ can be expressed as follows,*

$$\mathrm{P}[a \cdot X \geq c_1 + c_2 x \text{ and } b \cdot X \geq d_1 + d_2 x] = q_1(x) + \delta_1 \quad \text{if } x \in I_1,$$

$$\vdots$$

$$= q_k(x) + \delta_k \quad \text{if } x \in I_k,$$

*where, $|\delta_1|, \ldots, |\delta_k|$ are $O(\epsilon)$ and $q_1(x), \ldots, q_k(x)$ are polynomials of degree $O(-\log \epsilon)$ that can be constructed in $O(\log^2 \epsilon)$ time provided that $|a|$, $|b|$ and $a \cdot b$ are known.*

### Derandomization of Maximum Cut

Problems like Max Cut, Max Dicut and Max 2SAT [GW94], share the randomized strategy of using one randomly chosen plane to create a partition, and can be derandomized in essentially the same way. The following discussion looks at the Max Cut problem. Given a weighted undirected graph $G = (V, E)$, the Max Cut problem seeks to partition the set of vertexes, $V$, into two subsets such that the weight of edges that cross the partition is maximized. We derandomize the 0.878 approximation algorithm of Goemans and Williamson [GW94]. A relaxation of the Max Cut problem can be formulated as a semidefinite program. Let $n$ be the number of vertexes in the set $V$ and let there be an $n$-dimensional real vector, $x_u$ associated with each vertex $u \in V$. Let $w_{u,v}$ be the weight of edge $(u, v)$. Then the relaxed Max Cut formulation [GW94] is

$$\text{maximize } \frac{1}{2} \sum_{\substack{(u,v) \in E \\ u < v}} w_{u,v}(1 - x_u \cdot x_v), \text{ s.t. } \forall u \in V,\ x_u \in \mathbb{R}^n \text{ and } |x_u| = 1. \quad (4)$$

Let the superscript $*$ denote values that maximize the objective function, that is, $\forall u \in V$, if $x_u = x_u^*$ then the objective function is maximum. The randomized algorithm selects a normally distributed random vector $X$ from $\mathbb{R}^n$. All vertex vectors that form an acute angle with this random vector are placed in one set of the partition, while the ones that form an obtuse angle are placed in the other set of the partition. The cut is the set of edges that cross the partition. The expected size of the cut is

$$\sum_{(u,v) \in E} w_{u,v}\, \mathrm{P}[x_u^* \cdot X \geq 0 \text{ and } x_v^* \cdot X < 0] \geq (0.878 - \epsilon')W^*, \quad (5)$$

where, $W^*$ is the maximum value of the objective function in (4) and $\epsilon'$ captures the accuracy of solving the SDP in (4).

The derandomized algorithm fixes one dimension at a time. If the first $i - 1$ dimensions have been derandomized, i.e., $X_1 = x_1, \ldots, X_{i-1} = x_{i-1}$ then we minimize the summation,

$$\sum_{(u,v) \in E} w_{u,v}\, \mathrm{P}[x_u^* \cdot X \geq 0 \text{ and } x_v^* \cdot X < 0 : X_1 = x_1, \ldots, X_{i-1} = x_{i-1}, X_i = x], \quad (6)$$

over all values of $x$ and set $x_i$, the derandomized value of $X_i$, to that $x$ which gave the least sum. The process repeats till all the dimensions are derandomized. The minimization is done by approximating the probability with a polynomial, and then minimizing this polynomial. Since we are approximating the probability $n$ times, once for each dimension derandomized, we need an accuracy of $\epsilon/n$ for every approximation, so that the total error is no more than $\epsilon$. At the $i$th step, let the probability inside the summation in (6) be $p_{u,v}(x)$. We know from lemma 5 that $p_{u,v}(x)$ is approximable by some constant $k(u,v)$ sized set of polynomials of degree $\log n/\epsilon$, over different ranges of $x$,

$$p_{u,v}(x) \quad \approx q^1_{u,v}(x), \text{ if } x \in I^1_{u,v}$$

$$\vdots$$

$$\approx q^{k(u,v)}_{u,v}(x), \text{ if } x \in I^{k(u,v)}_{u,v}.$$

The range of $x$, i.e., $(-\infty, \infty)$, can be partitioned into $O(|E|)$ different intervals, $I_1, I_2, \dots I_{O(|E|)}$, such that in any interval $I_j$, $\forall x, y \in I_j$ the approximation of every $p_{u,v}(x)$ and $p_{u,v}(y)$ is the same polynomial. The intervals $I_j$ are simply defined by all the end points of all the $O(|E|)$ intervals, $I^k_{u,v}$. We proceed in a left to right order. For $I_1$ we sum all the polynomials. This first step takes $O(|E|)$ time. Since the degree of each polynomial is $O(\log n/\epsilon)$ we end up with a sum that is a polynomial of degree $O(\log n/\epsilon)$. Let this polynomial be $q_1$. We then proceed to the next interval, $I_2$, where one of the polynomials which was a good approximation in $I_1$ is not a good approximation any more. We discard this by subtracting it from $q_1$ and adding in its replacement (one that is a good approximation in interval $I_2$), giving us $q_2$. We compute the minima of $q_2$ in $I_2$ and then proceed to the next interval. The process repeats till all the intervals have been minimized. Out of the $O(|E| \log n/\epsilon)$ minimia, we select the one which is the least. The whole process takes $O(|E| \log^3 n/\epsilon)$ time. The logarithmic factor in our analysis is not tight. Recall that the time needed to construct each polynomial is $O(\log^2 n/\epsilon)$. In addition, we utilize the fact that given a polynomial of degree $d$ it is possible to find all its roots within an accuracy of $\delta$ in $O(d^2(\log^2 d - \log d \log \delta))$ time [P97].

The algorithm runs in $O(n^3 \log^3(n/\epsilon))$ time and guarantees that the cut found is within $\epsilon W^*$ of what is guaranteed by the randomized algorithm.

## 4    Conclusion

We have derived a simple pessimistic estimator for derandomizing the JL lemma based randomized projection algorithm. The construction is faster than known derandomizations. This kind of pessimistic estimator may yield insight into the kind of estimators needed to derandomize randomized projection algorithms for embedding $\ell_2$ into $\ell_1$, or the more general $\ell_2$ into $\ell_p$ using $\pm 1$ valued r.v.'s [MS80].

We have also presented a general technique for derandomizing SDP based approximation algorithms. When applied to Max Cut our algorithm is faster than known techniques. This technique is applicable to other SDP based approximations, such as Max 2SAT, Max Dicut, the coloring algorithm of [KMS94], the independent set algorithm of [AK95] and the max k-cut algorithm of [FJ95].

# References

[A01]     D. Achlioptas, Database-Friendly Random Projections, *Proceedings of the 20th ACM Symposium on Principles of Database Systems*, 274–281, 2001.

[AK95]    N. Alon and N. Kahale, Approximating the Independence Number via the $\theta$-Function, *Mathematical Programming*, 80(3):253–264, 1998.

[AS91]    N. Alon and J.H. Spencer, The Probabilistic Method, *Wiley Inter-science Series in Discrete Mathematics and Optimization*, 1991.

[AV99]    R.I. Arriaga and S. Vempala, An Algorithmic Theory of Learning: Robust Concepts and Random Projection, *Proceedings of the 40th IEEE Symposium on Foundations of Computer Science*, 616–623,1999.

[B91]     B. Berger, The Fourth Moment Method, *Proceedings of the 2nd Annual ACM SIAM Symposium on Discrete Algorithms*, 373–383, 1991.

[B85]     J. Bourgain, On Lipschitz Embedding of Finite Metric Spaces in Hilbert Space, *Israeli Journal of Mathematics*, 52:46–52, 1985.

[DG03]    S. Dasgupta and A. Gupta, An Elementary Proof of a Theorem of Johnson and Lindenstrauss, *Random Structures and Algorithms*, 22(1):60–65, 2003.

[EIO02]   L. Engebretsen, P. Indyk and R. O'Donnell, Derandomized Dimensionality Reduction with Applications, *Proceedings of the 13th Annual ACM SIAM Symposium on Discrete Algorithms*, 705–712, 2002.

[F70]     W. Feller, An Introduction to Probability Theory and its Applications, Volumes 1 & 2, 3rd Edition, Wiley Series in Probability and *Mathematical Statistics*, 1970.

[FM88]    P. Frankl and H. Maehara, The Johnson-Lindenstrauss Lemma and the Sphericity of Some Graphs, *Journal of Combinatorial Theory Series A*, 44(3):355–362, 1987

[FJ95]    A. Frieze and M. Jerrum, Improved Approximation Algorithms for Max k-Cut and Max Bisection, *Proceedings of the 3rd International Conference on Integer Programming and Combinatorial Optimization*, 1–13, 1995.

[GW94]    M. Goemans and D. Williamson, 0.878 Approximation Algorithms for MAX CUT and MAX 2SAT, *Proceedings of the 26th Annual Symposium on the Theory of Computing*, 422–431, 1994.

[IM98]    P. Indyk and R. Motwani, Approximate Nearest Neighbors: Towards Removing the Curse of Dimensionality, *Proceedings of the 30th Annual Symposium on the Theory of Computing*, 604–613, 1998.

[JL84]    W.B. Johnson and J. Lindenstrauss, Extensions of Lipschitz Mappings into Hilbert Space, *Contemporary Mathematics*, 26:189–206, 1984.

[KMS94]   D. Karger, R. Motwani and M. Sudan, Approximate Graph Coloring by Semidefinite Programming, *Proceedings of the 35th IEEE Symposium on Foundations of Computer Science*, 1–10, 1994.

[LLR94]   N. Linial, E. London and Y. Rabinovich, The Geometry of Graphs and Some of its Algorithmic Applications, *Proceedings of the 35th IEEE Symposium on Foundations of Computer Science*, 577–591, 1994.

[MR95]    S. Mahajan and H. Ramesh, Derandomizing Semidefinite Programming Based Approximation Algorithms, *Proceedings of the 36th IEEE Symposium on Foundations of Computer Science*, 162–169, 1995.

[MS80]    V.D. Milman and G. Schechtman, Asymptotic Theory of Finite Dimensional Spaces, *Lecture Notes in Mathematics. Springer-Verlag.* 1980.

[MR95b]   R. Motwani and P. Raghavan, Randomized Algorithms, *Cambridge University Press*, 1995.

[P97]     V.Y. Pan, Solving a Polynomial Equation: Some History and Recent Progress, *SIAM Review*, 39(2):187–220, 1997.

[R88]    P. Raghavan, Probabilistic Construction of Deterministic Algorithms: Approximating Packing Integer Problems, *Journal of Computer and System Sciences*, 37:130–143, 1988.

[RT87]   P. Raghavan and C.D. Thompson, Randomized Rounding: A Technique for Provably Algorithms and Algorithmic Proofs, *Combinatorica*, 7:365–374, 1987.

[S02]    D. Sivakumar, Algorithmic Derandomization via Complexity Theory, *Proceedings of the 34th Annual Symposium on the Theory of Computing*, 619–626, 2002.

# Subquadratic Algorithms for 3SUM

Ilya Baran, Erik D. Demaine, and Mihai Pătraşcu

MIT Computer Science and Artificial Intelligence Laboratory,
{ibaran, edemaine, mip}@mit.edu

**Abstract.** We obtain subquadratic algorithms for 3SUM on integers and rationals in several models. On a standard word RAM with $w$-bit words, we obtain a running time of $O(n^2 / \max\{\frac{w}{\lg^2 w}, \frac{\lg^2 n}{(\lg \lg n)^2}\})$. In the circuit RAM with one nonstandard $AC^0$ operation, we obtain $O(n^2 / \frac{w^2}{\lg^2 w})$. In external memory, we achieve $O(n^2/(MB))$, even under the standard assumption of data indivisibility. Cache-obliviously, we obtain a running time of $O(n^2 / \frac{MB}{\lg^2 M})$. In all cases, our speedup is almost quadratic in the parallelism the model can afford, which may be the best possible. Our algorithms are Las Vegas randomized; time bounds hold in expectation, and in most cases, with high probability.

## 1 Introduction

The 3SUM problem can be formulated as follows: given three sets $A, B, C$ of cardinalities at most $n$, determine whether there exists a triplet $(a, b, c) \in A \times B \times C$, with $a + b = c$. This problem has a simple $O(n^2)$-time algorithm, which is generally believed to be the best possible. Erickson [7] proved an $\Omega(n^2)$ lower bound in the restricted linear decision tree model. Many problems have been shown to be 3SUM-hard (reducible from 3SUM), suggesting that they too require $\Omega(n^2)$ time; see, for example, the seminal work of [8]. This body of work is perhaps the most successful attempt at understanding complexity inside **P**.

In this paper, we consider the 3SUM problem on integers and rationals. We consider several models of computation, and achieve $o(n^2)$ running times in all of them. Our algorithms use Las Vegas randomization. The only previously known subquadratic bound is an FFT-based algorithm which can solve 3SUM on integers in the range $[0, u]$ in $O(u \lg u)$ time [4–Ex. 30.1-7]. However, this bound is subquadratic in $n$ only when $u = o(n^2 / \lg n)$, whereas our algorithm improves on the simple $O(n^2)$ solution for all values of $n$ and $u$.

It is perhaps not surprising that 3SUM should admit slightly subquadratic algorithms. However, our solution requires ideas beyond just bit tricks, and we believe that our techniques are interesting in their own right. In addition, observe that in all cases our speedups are quadratic or nearly quadratic in the parallelism that the model can afford. If the current intuition about the quadratic nature of the 3SUM problem turns out to be correct, such improvements may be the best possible.

F. Dehne, A. López-Ortiz, and J.-R. Sack (Eds.): WADS 2005, LNCS 3608, pp. 409–421, 2005.
© Springer-Verlag Berlin Heidelberg 2005

*The Circuit RAM.* The transdichotomous RAM (Random Access Machine) assumes memory cells have a size of $w$ bits, where $w$ grows with $n$. Input integers must fit in a machine word, and so must the problem size $n$ (so that the entire memory can be addressed by word-size pointers); as a consequence, $w = \Omega(\lg n)$. Operations are allowed to touch a constant number of machine words at a time.

It remains to specify the operations and their costs. The circuit RAM allows any operation which has a polynomial-size (in $w$) circuit, with unbounded fan-in gates. The cost (time) of the operation is its depth; thus, unit-cost operations are those with an $AC^0$ implementation. The model tries to address the thorny issue of what is a "good" set of unit-cost operations. Allowing any operation from a complexity-theoretic class eliminates the sensitivity of the model to arbitrary choices in the instruction set. The depth of the circuit is a restrictive and realistic definition for the time it should take to execute the operation, making the model theoretically interesting. Previously, the circuit RAM was used to study the predecessor and dictionary problems, as well as sorting (see, e.g. [2, 3, 9]).

There are $AC^0$ implementations for all common arithmetic and boolean operations, except multiplication and division. Multiplication requires depth (in this model, time) $\Theta(\frac{\lg w}{\lg \lg w})$. However, the model allows other arbitrary operations, and we shall use this power by considering one nonstandard $AC^0$ operation. Note that the previous investigations of the model also considered nonstandard operations. Our operation, WORD-3SUM solves the 3SUM problem on small sets of small integers, in the sense that they can be packed in a word. For a fixed $s = \Theta(\lg w)$, WORD-3SUM takes three input words which are viewed as sets of $O(w/s)$ $s$-bit integers. The output is a bit which specifies whether there is a triplet satisfying $a + b = c \pmod{2}^s$ with $a, b, c$ in these sets, respectively.

It is easy to see that WORD-3SUM is in $AC^0$, because addition is in $AC^0$, and all $O((\frac{w}{\lg w})^3)$ additions can be performed in parallel. On a RAM augmented with this operation, we achieve a running time of $O(n^2 \cdot \frac{\lg^2 w}{w^2} + n \cdot \frac{\lg w}{\lg \lg w} + \text{sort}(n))$. The first term dominates for any reasonable $w$ and we essentially get a speedup of $\frac{w^2}{\lg^2 w}$. The term $O(n \frac{\lg w}{\lg \lg w})$ comes from evaluating $O(n)$ multiplicative hash functions, and $\text{sort}(n)$ is the time it takes to sort $n$ values ($\text{sort}(n) = O(n \lg \lg n)$ by [9]; note that this is also $O(n \lg w)$).

*The Word RAM.* Perhaps the most natural model for our problem is the common word RAM. This model allows as unit-time operations the usual arithmetic and bitwise operations, including multiplication. In this model, we achieve a running time of $O(n^2 / \max\{\frac{\lg^2 n}{(\lg \lg n)^2}, \frac{w}{\lg^2 w}\} + \text{sort}(n))$.

To achieve the speedup by a roughly $\lg^2 n$ factor, we use the algorithm for the circuit RAM. We restrict WORD-3SUM to inputs of $\varepsilon \lg n$ bits, so that it can be implemented using a lookup table of size $n^{3\varepsilon} = o(n)$, which is initialized in negligible time. One could hope to implement WORD-3SUM using bit tricks, not a lookup table, and achieve a speedup of roughly $w^2$. However, if the current conjecture about the hardness of 3SUM is true, any circuit for WORD-3SUM should require $\Omega(w^2 / \text{poly}(\lg w))$ gates (regardless of depth). On the other hand, the standard operations for the word RAM have implementations with

$O(w \operatorname{poly}(\lg w))$ gates (for instance, using FFT for multiplication), so we cannot hope to implement WORD-3SUM efficiently using bit tricks.

We can still take advantage of a higher word size, although our speedup is just linear in $w$. This result uses bit tricks, but, to optimize the running time, it uses a slightly different strategy than implementing WORD-3SUM.

*External Memory.* Finally, we consider the external-memory model, with pages of size $B$, and a cache of size $M$. Both these quantities are measured in data items (cells); alternatively, the cache has $M/B$ pages. As is standard in this model, we assume that data items are indivisible, so we do not try to use bit blasting inside the cells inside each page. If desired, our algorithms can be adapted easily to obtain a speedup both in terms of $w$ and in terms of $M$ and $B$.

In this model, we achieve a running time of $O(\frac{n^2}{MB}+\operatorname{sort}(n))$, where $\operatorname{sort}(n)$ is known to be $\Theta(\frac{n}{B}\lg_{M/B}\frac{n}{B})$. Note that even though the external-memory model allows us to consider $M$ data items at a time, reloading $\Omega(M)$ items is not a unit-cost operation, but requires $\Omega(M/B)$ page reads. Thus, it is reasonable that the speedup we achieve is only $MB$, and not, say, $M^2$.

*The Cache-Oblivious Model.* This model is identical to the external-memory model, except that the algorithm does not know $M$ or $B$. For a survey of results in this model, see [5]. Under the standard tall-cache assumption ($M = B^{1+\Omega(1)}$), we achieve a running time of $O(n^2/\frac{MB}{\lg^2 M} + \operatorname{sort}(n)\lg n)$. This bound is almost as good as the bound for external memory.

*Organization.* In Section 2, we discuss a first main idea of our work, which is needed as a subroutine of the final algorithm. In Section 3, we discuss a second important idea, leading to the final algorithm. These sections only consider the integer problem, and the discussion of all models is interwoven, since the basic ideas are unitary. Section 4 presents some extensions: the rational case, testing approximate satisfiability, and obtaining time bounds with high probability.

# 2   Searching for a Given Sum

We first consider the problem of searching for a pair with a given sum. We are given sets $A$ and $B$, which we must preprocess efficiently. Later we are given a query $\sigma$, and we must determine if there exists $(a, b) \in A \times B : a + b = \sigma$. The 3SUM problem can be solved by making $n$ queries, for all $\sigma \in C$.

Our solution is based on the following linear time algorithm for answering a query. In the preprocessing phase, we just sort $A$ and $B$ in increasing order. In the query phase, we simultaneously scan $A$ upwards and $B$ downwards. Let $i$ be the current index in $A$, and $j$ the current index in $B$. At every step, the algorithm compares $A[i] + B[j]$ to $\sigma$. If they are equal, we are done. If $A[i] + B[j] < \sigma$, we know that $A[t] + B[j] < \sigma$ for all $t \le i$ (because the numbers are sorted). Therefore, we can increment $i$ and we do not lose a solution. Similarly, if $A[i] + B[j] > \sigma$, $j$ can advance to $j - 1$.

**Lemma 1.** *In external memory, searching for a given sum requires $O(n/B)$ time, given preprocessing time $O(\text{sort}(n))$.*

*Proof.* Immediate, since the query algorithm relies on scanning.     □

Our goal now is to achieve a similar speedup, which is roughly linear in $w$, for the RAM models. The idea behind this is to replace each element with a hash code of $\Theta(\lg w)$ bits, which enables us to pack $P = O(\min\{\frac{w}{\lg w}, n\})$ elements in a word. A query maintains indices $i$ and $j$ as above. The difference is that at every step we advance either $i$ or $j$ by $P$ positions (one word at a time). We explain below how to take two chunks $A[i \mathrel{..} (i + P - 1)]$ and $B[(j - P + 1) \mathrel{..} j]$, both packed in a word, and efficiently determine whether there is a pair summing to $\sigma$. If no such pair exists, we compare $A[i + P - 1] + B[j - P + 1]$ to $\sigma$ (using the actual values, not the hash codes). If the first quantity is larger, we know that $A[q] + B[r] > \sigma$ for all $q \geq i + P - 1$ and $r \geq j - P + 1$. Also, we know that $A[q] + B[r] \neq \sigma$ for all $q \leq i + P - 1$ and $r \geq j - P + 1$. It follows that no value $B[r]$ with $r \geq j - P + 1$ can be part of a good pair, so we advance $j$ to $j - P$. Similarly, if $A[i + P - 1] + B[j - P + 1] < \sigma$, we advance $i$ to $i + P$.

## 2.1   Linear Hashing

We now describe how to test whether any pair from two chunks of $P$ values sums to $\sigma$. We hash each value into $s = \Theta(\lg w)$ bits (then, $P$ is chosen so that $P \cdot s \leq w$). In order to maintain the 3SUM constraint through hashing, the hash function must be linear. We use a very interesting family of hash functions, which originates in [6]. Pick a random *odd* integer $a$ on $w$ bits; the hash function maps $x$ to `(a*x)>>(w-s)`. This should be understood as C notation, with the shift done on unsigned integers. In other words, we multiply $x$ by $a$ on $w$ bits, and keep the high order $s$ bits of the result. This function is almost linear. In particular $h(x) \oplus h(y) \oplus h(z) \in h(x+y+z) \ominus \{0, 1, 2\}$, where circled operators are modulo $2^s$. This is because multiplying by $a$ is linear (even in the ring modulo $2^w$), and ignoring the low order $w - s$ bits can only influence the result by losing the carry from the low order bits. When adding three values, the carry across any bit boundary is at most 2.

Reformulate the test $x + y = \sigma$ as testing whether $z = x + y - \sigma$ is zero. If $z = 0$, we have $h(x) \oplus h(y) \oplus h(-\sigma) \in \{0, -1, -2\}$, because $h(0) = 0$ for any $a$. If, on the other hand, $z \neq 0$, we want to argue that $h(x) \oplus h(y) \oplus h(-\sigma) \in \{0, -1, -2\}$ is only true with small probability. We know that $h(x) \oplus h(y) \oplus h(-\sigma)$ is at most 2 away from $h(z)$. So if $h(z) \notin \{0, \pm 1, \pm 2\}$, we are fine. Since $z \neq 0$, it is equal to $b \cdot 2^c$, for odd $b$. Multiplying $b \cdot 2^c$ with a random odd number $a$ will yield a uniformly random odd value on the high order $w - c$ bits, followed by $c$ zeros. The hash function retains the high order $s$ bits. We now have the following cases:

$w - c > s$: $h(z)$ is uniformly random, so it hits $\{0, \pm 1, \pm 2\}$ with probability $\frac{5}{2^s}$.

$w - c = s$: the low order bit of $h(z)$ is one, and the rest are uniformly random. Then $h(z)$ can only hit $\pm 1$, and this happens with probability $\frac{2}{2^s - 1}$.

$w - c = s - 1$: the low order bit $h(z)$ is zero, the second lowest is one, and the rest are random. Then $h(z)$ can only hit $\pm 2$, and this happens with probability $\frac{2}{2^{s-2}}$.

$w - c \leq s - 2$: the two low order bits of $h(z)$ are zero, and $h(z)$ is never zero, so the bad values are never attained.

We have thus shown that testing whether $h(x) \oplus h(y) \oplus h(-\sigma) \in \{0, -1, -2\}$ is a good filter for the condition $x + y = \sigma$. If the condition is true, the filter is always true. Otherwise, the filter is true with probability $\frac{O(1)}{2^s}$. When considering two word-packed arrays of $P$ values, the probability that we see a false positive is at most $P^2 \cdot \frac{O(1)}{2^s}$, by a union bound over all $P^2$ pairs which could generate a false positive through the hash function. For $s = \Theta(\lg P) = \Theta(\lg w)$, this can be made $1 / \operatorname{poly}(P)$, for any desired polynomial.

## 2.2    Implementation in the RAM Models

We now return to testing if any pair from two chunks of $P$ values sums to $\sigma$. If for all pairs $(x, y)$, we have $h(x) \oplus h(y) \oplus h(-\sigma) \notin \{0, -1, -2\}$, then we know for sure that no pair sums to $\sigma$. On the circuit RAM, this test takes constant time: this is the WORD-3SUM operation, where the third set is $h(-\sigma) \ominus \{0, 1, 2\}$ (mod 2)$^s$ (so the third set has size 3).

If the filter is passed, we have two possible cases. If there is a false positive, we can afford time $\operatorname{poly}(P)$, because a false positive happens with $1 / \operatorname{poly}(P)$ probability. In particular, we can run the simple linear-time algorithm on the two chunks, taking time $O(P)$. The second case is when we actually have a good pair, and the algorithm will stop upon detecting the pair. In this case, we cannot afford $O(P)$ time, because this could end up dominating the running time of a query for large $w$. To avoid this, we find *one* pair which looks good after hashing in $O(\lg P) = O(\lg n)$ time, which is vanishing. We binary search for an element in the first chunk which is part of a good pair. Throw away half of the first word, and ask again if a good pair exists. If so, continue searching in this half; otherwise, continue searching in the other half. After we are down to one element in the first chunk, we binary search in the second. When we find a pair that looks good after hashing, we test whether the actual values sum to $\sigma$. If so, we stop, and we have spent negligible time. Otherwise, we have a false positive, and we run the $O(P)$ algorithm (this is necessary because there might be both a false positive, and a match in the same word). The expected running time of this step is $o(1)$ because false positives happen with small probability.

**Lemma 2.** *On a circuit RAM, searching for a given sum takes $O(n \cdot \frac{\lg w}{w})$ expected time, given preprocessing time $O(n \cdot \frac{\lg w}{\lg \lg w} + \operatorname{sort}(n))$.*

*Proof.* Follows from the above. Note that the preprocessing phase needs to compute $O(n)$ hash values using multiplication. $\qquad\square$

**Lemma 3.** *On a word RAM, searching for a sum takes $O(n \cdot \min\{\frac{\lg^2 w}{w}, \frac{\lg \lg n}{\lg n}\})$ expected time, given preprocessing time $O(\operatorname{sort}(n))$.*

*Proof.* All we have to do is efficiently implement the test for $h(x) \oplus h(y) \oplus h(-\sigma) \in \{0, -1, -2\}$ for all pairs $(x, y)$ from two word-packed sets (a special case of WORD-3SUM). We use a series of bit tricks to do this in $O(\lg w)$ time. Of course, if we only fill words up to $\varepsilon \lg n$ bits, we can perform this test in constant time using table lookup, which gives the second term of the running time.

Hash codes are packed in a word, with a spacing bit of zero between values. We begin by replacing each $h(A[q])$ from the first word with $-h(-\sigma) \ominus h(A[q])$. Consider a $(s + 1)$-bit quantity $z$ with the low $s$ bits being $-h(-\sigma)$, and the high bit being one. We multiply $z$ by a constant pattern with a one bit for every logical position in the packed arrays, generating $P$ replicas of $z$. The set high order bits of $z$ overlap with the spacing zero bits in the word-packed array. Now, we subtract the word containing elements of $A$ from the word containing copies of $z$. This accomplished a parallel subtraction modulo $2^s$. Because each $(s+1)$-st bit was set in $z$, we don't get carries between the elements. Some of these bits may remain set, if subtractions do not wrap around zero. We can AND the result with a constant cleaning pattern, to force these spacing bits to zero.

Now we have one word with $h(B[r])$ and one with $-h(-\sigma) \ominus h(A[q])$. We must test whether there is a value $-h(-\sigma) \ominus h(A[q])$ at most 2 greater than some $h(B[r])$. To do this, we concatenate the two words (this is possible if we only fill half of each word in the beginning). Then, we sort the $2P$ values in this word-packed array (see below). Now, in principle, we only have to test whether two consecutive elements in the sorted order are at distance $\leq 2$. We shift the word by $s + 1$ bits to the left, and subtract it from itself. This subtracts each element from the next one in sorted order. A parallel comparison with 2 can be achieved by subtracting the word from a word with $2P$ copies of $2^s + 2$, and testing which high order bits are reset. There are two more things that we need to worry about. First, we also need to check the first and the last elements in sorted order (because we are in a cyclic group), which is easily done in $O(1)$ time. Then, we need to consider only consecutive elements such that the first comes from $h(B[r])$ and the second from $-h(-\sigma) \ominus h(A[q])$. Before sorting, we attach an additional bit to each value, identifying its origin. This should be zero for $h(B[r])$ and it should be treated as the low order bit for sorting purposes (this insures that equal pairs are sorted with $h(B[r])$ first). Then, we can easily mask away the results from consecutive pairs which don't have the marker bits in a zero-one sequence.

To sort a word-packed array, we simulate a bitonic sorting network as in [1-Sec. 3]. It is known (and easy to see) that one step of a comparison network can be simulated in $O(1)$ time using word-level parallelism. Because a bitonic sorting network on $2P$ elements has depth $O(\lg P)$, this algorithm sorts a bitonic sequence in $O(\lg P) = O(\lg w)$ time. However, we must ensure that the original sequence is bitonic. Observe that we are free to pack the hash codes in a word in arbitrary order (we don't care that positions of the hash values in a word correspond to indices in the array). Hence, we can pack each word of $B$ in increasing order of hash codes, ensuring that the values $h(B[r])$ appear in increasing order. We can also pack $h(A[q])$ values in decreasing. When we apply a subtraction modulo $2^s$, the resulting sequence is always a cyclic shift of a monotonic sequence, which is bitonic (the definition of bitonicity allows cyclic shifts). Thus we can first sort the values coming from $A$ using the bitonic sorting network, then concatenate with the second word, and sort again using the bitonic sorting network. There is one small problem that this scheme introduces: when we find a pair which looks good after hashing, we do not know which real elements generated this pair, because hash values are resorted. But we can attach $O(\lg P) = O(\lg w)$ bits of additional information to each value, which is carried along through sorting and identifies the original element.     □

## 3     The General Algorithm

By running the query algorithm from the previous section $n$ times, for every element in $C$, we obtain a speedup which is roughly linear in the parallelism in the machine. The subproblem that is being solved in parallel is a small instance of 3SUM with $|C| = O(1)$. We can make better use of the parallelism if we instead solve 3SUM subproblems with $A$, $B$, and $C$ roughly the same size. The linear scan from the previous section does not work because it may proceed differently for different elements of $C$. Instead, we use another level of linear hashing to break up the problem instance into small subproblems. We hash all elements to $o(n)$ buckets and the linearity of the hash function ensures that for every pair of buckets, all of the sums of two elements from them are contained in $O(1)$ buckets. The algorithm then solves the subproblems for every pair of buckets.

### 3.1     A Hashing Lemma

We intend to map $n$ elements into $n/m$ buckets, where $m$ is specified below. Without loss of generality, assume $n/m$ is a power of two. The algorithm picks a random hash function, from the same almost linear family as before, with an output of $\lg(n/m)$ bits. A bucket contains elements with the same hash value. In any fixed bucket, we expect $m$ elements. Later, we will need to bound the expected number of elements which are in buckets with more than $O(m)$ elements. By a Markov bound, we could easily conclude that the expected number of elements in buckets of size $\geq mt$ decreases linearly in $t$. A sharper bound is usually obtained with $k$-wise independent hashing, but a family of (almost) linear hash functions cannot even achieve strong universality (2-independence): consider hashing $x$ and $2x$. Fortunately, we are saved by the following slightly unusual lemma:

**Lemma 4.** *Consider a family of universal hash functions $\{h : U \to [\frac{n}{m}]\}$, which guarantees that $(\forall)x \neq y : \Pr_h[h(x) = h(y)] \leq \frac{m}{n}$. For a given set $S$, $|S| = n$, let $\mathcal{B}(x) = \{y \in S \mid h(y) = h(x)\}$. Then the expected number of elements $x \in S$ with $|\mathcal{B}(x)| \geq t$ is at most $\frac{2n}{t-2m+2}$.*

*Proof.* Pick $x \in S$ randomly. It suffices to show that $p = \Pr_{h,x}[|\mathcal{B}(x)| \geq s] \leq \frac{2}{s-2m+1}$. Now pick $y \in S \setminus \{x\}$ randomly, and consider the collision probability

$q = \Pr_{x,y,h}[h(x) = h(y)]$. By universality, $q \leq \frac{m}{n}$. Let $q_h = \Pr_{x,y}[h(x) = h(y)]$ and $p_h = \Pr_x[|\mathcal{B}(x)| \geq s]$.

We want to evaluate $q_h$ as a function of $p_h$. Clearly, $\Pr[h(x) = h(y) \mid |\mathcal{B}(x)| \geq s] \geq \frac{s-1}{n}$. Now consider $S' = \{x \mid |\mathcal{B}(x)| < s\}$; we have $|S'| = (1 - p_h)n$. If $x \in S'$ and we have a collision, then $y \in S'$ too. By convexity of the square function, the collision probability of two random elements from $S'$ is minimized when the same number of elements from $S'$ hash to any hash code. In this case, $|\mathcal{B}(x)| \geq \lfloor \frac{|S'|}{n/m} \rfloor \geq (1 - p_h)m - 1$. So $\Pr[h(x) = h(y) \mid x \in S'] \geq \frac{(1-p_h)m-2}{n}$. Now $q_h \geq p_h \frac{s-1}{n} + (1-p_h)\frac{(1-p_h)m-2}{n} \geq \frac{1}{n}(p_h(s-1) + (1-2p_h)m - 2(1-p_h)) = \frac{1}{n}(p_h(s-2m+2) + m - 2)$. But we have $q = E[q_h] \leq \frac{m}{n}$, so $p_h(s - 2m + 2) + m - 2 \leq m$, which implies $p_h \leq \frac{2}{s-2m+2}$.    □

The lemma implies that in expectation, $O(\frac{n}{m})$ elements are in buckets of size greater than $3m$. It is easy to construct a universal family showing that the linear dependence on $t$ is optimal, so the analysis is sharp. Note that the lemma is highly sensitive on the constant in the definition of universality. If we only know that $\Pr_h[h(x) = h(y)] \leq \frac{O(1) \cdot m}{n}$, the probability that an element is in a bucket larger than $t$ decreases just as $\frac{m}{t}$ (by the Markov bound). Again, we can construct families showing that this result is optimal, so the constant is unusually important in our application. Fortunately, the universal family that we considered achieves a constant of 1.

## 3.2   The Algorithm

The idea of the algorithm is now simple. We hash the three sets $A, B, C$ separately, each into $n/m$ buckets. Consider a pair of buckets $(\mathcal{B}^A, \mathcal{B}^B)$ from $A$ and $B$, respectively. Then, there exist just two buckets $\mathcal{B}_1^C, \mathcal{B}_2^C$ of $C$, such $(\forall)(x,y) \in \mathcal{B}^A \times \mathcal{B}^B$, if $x + y \in C$, then $x + y \in \mathcal{B}_1^C \cup \mathcal{B}_2^C$. This follows by the almost linearity of our hash functions: if $x + y = z$, then $h(z) \in h(x) \oplus h(y) \oplus \{0, 1\}$.

The algorithm iterates through all pairs of buckets from $A$ and $B$, and for each one looks for the sum in two buckets of $C$. This corresponds to solving $n^2/m^2$ independent 3SUM subproblems, where the size of each subproblem is the total size of the four buckets involved. The expected size of each bucket is $m$, but this does not suffice to get a good bound, because the running times are quadratic in the size of the buckets. Here, we use the Lemma 4, which states that the expected number of elements which are in buckets of size more than $3m$ is $O(n/m)$. The algorithm caps all buckets to $3m$ elements, and applies the above reasoning on these capped buckets. The elements that overflow are handled by the algorithms from the previous section: for each overflowing element, we run a query looking for a pair of values from the other two sets, which satisfy the linear constraint. Thus, we have $n^2/m^2$ subproblems of worst-case size $O(m)$, and $O(n/m)$ additional queries in expectation.

**Theorem 5.** *The 3SUM problem can be solved in $O(n^2 \cdot \frac{\lg^2 w}{w^2} + n \cdot \frac{\lg w}{\lg \lg w} + \text{sort}(n))$ expected time on a circuit RAM.*

*Proof.* Choose $m = O(\min\{\frac{w}{\lg w}, \sqrt{n}\})$. We use a second level of hashing inside each bucket: replace elements by an $O(\lg w)$-bit hash value, and pack each bucket into a word. For buckets of $C$, we pack two values per element: $h(z)$ and $h(z) \ominus 1$. Now we can test whether there is a solution in a triplet of buckets in constant time, using WORD-3SUM. If we see a solution, we verify the actual elements of the buckets using $O(m^2)$ time. When we find a solution, we have an additive cost of $O(m^2) = O(n)$. The time spent on useless verification is $o(1)$ since the probability of a false positive is $1/\operatorname{poly}(w)$. We also need to run $O(n/\frac{w}{\lg w})$ queries in the structure of Lemma 2. □

**Theorem 6.** *The 3SUM problem can be solved in $O(n^2 / \max\{\frac{\lg^2 n}{(\lg \lg n)^2}, \frac{w}{\lg^2 w}\} + \operatorname{sort}(n))$ expected time on a word RAM.*

*Proof.* As above, but we only choose $m = O(\frac{\lg n}{\lg \lg n})$ and we implement WORD-3SUM using a lookup table. □

**Theorem 7.** *In external memory, 3SUM can be solved in expected time $O(\frac{n^2}{MB} + \operatorname{sort}(n))$.*

*Proof.* Choose $m = O(M)$, so that each subproblem fits in cache, and it can be solved without memory transfers. Loading a bucket into cache requires $\Theta(\frac{M}{B})$ page transfers, so the running time for all subproblems is $O(\frac{n^2}{MB})$. We must also run $O(\frac{n}{M})$ additional queries, taking $O(\frac{n}{B})$ time each, by Lemma 1. The startup phases only require time to sort. □

**Theorem 8.** *There is a cache-oblivious algorithm for the 3SUM problem running in expected time $O(n^2 \cdot \frac{\lg^2 M}{MB} + \operatorname{sort}(n) \lg n)$.*

*Proof.* This requires some variations on the previous approach. First, we hash all values using our almost linear hash functions into the universe $\{0, \dots, n-1\}$. We store a redundant representation of each of the 3 sets; consider $A$ for concreteness. We construct a perfect binary tree $\mathcal{T}^A$ over $n$ leaves (we can assume $n$ is a power of two). The hash code $h(x)$ of an element $x \in A$ is associated with leaf $h(x)$. For some node $u$ of $\mathcal{T}^A$, let $s(u)$ be the number of elements of $A$ with a hash code that lies below $u$. Also, let $\ell(u)$ be the number of leaves of $\mathcal{T}^A$ under $u$. Each node $u$ stores a set $S(u)$ with elements having hash codes under $u$, such that $|S(u)| \leq \min\{s(u), 3\ell(u)\}$. Let $v, w$ be the children of $u$. Then $S(v) \subseteq S(u), S(w) \subseteq S(u)$. If exactly one child has $s(\cdot) > 3\ell(\cdot)$, $u$ has a chance to store more elements than $S(v) \cup S(w)$. The additional elements are chosen arbitrarily. Now, the representation of $A$ consists of all nodes, with elements stored in them, listed in a preorder traversal. The set $S(\cdot)$ of each node is sorted by element values. Elements which appear for the first time in $S(u)$, i.e. they are in $S(u) \setminus (S(v) \cup S(w))$, are specially marked. Note that the representation of $A$ has size $O(n \lg n)$, and it is easy to construct in time $O(\operatorname{sort}(n) \cdot \lg n)$.

We now give a recursive procedure, which is given three vertices $v^A, v^B, v^C$ and checks whether there is a 3SUM solution in $S(v^A) \times S(v^B) \times S(v^C)$. It is easy to calculate the allowable interval of hash codes that are under a vertex. If, looking at the three intervals, it is mathematically impossible to find a solution, the procedure returns immediately. Otherwise, it calls itself recursively for all 8 combinations of children. Finally, it must take care of the elements which appear for the first time in one of $S(v^A), S(v^B), S(v^C)$. Say we have such an $x \in S(v^A)$ which does not appear in the sets of the children of $v^A$. Then, we can run the linear-scan algorithm to test for a sum of $x$ in $S(v^B) \times S(v^C)$. These sets are conveniently sorted.

As usual in the cache-oblivious model, we analyze the algorithm by constructing an ideal paging strategy (because real paging strategies are $O(1)$-competitive, under constant-factor resource augmentation). Consider the depth threshold where $\ell(v) \le \varepsilon \frac{M}{\lg M}$. Observe that the representation for $v$ and all nodes below it has size $O(\ell(v) \lg \ell(v)) = O(M)$ in the worst-case. Furthermore, these items appear in consecutive order. When the algorithm considers a triplet of vertices under the depth threshold, the pager loads all data for the vertices and their descendents into cache. Until the recursion under those vertices finishes, there need not be any more page transfers. Thus, the number of page transfers required by vertices below the threshold is $O((n/\frac{M}{\lg M})^2 \cdot \frac{M}{B}) = O(n^2 \frac{\lg^2 M}{MB})$. This is because the triples that we consider are only those which could contain a sum (due to the recursion pruning), and there are $O((n/\frac{M}{\lg M})^2)$ such triplets.

For a vertex $u$ above the threshold, we only need to worry about the elements which appear for the first time in $S(u)$. If we define each vertex at the threshold depth to be a bucket, in the sense of Lemma 4, these are elements which overflow their buckets. Thus, we expect $O(n/\frac{M}{\lg M})$ such elements. An element which is first represented at level $i$ is considered at most twice in conjunction with any vertex on level $i$. Each time, a linear scan is made through the set of such a vertex; the cost is one plus the number of full pages read. Then, handling such an element requires $O(\frac{n}{B} + n\frac{M}{\lg M})$. The first term comes from summing the full pages, and the second comes from the startup cost of one for each vertex on level $i$. In total for all elements that appear above the threshold, we spend time $O(n^2 \cdot \frac{\lg M}{M}(\frac{\lg M}{M} + \frac{1}{B}))$. Under the standard tall-cache assumption, $\frac{\lg M}{M} = o(\frac{1}{B})$, and we obtain the stated time bound.    □

## 4    Further Results

*The Rational Case.* Up to now, we have only discussed the integer case, but the rational case can be solved by reduction. Let $d$ be the least common denominator of all fractions. We replace each fraction $\frac{a}{b}$ by $adb^{-1}$, which is an integer. The new problem is equivalent to the old one, but the integers can be rather large (up to $2^{wn}$). To avoid this, we take all integers modulo a random prime between 3 and $O(wn^c \lg(wn))$, for some constant $c$. Consider a triplet of integers $x, y, z$. If $x + y = z$, certainly $x + y = z \pmod p$. If $x + y - c \ne 0$, the Chinese remainder theorem guarantees that $x + y - z = 0 \pmod{p}$ for less than $wn$ primes. By the density of primes, there are $\Theta(wn^c)$ primes in our range, so with high probability a non-solution does not become a solution modulo $p$. By a union bound over all triples, the prime does not introduce any false positive with high probability. Note that our primes have $O(\lg w + \lg n) = O(w)$ bits, so the problem becomes solvable by the integer algorithm. We also need to check that a reported solution really is a solution, but the time spent verifying false positives is $o(1)$. Also note that in $a_1 d b_1^{-1} + a_2 d b_2^{-1} = a_3 d b_3^{-1} \pmod{p}$ we can eliminate $d$, because $d \neq 0$, so we don't actually need to compute the least common denominator. We simply replace $\frac{a}{b}$ with $ab^{-1} \pmod{p}$. Inverses modulo $p$ can be computed in $O(\text{poly}(\lg n, \lg w))$ time.

*Bounds with High Probability.* We describe a general paradigm for making our bounds hold with high probability, for reasonable $w, M, B$ (when the overall time bounds are not too close to linear). The higher level of hashing is easy to handle: repeat choosing hash functions until the total number of elements overflowing their buckets is only twice the expectation. Rehashing takes roughly linear time, so repeating this step $O(\lg n)$ times (which is sufficient with high probability) is not significant. For the RAM, we have two more places where randomness is used. First, all $\frac{n^2}{m^2}$ pairs of buckets are checked by packing hash values in a word. The trouble is that these pairs are not independent, because the same hash function is used. However, after going through $n^\varepsilon$ buckets in the outer loop, we choose a different hash function and reconstruct the word-packed representation of all buckets. Reconstruction takes linear time, so this is usually a lower order term. However, the total running time is now the sum of $n^{1-\varepsilon}/m$ independent components, and we achieve a with high probability guarantee by a Chernoff bound. Note that each term has a strict upper bound of $\text{poly}(m)$ times its mean, so we need $n^{1-\varepsilon}/m$ to be bigger than a polynomial in $m$ to cover the possible variance of the terms. We also have expected time bounds in searching for a given sum. Since we perform $O(\frac{n}{m})$ queries, we can reuse the same trick: reconstruct the word-packed representations every $n^\varepsilon$ queries, giving $n^\varepsilon$ independent terms.

*Approximate Linear Satisfaction.* We can also approximate the minimum triplet sum in absolute value (i.e., $\min_{a,b,c} |a + b + c|$), within a $1 + \varepsilon$ factor. The running time increases by $O(\lg w)$ for any constant $\varepsilon$ (also, the dependence on $\frac{1}{\varepsilon}$ can be made logarithmic). We consider a threshold 3SUM problem: for a given $T$, the problem is to find a triplet with sum in the interval $[-T, T]$, or report that none exists. It is straightforward to verify that all algorithms from above can solve this more general problem in the same time bounds, when $T = O(1)$.

To approximate the minimum sum, we first test whether the sum is zero (classic 3SUM). If not, we want to find some $e \in [0, w/\lg(1 + \varepsilon)]$ such that there is a sum with absolute value less than $(1 + \varepsilon)^e$, and none with absolute value less than $(1 + \varepsilon)^{e-1}$. We show how to perform a binary search for $e$, which adds an $O(\lg w)$ factor to the running time. We must be able to discern the case when the minimum sum is at least $(1 + \varepsilon)L$ (for some $L$) versus when it is at most $L$. We divide all numbers by $(\varepsilon L)/6$ and take the floor. Then we use threshold 3SUM with $T = 6/\varepsilon + 3$. Suppose a triplet $x, y, z$ has sum in $[-L, L]$. Then the modified triplet has a sum whose absolute value is bounded by $L(6/(\varepsilon L)) + 3$; the term of 3 comes from a possible deviation by at most 1 for every floor operation. But this bound is exactly $T$, so we always find a good triplet. Suppose on the other hand that all triplets have a sum whose absolute value is at least $(1 + \varepsilon)L$. Then the modified triplets have sums with absolute values larger than $(1 + \varepsilon)L(6/(\varepsilon L)) - 3 > T$, so we never report a bad triplet. Note that this algorithm also works in the rational case, because we take a floor at an early stage; the only difference is that we must also allow $e$ to be negative, up to $-w/\lg(1 + \varepsilon)$.

## 5    Conclusions

An interesting open problem is whether it is possible to obtain a subquadratic algorithm for finding collinear triples of points in the plane (the original motivation for studying 3SUM). This problem seems hard for a fundamental reason: there are no results on using bit tricks in relation to slopes. For example, the best bound for static planar point location is still the comparison-based $O(\lg n)$, despite attempts by several researchers.

Another interesting question is what improvements are possible for the $r$-SUM problem. There, the classic bounds are $\widetilde{O}(n^{\lceil r/2 \rceil})$ by a meet-in-the-middle attack. Unfortunately, our techniques don't mix well with this strategy, and we cannot achieve a speedup that grows with $r$.

*Acknowledgments.* We thank Jeff Erickson and Rasmus Pagh for helpful discussions.

## References

1. Albers, S., Hagerup, T.: Improved parallel integer sorting without concurrent writing. In: Proc. 3rd ACM/SIAM Symposium on Discrete Algorithms (SODA). (1992) 463–472.
2. Andersson, A., Miltersen, P.B., Riis, S., Thorup, M.: Static dictionaries on $AC^0$ RAMs: Query time $\Theta(\sqrt{\log n/ \log \log n})$ is necessary and sufficient. In: Proc. 37th IEEE Symposium on Foundations of Computer Science (FOCS). (1996) 441–450.
3. Andersson, A., Miltersen, P.B., Thorup, M.: Fusion trees can be implemented with $AC^0$ instructions only. Theoretical Computer Science **215** (1999) 337–344.
4. Cormen, T.H., Leiserson, C.E., Rivest, R.L., Stein, C.: Introduction to Algorithms. 2nd edn. MIT Press and McGraw-Hill (2001).
5. Demaine, E.D.: Cache-oblivious algorithms and data structures. In: Lecture Notes from the EEF Summer School on Massive Data Sets, BRICS, University of Aarhus, Denmark. LNCS (2002) 39–46.
6. Dietzfelbinger, M.: Universal hashing and $k$-wise independent random variables via integer arithmetic without primes. In: Proc. 13th Symposium on Theoretical Aspects of Computer Science (STACS). (1996) 569–580.

7. Erickson, J.: Bounds for linear satisfiability problems. Chicago Journal of Theoretical Computer Science, 1999(8).

8. Gajentaan, A., Overmars, M.H.: On a class of $O(n^2)$ problems in computational geometry. Computational Geometry: Theory and Applications **5** (1995) 165–185.

9. Thorup, M.: Randomized sorting in $O(n \log \log n)$ time and linear space using addition, shift, and bit-wise boolean operations. Journal of Algorithms **42** (2002) 205–230. See also SODA'97.

# Near-Optimal Pricing in Near-Linear Time[*]

Jason D. Hartline[1] and Vladlen Koltun[2]

[1] Microsoft Research, Mountain View, CA, 94043
`hartline@microsoft.com`
[2] Computer Science Division, University of California, Berkeley, CA 94720-1776
`vladlen@cs.berkeley.edu`

**Abstract.** We present efficient approximation algorithms for a number of problems that call for computing the prices that maximize the revenue of the seller on a set of items. Algorithms for such problems enable the design of auctions and related pricing mechanisms [3]. In light of the fact that the problems we address are APX-hard in general [5], we design near-linear and near-cubic time approximation schemes under the assumption that the number of distinct items for sale is constant.

## 1   Introduction

Imagine a software provider that is about to release a new product in *student,* *standard,* and *professional* editions. How should the editions be priced? Naturally, high prices lead to more revenue per sale, whereas lower prices lead to more sales. There is also interplay between the prices on the different versions, e.g., a consumer willing to pay a high price for the professional version might be lured away by a bargain price on the standard version. *Given consumer preferences over a set of items for sale, how can the items be priced to give the optimal profit?*

Consumer preferences are rich in combinatorial structure and this leads to much of the difficulty in price optimization. General *combinatorial* preferences might have aspects of both *substitutability* and *complementarity*. As an example, a vacationer might like either a plane ticket to Hawaii together with a beachfront hotel, or a ticket to Paris with a hotel on Ile St. Louis. We address the pure complements case (i.e., *single-minded* consumers that want a set bundle of products for a specific price) and the pure substitutes case (i.e., *unit-demand* consumers that want exactly one unit of any of the offered products and have different valuations on the products). In the above software pricing scenario, it is natural to assume that consumers are unit-demand. See the following sections for definitions.

These variants of the pricing problem are considered by Guruswami et al. [5] who show that even in simple special cases, these pricing problems are APX-hard.

---

[*] Work on this paper by the second author has been supported by NSF Grant CCR-01-21555.

F. Dehne, A. López-Ortiz, and J.-R. Sack (Eds.): WADS 2005, LNCS 3608, pp. 422–431, 2005.

(That is, there is no polynomial time approximation scheme given standard complexity assumptions.) They give a logarithmic factor approximation algorithm and leave the problem of obtaining a better approximation factor open. Unfortunately, the logarithmic approximation does not yield any insight on questions like the pricing problem faced by the software company in our opening example; it would recommend selling all editions of the software at the same, albeit optimally chosen, price. Obviously, this defeats the purpose of making different editions of the software and is thus inadequate.

In this paper we consider the above pricing problems under the assumption that the number of distinct items for sale is constant. As is illustrated by the software pricing example, this assumption is pertinent to many real-life scenarios. A brute-force exact algorithm exists, but is exponential in the number of items for sale and generally impractical. We present approximation schemes with significantly superior running times. As the APX-hardness of the problems indicates, simplifying assumptions such as this are necessary for the achievement of such results.

In the *unlimited supply* case in which the seller is able to sell any number of units of each item, we give near-linear time approximation schemes for both the pure substitutes problem and the pure complements problem. Specifically, for unit-demand consumers a $(1 + \varepsilon)$-approximation is achieved in time $O\left(n \log\left(\frac{1}{\varepsilon} \log \frac{n}{\varepsilon}\right) + \left(\frac{1}{\varepsilon} \log \frac{n}{\varepsilon}\right)^{m(m+1)}\right)$ for $n$ consumers, $m = O(1)$ items, and an arbitrarily small $\varepsilon > 0$. For single-minded consumers a $(1 + \varepsilon)$-approximation is achieved in time $O\left(\left(n + \left(\frac{1}{\varepsilon} \log \frac{n}{\varepsilon}\right)^m\right) \log\left(\frac{1}{\varepsilon} \log \frac{n}{\varepsilon}\right)\right)$.

The more general *limited supply* case requires more care. In particular, we demand that the computed prices satisfy an explicit fairness criterion called *envy-freedom* [5] that makes sure that the prices are such that no item is oversold.[1] For unit-demand consumers we give a $(1 + \varepsilon)$-approximation algorithm for the limited supply (envy-free) pricing problem that runs in time $O(n^3 \log_{1+\varepsilon}^m n)$.

Part of the motivation for considering these pricing problems comes from auction mechanism design problems. In mechanism design it is pointedly assumed that the seller does not know the consumer preferences in advance. Instead, an auction must compute payments that encourage the consumers to reveal their preferences. Intuitively, however, understanding how to optimally price items given *known* preferences is necessary for the more difficult problem of running an auction when the preferences are unknown in advance. For the unlimited supply unit-demand scenario, Goldberg and Hartline [3] made this connection concrete by giving a reduction from the game theoretic auction design problem to the algorithmic price optimization problem. They left the problem of computing optimal or approximately optimal prices from known consumer preferences in polynomial time open. This is one of the questions addressed by our work.

---

[1] Envy-freedom is implicit in the definition of unlimited supply pricing problem since we require that the identical units of each item be sold at the same price and that each consumer pick their desired items after the prices are set.

This paper is organized as follows. In Sections 2 and 3 we describe the unlimited supply algorithms for the pure substitutes and pure complements problems, respectively. In Section 4 we formally define envy-freedom, give background material, and present the approximation algorithm for the limited supply pure substitutes problem. We conclude in Section 5 with a discussion of the difficulty in generalizing our approach to combinatorial preferences that contain both substitutes and complements.

# 2     Unlimited Supply, Unit-Demand Consumers

We assume that the seller has $m$ distinct items for sale, each available in *unlimited supply*. There are $n$ consumers each of whom wishes to purchase at most one item (i.e., there is *unit-demand*). We define a consumer's *valuation* for an item as the value assigned by the consumer to obtaining one unit of the item. For consumer $i$ and item $j$, let $v_{ij}$ denote this valuation. Given a price $p_j$ for item $j$, consumer $i$'s *utility* for this item is $u_{ij} = v_{ij} - p_j$. We assume that a consumer's only goal is to maximize this utility. Therefore, given a pricing of all items, $\mathbf{p} = (p_1, \ldots, p_m)$, consumer $i$ will choose one of the items $j$ such that $v_{ij} - p_j \geq v_{ij'} - p_{j'}$ for all $j' \neq j$, or no item if $v_{ij} < p_j$ for all $j$.

The pricing problem we address asks for computing, given the consumer valuations, a pricing that is *seller optimal*, i.e., the price vector that maximizes the sum of the prices of all the units of items sold, also called the seller's *profit*. For a given price vector $\mathbf{p}$, let $\text{Profit}_{\mathbf{p}}$ be the profit obtained from $\mathbf{p}$. Let $\bar{\mathbf{p}}$ be the price vector with the maximum profit and set $\text{OPT} = \text{Profit}_{\bar{\mathbf{p}}}$. The following assumption is natural in many context (See Section 4) and insures that $\text{Profit}_{\mathbf{p}}$ is well defined: a consumer indifferent between several items will choose following the discression of the pricing algorithm. For our goal of unlimited supply profit maximization we assume that indifferent consumers consumers choose the item with the higher price.

This problem was first posed in [3]. Both Guruswami et al. [5] and Aggarwal et al. [1] give logarithmic approximations and APX-hardness proofs that hold even for the special case where $v_{ij} \in \{0, 1, 2\}$. For the case where the number of distinct items for sale is constant we get the following result.

**Theorem 1.** *For any $\varepsilon < 1$, an envy-free pricing that gives profit at least* $\text{OPT}/(1 + \varepsilon)$ *can be computed in time* $O(n \log(Mm^2) + m^{2m+1} M^{m+1})$, *where* $M = O(m^m \log_{1+\varepsilon}^m \frac{n}{\varepsilon}))$. *For constant $m$, the running time is*

$$O\left(n \log\left(\frac{1}{\varepsilon} \log \frac{n}{\varepsilon}\right) + \left(\frac{1}{\varepsilon} \log \frac{n}{\varepsilon}\right)^{m(m+1)}\right).$$

In brief synopsis, the analysis proceeds as follows. We first show that there is an approximately optimal price vector $\tilde{\mathbf{p}}$ that satisfies $\text{Profit}_{\tilde{\mathbf{p}}} \geq (1 - \delta) \text{OPT}$ with the property that for all $j$, $\tilde{p}_j \in [\frac{\delta h}{n}, h]$ where $h = \max_{i,j} v_{ij}$ is the highest valuation of any user for any item. We then show that we can cover the space of price vectors $[\frac{\delta h}{n}, h]^m$ with a "small" set of price vectors $\mathcal{W}$, such that there exists $\mathbf{p} \in \mathcal{W}$ with $\mathrm{Profit}_{\mathbf{p}} \geq \mathrm{Profit}_{\tilde{\mathbf{p}}} /(1 + \delta)$. A brute-force search over $\mathcal{W}$ can find the optimal such vector $\mathbf{p} \in \mathcal{W}$ in time $O(nm\,|\mathcal{W}|)$ which is $O(n \log n)$ for constant $\varepsilon$ and $m$. A more clever search yields the stated runtime $O(n \log \log n)$. From a practical standpoint the brute-force approach may be more desirable due to its great simplicity of implementation.

**Lemma 1.** *There exists $\tilde{\mathbf{p}}$ with $\tilde{p}_j \in [\frac{\delta h}{n}, h]^m$ for all $j$ and $\mathrm{Profit}_{\tilde{\mathbf{p}}} \geq (1-\delta)\,\mathrm{OPT}$.*

*Proof.* First, we can assume that $\bar{\mathbf{p}}$ satisfies $\bar{p}_j \leq h$ as setting a price above the highest valuation cannot increase the profit. Indeed, the revenue from an item priced this way can only be zero. Consider $\tilde{\mathbf{p}}$ with $\tilde{p}_j = \max(\bar{p}_j, \frac{\delta h}{n})$. Let $J'$ be the set of items with $\tilde{p}_j = \bar{p}_j$ (all other items have price $\tilde{p}_j = \frac{\delta h}{n} > \bar{p}_j$). Any consumer that prefers an item from $J'$ under pricing $\bar{\mathbf{p}}$ prefers the same item under $\tilde{\mathbf{p}}$. This is because we kept the price of this preferred item fixed and only raised prices of other items. On the other hand, the total profit from consumers that preferred items with $\bar{p}_j < \frac{\delta h}{n}$ is at most $\delta h \leq \delta\,\mathrm{OPT}$. (Here we note that $\mathrm{OPT} \geq h$, since for $h = v_{ij}$, setting the price of all items to $h$ ensures that consumer $i$ will purchase item $j$, yielding profit at least $h$.) Even if we assume no revenue from these consumers under $\tilde{\mathbf{p}}$, the profit is still $\mathrm{Profit}_{\tilde{\mathbf{p}}} \geq (1 - \delta)\,\mathrm{OPT}$. $\qquad\square$

We define below a grid of prices $\mathcal{W}$ parameterized by $\delta > 0$ to fill a region slightly larger than $[\frac{\delta h}{n}, h]^m$ such that there is a grid point $\mathbf{p} \in \mathcal{W}$ that gives a profit close to the optimal price vector $\tilde{\mathbf{p}}$ for the region $[\frac{\delta h}{n}, h]^m$. For integer $0 \leq i < \log_{1+\delta} \frac{n}{\delta}$ and $0 \leq k < (2+\delta)m$, let $Z$ be the $\lceil \log_{1+\delta} \frac{n}{\delta} \rceil$ values of $Z_i = \frac{\delta h}{n}(1 + \delta)^i$ on the interval $[\frac{\delta h}{n}, h)$ and let $W_i$ be the $\lceil (2+\delta)m \rceil$ values of the form $Z_{i-1} + Z_{i-1}\frac{\delta k}{m}$ on the interval $[Z_{i-1}, Z_{i+1})$. Let $W = \bigcup_i W_i$ for $i$ with $Z_i \in Z$. Define the sets $\mathcal{W} = W^m$ and $\mathcal{Z} = Z^m$. Let $M = |\mathcal{W}| = \left(\lceil (2+\delta)m \rceil \lceil \log_{1+\delta} \frac{n}{\delta} \rceil\right)^m$.

**Lemma 2.** *For any $\tilde{\mathbf{p}} \in [\frac{\delta h}{n}, h]^m$, there exists $\mathbf{p} \in \mathcal{W}$ such that $\mathrm{Profit}_{\mathbf{p}} \geq \mathrm{Profit}_{\tilde{\mathbf{p}}} /(1 + \delta)$.*

*Proof.* Reindex the items so that $\tilde{p}_j \leq \tilde{p}_{j+1}$ for all $j$. For each $j$, let $Z_{i_j}$ (resp., $w_j$) be the price obtained by rounding $\tilde{p}_j$ down to the nearest value in $Z$ (resp., in $W_{i_j}$). Set $d_j = \frac{\delta}{m} Z_{i_j}$ and consider the price vector $\mathbf{p}$ defined with $p_j = w_j - jd_j \in W_{i_j}$. We show that $\mathbf{p} \in \mathcal{W}$ satisfies the statement of the lemma.

We claim that no consumer that prefers an item $j$ under $\tilde{\mathbf{p}}$ would prefer an item $j' < j$ under pricing $\mathbf{p}$ (i.e., one with a lesser price). In going from prices $\tilde{\mathbf{p}}$ to $\mathbf{p}$, the increase $\tilde{p}_j - p_j$ in a consumer's utility for item $j$, is higher than the increase $\tilde{p}_{j'} - p_{j'}$ for item $j'$. Given $j' + 1 \leq j$ we have:

$$\tilde{p}_j \geq w_j \qquad\qquad \tilde{p}_{j'} < w_{j'} + d_{j'}$$
$$\tilde{p}_j - p_j \geq jd_j \qquad\qquad \tilde{p}_{j'} - p_{j'} < (j' + 1)d_{j'}.$$

Since $d_j \geq d_{j'}$ and therefore $jd_j \geq (j' + 1)d_{j'}$, the desired inequality is established.

Since $p_j \geq \tilde{p}_j/(1+\delta)$ and no consumer changes preference to a cheaper item under $\mathbf{p}$, we have $\text{Profit}_{\mathbf{p}} \geq \text{Profit}_{\tilde{\mathbf{p}}}/(1+\delta)$.                            □

Combining Lemmas 1 and 2 and taking $\varepsilon = \Theta(\delta)$, we conclude that there exists $\mathbf{p} \in W$ such that $\text{Profit}_{\mathbf{p}} \geq \text{Profit}_{\tilde{\mathbf{p}}}/(1+\varepsilon)$.

It remains to show that the optimal price vector from $W$ can be found efficiently. The trivial algorithm enumerates all $\mathbf{p} \in W$ and computes the profit of each, which takes time $O(Mmn)$. Below is an improved algorithm that utilizes the assumption that $n \gg M$, which suggests the possibility of preprocessing the consumer valuations into a data structure that can be queried at each $\mathbf{p} \in W$.

A given price vector $\mathbf{p}$ divides $\mathbb{R}^m$ (which we view as the space of consumer valuation vectors) into $m+1$ regions, each corresponding to how a consumer valuation relates to the given prices. Given $\mathbf{p}$, each consumer prefers one of $m$ items or none. The boundaries of all the regions are delimited by $O(m^2)$ hyperplanes. For all $M$ price vectors in $W$ we have a total of $O(Mm^2)$ hyperplanes. We use the fact that an arrangement of $K$ hyperplanes in $\mathbb{R}^m$ can be preprocessed in time $O(K^m/\log^m K)$, such that a point location query can be performed in time $O(\log K)$ [2].

### Definition 1 (Unlimited Supply Pricing Algorithm).

1. *Preprocess the hyperplane arrangement defined by the $M$ price points into a point location data structure. Associate a counter with each arrangement cell, initially set to zero. Analysis: $O((Mm^2)^m/\log^m(Mm^2))$ time.*
2. *For each consumer, query the point location data structure with the consumer's valuation vector. Increment the counter associated with the located cell. Analysis: $O(n\log(Mm^2))$ time.*
3. *For each $\mathbf{p} \in W$, iterate over all arrangement cells. In each cell, multiply the value of the associated counter by the price yielded by the preferred item for that cell (if any), and add this to the profit associated with $\mathbf{p}$. Analysis: $O(Mm(Mm^2)^m)$ time.*
4. *Output $\mathbf{p} \in W$ with highest profit.*

The total runtime of the unlimited supply pricing algorithm is $O(n\log(Mm^2) + Mm(Mm^2)^m)$. Taking $m$ to be a constant, $M = O(\log_{1+\varepsilon}^m \frac{n}{\varepsilon})$ and we have a total runtime of $O(n\log\left(\frac{1}{\varepsilon}\log\frac{n}{\varepsilon}\right) + \left(\frac{1}{\varepsilon}\log\frac{n}{\varepsilon}\right)^{m(m+1)})$. This concludes the proof of Theorem 1.

## 3   Unlimited Supply, Single-Minded Consumers

As in the unit-demand case of the preceding section, we assume that the seller has $m$ distinct items available for sale, each in *unlimited supply*. Each of the $n$ consumers is interested in purchasing a particular bundle $S_i \subseteq \{1,\ldots,m\}$ of items. Denote consumer $i$'s valuation for their desired bundle by $v_i$ and let $h = \max_i v_i$. We assume that if the total cost of the items in $S_i$ is at most $v_i$, the consumer will purchase all of $S_i$, otherwise the consumer will purchase nothing.

Given the valuations $v_i$, we wish to compute a pricing that maximizes the seller's profit. Define $\text{Profit}_{\mathbf{p}}$, $\bar{\mathbf{p}}$, and OPT as in the previous section.

Guruswami et al. [5] give an algorithm for computing a logarithmic approximations to OPT and an APX-hardness proof that hold evens for the special case where $v_i = 1$ and $|S_i| \leq 2$. For the case where the number of distinct items for sale is constant we get the following result.

**Theorem 2.** *For any $\varepsilon < 1$, a pricing that gives profit at least $\text{OPT}/(1 + \varepsilon)$ can be computed in time $O((n + 2^m M) \log M)$, where $M = O(\log_{1+\varepsilon}^m \frac{n}{\varepsilon}))$. For constant $m$, the running time is*

$$O\left(\left(n + \left(\frac{1}{\varepsilon} \log \frac{n}{\varepsilon}\right)^m\right) \log \left(\frac{1}{\varepsilon} \log \frac{n}{\varepsilon}\right)\right).$$

**Lemma 3.** *There exists $\tilde{\mathbf{p}}$ with $\tilde{p}_j \in \{0 \cup [\frac{\delta h}{nm}, h]\}^m$ for all $j$ and $\text{Profit}_{\tilde{\mathbf{p}}} \geq (1 - \delta)\text{OPT}$.*

*Proof.* As in Lemma 1, we can assume that $\bar{\mathbf{p}}$ satisfies $\bar{p}_j \leq h$. Consider $\tilde{\mathbf{p}}$ with $\tilde{p}_j = 0$ if $\bar{p}_j \leq \frac{\delta h}{nm}$ and $\tilde{p}_j = \bar{p}_j$ otherwise. It is clear that the overall profit decreases by at most $\delta h$ when prices shift from $\bar{\mathbf{p}}$ to $\tilde{\mathbf{p}}$. It is easy to see that $\text{OPT} \geq h$ and thus $\text{Profit}_{\tilde{\mathbf{p}}} \geq (1 - \delta)\text{OPT}$.                    □

For integer $0 \leq i < \lceil \log_{1+\delta} \frac{nm}{\delta} \rceil$, define $Z$ to be the $\lceil \log_{1+\delta} \frac{nm}{\delta} \rceil$ values of $Z_i = \frac{\delta h}{nm}(1 + \delta)^i$ on the interval $[\frac{\delta h}{nm}, h)$, augmented by the value 0. Define $\mathcal{Z} = Z^m$. Let $M = |\mathcal{Z}| = O(\log_{1+\delta}^m(\frac{nm}{\delta}))$.

**Lemma 4.** *For any $\tilde{\mathbf{p}} \in [\frac{\delta h}{nm}, h]^m$, there exists $\mathbf{p} \in \mathcal{Z}$ such that $\text{Profit}_{\mathbf{p}} \geq \text{Profit}_{\tilde{\mathbf{p}}}/(1 + \delta)$.*

*Proof.* Let $\mathbf{p} \in \mathcal{Z}$ be the price vector obtained by taking the coordinates of $\tilde{\mathbf{p}}$ and rounding each of them down to the nearest value in $Z$. Since the price of any bundle under $\mathbf{p}$ is at least the price of this bundle under $\tilde{\mathbf{p}}$ divided by $1 + \delta$, we have $\text{Profit}_{\mathbf{p}} \geq \text{Profit}_{\tilde{\mathbf{p}}}/(1 + \delta)$.                    □

We conclude that by setting $\delta = \Theta(\varepsilon)$, there exists $\mathbf{p} \in \mathcal{Z}$ such that $\text{Profit}_{\mathbf{p}} \geq \text{Profit}_{\bar{\mathbf{p}}}/(1+\varepsilon)$. The following algorithm computes the optimal price vector from $\mathcal{Z}$.

**Definition 2 (Single-Minded Consumers Pricing Algorithm).**

1. *Compute the profit for each bundle and price vector, $(S, \mathbf{p}) \in 2^{\{1,\dots,m\}} \times \mathcal{Z}$:*
   (a) *Compute total cost of $S$ given $\mathbf{p}$: $\sum_{j \in S} p_j$.*
   (b) *For each $S$ sort $\mathbf{p} \in \mathcal{Z}$ by total cost. Associate a counter for each price vector in this sorted array.*
   (c) *For each consumer $i$, consider the sorted array associated with $S_i$ and increment the counter on the price vector in this array that has the highest total cost that is at most $v_i$.*

(d) The seller's profit for $S$ and $\mathbf{p}$ is the sum of the counters for price vectors with total cost at least that of $\mathbf{p}$ on $S$.

Analysis: $O(n \log M + 2^m M \log M)$

2. For each $\mathbf{p} \in \mathcal{Z}$, sum profits from all bundles. Analysis: $O(2^m M)$
3. Output optimal $\mathbf{p} \in \mathcal{Z}$.

The total runtime of the unlimited supply pricing algorithm is $O((n+2^m M) \log M)$. For $m = O(1)$, the running time is $O\left(\left(n + \left(\frac{1}{\varepsilon} \log \frac{n}{\varepsilon}\right)^m\right) \log \left(\frac{1}{\varepsilon} \log \frac{n}{\varepsilon}\right)\right)$. This concludes the proof of Theorem 2.

# 4   Limited Supply, Unit-Demand Consumers

We now consider the limited supply case of the unit-demand pricing problem. In this case the seller can only offer a limited number of units of each of the $m$ items available for sale. For limited supply pricing problems not all pricings lead to well defined outcomes. For example, an item might be priced so low that the demand for the item exceeds the supply. To avoid this problem, and to make our pricing algorithms useful in auction design problems, we restrict our attention to computing prices that are *envy-free*. We review envy-freedom as well as some concepts from Economics literature below.

## 4.1   Notation and Background

Index the items $1, \ldots, m$ and let $J$ represent the multiset of items the seller is offering. There are $n$ consumers each of whom wishes to purchase at most one item. Define a consumer's valuation for an item as the value assigned by the consumer to obtaining one unit of the item. For consumer $i$ and item $j$, let $v_{ij}$ denote this valuation and $V$ the matrix of consumer valuations. Given a price $p_j$ for item $j$, consumer $i$'s utility for this item (at its price) is $u_{ij} = v_{ij} - p_j$. Given a pricing of all items, $\mathbf{p} = (p_1, \ldots, p_m)$, consumer $i$ will choose one of the items that maximize their utility, i.e., an item $j$ such that $v_{ij} - p_j \geq v_{ij'} - p_{j'}$ for all $j' \neq j$, or no item if $v_{ij} < p_j$ for all $j$. We say that a consumer is *happy* with an item they would choose in this way. Under a particular pricing, we refer to the set of items that a consumer is happy with as the *demand set*. A consumer's utility under a pricing, denoted $u_i$ for consumer $i$, is the utility they would obtain for any item in this set.

A pricing $\mathbf{p}$ is *envy-free* if there is an assignment of items to consumers such that (a) no item is oversold, and (b) all consumers with positive utility under the pricing obtain an item in their demand set. Given the demand sets, computing such an assignment or determining that none exists is a simple bipartite matching problem.

The pricing problem we address is that of computing, when given the consumer valuations, the envy-free pricing that is seller optimal. This is the price vector that maximizes the seller's profit (i.e., the sum of the prices of all the units of items sold) subject to the condition that the pricing is envy-free.

Envy-free pricing for unit-demand problems is closely related to *Walrasian equilibria* [6]. When restricted to unit-demand problems Walrasian equilibria are precisely those pricings that are both envy-free and satisfy the additional condition that unsold items have price zero. Let $\mathrm{MWM}_V(J)$ denote the value of the maximum weighted matching on the bipartite graph induced by the consumers' valuations $V$ on item multiset $J$. Gul and Stacchetti [4] give the following algorithm for computing the Walrasian equilibrium that gives the seller the maximum profit.

**Definition 3 (Maximum Walrasian Prices Algorithm, MaxWalEq).** *Sell item $j$ at price $p_j = \mathrm{MWM}_V(J) - \mathrm{MWM}_V(J \setminus \{j\})$.*

It is easy to see that MaxWalEq can be computed in time $O(m^3n^3)$, since the initial maximum weighted matching solution $\mathrm{MWM}_V(J)$ can be computed in time $O(m^3n^3)$ and then $\mathrm{MWM}_V(J \setminus \{j\})$ can be computed for each $j \in \{1,\ldots,m\}$ that is matched in time $O(m^2n^2)$.

The difference between the seller optimal Walresian equilibrium (as computed by MaxWalEq) and the seller optimal envy-free prices is that in the former all unsold items must have price zero. This constraint results in a total seller profit for MaxWalEq that in general is lower than that obtained in the optimal envy-free pricing. In the special case that the optimal envy-free pricing happens to sell all available items then it gives a Walrasian equilibrum (since all unsold items have price zero). Thus, under the assumption that all items are sold in the optimal envy-free pricing, MaxWalEq computes this optimal pricing. We use this fact below to search for the optimal subset of items to sell, $J' \subseteq J$, using MaxWalEq to obtain prices for each subset.

## 4.2    The Algorithm

The following algorithm gives an exact optimal envy free pricing for the limited supply case. It runs in cubic time assuming that the total number of units for sale, $m'$, is constant.

**Definition 4 (Limited Supply Pricing Algorithm).** *For each subset $J'$ of the $J$ items for sale, compute MaxWalEq($J'$). Output the prices that give the highest profit.*

In the case that the number of units $m'$ for sale is constant, then we make $2^{m'}$ calls to MaxWalEq giving a total runtime of $O(m^3n^32^{m'})$, which is cubic for constant $m'$.

We next give give a simple algorithm for computing approximately optimal within a factor of $(1+\varepsilon)$ envy-free prices when the number of distinct items for sale is a constant (but with an arbitrary number of units of each item). This algorithm runs in time $O(m^3n^3 \log^m_{1+\varepsilon} n)$. First a lemma.

**Lemma 5.** *Consider multisets $J$ and $J'$, such that $J' \subset J$. For item $j \in J'$, let $p_j$ (resp., $p'_j$) be the price for $j$ in MaxWalEq($J$) (resp., in MaxWalEq($J'$)). Then $p'_j \geq p_j$.*

*Proof.* Given $J' \subset J$, we are arguing that $\mathrm{MWM}_V(J) - \mathrm{MWM}_V(J \setminus \{j\}) \leq \mathrm{MWM}_V(J') - \mathrm{MWM}_V(J' \setminus \{j\})$ for all $j \in J'$. We note in passing that this is equivalent to showing that $\mathrm{MWM}_V(\cdot)$ is a submodular function. We rearrange the statement to show that $\mathrm{MWM}_V(J) + \mathrm{MWM}_V(J' \setminus \{j\}) \leq \mathrm{MWM}_V(J') + \mathrm{MWM}_V(J \setminus \{j\})$. Letting $A = J \setminus \{j\}$ and $B = J'$ we have $A \cap B = J' \setminus \{j\}$ and $A \cup B = J$ making our goal to prove that $\mathrm{MWM}_V(A \cap B) + \mathrm{MWM}_V(A \cup B) \leq \mathrm{MWM}_V(A) + \mathrm{MWM}_V(B)$, the familiar definition of submodularity. We will show this for arbitrary sets $A$ and $B$. We proceed by showing that given the matched edges of $\mathrm{MWM}_V(A \cap B)$ and $\mathrm{MWM}_V(A \cup B)$ we can assign them to either matchings of $A$ or $B$. Of course $\mathrm{MWM}_V(A)$ will be at least the sum of the weights of edges assigned to the matching of $A$ and $\mathrm{MWM}_V(B)$ at least that for $B$ proving the result.

Consider putting the matchings $\mathrm{MWM}_V(A \cap B)$ and $\mathrm{MWM}_V(A \cup B)$ together; we get single edges, double edges, cycles, and paths. We now show how to use all the edges that make up these components to construct a matching of $A$ and a matching of $B$. Single edges incident on $A \setminus B$ go in the matching of $A$, single edges incident on $B \setminus A$ go in the matching of $B$, and double edges incident on $A \cap B$ go one to each of $A$ and $B$. A cycle can only have edges incident upon $\mathrm{MWM}_V(A \cap B)$; we assign its odd edges to the matching of $A$ and even edges to the matching of $B$ (say). As for paths, note that a path has every other edge from $\mathrm{MWM}_V(A \cap B)$ and $\mathrm{MWM}_V(A \cup B)$. Thus a path cannot have both endpoints at vertices representing items in $A \otimes B$ as it would then have even length requiring one of the endpoints to have an incident edge from $\mathrm{MWM}_V(A \cap B)$, a contradiction. Thus we can simply assign every other edge in each path to the matching of $A$ or $B$ making sure that if the path has an end point in $A \setminus B$ (resp., $B \setminus A$) then the incident edge on this vertex is assigned to the matching of $A$ (resp., $B$).                                                 □

**Definition 5 (General Approximate Pricing Algorithm, GAPA).** *For each subset $J'$ of the $J$ items for sale with the property that the multiplicity of each distinct item is either 0 or a power of $(1 + \varepsilon)$ (rounded down), compute* MaxWalEq$(J')$. *Output the prices that give the highest profit.*

We need only consider having at most $n$ copies of each item. There are at most $\lfloor \log_{1+\varepsilon} n \rfloor + 1$ multiplicities as described in the definition. Considering all possible combinations of these multiplicities across the distinct items gives at most $(1 + \log_{1+\varepsilon} n)^m$ calls to MaxWalEq. This gives a total runtime of $O(n^3 \log_{1+\varepsilon}^m n)$, which is near-cubic for a constant $m$.

We now show that this algorithm computes a $(1 + \varepsilon)$-approximation to the optimal envy-free pricing. Let $J^*$ be the set of items sold by the optimal envy-free pricing. Let $n_j$ be the number of items of type $j$ sold by $J^*$ and let $p_j$ be their prices. The optimal envy-free profit is thus $\sum_j n_j p_j$. Round the multiplicities of $J^*$ down to the nearest power of $(1 + \varepsilon)$ (or to zero) and note that GAPA gets at least the profit of MaxWalEq on this smaller set of items. Let $p'_j$ and $n'_j$ be the resulting prices and number of items sold. We have $n'_j \geq n_j/(1 + \varepsilon)$ and by Lemma 5, $p'_j > p_j$. Thus the profit of GAPA is at least $\sum_j n'_j p'_j \geq \sum_j n_j p_j/(1 + \varepsilon) = \mathrm{OPT}/(1 + \varepsilon)$.

## 5    Conclusion

In this paper we have shown that it is possible to approximate the optimal pricing in near-linear or near-cubic time for several natural pricing problems. One of the techniques we utilize is based on forming a small grid of prices that is guaranteed to contain an approximate optimum. Another technique is to guess how many of each item are sold in the optimal solution and use the seller optimal Walrasian equilibrium prices for this set of items.

It is our hope that these ideas can be extended to address the unlimited supply problem for general combinatorial consumers, described in [5]. There are several difficulties in proceeding in this direction. First, we have to show that a price vector in some smaller space, such as $(\{0\} \cup [\frac{\delta h}{n}, h])^m$ is close to optimal. This is not trivial, since when the items are complements, e.g., in the single-minded case, it is natural to round prices in $(0, \frac{\delta h}{n})$ down to zero; whereas when the items are substitutes, as in the unit-demand case, it is natural to round these prices up to $\frac{\delta h}{n}$. Second, a small grid has to be generated such that for any point in our reduced price space we can find a grid point that is almost as good. For single-minded consumers this task is simplified by the fact that each consumer only wants one particular product bundle, and rounding item prices down cannot cause them to switch to a drastically different set of products. For unit-demand consumers it is more complicated, and a locally linear grid has to be generated so that when we round down to a grid point we give up more for higher priced items, thus ensuring that a consumer only switches to a higher priced item. General combinatorial consumers, however, may want to switch from one set to a completely different set of items. These two sets may have different cardinality and consist of differently priced items. Further elaboration of our ideas seems necessary to address these challenges.

Another problem left for future work is the design of efficient algorithms for the limited supply case when the consumers are single-minded.

## References

1. G. Aggarwal, T. Feder, R. Motwani, and A. Zhu. Algorithms for multi-product pricing. In *Proc. of ICALP*, 2004.
2. B. Chazelle and J. Friedman. Point location among hyperplanes and unidirectional ray-shooting. *Comput. Geom. Theory Appl.*, 4:53–62, 1994.
3. A. Goldberg and J. Hartline. Competitive Auctions for Multiple Digital Goods. In *Proc. 9th European Symposium on Algorithms*, pages 416–427. Springer–Verlag, 2001.
4. F. Gul and E. Stacchetti. Walrasian Equilibrium with Gross Substitutes. *Journal of Economic Theory*, 87:95–124, 1999.
5. V. Guruswami, J. Hartline, A. Karlin, D. Kempe, K. Kenyon, and F. McSherry. On Profit Maximizing Envy-free Pricing. In *Proc. 16th Symp. on Discrete Algorithms*. ACM/SIAM, 2005.
6. L. Walras. *Elements of Pure Economics*. Allen and Unwin, 1954.

# Improved Approximation Bounds for Planar Point Pattern Matching*

Minkyoung Cho and David M. Mount

Department of Computer Science,
Institute for Advanced Computer Studies,
University of Maryland, College Park MD 20742, USA
{minkcho, mount}@cs.umd.edu

**Abstract.** We consider the well known problem of matching two planar point sets under rigid transformations so as to minimize the directed Hausdorff distance. This is a well studied problem in computational geometry. Goodrich, Mitchell, and Orletsky [GMO94] presented a very simple approximation algorithm for this problem, which computes transformations based on aligning pairs of points. They showed that their algorithm achieves an approximation ratio of 4. We consider a minor modification to their algorithm, which is based on aligning midpoints rather than endpoints. We show that this algorithm achieves an approximation ratio not greater than 3.13. Our analysis is sensitive to the ratio between the diameter of the pattern set and the Hausdorff distance, and we show that the approximation ratio approaches 3 in the limit. Finally, we provide lower bounds that show that our approximation bounds are nearly tight.

## 1  Introduction

Geometric pattern matching problem is a fundamental computational problem and has numerous applications in areas such as computer vision [MNM99], image or video compression [ASG02], model-based object recognition [HHW92], and computational chemistry [FKL+97]. In general, we are given two point sets, a *pattern set* $P$ and *background set* $Q$ from some geometric space. The goal is to compute the transformation $E$ from some geometric group that minimizes some distance measure from $EP$ to $Q$. Throughout, we will consider point sets in the Euclidean plane under rigid transformations (translation and rotation). Distances between two point sets $P$ and $Q$ will be measured by the *directed Hausdorff distance*, denoted $h(P,Q)$, which is defined to be

$$h(P,Q) \;=\; \max_{p \in P} \min_{q \in Q} \|pq\|,$$

---

* This work was supported by the National Science Foundation under grant CCR-0098151.

F. Dehne, A. López-Ortiz, and J.-R. Sack (Eds.): WADS 2005, LNCS 3608, pp. 432–443, 2005.

where $\|pq\|$ denotes the Euclidean distance between points $p$ and $q$. Throughout, unless otherwise specified, we use the term *Hausdorff distance* to denote the directed Hausdorff distance. Thus, our objective is to compute the rigid transformation $E$ that minimizes $h(EP, Q)$.

The problem of point pattern matching has been widely studied. We will focus on methods from the field of computational geometry. Excellent surveys can be found in [AG96] and [Vel01]. There are many variants of this problem, depending on the nature of the inputs, the allowable group of aligning transformations, and the distance function employed. A number of approaches have been proposed for matching points under the Hausdorff distance [AAR94, CGH+97, HKS93]. The computational complexity can be quite high. For example, the best-known algorithm for determining the translation and rotation that minimize the directed Hausdorff distance between sets $P$ and $Q$ of sizes $m$ and $n$, respectively, runs in $O(m^3 n^2 \log^2 mn)$ time [CGH+97].

Because of the high complexity of exact point pattern matching, approximation algorithms have been considered. Heffernan and Schirra [HS94] proposed an approximate decision algorithm that tests for congruence within a user-supplied absolute error $\epsilon$. Alt, Aichholzer and Rote [AAR94] presented a 1.43-factor approximation algorithm for planar point pattern matching under similarity transformations under the *bidirectional Hausdorff distance* (which is defined to be $\max(h(P,Q), h(Q,P))$). Neither of these is applicable to our problem, since the cost functions require that every point of $Q$ has a close match in $P$.

The starting point of this work is the alignment-based approximation algorithms given by Goodrich, Mitchell, and Orletsky [GMO94]. They presented a very simple approximation algorithm for a number of pattern matching formulations, including ours. This is arguably the simplest and most easily implemented algorithm for approximate pattern matching. It operates by computing the transformation that aligns a diametrical pair of points of $P$ with all possible pairs of $Q$, and then returning the transformation with the minimum Hausdorff distance. (We will present the algorithm below.) It runs in $O(n^2 m \log n)$ time, and they prove that it achieves an approximation ratio of 4, that is, it returns a transformation whose Hausdorff distance is at most a factor 4 larger than the optimal Hausdorff distance.

Another closely related piece of work is by Indyk, Motwani, and Venkatasubramanian [IMV99]. They considered the error model of Heffernan and Schirra. They presented an approximation algorithm whose running time is sensitive to the spread $\Delta$ of the point set, which is defined to be the ratio of the distances between the farthest and closest pairs of points. They showed that, for any fixed $\beta > 0$, an $(1 + \beta)$-approximation can be computed in time $O(n^{4/3} \Delta^{1/3} \log n)$. Cardoze and Schulman [CS98] presented a randomized approximation algorithm based on computing convolutions, whose running time is $O(n^2 \log n + \log^{O(1)} \Delta)$ for any fixed precision parameter and any fixed success probability. Although these approaches are asymptotically superior to that of Goodrich *et al.*, they lack its simplicity and ease of implementation.

In this paper, we reconsider the issue of the approximation ratio for the alignment-based algorithm of Goodrich, Mitchell, and Orletsky. We show that it is possible to improve on the approximation ratio of 4 without altering the simplicity of the algorithm. Our approach is identical in spirit and running time to theirs, but is based on a minor modification of the way in which the aligning transformation is computed. We call our modification *symmetric alignment* because it is based on aligning midpoints, rather than endpoints. We show that the resulting approximation ratio is never more than 3.13 (a quantity resulting from the numerical solution of an equation).

We also analyze the approximation ratio as a function of a parameter that depends on the closeness of the optimal match. Let $\rho$ be half the ratio of the diameter of $P$ to the minimum Hausdorff distance. We show that our algorithm achieves an approximation ratio of at most $3 + \frac{1}{\sqrt{3}\,\rho}$. In many applications $\rho$ can be quite large. Examples include document analysis and satellite image analysis, where the ratio of diameter of the pattern ranges from tens to hundreds of pixels, while the expected digitization error is roughly one pixel. For these applications, the approximation ratio will be quite close to 3. We also show that for sufficiently large $\rho$, the approximation factor is at least $3 + \frac{1}{10\rho^2}$.

It is worth noting that the more sophisticated approximation algorithm of Indyk, et al. [IMV99] uses the simple alignment algorithm as a subroutine. The running time of their algorithm has a cubic dependence on the approximation ratio of the alignment algorithm. So, an improvement in the approximation ratio has the effect of reducing the running time of their algorithm as well.

The remainder of the paper is organized as follows. In Section 2 we present the algorithm of [GMO94], which we call the *serial alignment algorithm* and review the derivation of its approximation bounds. Next, we will introduce the *symmetric alignment algorithm*, and derive a crude analysis of its approximation ratio. In Section 3, we will present a more accurate analysis of the approximation bound. Finally, in Section 4 we give concluding remarks.

## 2   The Serial and Symmetric Alignment Algorithms

In this section we present a description of the serial alignment algorithm of Goodrich, Mitchell, and Orletsky [GMO94] and review its approximation bound. We also described our modification of this algorithm, called symmetric alignment. Recall that $P = \{p_1, \ldots, p_m\}$ is the pattern set and $Q = \{q_1, \ldots, q_n\}$ is the background set to search, and we seek a rigid transformation of $P$ that minimizes the Hausdorff distance with $Q$.

The serial alignment algorithm proceeds as follows. Let $(p_1, p_2)$ denote a pair of points having the greatest distance in $P$, that is, a diametrical pair. For each distinct pair $(q_1, q_2)$ in $Q$, the algorithm computes a rigid transformation matching $(p_1, p_2)$ with $(q_1, q_2)$ as follows. First, it determines a translation that maps $p_1$ to $q_1$ and then a rotation about $p_1$ that aligns the directed line segment $\overrightarrow{p_1 p_2}$ with $\overrightarrow{q_1 q_2}$. The composition of these two transformations is then applied to the entire pattern set $P$, and the Hausdorff distance is computed. At the end, the algorithm returns the rigid transformation that achieves the minimum Hausdorff distance. The running time of this algorithm is $O(n^2 m \log n)$ because, for each of the $n(n-1)$ distinct pairs of $Q$, we compute the aligning transformation $E$ in $O(1)$ time, and then, for each point $p \in P$, we compute the distance from $E(p)$ to its nearest neighbor in $Q$. Nearest neighbors queries in a planar set $Q$ can be answered in time $O(\log n)$ after $O(n \log n)$ preprocessing [dBvKOS00]. Goodrich, Mitchell, and Orletsky prove that this algorithm achieves an approximation ratio of 4. For completeness let us summarize their argument.

**Lemma 1.** (Goodrich, Mitchell, and Orletsky [GMO94]) *The serial alignment algorithm has an approximation ratio of at most* 4.

*Proof.* For simplicity of notation, let us assume that $P$ has been presented to the algorithm in its optimal position with respect to $Q$. Let $h_{opt}$ denote the optimal Hausdorff distance from $P$ to $Q$. This means that for each point of $P$ there exists a point of $Q$ within distance $h_{opt}$. Now, we run the above algorithm and bound the maximum distance by which any point of $P$ is displaced relative to its original (optimal) position.

Clearly, in the process of translation, each point of the pattern set $P$ moves by at most $h_{opt}$. To determine the displacement due to rotation, it suffices to consider $p_2$ since it is the farthest point from $p_1$. The point $p_2$ was initially within distance $h_{opt}$ of its corresponding points of $Q$, denoted $q_2$, and the above translation moves it by an additional distance of at most $h_{opt}$. Thus, rotating $p_2$ into alignment with $q_2$ moves it by a distance of at most $2h_{opt}$. Therefore, given that it started within distance $h_{opt}$ of some point of $Q$, it follows that its contribution to the Hausdorff distance is at most $h_{opt} + 2h_{opt} + h_{opt} = 4h_{opt}$, and the approximation bound follows directly.                                    □

There are two obvious shortcomings with this algorithm and its analysis. The first is that the algorithm does not optimally align the pair $(p_1, p_2)$ with the pair $(q_1, q_2)$ with respect to Hausdorff distance. One would expect such an optimal alignment to provide a better approximation bound. This observation is the basis of our algorithm. The second shortcoming is that the analysis fails to account for the fact that the displacement distances due to translation and rotation are functions of the points' relative position to $p_1$ and $p_2$, and so share some dependency.

Next, we describe our approach, called *symmetric alignment*. The algorithm differs only in how the aligning transformation is computed, given the pairs $(p_1, p_2)$ in $P$ and $(q_1, q_2)$ in $Q$. Let $m_p$ and $m_q$ denote the respective midpoints of line segments $\overline{p_1 p_2}$ and $\overline{q_1 q_2}$. First, translate $P$ to map $m_p$ to $m_q$ and then rotate $P$ about $m_p$ to align the directed segments $\overrightarrow{p_1 p_2}$ with $\overrightarrow{q_1 q_2}$. Thus, the only difference is that we align and rotate around the midpoints rather than $p_1$. Observe that this alignment transformation minimizes the Hausdorff distance between the pairs $(p_1, p_2)$ and $(q_1, q_2)$.

Before giving our detailed analysis of the approximation bound, we establish a crude approximation bound, which justifies the benefit of symmetric alignment.

**Lemma 2.** *The symmetric alignment algorithm has an approximation ratio of at most* $(2 + \sqrt{3}) \approx 3.732$.

*Proof.* We will apply the same approach used in the analysis of serial alignment. As before, let us assume that $P$ has been presented to the algorithm in its optimal position with respect to $Q$. (Note that the algorithm's final alignment is independent of the initial point placement.) Let $v_1$ and $v_2$ denote the vector forms of $\overrightarrow{p_1 q_1}$ and $\overrightarrow{p_2 q_2}$, respectively. To match the midpoints $m_p$ and $m_q$, we translate the pattern set $P$ by $(v_1 + v_2)/2$. Let $T$ denote this translation. The distance between $T(p_1)$ and $q_1$ is $\|v_1 - (v_1 + v_2)/2\| = \|(v_1 - v_2)/2\| \leq h_{opt}$. By symmetry, the same bound applies to the distance from $T(p_2)$ to $q_2$. Therefore, $T(p_1)$ will move by at most $h_{opt}$ during the rotation process.

Because $(p_1, p_2)$ is a diametrical pair, it follows that all the points of $P$ lie in a lune defined by the intersection of two discs, both of radius $\|p_1 p_2\|$, centered at these points. Thus, no point of $P$ is farther from the midpoint $m_p$ than the apex of the lune, which is at distance $(\sqrt{3}/2)\|p_1 p_2\|$. Since this rotation moves $p_1$ (or equivalently $p_2$) by a distance of at most $h_{opt}$, and $p_1$ is within distance $(1/2)\|p_1 p_2\|$ of $m_p$, it follows that any point $p \in P$ is moved by at most $\sqrt{3} h_{opt}$. Given that $p$ started within distance $h_{opt}$ of some point of $Q$, its final contribution to the Hausdorff distance is at most $h_{opt} + \sqrt{3} h_{opt} + h_{opt} = (2 + \sqrt{3}) h_{opt} \approx 3.732 h_{opt}$.  □

This crude bound is already an improvement; however, the analysis suffers from the same problem as that for serial alignmnment in that it does not consider the geometric relationship between the translation and rotation vectors.

## 3  Main Results

In this section, we derive a tightly approximation bound for symmetric alignment. As before, let $h_{opt}$ denote the optimal Hausdorff distance between $P$ and $Q$ achievable under any rigid transformation of $P$. Let $A_{sym}(\rho)$ denote the approximate ratio for symmetric alignment.

**Theorem 1.** *Consider two planar point sets $P$ and $Q$ whose optimal Hausdorff distance under rigid transformations is $h_{opt}$. Let $\rho = \frac{1}{2} diam(P)/h_{opt}$, where $diam(P)$ denotes the diameter of $P$. Then the for all $\rho > 0$, the approximation ratio of symmetric alignment satisfies:*

$$A_{sym}(\rho) \leq \min\left(3 + \frac{1}{\sqrt{3}\,\rho}, \sqrt{4\rho^2 + 2\rho + 1}\right).$$

In typical applications where $\rho$ is relatively large, this shows that the approximation bound is nearly 3. The remainder of this section is devoted to proving this theorem. As usual, it will simplify notation to assume that $P$ has been transformed to its optimal placement with respect to $Q$, and we will bound the amount of additional displacement of each point of $P$ relative to this placement.

Without loss of generality we may scale space uniformly so that $h_{opt} = 1$, implying that for each point of $P$ there exists at least one point of $Q$ within distance 1. The algorithm starts by computing a pair $(p_1, p_2)$ of points of $P$ with the greatest distance. Let $(q_1, q_2)$ denote the corresponding pair of points of $Q$. From our scaling it follows that for $i \in \{1, 2\}$, $q_i$ lies within a unit disc centered at $p_i$. (See Fig. 1.) Let $m_p$ and $m_q$ denote the respective midpoints of the segments $\overline{p_1 p_2}$ and $\overline{q_1 q_2}$. Note that $\rho$ is just the distance from $m_p$ to either $p_1$ or $p_2$. Let $\alpha$ denote the absolute acute angle between the directed lines supporting the segments $\overrightarrow{p_1 q_1}$ and $\overrightarrow{p_2 q_2}$. Without loss of generality, we may assume this angle is acute. If $\rho > 1$, then the two unit discs do not intersect, and it follows easily that $0 \leq \sin \alpha \leq 1/\rho$.

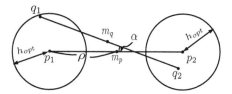

**Fig. 1.** The positions of the point sets prior to running the algorithm

Here is a brief overview of the proof. We establish two approximation bounds, each a function of $\rho$. One bound is better for small $\rho$ and the other better for high $\rho$. We shall show that the crossover value of these two functions, denoted $\rho^*$, is approximately 1.26. Due to space limitations, we only present the approximation bound for the more interesting case, when $\rho > \rho^*$.

We begin by considering the space of possible translations that align $m_p$ with $m_q$. It will make matters a bit simpler to think of translating $Q$ to align $m_q$ with $m_p$, but of course any bounds on the distance of translation will apply to case of translating $P$. The following lemma bounds the translation distance, which we will denote by $\|T\|$, as a function of $\rho$ and $\alpha$.

**Lemma 3.** *The symmetric alignment's translation transformation displaces any point of $P$ by a vector $T = T_\rho(\alpha)$ satisfying $\|T\| \leq \sqrt{1 - \rho^2 \sin^2 \alpha}$.*

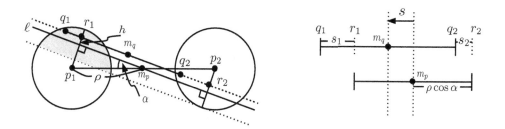

**Fig. 2.** The analysis of the midpoint translation

438     M. Cho and D.M. Mount

*Proof.* Consider a line passing through $m_p$ that is parallel to $\overline{q_1 q_2}$. Let $r_1$ and $r_2$ be the respective orthogonal projections of $p_1$ and $p_2$ onto the line $\overline{q_1 q_2}$ and let $s_1$ and $s_2$ denote the signed distances $\|\overrightarrow{r_1 q_1}\|$ and $\|\overrightarrow{r_2 q_2}\|$, respectively. Consider a coordinate system centered at $m_p$ whose positive $s$-axis is located along a line $\ell$ that is directed in the sense of $\overrightarrow{q_2 q_1}$, and whose positive $h$-axis is perpendicular to this and directed away from $\ell$ towards the line $\overline{q_1 q_2}$. In this coordinate system we have $m_p = (0,0)$, $q_1 = (s_1 + \rho \cos \alpha, h)$, and $q_2 = (s_2 - \rho \cos \alpha, h)$. Thus, $m_q = ((s_1+s_2)/2, h)$, and the translation vector $T$ is $m_q - m_p = m_q$. By straightforward calculations we have $|s_1| \leq \sqrt{1 - (\rho \sin \alpha + h)^2}$ and $|s_2| \leq \sqrt{(1 - (\rho \sin \alpha + h)^2}$. Therefore,

$$\|T\|^2 = s^2 + h^2 = \left(\frac{s_1 + s_2}{2}\right)^2 + h^2 = \frac{1}{2}(s_1^2 + s_2^2) - \frac{1}{4}(s_1 - s_2)^2 + h^2$$
$$\leq \frac{1}{2}(s_1^2 + s_2^2) + h^2 \leq \frac{1}{2}\{(1 - (\rho \sin \alpha + h)^2) + (1 - (\rho \sin \alpha - h)^2)\} + h^2$$
$$= 1 - \rho^2 \sin^2 \alpha.$$

Observe that the translation is maximized when $s_1 = s_2$, and this means that the line $\overline{q_1 q_2}$ passes through $m_p$.  $\square$

Next we consider the effect of rotation on the approximation error. Unlike translation we need to consider the placement of points in $P$ because the distance by which a point is moved by rotation is determined by both the rotation angle $\alpha$ and the distance from this point to the center of rotation. As mentioned in the proof of Lemma 2 every point of $P$ lies within a lune formed by the intersection of two discs of radius $2\rho$ centered at $p_1$ and $p_2$. (See Fig. 3.) The following lemma describes the possible displacements of a point of $P$ under the rotational part of the aligning transformation. This is done relative to a coordinate system centered at $m_p$, whose $x$-axis is directed towards $p_2$.

**Lemma 4.** *The symmetric alignment's rotation transformation displaces any point of $P$ by a vector $R = R_\rho(\alpha)$ satisfying $\|R\| \leq 2\sqrt{3}\rho \sin \frac{\alpha}{2}$.*

*Proof.* Let $p_\theta$ be any point of $P$ such that the signed angle from $\overrightarrow{m_p p_2}$ to $\overrightarrow{m_p p_\theta}$ is $\theta$. Let $p'_\theta$ denote the point after rotating $p_\theta$ counterclockwise about $m_p$ by

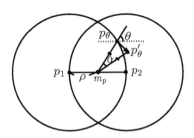

**Fig. 3.** The proof of rotation

angle $\alpha$. By simple trigonometry and the observation that $\triangle m_p p_\theta p'_\theta$ is isosceles, it follows that the length of the displacement $\overrightarrow{p_\theta p'_\theta}$, which we denote by $\|R\|$, is $\|m_p p_\theta\| \cdot 2 \sin \frac{\alpha}{2}$. The farthest point of the lune from $m_p$ is easily seen to be the apex, which is at distance $\sqrt{3}\rho$. Thus, $\|R\|$ is at most $2\sqrt{3}\rho \sin \frac{\alpha}{2}$. $\quad\square$

We are now ready to derive the approximation bound on the symmetric alignment algorithm by combining the bounds on the translational and rotational displacements. Recall that at the start of the algorithm, the points are assumed to placed in the optimal positions and that space has been scaled so that $h_{opt} = 1$. It follows that each point of $P$ has been displaced from its initial position within unit distance of a point of $Q$ to a new position that is now within distance $\|T_\rho(\alpha) + R_\rho(\alpha)\| + 1$ of some point of $Q$. For any fixed value of $\rho$ it follows that the approximation bound $A_{sym}(\rho)$ satisfies

$$A_{sym}(\rho) \leq \max_\alpha \|T_\rho(\alpha) + R_\rho(\alpha)\| + 1.$$

Recall that $0 \leq \sin \alpha \leq 1/\rho$.

Unfortunately, determining this length bound exactly would involve solving a relatively high order equation involving trigonometric functions. Instead, we will compute an upper bound by applying the triangle inequality to separate the translation and rotation components. From Lemmas 3 and 4 we see that $A_{sym}(\rho)$ is at most

$$\max_\alpha (\|T_\rho(\alpha)\| + \|R_\rho(\alpha)\|) + 1 = \max_\alpha \left( \sqrt{1 - \rho^2 \sin^2 \alpha} + 2\sqrt{3}\rho \sin \frac{\alpha}{2} \right) + 1.$$

Substituting $x = \cos \alpha$, it follows that the quantity to be maximized is

$$f_\rho(x) = \sqrt{1 - \rho^2(1 - x^2)} + \sqrt{6}\rho\sqrt{1 - x} + 1, \qquad \text{where } \sqrt{1 - \frac{1}{\rho^2}} \leq x \leq 1.$$

To find the maximum of $f_\rho$, we take the partial derivative with respect to $x$.

$$\frac{\partial f_\rho}{\partial x} = -\frac{\sqrt{3}\rho}{\sqrt{2}\sqrt{1 - x}} + \frac{\rho^2 x}{\sqrt{1 - \rho^2(1 - x^2)}}.$$

By our earlier assumption that $\rho \geq \rho^* \approx 1.26$, it follows by symbolic manipulations that $\partial f_\rho/\partial x = 0$ has a single real root for $\rho \geq \rho^*$, which is

$$x_0 = \frac{1}{6}\left( -1 + c_1(\rho) + \frac{1}{c_1(\rho)} \right),$$

$$\text{where} \quad c_1(\rho) = \frac{\rho^2}{(161\rho^6 + 18\rho^4\sqrt{c_2(\rho)} - 162\rho^4)^{1/3}}$$

$$\text{and} \quad c_2(\rho) = 80\rho^4 - 161\rho^2 + 81.$$

By an analysis of this derivative it follows that the function $f_\rho$ achieves its maximum value when $x = x_0$, and so the final approximation bound is $f_\rho(x_0) =$

$\sqrt{1 - \rho^2(1 - x_0^2)} + \sqrt{6}\rho\sqrt{1 - x_0} + 1$. By computing the partial derivation of $f_\rho(x)$ with respect to $\rho$ that, for any fixed $x$, this function is a monotonically decreasing function of $\rho$.

Unfortunately, this function is too complex to reason about easily. Nonetheless, we can evaluate it for any fixed value of $\rho$. The resulting approximation bound (together with the alternate bound for low $\rho$) is plotted as a function of $\rho$ in Fig. 4 below. The figure shows that as $\rho$ increases, the approximation bound converges to 3. The following result establishes this asymptotic convergence by proving a somewhat weaker approximation bound.

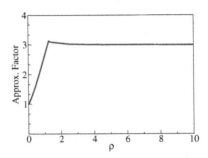

**Fig. 4.** The approximation bound for symmetric alignment as a function of $\rho$

**Lemma 5.** *If $\rho > \rho^*$ then the approximation bound of the symmetric alignment algorithm, $A_{sym}(\rho)$, is at most $3 + \frac{1}{\sqrt{3}\rho}$.*

*Proof.* Before presenting the general case, we consider the simpler limiting case when $\rho$ tends to $\infty$. Since $0 \leq \sin \alpha \leq 1/\rho$, in the limit $\alpha$ approaches 0. Using the fact of $\lim_{\alpha \to 0} \frac{\sin \alpha}{\alpha} = 1$ we have

$$A_{sym}(\infty) = \lim_{\rho \to \infty} A_{sym}(\rho) = \lim_{\rho \to \infty} \sqrt{1 - \rho^2 \sin^2 \alpha + 2\sqrt{3}\rho \sin \frac{\alpha}{2} + 1}$$

$$= \sqrt{1 - \rho^2 \alpha^2} + \sqrt{3}\rho\alpha + 1.$$

Let $x = \rho\alpha$. In the limit we have $0 \leq x \leq 1$ and so

$$A_{sym}(\infty) \leq \max_{0 \leq x \leq 1} \left( \sqrt{1 - x^2} + \sqrt{3}x + 1 \right).$$

It is easy to verify that $A_{sym}(\infty)$ achieves a maximum value of 3 when $x = \sqrt{3}/2$.

Next, we consider the general case. Since $\rho > \rho^*$ and $\sin \alpha \leq 1/\rho$, it follows that $0 \leq x < 1.16$. We have two cases, $0 \leq x \leq 1$ and $1 < x < 1.16$. We present only the first case, due to space limitations. To simplify $A_{sym}(\rho)$, we use a Taylor series expansion and the fact that $A_{sym}(\infty) \leq 3$.

$$A_{sym}(\rho) \leq \max_\alpha \left( \sqrt{1 - \rho^2 \sin^2 \alpha + 2\sqrt{3}\rho \sin \frac{\alpha}{2} + 1} \right)$$

$$\leq \max_\alpha \left( \sqrt{1 - \rho^2 (\alpha - \alpha^3/6)^2} + \sqrt{3}\rho\alpha + 1 \right).$$

$$A_{sym}(\rho) \leq \max_{\alpha} \left( \sqrt{1 - \rho^2 \left(\alpha - \alpha^3/6\right)^2} + \sqrt{3}\rho\alpha - \left(\sqrt{1 - \rho^2\alpha^2} + \sqrt{3}\rho\alpha + 1\right) + 3 + 1 \right)$$

$$= \max_{\alpha} \left( \sqrt{1 - \rho^2 \left(\alpha - \alpha^3/6\right)^2} - \sqrt{1 - \rho^2\alpha^2} + 3 \right).$$

Substituting $x = \rho\alpha$, let

$$g(x) = \sqrt{1 - x^2 \left(1 - \frac{x^2}{6\rho^2}\right)^2} - \sqrt{1 - x^2} + 3.$$

For all fixed $\rho$, this is a monotonically increasing function in $x$. Thus, since $x \leq 1$, this function achieves its maximum value at $x = 1$ of

$$g(1) = 3 + \sqrt{1 - \left(1 - \frac{1}{6\rho^2}\right)^2} = 3 + \sqrt{\frac{1}{3\rho^2} - \frac{1}{36\rho^4}} \leq 3 + \frac{1}{\sqrt{3}\rho}. \qquad \square$$

### 3.1  A Lower Bound for Symmetric Alignment

It is natural to wonder whether the upper bounds on the approximation ratio $A_{sym}$ proved in Theorem 1 are tight. Next, we prove that this approximation bound is close to tight. We show that for all sufficiently large $\rho$ the approximation factor is strictly greater than 3, and approaches 3 in the limit.

**Lemma 6.** *For all $\rho > 2A_{sym}$, there exists an input on which symmetric achieves an approximation factor of at least $3 + \frac{1}{10\rho^2}$.*

The remainder of this section is devoted to proving this. We consider two configurations of points. Consider a fixed value $\rho > 2A_{sym}$. We define the pattern point set $P = \{p_1, p_2, p_3, p_4\}$, where the first three points form an equilateral triangle of side length $2\rho$, and place $p_4$ at the midpoint of $\overline{p_1 p_2}$. (See Fig. 5.) By an infinitesimal perturbation of the points, we may assume that the pair $(p_1, p_2)$ is the unique diametrical pair for $P$.

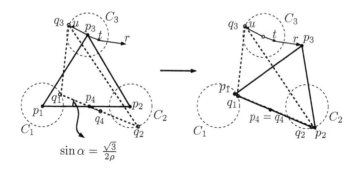

$$\sin \alpha = \frac{\sqrt{3}}{2\rho}$$

**Fig. 5.** The lower bound on $A_{sym}$

Next, to define the locations of the other background set $Q = \{q_1, q_2, q_3, q_4\}$, let $C_i$ denote a circle of unit radius centered $p_i$. Consider a line passing through $p_4$ (the midpoint of $\overline{p_1 p_2}$) forming an angle $\alpha$ with line $\overline{p_1 p_2}$, where $\sin \alpha = \frac{\sqrt{3}}{2\rho}$. Place $q_1$ and $q_2$ at the rightmost intersection points of the line and $C_1$ and $C_2$, respectively. Let $t$ and $r$ denote the translation and rotation displacement vectors resulting from symmetric alignment, respectively. (They are described below.) Let $q_3 = p_3 + u$, where $u$ is a unit length vector whose direction is chosen to be directly opposite that of $t + r$. Finally, place $q_4$ at the midpoint of $\overline{q_1 q_2}$. Observe that for each $p_i$, the corresponding point $q_i$ lies within distance 1.

For this placement, the directed Hausdorff distance is exactly 1, and so $h_{opt} \leq 1$. We run the symmetric alignment algorithm on $P$ and $Q$, and consider the displacement distance of $p_3$ relative to $q_3$. The fact that $\rho > 2A_{sym}$ and the presence of $p_4$ and $q_4$ imply that the final matching chosen by symmetric alignment results by aligning $(p_1, p_2)$ with $(q_1, q_2)$. The reason is that any other choice will result in $p_4$ having no point within distance $A_{sym}$, thus leading to a contradiction. (Details have been omitted due to space limitations.)

By adapting the analyses of Lemmas 3 and 4, it can be seen that this configuration achieves the worst case for these two lemmas. In particular, the length of the translation displacement is $\|t\| = \sqrt{1 - \rho^2 \sin^2 \alpha} = \sqrt{1 - (3/4)} = 1/2$, and the length of the rotation displacement is $\|r\| = 2\sqrt{3}\rho \sin \frac{\alpha}{2}$. Furthermore, it is easy to show that the angle between these two vectors is $\frac{\alpha}{2}$.

To complete the analysis, we decompose the rotation displacement vector $r$ into two components, $r_1$ is parallel to $t$ and $r_2$ is perpendicular to it, and apply the fact that the angle between these vectors is $\alpha/2$. After some algebraic manipulations, it follows that the squared magnitude of the final displacement is at least $4 + \frac{27}{64\rho^2}$, and under the assumption that $\rho > 2A_{sym} \geq 2$, the bound given in the statement of the lemma follows. Details have been omitted due to space limitations.

## 4    Concluding Remarks

We have presented a simple modification to the alignment-based algorithm of Goodrich, Mitchell, and Orletsky [GMO94]. Our modification has the same running time and retains the simplicity of the original algorithm. We have shown that, in contrast to the factor-4 approximation bound proved in [GMO94], our approach has an approximation ratio that is never larger than 3.13, and converges to 3 in the limit as the ratio $\rho$ between the pattern set diameter and the Hausdorff distance increases. We have also shown that the bound is nearly tight. This paper opens the question of what are the performance limits of approximations based on point alignments. It is natural to consider generalizations of this approach to higher dimensions, to matchings based on alignments of more than two points, and to other sorts of distance measures and transformation groups.

# References

[AAR94]      H. Alt, O. Aichholzer, and G. Rote. Matching shapes with a reference point. In *Proc. 10th Annu. ACM Sympos. Comput. Geom.*, pages 85–92, 1994.

[AG96]       H. Alt and L. J. Guibas. Discrete geometric shapes: Matching, interpolation, and approximation. Technical Report 96–11, Institut für Informatik, Freie Unversität Berlin, Berlin, Germany, 1996.

[ASG02]      M. Alzina, W. Szpankowski, and A. Grama. 2d-pattern matching image and video compression: Theory, algorithms, and experiments. *IEEE Trans. on Image Processing*, 11:318–331, 2002.

[CGH+97]     L. P. Chew, M. T. Goodrich, D. P. Huttenlocher, K. Kedem, J. M. Kleinberg, and D. Kravets. Geometric pattern matching under Euclidean motion. *Comput. Geom. Theory Appl.*, 7:113–124, 1997.

[CS98]       D. E. Cardoze and L. Schulman. Pattern matching for spatial point sets. In *Proc. 39th Annu. IEEE Sympos. Found. Comput. Sci.*, pages 156–165, 1998.

[dBvKOS00]   M. de Berg, M. van Kreveld, M. H. Overmars, and O. Schwarzkopf. *Computational Geometry: Algorithms and Applications*. Springer-Verlag, 2nd edition, 2000.

[FKL+97]     P. Finn, L. E. Kavraki, J. C. Latombe, R. Motwani, C. Shelton, S. Venkatasubramanian, and A. Yao. Rapid: Randomized pharmacophore identification for drug design. In *Proc. 13th Annu. ACM Sympos. Comput. Geom.*, pages 324–333, 1997.

[GMO94]      M. T. Goodrich, J. S. Mitchell, and M. W. Orletsky. Practical methods for approximate geometric pattern matching under rigid motion. In *Proc. 10th Annu. ACM Sympos. Comput. Geom.*, pages 103–112, 1994.

[HHW92]      J. E. Hopcroft, D. P. Huttenlocher, and P. C. Wayner. *Affine invariants for model-based recognition*. MIT Press, Cambridge, 1992.

[HKS93]      D. P. Huttenlocher, K. Kedem, and M. Sharir. The upper envelope of Voronoi surfaces and its applications. *Discrete Comput. Geom.*, 9:267–291, 1993.

[HS94]       P. J. Heffernan and S. Schirra. Approximate decision algorithms for point set congruence. *Comput. Geom. Theory Appl.*, 4:137–156, 1994.

[IMV99]      P. Indyk, R. Motwani, and S. Venkatasubramanian. Geometric matching under noise: Combinatorial bounds and algorithms. In *Proc. 8th Annual ACM-SIAM on Discrete Algorithm*, pages 457–465, 1999.

[MNM99]      D. M. Mount, N. S. Netanyahu, and J. Le Moigne. Efficient algorithms for robust point pattern matching. *Pattern Recognition*, 32:17–38, 1999.

[Vel01]      R. C. Veltkamp. Shape matching: Similarity measures and algorithms. Technical Report UU-CS 2001-03, Utrecht University: Information and Computing Sciences, Utrecht, The Netherlands, 2001.

# Author Index

# Lecture Notes in Computer Science

For information about Vols. 1–3509

please contact your bookseller or Springer

Vol. 3558: V. Torra, Y. Narukawa, S. Miyamoto (Eds.), Modeling Decisions for Artificial Intelligence. XII, 470 pages. 2005. (Subseries LNAI).

Vol. 3557: H. Gilbert, H. Handschuh (Eds.), Fast Software Encryption. XI, 443 pages. 2005.

Vol. 3556: H. Baumeister, M. Marchesi, M. Holcombe (Eds.), Extreme Programming and Agile Processes in Software Engineering. XIV, 332 pages. 2005.

Vol. 3555: T. Vardanega, A.J. Wellings (Eds.), Reliable Software Technology – Ada-Europe 2005. XV, 273 pages. 2005.

Vol. 3554: A. Dey, B. Kokinov, D. Leake, R. Turner (Eds.), Modeling and Using Context. XIV, 572 pages. 2005. (Subseries LNAI).

Vol. 3553: T.D. Hämäläinen, A.D. Pimentel, J. Takala, S. Vassiliadis (Eds.), Embedded Computer Systems: Architectures, Modeling, and Simulation. XV, 476 pages. 2005.

Vol. 3552: H. de Meer, N. Bhatti (Eds.), Quality of Service – IWQoS 2005. XVIII, 400 pages. 2005.

Vol. 3551: T. Härder, W. Lehner (Eds.), Data Management in a Connected World. XIX, 371 pages. 2005.

Vol. 3548: K. Julisch, C. Kruegel (Eds.), Intrusion and Malware Detection and Vulnerability Assessment. X, 241 pages. 2005.

Vol. 3547: F. Bomarius, S. Komi-Sirviö (Eds.), Product Focused Software Process Improvement. XIII, 588 pages. 2005.

Vol. 3546: T. Kanade, A. Jain, N.K. Ratha (Eds.), Audio- and Video-Based Biometric Person Authentication. XX, 1134 pages. 2005.

Vol. 3544: T. Higashino (Ed.), Principles of Distributed Systems. XII, 460 pages. 2005.

Vol. 3543: L. Kutvonen, N. Alonistioti (Eds.), Distributed Applications and Interoperable Systems. XI, 235 pages. 2005.

Vol. 3542: H.H. Hoos, D.G. Mitchell (Eds.), Theory and Applications of Satisfiability Testing. XIII, 393 pages. 2005.

Vol. 3541: N.C. Oza, R. Polikar, J. Kittler, F. Roli (Eds.), Multiple Classifier Systems. XII, 430 pages. 2005.

Vol. 3540: H. Kalviainen, J. Parkkinen, A. Kaarna (Eds.), Image Analysis. XXII, 1270 pages. 2005.

Vol. 3539: K. Morik, J.-F. Boulicaut, A. Siebes (Eds.), Local Pattern Detection. XI, 233 pages. 2005. (Subseries LNAI).

Vol. 3538: L. Ardissono, P. Brna, A. Mitrovic (Eds.), User Modeling 2005. XVI, 533 pages. 2005. (Subseries LNAI).

Vol. 3537: A. Apostolico, M. Crochemore, K. Park (Eds.), Combinatorial Pattern Matching. XI, 444 pages. 2005.

Vol. 3536: G. Ciardo, P. Darondeau (Eds.), Applications and Theory of Petri Nets 2005. XI, 470 pages. 2005.

Vol. 3535: M. Steffen, G. Zavattaro (Eds.), Formal Methods for Open Object-Based Distributed Systems. X, 323 pages. 2005.

Vol. 3534: S. Spaccapietra, E. Zimányi (Eds.), Journal on Data Semantics III. XI, 213 pages. 2005.

Vol. 3533: M. Ali, F. Esposito (Eds.), Innovations in Applied Artificial Intelligence. XX, 858 pages. 2005. (Subseries LNAI).

Vol. 3532: A. Gómez-Pérez, J. Euzenat (Eds.), The Semantic Web: Research and Applications. XV, 728 pages. 2005.

Vol. 3531: J. Ioannidis, A. Keromytis, M. Yung (Eds.), Applied Cryptography and Network Security. XI, 530 pages. 2005.

Vol. 3530: A. Prinz, R. Reed, J. Reed (Eds.), SDL 2005: Model Driven. XI, 361 pages. 2005.

Vol. 3528: P.S. Szczepaniak, J. Kacprzyk, A. Niewiadomski (Eds.), Advances in Web Intelligence. XVII, 513 pages. 2005. (Subseries LNAI).

Vol. 3527: R. Morrison, F. Oquendo (Eds.), Software Architecture. XII, 263 pages. 2005.

Vol. 3526: S. B. Cooper, B. Löwe, L. Torenvliet (Eds.), New Computational Paradigms. XVII, 574 pages. 2005.

Vol. 3525: A.E. Abdallah, C.B. Jones, J.W. Sanders (Eds.), Communicating Sequential Processes. XIV, 321 pages. 2005.

Vol. 3524: R. Barták, M. Milano (Eds.), Integration of AI and OR Techniques in Constraint Programming for Combinatorial Optimization Problems. XI, 320 pages. 2005.

Vol. 3523: J.S. Marques, N. Pérez de la Blanca, P. Pina (Eds.), Pattern Recognition and Image Analysis, Part II. XXVI, 733 pages. 2005.

Vol. 3522: J.S. Marques, N. Pérez de la Blanca, P. Pina (Eds.), Pattern Recognition and Image Analysis, Part I. XXVI, 703 pages. 2005.

Vol. 3521: N. Megiddo, Y. Xu, B. Zhu (Eds.), Algorithmic Applications in Management. XIII, 484 pages. 2005.

Vol. 3520: O. Pastor, J. Falcão e Cunha (Eds.), Advanced Information Systems Engineering. XVI, 584 pages. 2005.

Vol. 3519: H. Li, P. J. Olver, G. Sommer (Eds.), Computer Algebra and Geometric Algebra with Applications. IX, 449 pages. 2005.

Vol. 3518: T.B. Ho, D. Cheung, H. Liu (Eds.), Advances in Knowledge Discovery and Data Mining. XXI, 864 pages. 2005. (Subseries LNAI).

Vol. 3517: H.S. Baird, D.P. Lopresti (Eds.), Human Interactive Proofs. IX, 143 pages. 2005.

Vol. 3516: V.S. Sunderam, G.D.v. Albada, P.M.A. Sloot, J.J. Dongarra (Eds.), Computational Science – ICCS 2005, Part III. LXIII, 1143 pages. 2005.

Vol. 3515: V.S. Sunderam, G.D.v. Albada, P.M.A. Sloot, J.J. Dongarra (Eds.), Computational Science – ICCS 2005, Part II. LXIII, 1101 pages. 2005.

Vol. 3514: V.S. Sunderam, G.D.v. Albada, P.M.A. Sloot, J.J. Dongarra (Eds.), Computational Science – ICCS 2005, Part I. LXIII, 1089 pages. 2005.

Vol. 3513: A. Montoyo, R. Muñoz, E. Métais (Eds.), Natural Language Processing and Information Systems. XII, 408 pages. 2005.

Vol. 3512: J. Cabestany, A. Prieto, F. Sandoval (Eds.), Computational Intelligence and Bioinspired Systems. XXV, 1260 pages. 2005.

Vol. 3511: U.K. Wiil (Ed.), Metainformatics. VIII, 221 pages. 2005.

Vol. 3510: T. Braun, G. Carle, Y. Koucheryavy, V. Tsaousidis (Eds.), Wired/Wireless Internet Communications. XIV, 366 pages. 2005.